Advances in Modal Logic

Volume 3

Advances in Modal Logic

Volume 3

Editors

Frank Wolter
University of Leipzig, Germany

Heinrich Wansing
Dresden University of Technology, Germany

Maarten de Rijke
University of Amsterdam, The Netherlands

Michael Zakharyaschev
King's College London, UK

World Scientific
New Jersey • London • Singapore • Hong Kong

Published by

World Scientific Publishing Co. Pte. Ltd.
P O Box 128, Farrer Road, Singapore 912805
USA office: Suite 202, 1060 Main Street, River Edge, NJ 07661
UK office: 57 Shelton Street, Covent Garden, London WC2H 9HE

British Library Cataloguing-in-Publication Data
A catalogue record for this book is available from the British Library.

ADVANCES IN MODAL LOGIC — Vol. 3

ISBN 981-238-179-1

This book is printed on acid-free paper.

Printed in Singapore by World Scientific Printers (S) Pte Ltd

Contents

Preface

The present volume contains almost all invited and contributed papers of the third Advances in Modal Logic (AiML) workshop, University of Leipzig, October 4–7, 2000. Like the previous AiML workshops, *AiML 2000* brought together modal logicians from all over the world to present new results and to report on the latest developments in modal logic broadly conceived. Special efforts were made to deepen the interaction between the AiML community and the perhaps more computationally oriented Temporal Logic community by collocating *AiML 2000* and the International Conference on Temporal Logic, *ICTL 2000*. In addition, a session on Knowledge Representation and Modal Logic was organized around an invited survey lecture on Description Logics given by Giuseppe De Giacomo.

The complete list of invited speakers at *AiML 2000* was as follows:

- Nuel D. Belnap,
- Giuseppe De Giacomo,
- Stephane Demri,
- Silvio Ghilardi, and
- Krister Segerberg.

Twenty out of 38 submissions were selected for presentation as contributed papers. The *AiML 2000 Best Contributed Paper Award* was given to Rostislav E. Yavorsky for his paper "First-order logic of the standard proof predicate." A special highlight during the workshop was a ceremonial session in which the Faculty of Philosophy and Social Sciences of Leipzig University awarded a honorary doctoral degree to Nuel D. Belnap for his achievements in Non-classical but also in Modal Logic.

The Programme Committee for *AiML 2000* consisted of:

- Franz Baader,
- Greg Restall,

- Maarten de Rijke,
- Heinrich Wansing,
- Frank Wolter, and
- Michael Zakharyaschev.

We would like to thank our fellow PC members and all the colleagues who, in addition to the editors of this volume, acted as referees, namely:

- Natasha Alechina,
- Hajnal Andréka,
- Carlos Areces,
- Nuel Belnap,
- Johan van Benthem,
- Alexander Chagrov,
- Henry Chinaski,
- Marta Cialdea Mayer,
- Rosella Gennari,
- Giuseppe De Giacomo,
- Erich Grädel,
- Bernhard Heinemann,
- Wiebe van der Hoek,
- Marcus Kracht,
- Natasha Kurtonina,
- Agnes Kurucz,
- Gerhard Lakemeyer,
- Carsten Lutz,
- Fabio Massacci,
- Larry Moss,
- Hans Jürgen Ohlbach,
- Marc Pauly,
- Greg Restall,
- Ulrike Sattler,
- Valentin Shehtman,
- Maarten Stol,
- Holger Sturm,
- Nobu-Yuki Suzuki,
- Eugenia Ternovskaia,
- Stephan Tobies,
- Yde Venema, and
- Stefan Wölfl.

Moreover, we gratefully acknowledge generous financial support from the Deutsche Forschungsgemeinschaft (DFG), the Department of Computer Science at Leipzig University, the Department of Logic and Philosophy of Science at Leipzig University, the Graduate Programme in Knowledge Representation and the research group Communicative Understanding at Leipzig University.

During *AiML 2000*, a number of changes to the structure of the AiML initiative have been implemented. The AiML Steering Committee has been extended and now consists of Mark Reynolds, Maarten de Rijke, Renate Schmidt, Nobu-Yuki Suzuki, Heinrich Wansing, Frank Wolter, and Michael Zakharyaschev. Maarten de Rijke has been appointed president of AiML and Heinrich Wansing was made vice-president. Further, in addition to the AiML workshops, now continued as AiML conferences in even years, smaller workshops and less formal meetings will be organized in odd years. Information concerning AiML is available at the following website: `http://www.aiml.net`.

F. Wolter, H. Wansing, M. de Rijke, and M. Zakharyaschev
Leipzig, Dresden, Amsterdam, and London

1

On Arithmetical Completeness of First-Order Logics of Provability

ROSTISLAV E. YAVORSKY

ABSTRACT. In this paper we consider first-order extensions of the modal provability logic GL and the operational logic of proofs $\mathcal{LP}$. It is proved that for the so-called binding provability operator, QGL^b, a natural first-order extension of GL, is arithmetically complete. The same holds for the first-order logic of proofs with monotonicity axiom $\mathcal{QLP}^b$.

For the first-order modal logic of provability QGL^b we also establish completeness with respect to a Kripke-style semantics.

1 Introduction

The idea to study provability by means of modal logic was suggested in Orlov 1928 and Gödel 1933. Independently they wrote down the modal system coinciding with $S4$ that describes the intuitive notion of provability. Later it turned out that $S4$ is incorrect with respect to the direct interpretation of $\Box$ as provability in a given formal system, say Peano arithmetic PA or set theory. The reason is that the reflexivity axiom $\Box A \to A$ contradicts the Gödel incompleteness theorem. In Solovay 1976 it was proven that the set of modal formulas provable in PA under every arithmetical interpretation coincides with the modal logic GL. The complete classification of propositional modal provability logics is given in Beklemishev 1999.

The study of the *first-order modal logic of provability* has been suggested in Boolos 1979. The usual definition of first-order modal provability logic $QPL_T(M)$ of a theory T with respect to a metatheory M is the following (here T and M are arithmetical theories). Consider an extension of the pure predicate language with the modal operator $\Box$.

Advances in Modal Logic, Volume 3
F. Wolter, H. Wansing, M. de Rijke, and M. Zakharyaschev, eds.

An arithmetical interpretation f maps every pure predicate formula to an arithmetical formula with the same set of free variables, it commutes with boolean connectives and quantifiers, and

$$f(\Box A(x_1,\dots,x_k)) \rightleftharpoons \mathit{Prov}_T(\ulcorner f(A(\dot{x}_1,\dots,\dot{x}_k))\urcorner).$$

Here Prov_T is a standard provability formula, and $\ulcorner f(A(\dot{x}_1,\dots,\dot{x}_k))\urcorner$ is a primitive recursive term for the function which for every k-tuple of natural numbers $m_1,\dots,m_k$ returns the Gödel number of the formula $f(A(\overline{m_1},\dots,\overline{m_k}))$. Then, $QPL_T(M)$ is the set of all modal formulas whose interpretations are all provable in M.

It was shown in Artemov 1980 and Vardanyan 1986 that for many natural choices of T and M the modal provability logic $QPL_T(M)$ is not effectively axiomatizable (see de Jongh and Japaridze 1998 for a recent overview of the subject).

In this paper we consider the first-order modal language with a so-called binding modal operator. We suppose that all individual variables in the scope of the modal operator $\Box$ are bound. So, $\Box A(x_1,\dots,x_k)$ is a closed formula even if $x_1,\dots,x_k$ are free in the formula $A(x_1,\dots,x_k)$. The definition of an arithmetical interpretation should be corrected in the following way:

$$f(\Box A(x_1,\dots,x_k)) \rightleftharpoons \mathit{Prov}(\overline{\ulcorner f(A(x_1,\dots,x_k))\urcorner}),$$

where $\ulcorner f(A(x_1,\dots,x_k))\urcorner$ denotes the Gödel number of the formula, and $\overline{\ulcorner f(A(x_1,\dots,x_k))\urcorner}$ is the corresponding numeral.

In section 2 we prove that the corresponding first-order extension QGL^b of logic GL is arithmetically complete. We also establish completeness of QGL^b with respect to an appropriate Kripke-style semantics.

The study of *logics of proofs* with the atoms "t is a proof of F" was initiated in Artemov 1994. In Artemov 1998 the operational logic of proofs $\mathcal{LP}$ was presented and it was established that every theorem of the modal logic $S4$ admits a reading in $\mathcal{LP}$ as a statement about explicit provability. This completed the efforts in Kolmogoroff 1932 and Gödel 1933, Gödel 1995 to provide a Brouwer – Heyting – Kolmogorov style classical provability semantics for intuitionistic logic.

In Artemov and Yavorskaya-Sidon 1999 the study of first-order logics of proofs was started. It was proven that in many natural cases the corresponding arithmetically complete logics are not effectively axiomatizable.

In section 3 we consider the first-order logic $\mathcal{QLP}^b$ of the standard multiple conclusion proof predicate. Again, an important point is that we consider a language with a binding proof predicate. Arithmetical completeness of $\mathcal{QLP}^b$ is established.

2 Modal provability logic

2.1 Definition of QGL^b

The language of QGL^b is an extension of the pure predicate language by one unary modal operator $\Box$ with the following condition: we assume that $\Box$ binds all free occurrences of individual variables in a formula i.e., all occurrences of individual variables are considered to be bound in the formula $\Box A$ independently of the structure of the formula A.

We say that F is *a boxed formula* if it is of the form $\Box G$ for some formula G. Below, $d(F)$ denotes the maximal depth of nested $\Box$-es in the formula F.

Definition 2.1 The logic QGL^b is axiomatized over the first-order calculus in the extended language by the following axioms and rules of inference (below A, B stand for arbitrary formulas of the language QGL^b and x is any individual variable):

1. $\Box(A \to B) \to (\Box A \to \Box B)$
2. $\Box A \to \Box\Box A$
3. $\Box(\Box A \to A) \to \Box A$
4. $\Box A \to \Box \forall x A$

The rules of inference are: $\dfrac{A, A \to B}{B}$, $\dfrac{A}{\forall x A}$, $\dfrac{A}{\Box A}$.

We say that a formula A is in *normal form* iff for every subformula $\Box B$ of the formula A the formula B has no free occurrences of individual variables. It follows immediately from the Axioms 4 and 1 that every formula of the language QGL^b is equivalent to some formula in normal form.

2.2 Kripke-style semantics for QGL^b

Suppose K is a non-empty set, and $\prec$ is a binary relation on K. The relation $\prec$ is called converse well-founded if there are no infinite increasing chains of the kind $x_1 \prec x_2 \prec x_3 \prec \ldots$ in K. For finite K a transitive binary relation is converse well-founded if and only if it is irreflexive.

Definition 2.2 A Kripke model $\mathcal{K}$ is a triple $\langle \mathbf{K}, \prec, \{M_i\}_{i\in\mathbf{K}}\rangle$ where

$\mathbf{K}$ is a non-empty set of possible worlds,

$\prec$ is a transitive converse well-founded relation on $\mathbf{K}$,

$\{M_i\}$ $(i \in \mathbf{K})$ is a family of first-order structures labeled by elements of $\mathbf{K}$.

We define the forcing relation for formulas of the language QGL^b in every world $i \in \mathbf{K}$ in the following way.

- For pure predicate formulas $i \Vdash F$ iff $M_i \models F$;
- $i \Vdash \Box F$ iff $\forall j(i \prec j \Rightarrow j \Vdash F)$;
- Suppose that the forcing relation is already defined for all boxed subformulas of a formula F. Let F' be the result of substituting the propositional constant $\top$ for every true $\Box$-formula and $\bot$ for every false $\Box$-formula in a given world. Then we define the forcing relation for the formula F as follows:
$$i \Vdash F \text{ iff } M_i \models F'.$$

We say that $\mathcal{K} \models F$ (the formula F is true in the model K) if for all $i \in \mathbf{K}$ one has $i \Vdash F$.

Theorem 2.3 *$QGL^b \vdash A$ iff A is true in all Kripke models with a finite set of possible worlds.*

Proof. ($\Rightarrow$) Straightforward induction on the length of derivation in QGL^b.

($\Leftarrow$) Suppose $QGL^b \nvdash A$. Let Ω_A denote the following set of formulas: $A, \neg A \in \Omega_A$, and for every boxed subformula $\Box B$ of the formula A the formulas B, $\neg B$, $\Box B$ and $\neg\Box B$ are included in Ω_A.

Let $W \subset \Omega_A$. The set W is called *a maximal consistent subset* of Ω_A if it is QGL^b-consistent, i.e. $QGL^b \nvdash \neg(\bigwedge W)$, and for every formula B in Ω_A either B or its negation belongs to W (where the negation of a formula $\neg F$ is the formula F).

Note the following property of maximal consistent subsets of Ω_A. We skip the proof for reasons of triviality.

Lemma 2.4 *Every consistent subset of Ω_A can be extended to some maximal consistent subset.*

Let $\mathbf{K} = \{W_1, \ldots, W_n\}$ be the set of all maximal consistent subsets of Ω_A. We define the accessibility relation $\prec$ as follows: $W_i \prec W_j$ if for any boxed formula $\Box B \in \Omega_A$ if $\Box B \in W_i$ then $B, \Box B \in W_j$, and there exists a formula $\Box C \in \Omega_A$ such that $\neg\Box C \in W_i$ and $\Box C \in W_j$.

To complete the definition of $\mathcal{K}$ we have to describe a family $\{M_i\}$ ($i \in \mathbf{K}$) of first-order structures.

For every $W_i \in \mathbf{K}$ we define a set $\widetilde{W_i}^{(i)}$ of pure predicate formulas in the following way. Every formula $F \in W_i$ is translated to a pure predicate formula $\widetilde{F}^{(i)}$ as follows: if F is a pure predicate formula, the translation returns F, if $\Box B \in W_i$ then every occurrence of $\Box B$ in F is interpreted as $\top$, otherwise, if $\neg\Box B \in W_i$ then every occurrence of $\Box B$ is interpreted as $\bot$. (Clearly, it is sufficient to translate the most outward occurrences of $\Box$-es in F.) It is clear that the set $\widetilde{W_i}^{(i)}$ is consistent whenever W_i is consistent. So, according to the completeness theorem there exists a first-order structure M_i such that $M_i \models \widetilde{W_i}^{(i)}$.

Now we are going to show that the model $\mathcal{K} = \langle \mathbf{K}, \prec, \{M_i\}_{i \in \mathbf{K}} \rangle$ is the desired one, namely $\mathcal{K} \not\models A$.

Lemma 2.5 *For any formula $F \in \Omega_A$ one has*

$$F \in W_i \text{ iff } W_i \Vdash F.$$

Proof. We prove the lemma by joint induction on $d(F)$ and the length of F. Suppose $d(F) = 0$. The formula F has no occurrences of $\Box$-es, so $\widetilde{F}^{(i)}$ coincides with F. Thus, according to the definition of M_i one has $M_i \models F$ if $F \in W_i$ and $M_i \models \neg F$ otherwise.

Let now $d(F) > 0$. Consider the following two subcases.

Case 1; $F = \Box B$. If $\Box B \in W_i$ then according to the definition of the relation $\prec$ one has

$$\forall j (W_i \prec W_j \Rightarrow B \in W_j).$$

By the induction hypothesis $B \in W_j$ iff $W_j \Vdash B$, so $W_i \Vdash \Box B$.

If $\Box B \notin W_i$ then $\neg \Box B \in W_i$. We are going to show that there exists j such that $W_i \prec W_j$ and $W_j \Vdash \neg B$. Let $\Box D_1, \ldots, \Box D_k$ be an exhaustive enumeration without repetitions of all boxed formulas that belong to W_i. We state that the set $\{\Box D_1, \ldots, \Box D_k, \Box B, \neg B\}$ is consistent. Indeed, suppose it is not so:

$$QGL^b \vdash \neg(\Box D_1 \wedge \cdots \wedge \Box D_k \wedge \Box B \wedge \neg B).$$

Then one has

$$QGL^b \vdash \Box D_1 \wedge \cdots \wedge \Box D_k \to (\Box B \to B).$$

Using the modal necessity rule and Axioms 1, 2 one obtains

$$QGL^b \vdash \Box D_1 \wedge \cdots \wedge \Box D_k \to \Box(\Box B \to B).$$

Then, using Axiom 3 we have

$$QGL^b \vdash \Box D_1 \wedge \cdots \wedge \Box D_k \to \Box B.$$

Thus, we conclude that the set $\{\Box D_1, \ldots, \Box D_k, \neg \Box B\}$ is inconsistent, so W_i is inconsistent too; a contradiction.

So, there exists a maximal consistent subset W_j such that $W_i \prec W_j$ and $\neg B \in W_j$. Since $d(\neg B) < d(\Box B)$ by the induction hypotheses we conclude that $W_j \Vdash \neg B$.

Case 2; F is not a boxed formula. By the induction hypothesis the lemma holds for all boxed subformulas of F, since the length of a proper subformula is less than the length of the formula. Thus, F' is defined and one obtains

$$W_i \Vdash F' \leftrightarrow \widetilde{F}^{(i)}.$$

According to the definition of the forcing relation one has

$$W_i \Vdash F \Leftrightarrow W_i \Vdash F'.$$

On the other hand, $W_i \Vdash \widetilde{F}^{(i)}$ iff $\widetilde{F}^{(i)} \in \widetilde{W_i}^{(i)}$ iff $F \in W_i$. ⊣

Let us now complete the proof of the theorem. Since the set $\{\neg A\}$ is consistent, there exists a maximal consistent subset W_i such that $\neg A \in W_i$. According to Lemma 2.5 one can find a model such that $W_i \Vdash \neg A$. So, $\mathcal{K} \not\models A$. ⊣

2.3 Arithmetical completeness of QGL^b

An arithmetical interpretation $f = (T, *)$ of the language QGL^b has the following parameters:

- T is a recursively enumerable extension of PA such that all theorems of T are true in the standard model. We suppose that for every such a theory the standard provability formula $Prov_T(x)$ which enumerates all theorems of the theory T is given.
- $*$ is an arithmetical evaluation of all atomic predicate formulas by arithmetical formulas with the same set of free variables.

The interpretation f can be extended in natural way to all formulas of the language QGL^b. We stipulate that f commutes with boolean connectives, quantifiers and substitution of individual variables, i.e. if $f(Q(x)) = \varphi(x)$ then $f(Q(y)) = \varphi(y)$. For boxed formulas we define

$$f(\Box A) \rightleftharpoons Prov_T(\overline{\ulcorner f(A) \urcorner}),$$

where $\ulcorner f(A) \urcorner$ is a Gödel number of the formula $f(A)$ and $\overline{\ulcorner f(A) \urcorner}$ is the corresponding numeral. So, even in case A has occurrences of free individual variables the formula $f(\Box A)$ is bounded (as well as the formula $\Box A$ in the language QGL^b). In what follows we will usually omit the line over Gödel numbers to make reading easier.

Theorem 2.6 *Let A be an arbitrary formula of the language QGL^b. Then $QGL^b \vdash A$ iff for every arithmetical interpretation $f = (T, *)$ one has $T \vdash f(A)$.*

Proof. ($\Rightarrow$) Straightforward induction on the length of derivation of the formula A in QGL^b.

($\Leftarrow$) Suppose $QGL^b \nvdash A$. Consider the Kripke model $\mathcal{K}$ constructed in Theorem 2.3 such that $\mathcal{K} \not\models A$. Without loss of generality one can assume that $W_1 \nVdash A$.

According to the arithmetical formalization of the completeness theorem (see Hilbert and Bernays 1968) every satisfiable pure predicate formula is arithmetically satisfiable. Note that the sets of pure predicate formulas $\widetilde{W_1}^{(1)}, \dots, \widetilde{W_n}^{(n)}$ defined above are consistent, so there exist arithmetical interpretations $f_1, \dots, f_n$ of the pure predicate language such that formulas

$$f_1(\bigwedge \widetilde{W_1}^{(1)}), \dots, f_n(\bigwedge \widetilde{W_n}^{(n)})$$

are true in the standard model of arithmetic.

In order to define the desirable arithmetical interpretation f of the language QGL^b we need to choose a theory T and an evaluation $*$. Let

$$T \rightleftharpoons PA + f_1(\bigwedge \widetilde{W_1}^{(1)}) + \dots + f_n(\bigwedge \widetilde{W_n}^{(n)})$$

Thus, T is an extension of PA with a finite set of new axioms that are true in the standard model. Hence, all theorems of the theory T are true in the standard model too.

Now we have to define the evaluation $*$. We use here the standard Solovay-style embedding of Kripke models into arithmetic.

Lemma 2.7 (Solovay 1976) *There exist arithmetical sentences* σ_0, $\sigma_1, \dots, \sigma_n$ *such that*

1. $T \vdash \sigma_0 \vee \sigma_1 \vee \dots \vee \sigma_n$;
2. *for* $0 \le i < j \le n$ *one has* $T \vdash \neg\sigma_i \vee \neg\sigma_j$;
3. *for* $1 \le i, j \le n$ *if* $W_i \prec W_j$ *then* $T \vdash \sigma_i \to \neg Prov_T(\ulcorner \neg\sigma_j \urcorner)$;
4. *for* $1 \le i \le n$ *one has* $T \vdash \sigma_i \to Prov_T(\ulcorner \bigvee_{W_i \prec W_j} \sigma_j \urcorner)$;
5. *all* σ_i *are consistent, i.e.* $T \nvdash \neg\sigma_i$.

We define the evaluation $*$ in the following way. Let P be any atomic formula in the pure predicate language. Put

$$P^* \rightleftharpoons (\sigma_1 \wedge f_1(P)) \vee \dots \vee (\sigma_n \wedge f_n(P)).$$

Lemma 2.8 *For any pure predicate formula* Q *one has for* $1 \le i \le n$

$$T \vdash \sigma_i \to (f(Q) \leftrightarrow f_i(Q)).$$

Proof. Easily follows from the first and the second properties of the sentences σ_i. $\dashv$

Now we are going to verify that the interpretation f defined in this way is the desired one.

Lemma 2.9 *Let Ω_A be the set of formulas defined in the proof of Theorem 2.3, let $F \in \Omega_A$. Then for any i such that $1 \leq i \leq n$ one has*

1. *if $F \in W_i$ then $T \vdash \sigma_i \to f(F)$;*
2. *if $F \notin W_i$ then $T \vdash \sigma_i \to \neg f(F)$;*

Proof. Joint induction on $d(F)$ and the length of the formula.

Suppose $d(F) = 0$. Then the formula F has no occurrences of $\Box$. The statement of the lemma follows immediately from Lemma 2.8 and the definition of the theory T.

Let now $d(F) > 0$. Consider the following two cases.

Case 1; $F = \Box B$.

1.1. If $\Box B \in W_i$ then $\forall W_j(W_i \prec W_j \Rightarrow W_j \Vdash B)$. Since $d(B) < d(\Box B)$ one can apply the induction hypothesis, so one has

$$T \vdash \bigvee_{W_i \prec W_j} \sigma_j \to f(B).$$

Hence,

$$T \vdash Prov_T(\ulcorner \bigvee_{W_i \prec W_j} \sigma_j \urcorner) \to Prov_T(\ulcorner f(B) \urcorner).$$

Thus, according to the 4-th item of Lemma 2.7 one get

$$T \vdash \sigma_i \to Prov_T(\ulcorner f(B) \urcorner).$$

1.2. If $\Box B \notin W_i$ then $\exists W_j(W_i \prec W_j$ and $W_j \Vdash \neg B)$. By the induction hypothesis one has

$$T \vdash \sigma_j \to \neg f(B).$$

Hence,

$$T \vdash \neg Prov_T(\ulcorner \neg \sigma_j \urcorner) \to \neg Prov_T(\ulcorner f(B) \urcorner).$$

According to the 3-rd item of Lemma 2.7 one get

$$T \vdash \sigma_i \to \neg Prov_T(\ulcorner f(B) \urcorner).$$

Case 2; F is not a boxed formula. According to the induction hypothesis statements (1) and (2) of the lemma hold for all proper subformulas of the formula F. Therefore, the following holds:

$$T \vdash \sigma_i \to (f(F) \leftrightarrow f(\widetilde{F}^{(i)})).$$

Note that $\widetilde{F}^{(i)}$ is a pure predicate formula, so according to Lemma 2.8

$$T \vdash \sigma_i \to (f(F) \leftrightarrow f_i(\widetilde{F}^{(i)})).$$

On the other hand, according to the definition of the theory T one has

$$\text{if } F \in W_i \text{ then } T \vdash \widetilde{F}^{(i)},$$
$$\text{if } F \notin W_i \text{ then } T \vdash \neg\widetilde{F}^{(i)}.$$

This completes the proof of the lemma. ⊣

To complete the proof of the theorem recall that $\neg A \in W_1$ and $T \nvdash \neg\sigma_1$. $\dashv$

3 First-order logic of the standard proof predicate

3.1 Definition of the logic $\mathcal{QLP}^b$

The language of the first-order logic of proofs $\mathcal{QLP}^b$ is an extension of the pure predicate language with an infinite set of proof variables $p_1, p_2, \ldots$, the proof predicate $[\![t]\!]F$, symbols for operations on proofs $\cdot$, $+$, $!$, $?$, and an infinite set of unary operation symbols $g_1, g_2, \ldots$ for the operations of generalization over the corresponding individual variables $x_1, x_2, \ldots$.

Formulas and proof terms of the language $\mathcal{QLP}^b$ are constructed as follows:

- if Q is any n-ary predicate symbol and $y_1, \ldots, y_n$ are individual variables then $Q(y_1, \ldots, y_n)$ is a formula;
- if A, B are formulas then $\neg A$, $(A \wedge B)$, $(A \vee B)$, $(A \rightarrow B)$ are formulas too;
- if A is a formula and x_i is an individual variable then $\forall x_i A$ is a formula and x_i is excluded from the set of free variables in A;
- proof variables $p_1, p_2, \ldots$ are proof terms;
- if s, t are proof terms then $(s \cdot t)$, $(s + t)$, $!t$, $g_i(t)$ are proof terms too;
- if t is a proof term and A is a formula then $?(A, t)$ is a proof term;
- if A is a formula and t is a proof term then $[\![t]\!]A$ is a (quasiatomic) formula with no free variables, i.e. the proof operator binds all free occurrences of individual variables in A.

Definition 3.1 The logic $\mathcal{QLP}^b$ is defined by the following axioms and rules of inference over the first-order calculus in the language $\mathcal{QLP}^b$. (Below A, B, A_1, $\ldots$, A_n are arbitrary formulas of the language $\mathcal{QLP}^b$, s, t, t_1, $\ldots$, t_n are proof terms, and x_i is any individual variable. In the monotonicity axiom the notation $A(t)$ means that the term t explicitly occurs in the formula A.)

Operational axioms:

1. $[\![s]\!](A \rightarrow B) \rightarrow ([\![t]\!]A \rightarrow [\![s \cdot t]\!]B)$ (composition),
2. $([\![t]\!]A) \vee ([\![s]\!]A) \rightarrow [\![t + s]\!]A$ (non-deterministic choice),
3. $[\![t]\!]A \rightarrow [\![!t]\!]([\![t]\!]A)$ (verification),
4. $\neg[\![t]\!]A \rightarrow [\![?(A, t)]\!](\neg[\![t]\!]A)$ (negative introspection),
5. $[\![t]\!]A \rightarrow [\![g_i(t)]\!]\forall x_i A$ (generalization),

General axioms:

6. $[\![t]\!]A \to A$ (reflexivity),
7. $\neg([\![t_1]\!]A_1(t_2) \wedge [\![t_2]\!]A_2(t_3) \wedge \ldots \wedge [\![t_n]\!]A_n(t_1))$ (monotonicity).

The rules of inference: $\dfrac{A, A \to B}{B}$, $\dfrac{A}{\forall x_i A}$.

Concerning the operational version of the negative introspection axiom see Artemov et al. 1999. The monotonicity axiom was introduced in Artemov 1994. It formalizes the following property of the standard proof predicate. If m_1 is the Gödel number of a proof of a formula $F(m_2)$ then m_1 is greater than m_2. So, the monotonicity axiom is an analog of the following valid principle:

$$\neg((t_1 > t_2) \wedge (t_2 > t_3) \wedge \ldots \wedge (t_n > t_1)).$$

The system $\mathcal{QLP}^b$ has first appeared in Yavorsky 2000.

3.2 Arithmetical semantics for $\mathcal{QLP}^b$

Let T be any consistent extension of PA with a decidable set of axioms. We suppose that a certain Gödel numbering of the arithmetical language is fixed, so for every such theory we may consider the following multi-conclusion version of the standard proof predicate:

> "x is a code of a finite set of derivations in T, and y is a number of a formula, proved by one of these derivations".

The corresponding arithmetical formula is denoted by $PROOF_T(x, y)$.

An arithmetical interpretation f_T of the language $\mathcal{QLP}^b$ is organized in the following way:

- every atomic formula is interpreted by an arithmetical formula with the same set of free variables;
- f_T commutes with boolean connectives, quantifiers and substitution of variables, i.e. if $f_T(Q(x)) = \varphi(x)$ then $f_T(Q(y)) = \varphi(y)$;
- every proof variable is interpreted by a natural number;
- total computable functions unary $!'$, $\{g'_i\}_{i=1}^{\infty}$ and binary $\cdot'$, $+'$, $?'$ are defined, which satisfy the operational Axioms 1–5 of the logic $\mathcal{QLP}^b$, i.e.
  - ($\cdot$) if both $PROOF_T(n, \ulcorner\varphi \to \psi\urcorner)$ and $PROOF_T(m, \ulcorner\varphi\urcorner)$ hold then $PROOF_T(n \cdot' m, \ulcorner\psi\urcorner)$ holds too;
  - (+) if $PROOF_T(n, \ulcorner\varphi\urcorner)$ or $PROOF_T(m, \ulcorner\varphi\urcorner)$ holds then $PROOF_T(n +' m, \ulcorner\varphi\urcorner)$ holds too;
  - (!) if $PROOF_T(n, \ulcorner\varphi\urcorner)$ holds then $PROOF_T(!'(n), \ulcorner PROOF_T(\overline{n}, \overline{\ulcorner\varphi\urcorner})\urcorner)$ holds too;
  - (?) if the relation $PROOF_T(n, \ulcorner\varphi\urcorner)$ does not hold true then one has $PROOF_T(?'(\ulcorner\varphi\urcorner, n), \ulcorner\neg PROOF_T(\overline{n}, \overline{\ulcorner\varphi\urcorner})\urcorner)$;

(g_i) if $PROOF_T(n, \ulcorner\varphi\urcorner)$ holds then $PROOF_T(g_i'(n), \ulcorner\forall x_i\varphi\urcorner)$ is true too;

for any natural numbers i, m, n and any arithmetical formulas φ and ψ ;

- using these functions the interpretation of proof terms is computed;
- $f_T([\![t]\!]A) \rightleftharpoons PROOF_T(\overline{f_T(t)}, \overline{\ulcorner f_T(A)\urcorner})$.

Thus, every formula of the language $\mathcal{QLP}^b$ is interpreted by an arithmetical formula with the same set of free variables, and every proof term is interpreted by a natural number.

3.3 Arithmetical completeness of $\mathcal{QLP}^b$

Below **N** denotes the standard model of arithmetic.

Theorem 3.2 *Let A be any formula in the language $\mathcal{QLP}^b$. Then*

a) if $\mathcal{QLP}^b \vdash A$ then for every theory T and interpretation f_T one has $T \vdash f_T(A)$;

b) if $\mathcal{QLP}^b \nvdash A$ then there exists an extension T of PA by a finite set of true sentences and an interpretation f_T such that $\mathbf{N} \models \neg f_T(A)$.

Proof. a) Straightforward induction on the proof of A in $\mathcal{QLP}^b$.

b) Let *Terms*(A) denote the set of all proof terms occurring in A. Let X_A be the minimal set of quasiatomic formulas such that:

1. all quasiatomic subformulas of A belong to X_A;
2. if $(s \cdot t) \in Terms(A)$ and $[\![s]\!](B \to C) \in X_A$, $[\![t]\!]B \in X_A$ then $[\![s \cdot t]\!]C \in X_A$;
3. if $(s + t) \in Terms(A)$ and either $[\![s]\!]B \in X_A$, or $[\![t]\!]B \in X_A$ then $[\![s + t]\!]B \in X_A$;
4. if $(!t) \in Terms(A)$ and $[\![t]\!]B \in X_A$ then $[\![!t]\!][\![t]\!]B \in X_A$;
5. if $g_i(t) \in Terms(A)$ and $[\![t]\!]B \in X_A$ then $[\![g_i(t)]\!](\forall x_i B) \in X_A$;
6. if $?(B, t) \in Terms(A)$ then $[\![?(B, t)]\!](\neg[\![t]\!]B) \in X_A$.

One can show that for every $t \in Terms(A)$ the set $\{A \mid [\![t]\!]A \in X_A\}$ is finite. So, X_A is finite too.

Let K_A denote the conjunction of all proper axioms of $\mathcal{QLP}^b$ which contain only quasiatomic formulas from X_A. For Axiom 6 we take its universal closure.

Consider now a translation $\star$ of the language $\mathcal{QLP}^b$ into the pure predicate language which substitutes the corresponding fresh propositional variables $S_{[\![t]\!]F}$ for all occurrences of quasiatomic formulas $[\![t]\!]F$ (all pure predicate formulas are left unchanged). We stipulate that $\star$ commutes with boolean connectives, quantifiers and substitution of individual variables.

One can easily verify that the implication $K_A^\star \to A^\star$ is not provable in the first-order calculus, since $\mathcal{QLP}^b \nvdash A$. So, according to the arithmetical formalization of the completeness theorem for the first-order calculus (see Hilbert and Bernays 1968), there exists an arithmetical interpretation h of the pure predicate language, such that the formula $h(K_A^\star \to A^\star)$ is false in the standard model of arithmetic $\mathbf{N}$, i.e. $\mathbf{N} \models h^\star(K_A) \wedge \neg h^\star(A)$.

The composition of $\star$ and h (denoted by $h^\star$) is arranged very similar to the desirable interpretation f_T. Indeed, it assigns to every atomic predicate formula some arithmetical formula with the same set of free variables, it commutes with boolean connectives, quantifiers and substitution of variables, and $\mathbf{N} \models \neg h^\star(A)$.

Without loss of generality one may assume that for every propositional variable S_i either $h(S_i) = \top$ or $h(S_i) = \bot$ for $\top \rightleftharpoons (0 = 0)$, $\bot \rightleftharpoons (0 = 1)$. So, for every quasiatomic formula $[\![t]\!]F$ one has either $h^\star([\![t]\!]F) = \top$ or $h^\star([\![t]\!]F) = \bot$.

Recall that for every quasiatomic formula $[\![t]\!]F$ from X_A the implication $[\![t]\!]F \to F$ is included into K_A. So, if $h^\star([\![t]\!]F) = \top$ then $\mathbf{N} \models h^\star(F)$.

Now one can define the desired extension T of PA:

$$T = PA + \{h^\star(F) \mid [\![t]\!]F \in X_A \text{ and } h^\star([\![t]\!]F) = \top\}.$$

It follows from the observation above that all the additional axioms are true in $\mathbf{N}$.

Now one need to define the interpretation f_T. It will be done in three steps.

Step 1. Put $f_T(Q_i) \rightleftharpoons h^\star(Q_i)$ for all atomic formulas. Thus, f_T is defined for all pure predicate formulas since it commutes with boolean connectives and quantifiers. It is clear that for all these formulas the following condition holds:

$$PA \vdash f_T(F) \leftrightarrow h^\star(F) \tag{3}$$

Without loss of generality one can assume that the defined interpretation is injective. Otherwise, put $f_T(Q_i) \rightleftharpoons h^\star(Q_i) \wedge (i = i)$.

The condition of injectivity is not crucial for the proof, but it will be useful to simplify the definition of the functions in Step 3.

Step 2. Now we are going to define the interpretation f_T for all proof terms from *Terms*(A).

Consider a partial order on *Terms*(A) which is the transitive closure of the following relation: $t_1 \prec t_2$ iff for some formula $B(t_1)$ one has $[\![t_2]\!]B(t_1) \in X_A$ and $h^\star([\![t_2]\!]B(t_1)) = \top$. This order is well founded, since the monotonicity axiom for all such formulas is included in K_A.

Consider a term t_0, which is minimal with respect to this order relation. The set

$$Proved(t_0) \rightleftharpoons \{F \mid [\![t_0]\!]F \in X_A \text{ and } h^\star([\![t_0]\!]F) = \top\}$$

contains only pure predicate formulas, therefore the interpretation f_T is already defined for them. On the other hand, according to the definition of T and condition (3) the results of translation of all these formulas are equivalent in *PA* to the corresponding axioms of T. So, there are infinitely many proofs of the set $f_T(Proved(t_0))$ in T. We define $f_T(t_0)$ to be minimum of the Gödel numbers of such proofs.

Thus we have extended the set of $\mathcal{QLP}^b$-formulas for which the interpretation f_T is defined. It is clear that for all such formulas condition (3) still holds.

Let now $t_i \in Terms(A)$ be a minimal proof term (with respect to the defined order relation) such that $f_T(t_i)$ is not defined. Then, for every formula $F \in Proved(t_i)$ the interpretation $f_T(F)$ is already defined. According to property (3) all formulas from $f_T(Proved(t_i))$ are equivalent in *PA* to some axioms of T. So, they are provable in T. We define $f_T(t_i)$ to be minimum of the Gödel numbers of such proofs which were not assigned to proof terms from *Terms*(A) before.

Following this procedure we define $f_T(t)$ for all other terms occurring in the formula A.

After Step 2 terminates one has the interpretation f_T defined for all formulas containing occurrences of terms from *Terms*(A) only. One can easily check that f_T remains injective and for all such formulas condition (3) holds. In particular,

$$PA \vdash f_T(A) \leftrightarrow h^\star(A).$$

So, for the partially defined f_T one has

$$\mathbf{N} \models \neg f_T(A). \tag{4}$$

Step 3. Now one has to define the interpretation for all the other proof variables and to provide appropriate computable functions for the operations on proofs. These functions must satisfy all operational axioms and preserve the definition of the interpretation f_T for proof terms from *Terms*(A). This can be done easily, since f_T is injective on all terms from *Terms*(A) and all subformulas of A.

Note that at this point the inverse function f_T^{-1} is computable too. Moreover, for any natural number n one can effectively determine whether there exists a proof term t such that $f_T(t) = n$ and find this term in case the answer is positive. In particular, one can effectively determine whether the considered number is an image of a proof term from *Terms*(A).

Let k be a Gödel number of a proof such that $k \notin f_T(Terms(A))$. Put $f_T(p_i) \rightleftharpoons k$ for all proof variables not occurring in the formula A.

If both natural numbers n_1, n_2 are images of some proof terms $t_1, t_2 \in Terms(A)$ such that $(t_1 \cdot t_2) \in Terms(A)$ then put $n_1 \cdot' n_2 \rightleftharpoons f_T(t_1 \cdot t_2)$, otherwise define $n_1 \cdot' n_2$ in an arbitrary computable way such that the result does not get into the set $f_T(Terms(A))$ and the corresponding property of the operation $\cdot'$ is satisfied.

The operations $!', g_i', +'$ are treated similarly.

If natural numbers n_1, n_2 are such that $n_1 = \ulcorner f_T(B) \urcorner$, $n_2 = f_T(t)$, and $?(B,t) \in Terms(A)$ then put $?'(n_1, n_2) \rightleftharpoons f_T(?(B,t))$, otherwise define $?'(n_1, n_2)$ in an arbitrary computable way such that the result does not get into the set $f_T(Terms(A))$ and the corresponding property of the operation $?'$ is satisfied.

It is clear that these definitions of operations do preserve the definition of $f_T(A)$ given at Step 2.

To complete the proof one has to check whether the defined operations satisfy the corresponding operational Axioms 1–5. The definition of each operation consists of two cases. For the second case verification is trivial. For the first case it follows from the definitions of the set X_A and the conjunction K_A. Recall that K_A includes operational axioms for all terms from $Terms(A)$, and because of condition (3) these properties of operations are fulfilled under the interpretation f_T.

According to (4) one has $\mathbf{N} \models \neg f_T(A)$. $\dashv$

Comments

Here I would like to stress out two key points in the formulation of the completeness theorems. Namely, (1) why it is necessary to consider binding interpretations, and (2) why one should consider all arithmetically correct r.e. extensions of *PA*.

The main idea was to formulate the logic and the completeness theorem in such a way that both S. Artemov-V. Vardanyan's argument and T. Yavorskaya-Sidon's argument for non-axiomatizability failed.

Detailed inspection of the proof of non-axiomatizability theorems in Artemov 1980 and Vardanyan 1986 shows that the crucial point there is the possibility to express decidability in terms of provability. Consider the following scheme:

$$\forall x((P(x) \to Prov_T[P(x)])\&(\neg P(x) \to Prov_T[\neg P(x)])).$$

Whenever it is valid the formula $P(x)$ is decidable in T. On the other hand, if one considers the so-called binding interpretations the argument fails since one needs the variable x to be free in the formula $Prov_T[P(x)]$.

The second argument (see Artemov and Yavorskaya-Sidon 1999) is based on the following observation. For every recursively enumerable arithmetical theory T the set of pure predicate formulas that are not provable in T under any interpretation is strictly Π^0_1. It is clear that for every such a formula F the formula $\neg[\![p]\!]F$ is valid under any T-interpretation (the same holds for the formula $\neg\Box F$). So for every particular theory T the logic $\mathcal{QLP}^b(T)$ is not recursively enumerable. But if one considers intersection of all such logics for different arithmetically correct extensions of PA then the argument fails. Indeed, according to the arithmetical formalization of the Gödel completeness theorem every satisfiable formula in the pure predicate language is arithmetically satisfiable i.e. is true in the standard model of arithmetic under some interpretation. So, every such formula is provable under some interpretation in an appropriate extension of PA.

Acknowledgments

I am very grateful to my scientific advisor Professor Sergei N. Artemov for his permanent care and encouragement. Also, I am very thankful to Vladimir N. Krupski and Tatiana Yavorskaya-Sidon for helpful discussion on the subject.

References

Artemov, Sergei. 1980. Arithmetically complete modal theories (in Russian). *Semiotika i Informatika* 14:115–133. English translation, American Mathematical Society Translations (2), vol. 135 (1987), pp.39–54.

Artemov, Sergei. 1994. Logic of Proofs. *Annals of Pure and Applied Logic* 67:29–59.

Artemov, Sergei. 1998. Explicit provability: the intended semantics for intuitionistic and modal logic. *Techn. Rep. Mathematical Science Institute, Cornell University* (98-10). Available at http://www.math.cornell.edu/~artemov/publ.html.

Artemov, Sergei, Evgeni Kazakov, and Dmitri Shapiro. 1999. Logic of knowledge with justifications. *Techn. Rep. Mathematical Science Institute, Cornell University* (99-12).

Artemov, Sergei, and Tatiana Yavorskaya-Sidon. 1999. On the first order logic of proofs. *Techn. Rep. Mathematical Science Institute, Cornell University* (99-11).

Beklemishev, Lev. 1999. Classification of Propositional Provability Logics. In *Amer. Math. Soc. Transl. (2)*. 1–56. American Mathematical Society.

Boolos, Georg. 1979. *The unprovability of consistency: an essay in modal logic.* Cambridge: Cambridge University Press.

de Jongh, Dick, and Georgi Japaridze. 1998. The Logic of Provability. In *Handbook of proof theory*, ed. Samuel R. Buss. 475–546. Amsterdam: Elsevier.

Gödel, Kurt. 1933. Eine Interpretation des intuitionistischen Aussagenkalkuls. *Ergebnisse Math. Colloq.* 4:39–40.

Gödel, Kurt. 1995. Vortrag bei Zilsel (1938). In *Kurt Goedel Collected Works*, ed. Solomon Feferman. Oxford University Press.

Hilbert, David, and Paul Bernays. 1968. *Grundlagen der Mathematik*. Berlin: Springer–Verlag.

Kolmogoroff, Andrei. 1932. Zur Deutung der intuitionistischen Logik. *Math. Ztschr.* 35:58–65.

Orlov, I. E. 1928. The calculus of compatibility of propositions. *Mathematics of the USSR, Sbornik* 35:263–286. (in Russian).

Solovay, Robert M. 1976. Provability interpretations of modal logic. *Israel Journal of Mathematics* 25:287–304.

Vardanyan, Valeri A. 1986. Arithmetical complexity of predicate logics of provability and their fragments (in Russian). *Doklady Akademii Nauk SSSR* 288(1):11–14. English translation, Soviet Mathematics Doklady, vol. 34 (1986) pp. 384-387.

Yavorsky, Rostislav E. 2000. On the Logic of the Standard Proof Predicate. In *Lecture Notes in Computer Science*, ed. Peter G. Clote and Helmut Schwichtenberg. 527–541. Berlin Heidelberg New York: Springer. 14th International Workshop, CSL 2000, Annual Conference of the EACSL.

Rostislav E. Yavorsky
Steklov Mathematical Institute
Gubkina 8, Moscow,
117966, RUSSIA
E-mail: `rey@mi.ras.ru`
URL: `http://genesis.mi.ras.ru/~rey/`

2

From Description to Hybrid Logics, and Back

CARLOS ARECES AND MAARTEN DE RIJKE

ABSTRACT. Building on work by Schild, De Giacomo and Lenzerini, we establish a tight connection between description logics and hybrid logics. The main aim of the paper is to provide a modal perspective on some of the distinguishing features of description logic. In particular, by working in a hybrid logic setting we are able to develop a model-theoretic understanding of both assertional and terminological information. We also show how to use the connection between description and hybrid logics to transfer results on complexity and expressive power from one to the other.

1 Introduction

Nearly a decade ago, Schild (1991) observed and exploited the close correspondence between description logics and modal languages. He used it as a bridge to transfer complexity results and axiomatizations from modal logics to description logics, but noticed that the correspondence can only be established at the level of *concept satisfiability*. Basic modal logic is not expressive enough to account for either A-Box reasoning or inference in the presence of definitions (non-empty T-Boxes). Also, some very expressive description languages include constructions for building complex roles such as intersection, converse, and even transitive closure. By lifting the correspondence to Converse Propositional Dynamic Logic (CPDL, Fischer and Ladner 1979), Schild accounted for these constructions and for inference from non-empty T-Boxes. De Giacomo and Lenzerini (1994) extended these results by encoding A-Box reasoning in CPDL. While embeddings of description logics into CPDL have proved useful, they have two important disadvantages. Complexity-wise, the lo-

Advances in Modal Logic, Volume 3
F. Wolter, H. Wansing, M. de Rijke, and M. Zakharyaschev, eds.

Constructor	Syntax	Semantics
concept name	C	$C^{\mathcal{I}}$
top	$\top$	$\Delta^{\mathcal{I}}$
negation ($\mathcal{C}$)	$\neg C$	$\Delta^{\mathcal{I}} \setminus C^{\mathcal{I}}$
conjunction	$C_1 \sqcap C_2$	$C_1^{\mathcal{I}} \cap C_2^{\mathcal{I}}$
disjunction ($\mathcal{U}$)	$C_1 \sqcup C_2$	$C_1^{\mathcal{I}} \cup C_2^{\mathcal{I}}$
universal quant.	$\forall R.C$	$\{d_1 \mid \forall d_2 \in \Delta^{\mathcal{I}}.(R^{\mathcal{I}}(d_1, d_2) \rightarrow d_2 \in C^{\mathcal{I}})\}$
existential quant. ($\mathcal{E}$)	$\exists R.C$	$\{d_1 \mid \exists d_2 \in \Delta^{\mathcal{I}}.(R^{\mathcal{I}}(d_1, d_2) \wedge d_2 \in C^{\mathcal{I}})\}$
one-of ($\mathcal{O}$)	$\{a_1, \ldots, a_n\}$	$\{d \mid d = a_i^{\mathcal{I}}$ for some $a_i\}$
role filler ($\mathcal{B}$)	$\exists R.\{a\}$	$\{d \mid R^{\mathcal{I}}(d, a^{\mathcal{I}})\}$
role name	R	$R^{\mathcal{I}}$
inverse roles ($\mathcal{I}$)	R^{-1}	$\{(d_1, d_2) \mid R^{\mathcal{I}}(d_2, d_1)\}$

TABLE 1 Common operators of description logics.

cal satisfiability problem of CPDL (i.e., the problem of finding, given a CPDL formula ϕ, a model $\mathcal{M}$ and a state m such that $\mathcal{M}, m \Vdash \phi$) is already EXPTIME-complete, and this prohibits sharp complexity results. And perhaps more crucially if our main aim is to understand the general behavior of description logics, the model theory of CPDL is complex, because the Kleene star (and hence a weak notion of induction) needs to be taken into consideration.

In this paper, we replace CPDL by hybrid languages and in this way shed new light on the issues above. Our main aim is to establish a very tight connection between description and hybrid logics, and to show how it provides a modal perspective on description logics, by means of a number of transparent model-theoretic ideas. We also indicate some possible uses of this connection in terms of results on complexity, expressive power, and meta-logical properties like interpolation and Beth definability.

We start by providing some background. We then recall relevant work by Schild, and De Giacomo and Lenzerini. After that we set up the link between hybrid and description logics, and exploit it.

1.1 Description Logic

Description logics (DLs) are a family of formal languages with a clearly specified semantics, usually in terms of first-order models, together with inference mechanisms to account for knowledge classification. One of the main aims of the field is to identify fragments of first-order logic that are able to capture the features needed for representing a particular problem domain, and which still admit efficient reasoning algorithms.

Let $\mathsf{CON} = \{C_1, C_2, \ldots\}$ be a countable set of *atomic concepts*, $\mathsf{ROL} = \{R_1, R_2, \ldots\}$ a countable set of *atomic roles*, and $\mathsf{IND} = \{a_1, a_2, \ldots\}$ a countable set of *individuals*. For CON, ROL, IND, all pairwise disjoint, $\mathcal{S} = \langle \mathsf{CON}, \mathsf{ROL}, \mathsf{IND} \rangle$ is a *signature.* An *interpretation* $\mathcal{I}$ for $\mathcal{S}$ is a tuple $\mathcal{I} = \langle \Delta^{\mathcal{I}}, \cdot^{\mathcal{I}} \rangle$, where $\Delta^{\mathcal{I}}$ is a non-empty set, and $\cdot^{\mathcal{I}}$ assigns elements $a_i^{\mathcal{I}} \in \Delta^{\mathcal{I}}$ to constants a_i, subsets $C_i^{\mathcal{I}} \subseteq \Delta^{\mathcal{I}}$ to atomic concepts C_i, and relations $R_i^{\mathcal{I}} \subseteq \Delta^{\mathcal{I}} \times \Delta^{\mathcal{I}}$ to atomic roles R_i. The atomic symbols in a DL signature can be combined by means of *concept* and *role constructors*, to form complex expressions. Table 1 defines the constructors for the DLs we will discuss, together with their semantics. It is customary to define systems by postfixing the names of some basic description languages like $\mathcal{AL}$ or $\mathcal{FL}$ with the names of the added operators from Table 1. In this paper, we will be interested in languages having full Boolean expressivity and hence focus on $\mathcal{ALC}$ and its extensions. The most expressive language we will deal with is $\mathcal{ALCOI}$, that is $\mathcal{ALC}$ extended with the "one-of" operator and converse roles (this language can trivially encode the "role-filler" operator $\mathcal{B}$).

In DLs we want to perform inferences given certain background knowledge. Let $\mathcal{L}$ be any description logic, a *knowledge base* Σ in $\mathcal{L}$ is a pair $\Sigma = \langle T, A \rangle$ such that T is the T(erminological)-Box: a finite, possibly empty, set of expressions of the form $C_1 \sqsubseteq C_2$, where C_1, $C_2 \in \mathsf{CON}(\mathcal{L})$ ($C_1 \doteq C_2$ is short for '$C_1 \sqsubseteq C_2$ and $C_2 \sqsubseteq C_1$'). Formulas in T are called *terminological axioms.* In addition, A is the A(ssertional)-Box: a finite, possibly empty, set of expressions of the forms $a : C$ or $(a, b) : R$ where C is in $\mathsf{CON}(\mathcal{L})$, R is in $\mathsf{ROL}(\mathcal{L})$ and a, b are individuals. Formulas in A are called *assertions.*

Let $\mathcal{I}$ be an interpretation and ϕ a terminological axiom or assertion. Then $\mathcal{I}$ *models* ϕ (notation: $\mathcal{I} \models \phi$) if $\phi = C_1 \sqsubseteq C_2$ and $C_1^{\mathcal{I}} \subseteq {C_2}^{\mathcal{I}}$, or $\phi = a{:}C$ and $a^{\mathcal{I}} \in C^{\mathcal{I}}$, or $\phi = (a,b){:}R$ and $(a^{\mathcal{I}}, b^{\mathcal{I}}) \in R^{\mathcal{I}}$. If $\Sigma = \langle T, A \rangle$ is a knowledge base and $\mathcal{I}$ an interpretation, then $\mathcal{I}$ models Σ (notation: $\mathcal{I} \models \Sigma$) if for all $\phi \in T \cup A, \mathcal{I} \models \phi$. Given a knowledge base Σ and a terminological axiom or assertion ϕ, we write $\Sigma \models \phi$ if for all models $\mathcal{I}$ of Σ we have $\mathcal{I} \models \phi$. All standard description logic reasoning tasks (like subsumption or instance checking) can be defined in terms of this relation.

1.2 Hybrid Logic

Modal formulas are evaluated at a given *state* in a model, and their truth values depend on the value of formulas at some related *states.* Yet, nothing in modal syntax gets to grips with the *states* themselves. Hybrid languages are modal languages which solve this "reference problem" by introducing special symbols, called *nominals,* to explicitly name the states in a model.

The basic hybrid language is $\mathcal{H}_{\mathsf{N}}$, basic modal logic extended with nominals. Further extensions are named by listing the added operators.

The most expressive system we will discuss is $\mathcal{H}_{\mathsf{N}}(\langle R^{-1}\rangle, \mathsf{E}, @)$, the basic hybrid language extended with the converse and existential modalities, and the @ operator. More precisely, let $\mathsf{REL} = \{R_1, R_2, \ldots\}$ be a countable set of *relation symbols*, $\mathsf{PROP} = \{p_1, p_2, \ldots\}$ a countable set of *proposition letters*, and $\mathsf{NOM} = \{i_1, i_2, \ldots\}$ a countable set of *nominals*. $\mathsf{ATOM} = \mathsf{PROP} \cup \mathsf{NOM}$ is the set of *atoms*. The formulas of the hybrid language $\mathcal{H}_{\mathsf{N}}(\langle R^{-1}\rangle, \mathsf{E}, @)$ in the signature $\langle \mathsf{REL}, \mathsf{PROP}, \mathsf{NOM}\rangle$ are

$$\mathsf{FORMS} := \top \mid a \mid \neg\phi \mid \phi_1 \wedge \phi_2 \mid \langle R\rangle\phi \mid \langle R^{-1}\rangle\phi \mid \mathsf{E}\phi \mid @_i\phi,$$

where $a \in \mathsf{ATOM}$, $R \in \mathsf{REL}$, $i \in \mathsf{NOM}$, and ϕ, ϕ_1, $\phi_2 \in \mathsf{FORMS}$.

A *hybrid model* $\mathcal{M}$ is a triple $\mathcal{M} = \langle M, \{R_i\}, V\rangle$ where M is a non-empty set, $\{R_i\}$ is a set of binary relations on M, and $V : \mathsf{PROP} \cup \mathsf{NOM} \to Pow(M)$ is such that for all nominals $i \in \mathsf{NOM}$, $V(i)$ is a singleton subset of M. Let $\mathcal{M} = \langle M, \{R_i\}, V\rangle$ be a model and $m \in M$. The interesting cases of the *satisfiability relation* are as follows: $\mathcal{M}, m \Vdash a$ iff $m \in V(a)$, $a \in \mathsf{ATOM}$; $\mathcal{M}, m \Vdash \langle R\rangle\phi$ iff $\exists m' \,(R(m, m')\ \&\ \mathcal{M}, m' \Vdash \phi)$; $\mathcal{M}, m \Vdash \langle R^{-1}\rangle\phi$ iff $\exists m' \,(R(m', m)\ \&\ \mathcal{M}, m' \Vdash \phi)$; $\mathcal{M}, m \Vdash \mathsf{E}\phi$ iff $\exists m' \,(\mathcal{M}, m' \Vdash \phi)$; and $\mathcal{M}, m \Vdash @_i\phi$ iff $\mathcal{M}, m' \Vdash \phi$, where $V(i) = \{m'\}$, $i \in \mathsf{NOM}$.

We write $\mathcal{M} \Vdash \phi$ iff for all $m \in M$, $\mathcal{M}, m \Vdash \phi$. This notion extends to sets of formulas in the standard way. A formula ϕ is *satisfiable* if there is a model $\mathcal{M}$ and a state $m \in M$ with $\mathcal{M}, m \Vdash \phi$. A formula ϕ is *valid* if for all models $\mathcal{M}$, $\mathcal{M} \Vdash \phi$. ϕ is a *local consequence* of a set of formulas T (notation: $T \models^{loc} \phi$), if for all models $\mathcal{M}$ and points $m \in M$, $\mathcal{M}, m \Vdash T$ implies $\mathcal{M}, m \Vdash \phi$; ϕ is a *global consequence* of a set of formulas T (notation: $T \models^{glo} \phi$), if for all models $\mathcal{M}$, $\mathcal{M} \Vdash T$ implies $\mathcal{M} \Vdash \phi$. When T is the empty set, we have $\{\} \models^{glo} \phi$ iff $\{\} \models^{loc} \phi$, and simply write $\models \phi$.

2 Schild's Terminologies

It is easy to map concepts in $\mathcal{ALC}$ into formulas of CPDL, while preserving satisfiability — actually, basic poly-modal logic is enough. Just define the translation $\cdot^t$ by putting $(C_i)^t = p_i$, for C_i an atomic concept; $(\neg C)^t = \neg(C^t)$; $(C \sqcap D)^t = C^t \wedge D^t$; and $(\exists R.C)^t = \langle R\rangle C^t$. It is clear that $\cdot^t$ preserves satisfiability. But we need further expressive power to account for T-Box and A-Box reasoning.

Consider the signature $\mathcal{S} = \langle\{C_1, C_2\}, \{R\}, \{a\}\rangle$ and the interpretations $\mathcal{I}_1 = \langle\{m_1, m_2\}, \cdot^{\mathcal{I}_1}\rangle$ and $\mathcal{I}_2 = \langle\{m_3, m_4, m_5\}, \cdot^{\mathcal{I}_2}\rangle$ where $C_1^{\mathcal{I}_1} = \{m_1\}$, $C_2^{\mathcal{I}_1} = \{m_1, m_2\}$, $R^{\mathcal{I}_1} = \{\}$, $a^{\mathcal{I}_1} = m_1$; and $C_1^{\mathcal{I}_2}\{m_4\}$, $C_2^{\mathcal{I}_2} = \{m_3\}$, $R^{\mathcal{I}_2} = \{\}$, $a^{\mathcal{I}_2} = m_5$. Clearly, $\mathcal{I}_1$ models both $C_1 \sqsubseteq C_2$ and $a{:}C_1$ while $\mathcal{I}_2$ models neither. On the other hand, when we consider $\mathcal{I}_1$ and

$\mathcal{I}_2$ as modal models, the relation $\{(m_2, m_3)\}$ is a bisimulation, showing that there are no basic modal formula capturing these notions.

Observe that terminological axioms such as $C_1 \sqsubseteq C_2$ and $a : C_1$ express global properties: they are true in a point in a model iff they are true in all elements of the model. Schild (1991) accounts for the global nature of terminological axioms by using the collapsed model property of CPDL (any satisfiable CPDL formula is satisfiable in a connected model) and the availability of the Kleene star. Due to the former, we can ignore states which are not reachable by a finite sequence of backwards and forwards transitions. And due to the Kleene star we can "compress" these transitions into a single step. Formally, extend $\cdot^t$ by putting $(C \sqsubseteq D)^t = (C^t \to D^t)$. And for a finite set of terminological axioms T, let T^t be $\bigwedge \phi_i^t$, where $\phi_i \in T$. Now, let $T \cup \{\phi\}$ be a finite set of terminological axioms, and let $R_1, \ldots, R_n$ be all the roles in $T \cup \{\phi\}$. Then $\langle T, \{\}\rangle \models \phi$ iff $\models [(R_1 \cup R_1^{-1} \cup \cdots \cup R_n \cup R_n^{-1})^*]T^t \to \phi^t$.

As Schild remarks, this translation would not work for an infinite T. On the one hand, T might contain an infinite number of roles, but even in the case of a finite signature, PDL is not compact (Harel 1984, Theorem 2.15), hence inference from infinite sets does not coincide with inference from finite sets. More importantly, lack of compactness has a striking effect on the complexity of the consequence problem, which becomes highly undecidable, an indication that PDL is not computationally well behaved. The computational problems caused by the Kleene star have been well investigated both in the modal and description logic community (Ladner 1977, Halpern and Moses 1992, Sattler 1996, Horrocks and Gough 1997); and authors like Sattler, and Horrocks and Gough have argued that the ability to define a role as transitive often suffices in applications, instead of the full power of transitive closure.

What is going on here? How can we understand terminological axioms in modal terms, using the kind of model-theoretic tools that have been used to analyze modal logic itself? Before we answer this question we highlight one more distinguishing feature of DLs.

3 De Giacomo's Individuals

Accounting for assertional information in CPDL is more complicated than encoding terminological axioms. Below we present a simplified version of a translation proposed by De Giacomo and Lenzerini (1994); the latter enforce the unique name assumption and also deal with complex structure on roles (union, composition, transitive closure, etc.) which makes for the additional complexity. Here, we only discuss the handling of individuals.

Extend the translation $\cdot^t$ defined in Section 2 to assertions by defining $(a : C)^t = p_a \to C^t$, and $((a, b) : R)^t = p_a \to \langle R \rangle p_b$, where p_a and p_b are propositional symbols. Let A be a finite set of assertions, define A^t as $\bigwedge \phi_i^t$ for $\phi_i \in A$. The problem now is that in translating *individuals* as *propositions* in CPDL we have lost the information that individuals denote a single element in the domain. Hence, we have to explicitly force these symbols to behave as individuals. Let $\Sigma = \langle T, A \rangle$ be a knowledge base, $R_1, \ldots, R_n$ the roles appearing in Σ, $a_1, \ldots, a_m$ the individuals mentioned in Σ, and let $\mathsf{SF}(\phi)$ be the set of all subformulas of ϕ. Let $[U]$ stand for $[(R_1 \cup R_1^{-1} \cup \cdots \cup R_n \cup R_n^{-1})^*]$, and let S be a role not in Σ. Let Σ^t be

$$[S][U](A^t \wedge T^t) \wedge$$
$$\bigwedge_{1 \leq i \leq m} \left(\langle S \rangle p_{a_i} \wedge \left(\bigwedge_{\psi \in \mathsf{SF}(T^t \wedge A^t)} [S](\langle U \rangle (p_{a_i} \wedge \psi) \to [U](p_{a_i} \to \psi)) \right) \right).$$

It can be shown that Σ is consistent if and only if Σ^t is satisfiable. This is enough because in sufficiently expressive languages like the ones we consider in this paper, all standard reasoning tasks can be reduced to knowledge base consistency.

As remarked by Horrocks et al. (2000), De Giacomo's translation is probably too involved and costly to provide effective decision methods. It is also difficult to extract theoretical results from it, except for the general complexity results presented in De Giacomo and Lenzerini (1994). As we already remarked, the model theory of CPDL is intricate because of the inductive nature of the Kleene star, and the cryptic translation provides little help on simplifying things out.

4 Universal Statements and Individuals

The two features of DLs that most modal logicians would probably find hard to understand in terms of the model-theoretic notions they are used to (bisimulations, axiomatizations, etc.), are the use of global information and of information about individuals.

One way to understand the issue of global information is by paying special attention to the standard notions of consequence used in description and modal logics. In Section 1 we introduced two notions of consequence, a *local* one and a *global* one which differ in the way we quantify over states in the model. To understand what is going on, it is helpful to discuss consequence in first-order terms. Given a set of formulas $\Gamma \cup \{\phi\}$ which might contain free variables, the way we define the quantification over models $\mathcal{M}$ and first-order assignments g becomes meaningful. We can either require

(Global) $\forall \mathcal{M}(\forall g \mathcal{M} \models \Gamma[g] \Rightarrow \forall g. \mathcal{M} \models \phi[g])$, or

(Local) $\forall \mathcal{M} \forall g(\mathcal{M} \models \Gamma[g] \Rightarrow \mathcal{M} \models \phi[g])$

The global consequence relation is the one familiar from first-order logic, but first-order consequence is usually analyzed over sets of *sentences*, and in this case both the global and the local notions coincide. Modal languages are usually equivalent to first-order formulas with *free variables*, and choosing one of the two possibilities becomes an issue. In line with the general local perspective of modal logics (evaluation of a formula *at a state* in the model), the local notion of consequence is the most natural. But the presence of the E modality in hybrid languages makes things simpler, as we can easily interdefine local and global consequence. Let A (the universal modality) be the dual of $\Diamond$ (i.e., $\mathsf{A}\phi := \neg\mathsf{E}\neg\phi$), then

$$\Gamma \models^{glo} \phi \text{ iff } \{\mathsf{A}\psi \mid \psi \in \Gamma\} \models^{loc} \phi. \quad (1)$$

Goranko and Passy (1992) study properties of languages containing E, and prove that global properties of a logic $\mathcal{L}$ correspond to local properties of the logic $\mathcal{L}^{\mathsf{E}}$ which arises from $\mathcal{L}$ by adding E. In particular, for basic modal logics, global decidability, global finite model property, and global completeness of a logic $\mathcal{L}$ are equivalent to their local versions for $\mathcal{L}^{\mathsf{E}}$. This result can be extended to many hybrid languages.

Theorem 4.1 *Let the property P be either decidability, finite model property, or axiomatic completeness, and let $\mathcal{L}$ be any sublanguage of $\mathcal{H}_{\mathsf{N}}(\langle R^{-1}\rangle, @)$. Then $\mathcal{L}$ has P globally iff $\mathcal{L}^{\mathsf{E}}$ has P locally.*

Going back to description languages, notice that if we use $\models^{glo}$ instead of $\models^{loc}$, then basic modal logic *is* enough to encode terminological axioms, as the following equivalence holds: $\langle T, \{\}\rangle \models \phi$ iff $T^t \models^{glo} \phi^t$. By using (1) we can move to $\langle T, \{\}\rangle \models \phi$ iff $\{\mathsf{A}(T^t)\} \models^{loc} \phi^t$. And given that the local consequence relation satisfies the deduction theorem, we obtain

$$\langle T, \{\}\rangle \models \phi \text{ iff } \models \mathsf{A}(T^t) \rightarrow \phi^t.$$

By Theorem 4.1, then, we can study logical properties of *inference from non-empty knowledge bases* through *local properties of satisfiability* of languages containing E. If the logic is compact, we can perform this reduction even for infinite T-Boxes, but, most importantly, by replacing the Kleene star with the existential modality we obtain a language with a much better behaved and understood model theory.

Let us now turn to the second feature of DLs that many modal logicians find puzzling from a model-theoretic point of view: A-Box information.

The main difficulty with the extended translation $(\cdot)^t$ defined in Section 3 is in forcing propositional symbols in CPDL to behave as individuals. If we use hybrid logics instead, we can simply use *nominals.* A-Box statements can then be straightforwardly accounted for, as $a:C$ simply becomes $@_a C^t$ and $(a,b) : R$ can be translated as $@_a\langle R\rangle b$. The whole idea of nominals seems to fit neatly in the DLs perspective, so much so that many ideas and techniques from hybrid logics, involving the direct use of nominals, have been already taken up by the description logic community (see, e.g., Tobies 2000).

Given the presence of the existential modality and nominals, hybrid logics are very well suited to provide a modal perspective on description languages that is able to deal with full terminological and assertional reasoning. Consider the following translation $\cdot^h$ taking concepts, terminological axioms and assertions to hybrid formulas:

$$
\begin{aligned}
(C_i)^h &= p_i,\ (C_i \text{ atomic}) & (\{a_1,\dots,a_n\})^h &= a_1 \vee \dots \vee a_n\\
(\neg C)^h &= \neg(C^h) & (\exists R.\{a\})^h &= \langle R\rangle a\\
(C \sqcap D)^h &= C^h \wedge D^h & (C \sqsubseteq D)^h &= \mathsf{A}(C^h \to D^h)\\
(\exists R.C)^h &= \langle R\rangle C^h & (a:C)^h &= @_a C^h\\
(\exists R^{-1}.C)^h &= \langle R^{-1}\rangle C^h & ((a,b):R)^h &= @_a\langle R\rangle b.
\end{aligned}
$$

Theorem 4.2 *Let $\Sigma = \langle T, A\rangle$ be a knowledge base in $\mathcal{ALCOI}$, and ϕ a terminological axiom or an assertion, then $\langle T, A\rangle \models \phi$ iff $\models (\bigwedge_{\psi\in T}\psi^h \wedge \bigwedge_{\psi\in A}\psi^h) \to \phi^h$.*

The proof is obvious (and the connection between the two languages is stronger than with CPDL), as any model of $\langle T, A\rangle$ and ϕ can be viewed as a model of $(\bigwedge_{\psi\in T}\psi^h \wedge \bigwedge_{\psi\in A}\psi^h) \to \phi^h$ and vice versa.

Hybrid logics have already been proposed as natural counterparts of description languages. Blackburn and Tzakova (1998) propose using hybrid languages to embed description logics, highlighting the connection between assertional information and nominals, and the use of the existential modality to encode terminological axioms. But they introduce undecidable hybrid languages for this account, arguing that the increase in expressive power of these languages is an advantage. Instead, our translation tries to remain as faithful as possible to the original description language, and pays special attention to decidability issues.

What kind of expressive power is needed to encode the different languages and reasoning tasks? For example, the existential modality is required only for translating terminological axioms, while @ is only used for assertions. The following lists a number of relevant hybrid languages:

- $\mathcal{H}_{\mathsf{N}}(\langle R^{-1}\rangle, @, \mathsf{E})$, in which the full translation of $\mathcal{ALCOI}$ with non-empty T-Boxes and A-Boxes can be made.
- $\mathcal{H}_{\mathsf{N}}(\langle R^{-1}\rangle, @)$, in which only inferences in terms of empty T-Boxes can be performed.
- $\mathcal{H}(\langle R^{-1}\rangle, @, @\Diamond, \mathsf{E})$, in which we only allow nominals to appear as subindices of @ and in the construction $@_a\langle R\rangle b$ or $@_a\langle R^{-1}\rangle b$, and hence we can translate neither the one-of operator $\mathcal{O}$ nor role fillers $\mathcal{B}$.
- $\mathcal{H}(\langle R^{-1}\rangle, @, @\Diamond)$, the "local" version of the language above, where we work with empty T-Boxes.

We have defined each of the logics above to be expressive enough to permit the encoding of certain specific DLs. But it is also important to determine if and how we have *extended* the expressive power of the source language with the move into these hybrid languages. The general answer is: we have incorporated Boolean structure into the knowledge base, and allowed explicit interaction among T-Box definitions, A-Box assertions and concepts.

As an example, we will show how to recast satisfiability of a formula in $\mathcal{H}_{\mathsf{N}}(\langle R^{-1}\rangle, @, \mathsf{E})$, in terms of satisfiability in the corresponding description logic. It will be useful to first discuss a normal form for hybrid formulas. Take a formula ϕ in $\mathcal{H}_{\mathsf{N}}(\langle R^{-1}\rangle, @, \mathsf{E})$. We start by translating ϕ into propositional normal form. Then we use the following equivalences to "push out" the E and A operators from inside the other modalities (we use Q to range over E and A):

$$\begin{array}{rcl|rcl}
[R]\mathsf{Q}\psi & \leftrightarrow & [R]\bot \vee \mathsf{Q}\psi & @_s\mathsf{Q}\psi & \leftrightarrow & \mathsf{Q}\psi \\
[R](\theta \vee \mathsf{Q}\psi) & \leftrightarrow & [R]\theta \vee \mathsf{Q}\psi & @_s(\theta \vee \mathsf{Q}\psi) & \leftrightarrow & @_s\theta \vee \mathsf{Q}\psi \\
[R](\theta \wedge \mathsf{A}\psi) & \leftrightarrow & [R]\theta \wedge [R]\mathsf{A}\psi & @_s(\theta \wedge \mathsf{Q}\psi) & \leftrightarrow & @_s\theta \wedge \mathsf{Q}\psi \\
[R](\theta \wedge \mathsf{E}\psi) & \leftrightarrow & [R]\bot \vee ([R]\theta \wedge \mathsf{E}\psi) & & &
\end{array}$$

Similar equivalences hold for the dual modalities $\langle R\rangle$ (@ is self dual). For pushing out @ we have

$$\begin{array}{rcl|rcl}
[R]@\psi & \leftrightarrow & [R]\bot \vee @_s\psi & @_s@_t\psi & \leftrightarrow & @_t\psi \\
[R](\theta \vee @_s\psi) & \leftrightarrow & [R]\theta \vee @_s\psi & @_s(\theta \vee @_t\psi) & \leftrightarrow & @_s\theta \vee @_t\psi \\
[R](\theta \wedge @_s\psi) & \leftrightarrow & [R]\theta \wedge [R]@_s\psi & @_s(\theta \wedge @_t\psi) & \leftrightarrow & @_s\theta \wedge @_t\psi.
\end{array}$$

And similarly for the @ operators appearing under $\langle R\rangle$. We arrive then to the following result.

Proposition 4.3 *Let ϕ be a hybrid formula, then ϕ is equivalent to a formula ϕ' where subformulas of the form $\mathsf{E}\psi$, A and $@_i\psi$ (if any) occur* *only at modal depth* 0. *In particular,* ϕ' *can be taken to be*

$$\bigwedge_{l\in L}\Big(\bigvee_{m\in M}\mathsf{A}\rho_{(l,m)} \vee \mathsf{E}\sigma_l \vee \bigvee_{i\in\mathsf{NOM}(\phi)} @_i\nu_{(l,i)} \vee \tau_l\Big),$$

for some (possibly empty) index sets L, M, *where* $\rho_{(l,m)}$, σ_l, $\nu_{l,i}$ *and* τ_l *contain neither* E *nor* $@$.

By allowing negations in the T-Box we can encode validity of formulas in $\mathcal{H}_{\mathsf{N}}(\langle R^{-1}\rangle, @, \mathsf{E})$ as instance checking as follows. Define *Boolean knowledge bases* as pairs $\Sigma = \langle T, A\rangle$ where T is a set of Boolean combinations of terminological axioms, and A a set of Boolean combinations of assertions. In the description logic community, Boolean knowledge bases have been considered in the setting of the CLASSIC system (Borgida et al. 1989), while authors with a modal logic background, such as Wolter and Zakharyaschev (2000), have also considered this kind of knowledge bases.

Take $\phi \in \mathcal{H}_{\mathsf{N}}(\langle R^{-1}\rangle, @, \mathsf{E})$ in the normal form described in Proposition 4.3. For $l \in L$, define the knowledge base $\Sigma^l_\phi = \langle T^l_\phi, A^l_\phi\rangle$ by putting

$$\begin{aligned} T^l_\phi &= \{\neg(\top \sqsubseteq \rho^{h^{-1}}_{(m,l)}) \mid m \in M\} \cup \{\top \sqsubseteq \neg\sigma^{h^{-1}}_l\}, \text{ and}\\ A^l_\phi &= \{i : \neg\nu^{h^{-1}}_{(l,i)} \mid i \in \mathsf{NOM}(\phi)\}, \end{aligned}$$

where the mapping $\cdot^{h^{-1}}$ is the backwards translation from the hybrid language into $\mathcal{ALCOI}$ that sends Boolean and modal operators to the corresponding description logic ones and using singleton one-of sets $\{i\}$ for translating nominals.

Theorem 4.4 *For any formula* ϕ *in* $\mathcal{H}_{\mathsf{N}}(\langle R^{-1}\rangle, @, \mathsf{E})$, *let* $a \notin \mathsf{NOM}(\phi)$, *then* ϕ *is valid iff for all* $l \in L$, $\Sigma^l_\phi \models a : \tau^{h^{-1}}_l$.

Interestingly, even if we allow Boolean knowledge bases, we cannot recast validity of hybrid formulas as inference in terms of a *unique* knowledge base. This is because the separation between terminological axioms, assertions and simple concepts imposes syntactic restrictions which don't exist in hybrid logic. Trivially, if the index set L above is a singleton, then a unique knowledge base is sufficient. I.e., we can precisely characterize the fragment of $\mathcal{H}_{\mathsf{N}}(\langle R^{-1}\rangle, @, \mathsf{E})$ that captures the expressivity of $\mathcal{ALCOI}$ with Boolean knowledge bases.

As we will see in Section 5, allowing the extra flexibility that Boolean knowledge bases offer does not modify the complexity class in which the reasoning tasks fall (for the languages we are considering), but it increases expressive power and has an impact on meta-logical properties like interpolation and Beth definability.

5 Exploiting the Connection

The links between hybrid logics and DLs discussed in the previous sections are so strong, that we can immediately start harvesting by interpreting results from one field in the other. This is what we will do now, from a number of perspectives, including complexity, expressive power and meta-logical properties.

5.1 Complexity

We start by exporting complexity results for hybrid logics to DLs. We need to pay attention to the difference between local and global notions. For a modal language, we can distinguish between the local-*Sat* problem (given a formula ϕ, does there exist a model $\mathcal{M}$ and $m \in M$ with $\mathcal{M}, m \Vdash \phi$?), and the global-*Sat* problem (is there a model $\mathcal{M}$ with $\mathcal{M} \Vdash \phi$?). If the logic contains the E modality, the problems coincide, as we argued in Section 4.

First, we consider the "pure future" fragments of the hybrid languages defined in Section 4, i.e., we only consider formulas without the $\langle R^{-1}\rangle$ operator. The local-*Sat* problem for $\mathcal{H}_{\mathsf{N}}(@)$ is PSPACE-complete (Areces et al. 1999b). This result also settles the complexity of the language $\mathcal{H}(@, @\Diamond)$, because this language contains the basic modal language. As a corollary of the EXPTIME-completeness of CPDL, we obtain an EXPTIME upper bound for the local-*Sat* problem for $\mathcal{H}_{\mathsf{N}}(@, \mathsf{E})$. It follows from Spaan's (1993) results on the EXPTIME-completeness of modal logics with the existential modality, that both $\mathcal{H}(@, @\Diamond, \mathsf{E})$ and $\mathcal{H}_{\mathsf{N}}(@, \mathsf{E})$ are EXPTIME-complete.

Switching to the DL perspective, the results above imply that it is the move from empty T-Boxes to full T-Boxes that modifies complexity. And this does not depend on our extension to "Boolean" knowledge bases, as the same complexity results obtain when we restrict ourselves to standard knowledge bases. The one-of operator $\mathcal{O}$ and role fillers $\mathcal{B}$ offer more expressivity at no cost (up to a polynomial). At this very point the encoding of DLs into hybrid languages, instead of CPDL, works to our advantage, as we can identify cases with a PSPACE upper bound.

Theorem 5.1

1. *Instance checking for Boolean knowledge bases with empty T-Boxes is PSPACE-complete for the language $\mathcal{ALCO}$.*
2. *Instance checking for Boolean knowledge bases is solvable in EXPTIME (hence EXPTIME-complete) for the language $\mathcal{ALCO}$.*

Notice that we don't need to restrict to empty A-Boxes in item 1, and recall that the complexity results for instance checking extend to all standard reasoning tasks like knowledge base consistency, or subsumption checking.

What about the $\langle R^{-1}\rangle$ operator? Adding just *one* nominal to basic temporal logic moves the complexity of the local-*Sat* problem from PSPACE-hard to EXPTIME-hard. The known EXPTIME upper bound for CPDL plus nominals and E (Areces et al. 1999b) also covers the language $\mathcal{H}_{\mathsf{N}}(\langle R^{-1}\rangle, @, \mathsf{E})$; hence, the local-*Sat* problems of $\mathcal{H}_{\mathsf{N}}(\langle R^{-1}\rangle, @)$, $\mathcal{H}(\langle R^{-1}\rangle, @, @\Diamond, \mathsf{E})$ and $\mathcal{H}_{\mathsf{N}}(\langle R^{-1}\rangle, @, \mathsf{E})$ are EXPTIME-complete.

A PSPACE upper bound for $\mathcal{H}(\langle R^{-1}\rangle, @, @\Diamond)$ is easy to establish by using the fact that @ operators need only appear at modal depth 0.[1] We give a sketch of the proof. To avoid confusion we will write $@_i\langle R_r\rangle j$ as $R_r(i,j)$. Let

$$\phi = \bigwedge_{l\in L}\left(\bigvee_{i\in\mathsf{NOM}} @_i\nu_{(l,i)} \vee \bigvee T_l \vee \sigma_l\right),$$

where each T_l is a collection of formulas of the form $R_r(i,j)$ or $\neg R_r(i,j)$, and $\nu_{(l,i)}, \sigma_l$ contain neither @ nor nominals. As PSPACE = NPSPACE, non-deterministically choose from each conjunct of ϕ the disjunct satisfied by a model of ϕ. Collect these choices in a set called CHOICE. Now, for each i, let $S_i = \{\phi \mid @_i\phi \in \mathsf{CHOICE}\}$, and create a polynomial model satisfying S_i at the point m_i (notice that all formulas in S_i are basic temporal formulas and hence a PSPACE model can be constructed). Similarly, create a polynomial model for all formulas in CHOICE which are not @-formulas. Let $\mathcal{M}$ be the disjoint union of all these models. Finally, if $R_r(i,j) \in \mathsf{CHOICE}$, add the pair (m_i, m_j) to R_r. The model of ϕ obtained in this way has size polynomial in $|\phi|$.

With the translation into CPDL it would of course be impossible to evaluate the difference made by the presence or absence of the $\langle R^{-1}\rangle$ operator in terms of complexity.

Theorem 5.2

1. *Instance checking for Boolean knowledge bases with empty T-Boxes is solvable in PSPACE (hence PSPACE-complete) for the language $\mathcal{ALCBI}$.*
2. *Instance checking for knowledge bases with empty T- and A-Boxes is EXPTIME-hard for the language $\mathcal{ALCOI}$.*

[1] Notice that we cannot directly use the normal form of Proposition 4.3 here, as the formula ϕ' in normal form can be exponentially larger than the original ϕ (already the first step in the transformation is to translate ϕ in propositional normal form). But pulling out the @ modality with the help of new propositional symbols is simple.

3. Instance checking for Boolean knowledge bases is solvable in EXPTIME (hence EXPTIME-complete) for the language $\mathcal{ALCOI}$.

The complexity results listed so far were based on importing hybrid logic results into DL. The EXPTIME-hardness result for $\mathcal{H}_{\mathsf{N}}(\langle R^{-1}\rangle)$ (basic temporal logic with at least one nominal) contrast sharply with the good complexity behavior of $\mathcal{H}_{\mathsf{N}}(@)$. For example, if we move to the class of transitive models, even $\mathcal{H}_{\mathsf{N}}(@, \mathsf{E})$ is PSPACE-complete (meaning that there are PSPACE algorithms even for inference from non-empty T-Boxes, when restrictions are set on roles), while $\mathcal{H}_{\mathsf{N}}(\langle R^{-1}\rangle)$ remains obstinately in EXPTIME. Results concerning the complexity of hybrid logics in different classes of models are investigated in (Areces et al. 2000). When translated in DL terms, one of the main results in that paper implies that instance checking for Boolean knowledge bases in $\mathcal{ALCOI}$ can be solved in PSPACE if we consider only transitive trees as models.

Here we have mainly exported complexity results from hybrid logics to description logics, but of course the other direction is also open: known complexity results from DLs can be translated in hybrid terms. For example, little is known about the extension of hybrid languages with counting, while counting is widely used in DLs. Further, there is a "folklore" result which states that instance checking for $\mathcal{ALC}$ with T-Boxes restricted to simple and acyclic terminological axioms is PSPACE-complete; this implies that when syntactic restrictions are imposed on the use of E, we can avoid EXPTIME-hardness for the local-*Sat* problem of $\mathcal{H}(@, @\Diamond, \mathsf{E})$. In this respect, Lutz (1999a, 1999b) provides the first detailed complexity analysis of inference from simple, acyclic T-Boxes. Interestingly, the restriction to simple, acyclic T-Boxes does not always preserve complexity: instance checking in $\mathcal{ALCF}$ ($\mathcal{ALC}$ extended with features, feature agreement and feature disagreement) is PSPACE-complete for empty T-Boxes, but it turns NEXPTIME-complete even when only simple, acyclic T-Boxes are allowed.

5.2 Expressive Power

We now consider expressive power, and we do so by taking advantage of hybrid bisimulations. Bisimulations are binary relations on the domain of hybrid models. Kurtonina and de Rijke (1999) provide a detailed analysis of the expressive power of DLs by means of (bi-)simulations, but their results only address the expressive power of *concepts*. In this section, we will instead study the expressive power offered by full knowledge bases.

Let $\mathcal{M} = \langle M, \{R_r^{\mathcal{M}}\}, V^{\mathcal{M}}\rangle$ and $\mathcal{N} = \langle N, \{R_r^{\mathcal{N}}\}, V^{\mathcal{N}}\rangle$ be two hybrid models. For $i \in \mathsf{NOM}$, let $i^{\mathcal{M}}$ be the denotation of i in $\mathcal{M}$ and similarly for $i^{\mathcal{N}}$. Let $\sim$ be a non-empty binary relation on $M \times N$, and consider the following properties on $\sim$ in addition to the conditions for bisimulation for a basic temporal language (Blackburn et al. 2001):

(@) For all nominals i in NOM, $i^{\mathcal{M}} \sim i^{\mathcal{N}}$.
(@◇) Let i, j be nominals in NOM, then $R_r(i^{\mathcal{M}}, j^{\mathcal{M}})$ iff $R_r(i^{\mathcal{N}}, j^{\mathcal{N}})$.
(E) $\sim$ is total and surjective.

A bisimulation $\sim$ for a basic temporal language is a $\mathcal{H}(\langle R^{-1}\rangle, @, @\Diamond)$-bisimulation if it satisfies the conditions **(@)** and **(@◇)**. Further, $\sim$ is a $\mathcal{H}_{\mathsf{N}}(\langle R^{-1}\rangle, @)$-bisimulation if bisimilar states agree on all nominals (and in this case **(@◇)** can be derived from the others). $\mathcal{H}(\langle R^{-1}\rangle, @, @\Diamond, \mathsf{E})$- and $\mathcal{H}_{\mathsf{N}}(\langle R^{-1}\rangle, @, \mathsf{E})$-bisimulations are obtained from $\mathcal{H}(\langle R^{-1}\rangle, @, @\Diamond)$- and $\mathcal{H}_{\mathsf{N}}(\langle R^{-1}\rangle, @)$-bisimulations, respectively, by requiring, in addition, condition **(E)**.

Proposition 5.3 *Let $\mathcal{H}$ be one of $\mathcal{H}(\langle R^{-1}\rangle, @, @\Diamond)$, $\mathcal{H}_{\mathsf{N}}(\langle R^{-1}\rangle, @)$, $\mathcal{H}(\langle R^{-1}\rangle, @, @\Diamond, \mathsf{E})$ and $\mathcal{H}_{\mathsf{N}}(\langle R^{-1}\rangle, @, \mathsf{E})$. Let $\mathcal{M} = \langle M, \{R_r^{\mathcal{M}}\}, V^{\mathcal{M}}\rangle$ and $\mathcal{N} = \langle N, \{R_r^{\mathcal{N}}\}, V^{\mathcal{N}}\rangle$, and let $\sim$ be an $\mathcal{H}$-bisimulation between $\mathcal{M}$ and $\mathcal{N}$. Then for $m \in M, n \in N$, and ϕ in $\mathcal{H}$, $m \sim n$ implies $\mathcal{M}, m \Vdash \phi$ iff $\mathcal{N}, n \Vdash \phi$.*

For two logics $\mathcal{H}$ and $\mathcal{H}'$ we write $\mathcal{H} \preccurlyeq \mathcal{H}'$ if there is a translation $\cdot^* : \mathcal{H} \to \mathcal{H}'$, such that for each formula ϕ, for any model $\mathcal{M}$ and state m in $\mathcal{M}$, $\mathcal{M}, m \Vdash \phi$ if and only if $\mathcal{M}, m \Vdash \phi^*$. We write $\mathcal{H} \prec \mathcal{H}'$ if $\mathcal{H} \preccurlyeq \mathcal{H}'$ and not $\mathcal{H}' \preccurlyeq \mathcal{H}$. It is immediate that $\mathcal{H}(\langle R^{-1}\rangle, @, @\Diamond) \preccurlyeq \mathcal{H}_{\mathsf{N}}(\langle R^{-1}\rangle, @)$ and $\mathcal{H}(\langle R^{-1}\rangle, @, @\Diamond, \mathsf{E}) \preccurlyeq \mathcal{H}_{\mathsf{N}}(\langle R^{-1}\rangle, @, \mathsf{E})$. More interestingly, each of the relations is actually strict, which can be shown by means of bisimulations. In DL terms this means, for instance, that the one-of operator $\mathcal{O}$ does increase the expressive power of the language, both with full and empty T-Boxes.

The relation between $\mathcal{H}_{\mathsf{N}}(\langle R^{-1}\rangle, @)$ and $\mathcal{H}(\langle R^{-1}\rangle, @, @\Diamond, \mathsf{E})$ is more complex. Using bisimulations, we can prove both that $\mathcal{H}_{\mathsf{N}}(\langle R^{-1}\rangle, @) \not\preccurlyeq \mathcal{H}(\langle R^{-1}\rangle, @, @\Diamond, \mathsf{E})$ and $\mathcal{H}(\langle R^{-1}\rangle, @, @\Diamond, \mathsf{E}) \not\preccurlyeq \mathcal{H}_{\mathsf{N}}(\langle R^{-1}\rangle, @)$. Nevertheless, $\mathcal{H}(\langle R^{-1}\rangle, @, @\Diamond, \mathsf{E})$ is at least as expressive as $\mathcal{H}_{\mathsf{N}}(\langle R^{-1}\rangle, @)$ if we are only interested in satisfiability (and not in the existence of an equivalent formula).

Proposition 5.4 *There exists a translation $\cdot^*$ from $\mathcal{H}_{\mathsf{N}}(\langle R^{-1}\rangle, @)$ into $\mathcal{H}(\langle R^{-1}\rangle, @, @\Diamond, \mathsf{E})$ such that such that for all $\phi \in \mathcal{H}_{\mathsf{N}}(\langle R^{-1}\rangle, @)$, ϕ is satisfiable iff ϕ^* is satisfiable.*

Such comparisons of the expressive power of hybrid logics easily translate to description languages. For two description languages $\mathcal{L}_1$ and $\mathcal{L}_2$, define $\mathcal{L}_1 \preccurlyeq \mathcal{L}_2$ if for any knowledge bases Σ in $\mathcal{L}_1$ there is a knowledge base

Σ' in $\mathcal{L}_2$ such that for all interpretations $\mathcal{I}$, $\mathcal{I} \models \Sigma$ iff $\mathcal{I} \models \Sigma'$. Now, the formulas used to separate the languages can easily be recast as assertions or terminological definitions, and similarly for the translation used in the proof of Proposition 5.4.

The notions of bisimulation we have defined not only separate the fragments of first-order logic which corresponds to the hybrid logics we have been discussing, they also *characterize* them. For $\mathcal{H}$ any of our hybrid languages, we say that a first-order formula $\alpha(x)$ in the first-order language over $\langle \mathsf{REL} \cup \{P_j \mid p_j \in \mathsf{PROP}\}, \mathsf{NOM}, \{x, y\}\rangle$ is *invariant for $\mathcal{H}$-bisimulations* if for all models $\mathcal{M}$ and $\mathcal{N}$, and all states m in $\mathcal{M}$, n in $\mathcal{N}$, and all $\mathcal{H}$-bisimulations $\sim$ between $\mathcal{M}$ and $\mathcal{N}$ such that $m \sim n$, we have $\mathcal{M} \models \alpha(x)[m]$ iff $\mathcal{N} \models \alpha(x)[n]$.

Theorem 5.5 *Assume that $\mathcal{H}$ is one of the following: $\mathcal{H}(\langle R^{-1}\rangle, @, @\Diamond)$, $\mathcal{H}_{\mathsf{N}}(\langle R^{-1}\rangle, @)$, $\mathcal{H}(\langle R^{-1}\rangle, @, @\Diamond, \mathsf{E})$ or $\mathcal{H}_{\mathsf{N}}(\langle R^{-1}\rangle, @, \mathsf{E})$. Then, a first-order formula $\alpha(x)$ over the signature $\langle \mathsf{REL} \cup \{P_j \mid p_j \in \mathsf{PROP}\}, \mathsf{NOM}, \{x, y\}\rangle$ is invariant for $\mathcal{H}$-bisimulations iff it is equivalent to the hybrid translation of a hybrid formula in $\mathcal{H}$.*

We have only scratched the surface on expressivity issues. For example, definability results for hybrid languages, like the ones presented here and those in (de Rijke 1992, Gargov and Goranko 1993, de Rijke and Sturm 2001) shed light on the kinds of models that can be captured by means of knowledge bases of a given description language. More generally, Gargov and Goranko (1993) discuss transfer results when moving from basic modal languages to languages with nominals, while Goranko and Passy (1992) give a similar analysis for the extension with the existential modality. These results are closely related to the move from empty knowledge bases to non-empty A- and T-Boxes, respectively.

5.3 Interpolation and Beth Definability

In (Areces et al. 1999a) results concerning the interpolation and Beth definability properties for a variety of hybrid languages are provided. What is the role of these two properties in the setting of description logics? Let's first introduce some notation. For $\Sigma = \langle T, A\rangle, \Sigma' = \langle T', A'\rangle$ two knowledge bases, let $\Sigma \cup \Sigma'$ be $\langle T \cup T', A \cup A'\rangle$, and $\Sigma[C/D]$ be the knowledge base obtained from Σ by replacing each occurrence of the concept C by D. Now, suppose that for a given knowledge base Σ the following holds,

(2) $\Sigma[C/D_1] \cup \Sigma[C/D_2] \models D_1 \doteq D_2$ for some $D_1, D_2 \notin \mathsf{CON}(\Sigma)$.

Notice that this equation need not be true for all knowledge bases Σ and concepts C. Actually, (2) implies that Σ encodes enough information concerning C to provide a complete—though not necessarily explicit—definition. Now, if the (global) Beth definability property holds for the language of Σ, then there actually exists an explicit definition of C. I.e., there is a concept D not involving C such that $\Sigma \models C \doteq D$. Given that description languages take definitions very seriously (partial and complete definitions are exactly the content of T-Boxes), the Beth definability property (i.e., the capacity of the language to turn implicit definitions into explicit ones) seems highly relevant.

There doesn't seem to be a uniform direct way of proving or disproving Beth definability. The standard approach to establishing the property is via a detour through interpolation. In first-order and modal languages, the (arrow) interpolation property implies the Beth definability property and the same relation holds for the hybrid languages we have introduced.

Hence, positive interpolation results for hybrid languages translate into nice definability properties of the corresponding description language. Unfortunately, for languages where nominals appear free in formulas, and which do not provide a binding mechanism, failure of arrow interpolation seems to be the norm. In particular, (Areces et al. 1999a) provides counter-examples to the arrow interpolation property for the basic modal language extended with nominals $\mathcal{H}_{\mathsf{N}}(@)$. The extensions of this language with the $\langle R^{-1}\rangle$ operator fare no better, and adding the E operator doesn't help either. Hence, in all these cases, the standard road to to establish Beth definability is closed for us. Interestingly, the counter-examples to arrow interpolation obtained are based on counting arguments. Because the language is not expressive enough to bound the number of successors of a given state we can establish bisimulations between points with different number of successors and use this to prove failure of the interpolation property. The language extended with counting operators (even unqualified counting) would destroy the bisimilarity and hence invalidate the counter-examples, and perhaps restore interpolation and hence Beth definability.

The case is different for $\mathcal{H}(@, @\Diamond)$ and $\mathcal{H}(\langle R^{-1}\rangle, @, @\Diamond)$. As we will now show, we can extend the constructive method for establishing arrow interpolation presented in (Kracht 1999, Section 3.8), to handle @ and $@\Diamond$. Again, we will use the normal form introduced in Proposition 4.3.

Theorem 5.6 *$\mathcal{H}(@, @\Diamond)$ and $\mathcal{H}(\langle R^{-1}\rangle, @, @\Diamond)$ have arrow interpolation.*

Given that arrow interpolation implies global Beth definability for these languages, implicit definitions in $\mathcal{H}(@, @\Diamond)$ can be turned into explicit definitions. In an attempt to transfer this property to the description logic counterpart of $\mathcal{H}(@, @\Diamond)$, we would proceed as follows. Suppose a knowledge base $\Sigma = \langle T, A \rangle$ in $\mathcal{ALC}$ satisfies the conditions in (2). Then we can translate Σ into a theory Θ of $\mathcal{H}(@, @\Diamond)$ (as we are using global consequence this time we don't need E), and obtain $\Theta[p_C/p_{D_1}] \cup \Theta[p_C/p_{D_2}] \models^{glo} p_{D_1} \leftrightarrow p_{D_2}$. Applying Beth definability for $\mathcal{H}(@, @\Diamond)$ we obtain a formula δ such that $\Theta \models^{glo} \delta \leftrightarrow p_C$. Now, δ is an explicit definition of C, but it is in the *full* language $\mathcal{H}(@, @\Diamond)$, i.e., it might contain subformulas of the form $@_i\psi$ and $@_i\Diamond j$. Because of the syntactic restrictions imposed by the division into T- and A-Box information it will not always be possible to translate δ into a concept in $\mathcal{ALC}$. To see an example, suppose δ is of the form $@_i\nu \vee \psi$. Hence we will have that $\Sigma \models (@_i\nu \to (p_C \leftrightarrow \top)) \wedge (@_i\neg\nu \to (p_C \leftrightarrow \psi))$. That is, we obtain a definition of C that is conditional on assertional information.

More generally, we first write δ in normal form to obtain

$$\Theta \models^{glo} \left(\bigwedge_{l \in L} (\bigvee_{i \in \mathsf{NOM}} @_i\nu_{(i,l)}) \vee \tau_l \right) \leftrightarrow p_C.$$

Notice that for a hybrid formula ψ and $@_i\nu \in \mathsf{SF}(\psi)$ such that $@$ does not appear in ν, ψ is equivalent to $(@_i\nu \to \psi[@_i\nu/\top]) \wedge (@_i\neg\nu \to \psi[@_i\nu/\bot])$. By iterating this rewriting on the formula $(\bigwedge_{l \in L}(\bigvee_{i \in \mathsf{NOM}} @_i\nu_{(i,l)}) \vee \tau_l) \leftrightarrow p_C$, we finally obtain a series of definitions of C in terms of concepts of $\mathcal{ALC}$, but conditional on assertional information to be inferred from Σ.

There is an interesting connection between the Beth definability property and acyclic definitions in T-Boxes. The latter restriction was aimed at avoiding the introduction of circular concepts, i.e., concepts defined in terms of themselves. This kind of concepts, it was argued, called for some kind of fixed point semantics which would be computationally expensive (Baader 1990, Nebel 1990). But if the language has the Beth definability property, any concept implicitly defined in a knowledge base also has an explicit definition *without self reference*. Hence, considering only acyclic definitions does not carry any loss of expressivity.

6 Conclusions and Further Directions

Nearly a decade after Schild started exploring the connection between modal and description logic, we have made another step forwards in finding a precise modal logical counterpart of DLs. One of the key points of DLs is their use of non-empty T-Boxes and A-Boxes; we have shown how hybrid languages provide simple mechanisms to deal with them and to understand their inter-relations. We have illustrated some of the possibilities by means of examples. Our analysis of DLs in terms of hybrid logics has shown that we can reason with non-empty A-Boxes (and empty T-Boxes) in PSPACE; we have also shown how to capture the expressive power offered by T-Boxes, A-Boxes, and how to transfer meta-logical properties such as interpolation and Beth definability.

It is striking how description and hybrid logics are similar and different at the same time, like twin brothers raised separately. Because the connection is so tight, modal logicians can use hybrid logic both as an entrypoint to DLs, and as a means for understanding them. In particular, we can investigate meta-logical properties on the "hybrid logic side" (which appears to be more amenable to standard model-theoretic techniques), and these results can help us understand the behavior of description languages.

We have only investigated some of the possibilities of this two-way interchange, but there are many others of course. One direction, for example, concerns the classical hybrid topic of ***binders and variables*** (Blackburn and Seligman, 1995). We have focused on "weak" hybrid languages which remain close to the basic DLs. But a natural step from the hybrid point of view is to regard nominals not as names but as ***variables*** over individual states, and to add quantifiers. Undecidability quickly shows up in this setting, but syntactic restriction can restore decidability, while providing interesting new concepts when introduced in a description language. A different direction concerns *counting*. Graded or counting modalities $\langle n \rangle \phi$ restrict the number of possible successors satisfying ϕ that a state has in a model. While their theory is not so well developed, qualifying number restrictions are actively used in description formalisms, as they provide important modeling power. Recent work by Tobies (2000) provides promising new complexity results.

Different enough to make comparisons interesting, but similar enough to allow for extensive traffic of results, extensions and variations, description logics and hybrid logics form an interesting pair. We hope that this paper paves the way for further cross-fertilization.

Acknowledgements

We would like to thank Carsten Lutz, Ulrike Sattler, Stephan Tobies, and the anonymous referees for their valuable comments. Maarten de Rijke was supported by the Spinoza project 'Logic in Action' and by grants from the Netherlands Organization for Scientific Research (NWO), under project numbers 612.069.006, 365-20-005, and 220-80-001.

References

Areces, C., P. Blackburn, and M. Marx. 1999a. Hybrid logics: characterization, interpolation and complexity. Technical Report CLAUS-Report 104, http://www.coli.uni-sb.de/cl/claus. Computerlinguistik, Universität des Saarlandes. To appear in the *Journal of Symbolic Logic*.

Areces, C., P. Blackburn, and M. Marx. 1999b. A road-map on complexity for hybrid logics. In *Computer Science Logic*, ed. J. Flum and M. Rodríguez-Artalejo, 307–321. LNCS, No. 1683. Springer. Proceedings of the 8th Annual Conference of the EACSL, Madrid, September 1999.

Areces, C., P. Blackburn, and M. Marx. 2000. The computational complexity of hybrid temporal logics. *Logic Journal of the IGPL* 8(5):653–679.

Baader, F. 1990. Terminological cycles in KL-ONE-based knowledge representation languages. In *Proceedings of the 8th National Conference on Artificial Intelligence (AAAI-90)*, 621–626.

Blackburn, P., M. de Rijke, and Y. Venema. 2001. *Modal Logic*. Cambridge University Press.

Blackburn, P., and J. Seligman. 1995. Hybrid languages. *Journal of Logic, Language and Information* 4(3):251–272. Special issue on decompositions of first-order logic.

Blackburn, P., and M. Tzakova. 1998. Hybridizing concept languages. *Annals of Mathematics and Artificial Intelligence* 24:23–49.

Borgida, A., R. Brachman, D. McGuinness, and L. Alperin Resnick. 1989. CLASSIC: a structural data model for objects. In *Proceedings of the ACM SIGMOD International Conference of Management of Data*, 59–67.

De Giacomo, G., and M. Lenzerini. 1994. Boosting the correspondence between description logics and propositional dynamic logics. In *Proceedings of the 12th National Conference on Artificial Intelligence (AAAI'94)*, 205–212.

de Rijke, M. 1992. The modal logic of inequality. *Journal of Symbolic Logic* 57(2):566–584.

de Rijke, M., and H. Sturm. 2001. Global definability in basic modal logic. In *Essays on Non-classical Logic*, ed. H. Wansing. World Scientific Publishers.

Fischer, M., and R. Ladner. 1979. Propositional dynamic logic of regular programs. *Journal of Computer and System Sciences* 18(2):194–211.

Gargov, G., and V. Goranko. 1993. Modal logic with names. *Journal of Philosophical Logic* 22(6):607–636.

Goranko, V., and S. Passy. 1992. Using the universal modality: gains and questions. *Journal of Logic and Computation* 2:5–30.

Halpern, J., and Y. Moses. 1992. A guide to completeness and complexity for modal logics of knowledge and belief. *Artificial Intelligence* 54:319–379.

Harel, D. 1984. Dynamic logic. In *Handbook of Philosophical Logic. Vol. II*, ed. D. Gabbay and F. Guenthner. Synthese Library, Vol. 165, 497–604. Dordrecht: D. Reidel Publishing Co.

Horrocks, I., and G. Gough. 1997. Description logics with transitive roles. In *Proceedings of the International Workshop on Description Logics*, ed. M. Rousset, R. Brachmann, F. Donini, E. Franconi, I. Horrocks, and A. Levy, 25–28. Gif sur Yvette, France.

Horrocks, I., U. Sattler, and S. Tobies. 2000. Practical reasoning for very expressive description logics. In *Methods for Modalities, 1*, ed. C. Areces, E. Franconi, R. Goré, M. de Rijke, and H. Schlingloff, 239–264. Logic Journal of the IGPL.

Kracht, M. 1999. *Tools and Techniques in Modal Logic.* Amsterdam: North-Holland Publishing Co.

Kurtonina, N., and M. de Rijke. 1999. Expressiveness of concept expressions in first-order description logics. *Artificial Intelligence* 107(2):303–333.

Ladner, R. 1977. The computational complexity of provability in systems of modal propositional logic. *SIAM Journal of Computing* 6(3):467–480.

Lutz, C. 1999a. Complexity of terminological reasoning revisited. In *Proceedings of the 6th International Conference on Logic for Programming and Automated Reasoning LPAR'99*, 181–200. Lecture Notes in Artificial Intelligence. Springer-Verlag.

Lutz, Carsten. 1999b. On the complexity of terminological reasoning. LTCS-Report 99-04. LuFg Theoretical Computer Science, RWTH Aachen, Germany.

Nebel, B. 1990. Terminological cycles: semantics and computational properties. In *Principles of Semantic Networks*, ed. J. Sowa. 331–361. Los Altos: Morgan Kaufmann.

Sattler, U. 1996. A concept language extended with different kinds of transitive roles. In *20. Deutsche Jahrestagung für Künstliche Intelligenz*, ed. G. Görz and S. Hölldobler. Lecture Notes in Artificial Intelligence, No. 1137. Springer Verlag.

Schild, K. 1991. A correspondence theory for terminological logics. In *Proceedings of the 12th IJCAI*, 466–471.

Spaan, E. 1993. *Complexity of Modal Logics.* Doctoral dissertation, Institute for Logic, Language and Computation, University of Amsterdam.

Tobies, S. 2000. The complexity of reasoning with cardinality restrictions and nominals in expressive description logics. *Journal of Artificial Intelligence Research.* To appear.

Wolter, F., and M. Zakharyaschev. 1999. Modal description logics: modalizing roles. *Fundamenta Informaticae* 39:411–438.

Carlos Areces and Maarten de Rijke
ILLC, University of Amsterdam
Nieuwe Achtergracht 166
1018 WV Amsterdam
The Netherlands
E-mail: {carlos, mdr}@science.uva.nl
URL: http://www.science.uva.nl/~{carlos, mdr}

3

Double Time References: Speech-act Reports as Modalities in an Indeterminist Setting

NUEL D. BELNAP

ABSTRACT. The background is a theory that describes agents and their choices in the branching-time representation of indeterminism, and that describes a language appropriate for use in such a setting. The problem of this paper is to clarify the meaning, in the context of indeterminism, of speech acts such as promising, ordering, advising, asserting, and betting. Such speech acts often have a sentence that serves as a "declarative core," whose truth-conditional semantics in branching time contributes to the meaning of the speech act. Direct-discourse reports of such speech acts can be treated as a kind of "modal" connective, e.g. "a_1 promised a_2 to make 'A' true." A key feature of the contribution of the declarative core is made in terms of its (*now*) being definitely settled at a later moment (in the branching tree) that the declarative core of the speech act was true (*then*) at the moment of the speech act. That is the "double time reference" that needs clarification.

This paper takes some steps toward understanding the causal structure of speech acts considered as taking place in an indeterminist world; that is, in a world such that at any moment, while the past is fixed, the future is open to multiple possibilities. In the course of this exploration, it helps to introduce some "double time reference" ideas that are confusing and difficult to express in idiomatic English.

Advances in Modal Logic, Volume 3
F. Wolter, H. Wansing, M. de Rijke, and M. Zakharyaschev, eds.

1 Assumptions about indeterminism and actions

The study is based on the "theory of agents and choices in branching time," the elements of which are presented in detail in the forthcoming Belnap et al. 2001 (*FF*).[1] Before getting on to double time references for speech acts in an indeterminist setting, I note some of the assumptions on which that theory is based.

- There is no real action without real choice. (Evidently this assumption implies that we seek no application to "action" in the sense of "action of the starter motor.")
- There is no real choice without the reality of alternative possible choices facing the agent. Each of these possibilities is, before the moment of choice, as real as any other. It is true and important that at most one of these possibilities will be realized. It is equally true and equally important that none of these possibilities is a ghostly image of some specially distinguished one among them that some philosopher might label "the actual choice." This form of "actualism," we argue, is a bad idea.
- Since the alternative possibilities are incompatible, they cannot inhabit the same linear time structure, nor even the same space-time structure. *FF* runs with the (obviously rudimentary) idea that the causal structure of our world is neither linear nor Minkowski-like, but branching. We postulate for the study of agency and other indeterminist phenomena a causal structure that branches toward the future, but that always leaves behind it a settled past. As jargon, we call each node of the tree a *moment*, and we call each maximal chain of the tree a *history*.
- Given that our world exhibits (something like) this structure, a structure that is usually if unhappily called "branching time," care is called for when we come to think about a language spoken by its denizens. Let us call these folks (whom we believe to include ourselves) the "tree dwellers," so that their language is "the language of the tree dwellers," or *LTD*.
- Combining Tarski and Prior and Thomason and Kaplan, we argue that for understanding the language of the tree dwellers, it is not optional, but necessary, to see the truth value of sentences as dependent on a number of parameters, as follows.

[1]A substantial portion of the theory has been given initial presentation in the following publications: Belnap 1991, 1996a, 1996b; Belnap and Perloff 1988, 1992, 1993; Perloff 1991, 1995; Xu 1994a, 1994b, 1994c, 1995a, 1995b; Horty 2000; Belnap and Bartha 1995; and Belnap and Green 1994.

1. As always, there are the underlying structure and the interpretation of individual and predicate constants, which we put together into the "model." The structure may be conceived as a tuple $\langle$*Tree*, $\leqslant$, *Agent*, *Choice*$\rangle$, where *Tree* is a set of moments representing our entire possibility-filled world, $\leqslant$ is the branching causal order on *Tree*, *Agent* is the set of all agents, and *Choice* represents the possible choices open to each agent at each moment. For each agent, a, and moment, m, we assume that $Choice_m^a$ is a partition of the set of histories to which m belongs. We write $Choice_m^a(h)$ for the (unique) possible choice for agent a at m to which h belongs. On the theory of agents and choices in branching time, if $m \in h$, then $Choice_m^{a_1}(h) \cap Choice_m^{a_2}(h)$ is a *joint choice* for $\{a_1, a_2\}$. For this paper, we think of there being a fixed *standard model*, and suppress its mention.
2. We need the Tarski-inspired assignment of values to the variables in order to treat quantifiers in a lucid fashion. Since we do not in this paper deal with quantifiers, we also suppress this parameter.
3. We need moments in the causal order so that we can, as we learned from Prior 1957, get our tense modalities straight. The semantics of *LTD* invokes an auxiliary *moment of evaluation* for this purpose.
4. From Kaplan 1989 comes the need for the context of utterance both to fix indexicals and the like, and also as giving a starting point when evaluating stand-alone sentences. In this paper, because of our particular concern with speech acts, we will need three context elements: the *speaker* of the context, the *audience* of the context (both agents), and the *moment* of the context. The latter must be carefully distinguished from the auxiliary moment of evaluation that is used in connection with tense modalities.
5. Lastly, crucially, and less obviously, from Thomason 1970, 1984, after Prior 1967, pp. 126–127, we learn that to make sense out of future-pointing temporal constructions in a branching structure, we must also relativize truth to histories. More colloquially, we make rigorous the evident fact that in the presence of indeterminism, the truth of "Themistocles will choose to fight a sea battle tomorrow" depends on how the future works itself out. For this purpose we invoke an auxiliary *history of evaluation* parameter.

In sum, by combining Prior–Thomason's branching-histories semantics with Kaplan's treatment of context, we are left with these semantic parameters to which to relativize truth: three elements of context (speaker, audience, and moment) and two auxiliary parameters to help with tense and modality (the moment and history of evaluation). We assemble these five parameters of truth as follows into a *point of evaluation*:

$$\langle a_1, a_2, m_c, m/h \rangle \tag{1}$$

The first three are the speaker, audience, and moment of context, while the last two are the auxiliary moment of evaluation and the auxiliary history of evaluation for use with tenses and modalities. It is required that $m \in h$, which is one reason the notation binds these two parameters together.

We abbreviate our language of truth and denotation for *LTD* as follows:

> $a_1, a_2, m_c, m/h \vDash A$ for "the sentence, A, is true at the point $\langle a_1, a_2, m_c, m/h \rangle$," and $Val_{a_1,a_2,m_c,m/h}(t)$ for "the denotation of the term, t, at the point $\langle a_1, a_2, m_c, m/h \rangle$." (2)

Here are a few suggestive semantic analyses.

- $a_1, a_2, m_c, m/h \vDash$ *Was:A* iff $\exists m_1[m_1 \in h$ and $m_1 < m$ and $a_1, a_2, m_c, m_1/h \vDash A]$.
- $a_1, a_2, m_c, m/h \vDash$ *Will:A* iff $\exists m_1[m_1 \in h$ and $m < m_1$ and $a_1, a_2, m_c, m_1/h \vDash A]$.
- $a_1, a_2, m_c, m/h \vDash$ *Sett:A* iff $\forall h_1[(m \in h_1$ and h_1 is a history in *Tree*$) \to a_1, a_2, m_c, m/h_1 \vDash A]$.

Part of the theory consists in endowing *LTD* with a modal logic of agency. The modality at issue is written "t sees to it that A," where t is a term denoting an agent, and where A is any sentence. We abbreviate the seeing-to-it-that modality by [t *stit:* A]. Its semantic analysis makes stit depend on choice, which is an element of the model. Then, where we let $a = Val_{a_1,a_2,m_c,m/h}(t)$, we may say

> $a_1, a_2, m_c, m/h \vDash [t$ *stit:* $A]$ iff (i) $a_1, a_2, m_c, m/h_1 \vDash A$ for every h_1 belonging to $Choice^a_m(h)$, and (ii) $a_1, a_2, m_c, m/h_1 \nvDash A$ for some h_1 to which m belongs. (3)

This gives the truth conditions for the "deliberative" stit, which we choose to employ instead of the "achievement" stit that is also defined in *FF*.

2 Some speech acts

The topic of speech acts was introduced by John Austin, most especially in his 1962 book, *How to do things with words.* It has been pushed forward by e.g. Searle and Vanderveken 1985. When is an utterance a speech act? The speech-act community tends to follow the thought that a speech act such as an order involves a certain intention on the part of the speaker to cause the audience to act in certain ways. The following considerations, while by no means offered as argumentative or exclusionary, are intended to guide us to see a different way in which speech acts are set into the causal structure of our world. First of all, it is not so common to consider speech acts in the context of a wider theory of actions. Since a "speech act" is, however, an "act," it has to have some *differential consequences.* It has to make a difference in our world. As a result of the act, things in the future have to go one way rather than another. That is where indeterminism comes in. Second, since a "speech act" is a bit of speech, not mere rumbling sounds, nor always attention-getting shouts, the *semantics* of the utterance should be involved. That is why we need to pay careful attention to the semantics of a language such as *LTD* designed to be spoken in an indeterminist world.

This paper considers how speech acts look if the changes caused by a speech act are *publicly normative* rather than either directly physical or mental, and if the specification of the new norms created by a speech act requires reference to the *semantic* understanding of an utterance as it bears on an indeterministic future. Here are some quick examples.

- A *promisor* changes the world by making *commitments* to take certain steps. Since whether a promise will be kept depends on later choices, the promise does not guarantee those choices, and there is a built-in reference to norms governing an indeterministic future.
- Someone who gives an *order* puts *requirements* on the choices—the future choices—of someone else.
- Someone who gives *advice commits* herself to giving reasons, if they are subsequently requested. Furthermore, the adviser *commits* herself to accepting blame or making amends or explanations or excuses, etc., should things later go wrong.
- *Assertion* is more difficult, partly because it is multiform. Brandom has given a rich account of assertions that describes them in terms of the commitments and entitlements they create. For example, the assertee is *entitled* to *require* the assertor to defend the assertion, and, should the assertee act in reliance on the assertion, to *require* compensation if things go wrong. In some cases, the assertor *deserves* credit or discredit depending on whether the assertion turns out to have been true or false. At a minimum, the assertor is *vindicated* in the one case and *impugned* in the other.
- A typical *bet* between two parties says that the loser *owes* a certain sum to the winner. The indebtedness is guaranteed by the bet, but the actual payoff cannot be guaranteed by the bet, not even conditionally, since it requires later choices.

- Someone who *invites an offer commits* herself to entertaining any forthcoming offer with at least minimum care. Someone who *makes an offer empowers* the offeree to *accept*, and in case of acceptance, *commits* himself to making subsequent choices in accord with the terms of the resultant contract. Of course the acceptance of an offered contract also lays *requirements* on the accepter.

From these examples, it is easy to see how natural it is to attribute the creation of norms; nevertheless, this territory is evidently a swamp. The current project is to see if placing speech acts in the causal structure represented by branching time can help us to see some features of the murky situation that we might otherwise overlook.

3 Uttering

We prepare for speech acts by studying "uttering." We take uttering to be an act that may or may not be a *speech* act; in part this means that uttering may or may not be an act whose accomplishment *requires* that the utterance be subject to semantic analysis.

3.1 Mere uttering

Not every act of e.g. promising need involve utterance, but presumably every *speech* act does so. We suppose that *LTD* carries the idea of uttering with

$$Utters(t_1, t_2, \text{`}A\text{'}), \qquad (4)$$

to be read in English as "t_1 utters 'A' to t_2." The grammar of *LTD* requires that t_1 and t_2 be terms of *LTD*, and that A be a sentence of *LTD* without free variables.[2] The semantic interpretation of (4) treats it in effect like a 3-place predicate, so that whether

$$a_1, a_2, m_c, m_0/h \vDash Utters(t_1, t_2, \text{`}A\text{'}) \qquad (5)$$

depends entirely on $Val_{a_1,a_2,m_c,m_0/h}(t_1)$, on $Val_{a_1,a_2,m_c,m_0/h}(t_2)$, on the identity (but not the semantic content) of A, and, as for any predicate of *LTD*, on m_0/h.[3] We suppose as a necessary condition for the truth of (4) that the two terms denote agents:

[2] Notation. *We* use "A" to name a sentence, whereas *they* (the tree dwellers) use "'A'." *They* use single quotes carefully, whereas *we* use double quotes casually. Furthermore, in this paper we restrict *LTD* by supposing that the *LTD* quote-name, 'A', is the only way in which *LTD* names the *LTD* sentence, A. The restriction to quote-names avoids the threat of circularity via self-reference that might otherwise arise. "Paradoxes" would threaten if we permitted arbitrary terms denoting sentences. In a treatment of a language closer to our own, we should admit the circularity and squarely meet the threat, invoking rules of revision in order to cope with it.

[3] Merely as an aid to the eye, we use the symbol "m_0" instead of "m" for a moment of uttering. Note that m_0 in (5) is a "moment of evaluation," not "the moment of context." In (5), the moment m_c represents the context in which the

Assumption 3.1 (Terms denote agents) a_1, a_2, m_c, $m_0/h \models$ *Utters*$(t_1, t_2, \text{'}A\text{'}) \rightarrow Val_{a_1,a_2,m_c,m_0/h}(t_1) \in$ *Agent* and $Val_{a_1,a_2,m_c,m_0/h}(t_2) \in$ *Agent.*

An uttering involves speech production by the utterer, and to utter is to act. But it is not only the utterer who serves as an actor in this transaction. We also build in the Thomson 1990 requirement of "uptake" at this early stage by supposing that the utteree acts as well, at the very least by "lending an ear" to the utterer. We carry this formally by supposing that there is *joint* action by utterer and utteree. The agentiveness of uttering is therefore properly carried by the following semantic requirement:

Assumption 3.2 (Agentiveness of uttering) a_1, a_2, m_c, $m_0/h \models$ *Utters*$(t_1$, t_2, 'A') iff a_1, a_2, m_c, $m_0/h \models [\{t_1, t_2\}$ *stit:* *Utters*$(t_1$, t_2, 'A')].[4]

Stit theory allows us to treat stit in this formula as either the achievement stit or the deliberative stit. The deliberative stit is in some respects simpler to use, and we shall do so, but there is a little complication. The reason is that it is natural to think of *Utters*$(t_1$, t_2, 'A') as history-independent—as not depending on what happens next. Since we are treating it as a deliberative stit, however, *Utters*$(t_1$, t_2, 'A') is open at the "moment of uttering" m_0, depending (in the logical sense) as it does on the choice to utter. That is, *Utters*$(t_1$, t_2, 'A') is true at m_0/h depending on whether the historical continuation, h, does or does not represent a history on which the utterance was made. This will become easier to follow if we reify utterances—not indeed in order to enrich our serious ontology, but just to give us convenient ways of speaking. In expressing the idea of the following definition, we pretend for convenience that for every agent, a, there is a distinguished *LTD* term, which we call a^*, that "rigidly" denotes a independent of any context or auxiliary parameters.

Definition 3.3 (Reified utterance) A 5-tuple, $\langle a_1, a_2, m_0, H, A\rangle$, is *an utterance* iff H is a possible joint choice for $\{a_1, a_2\}$ at m_0 (so that H is a set of histories all of which contain m_0), and for some (or all—it makes no difference) agents a_3 and a_4, we have that a_3, a_4, m_0, $m_0/h \models$ *Utters*$(a_1{}^*$, $a_2{}^*$, 'A') for every h in H. When $\langle a_1, a_2, m_0, H, A\rangle$ is an utterance, a_1 is defined as its *utterer* or *speaker*, a_2 as its *utteree*

---

report of uttering occurs, whereas m_0 represents the moment at which the *uttering* is described as occurring.

[4]Strictly speaking, the stit should be a "strict stit," which requires that each agent be essential to the outcome.

or *audience*, m_0 as the *moment of utterance*, H as the *outcome* of the utterance, and A as *the sentence uttered.*

Here by "utterance" we refer not to something uttered, but to the very act of uttering. It is good to think of an utterance in this sense as a *transition* from m_0 to H, and to think of H as representing the proposition that the utterance occurs. The uttered sentence will contain no free variables.

3.2 Truly uttering

It makes sense to say that someone uttered A truly to someone else, even if they have *not* performed a speech act; but one cannot unpack "uttered truly" by a simple conjunction of a direct-discourse report of uttering conjoined with a truth claim about A (or with A itself).[5] For sanity, it is necessary to evaluate the report of the utterance in one context, and the uttered sentence, A, in the context in which it was uttered. These are not likely to be the same. There are three elements of context: speaker, audience, and moment of context. Each must be *shifted*—which is at this point a key idea. The speaker and the audience must be shifted to those explicitly mentioned in the sentence reporting the uttering, and the moment of context must be shifted to the moment of evaluation of the utterance report. We therefore introduce a new construction, *Truly utters*$(t_1, t_2, `A\text{'})$. We better write the semantics of this form before we forget how all that shifting goes.

Definition 3.4 (Truly utters) $a_1, a_2, m_c, m_0/h \vDash \textit{Truly utters}(t_1, t_2,$ $`A\text{'})$ iff both of the following: $a_1, a_2, m_c, m_0/h \vDash \textit{Utters}(t_1, t_2, `A\text{'})$ and $Val_{a_1,a_2,m_c,m_0/h}(t_1), Val_{a_1,a_2,m_c,m_0/h}(t_2), m_0, m_0/h \vDash A$.

That is, since we evaluate the utterance-report, *Utters*$(t_1, t_2, `A\text{'})$, at

$$a_1, a_2, m_c, m_0/h,$$

we must evaluate the truth value of the uttered sentence, A, at

$$Val_{a_1,a_2,m_c,m_0/h}(t_1), Val_{a_1,a_2,m_c,m_0/h}(t_2), m_0, m_0/h$$

Observe that not only is the latter point context-initialized (by shifting the moment of context to be exactly the same as the moment of evaluation), but the speaker and audience of the context have been properly identified as the speaker and audience of the utterance.

[5]In his unpublished 1999 working paper, "Indexicals in direct and indirect discourse," M. Weiner (University of Pittsburgh) solves the problem of articulating "truly uttering." The present solution is technically a little different, but not at bottom. I am much indebted to Weiner for access to his paper, and for numerous conversations.

It is worth remarking in addition that although *uttering* is an act, we are by no means suggesting that *truly uttering* is invariably an act. The obvious point is that uttering A is not enough to guarantee the truth of A, so that even when you can choose to utter A, you may well not be able to choose to speak the truth. For example, yesterday you truly uttered to me, "Tomorrow I will win the Irish Sweepstakes." The utterance, but not the truth of the utterance, was within your power of choice. Thus, truly uttering is not invariably a speech act.

As we noted above, $Utters(t_1, t_2, \text{'}A\text{'})$ behaves like a standard 3-place predicate. $Truly\ utters(t_1, t_2, \text{'}A\text{'})$, however, does not; sometimes we call it a "quasi-3-place predicate." Its first two places are filled with terms treated semantically in the standard way, but in contrast to mere uttering, the evaluation of the truly-uttering construction depends not just on the identity of A, but on its semantic content. In this regard, $Truly\ utters(t_1, t_2, \text{'}A\text{'})$ works something like a standard modal connective in its third position. There is, however, a critical systematic difference here that arises from the fact that $Truly\ utters(t_1, t_2, \text{'}A\text{'})$ is a "direct discourse modality." The difference is this: Unlike the case for standard "indirect discourse modalities," when you evaluate the sentential argument, A, of $Truly\ utters(t_1, t_2, \text{'}A\text{'})$, you need to shift context.

4 Normal form of speech acts

In order to (over-) simplify, we suppose that every speech act has an agent as speaker, an agent as audience, and, in direct speech, involves a declarative sentence as expressing what is sometimes unhappily called the "content" of the speech act. We may suppose that every speech act can be described in a certain normal form. In our simple idealization, this idea is carried by a moment-dependent quasi-3-place agentive verb such as the following, with "*Speech-act*" being appropriately replaced, depending on the particular speech act at issue:

$$Speech\text{-}act(t_1, t_2, \text{'}A\text{'}), \tag{6}$$

We will discuss the following special cases of (6).

Example 4.1 (Some speech acts) In the informal English readings below, with respect to which we ask your charity, we suppose that the speaker is female and the audience is male.

- $Promise(t_1, t_2, \text{'}A\text{'})$. t_1 promises t_2 to see to it that 'A' is true.[6]
- $Order(t_1, t_2, \text{'}A\text{'})$. t_1 orders t_2 to see to it that 'A' is true.

[6] Here and in the informal English readings below, we do not intend that "sees to it that" be symbolized with the formal stit described at the end of section 1. As we begin to see in section 6.2, what is promised is more complicated than a simple stit.

- *Advise*$(t_1, t_2,$ 'A'$)$. t_1 advises t_2 to see to it that 'A' is true.
- *Assert*$(t_1, t_2,$ 'A'$)$. t_1 asserts to t_2 that 'A' is true.
- *Bet \$3/\$5*$(t_1, t_2,$ 'A'$)$. t_1 bets t_2 her \$3 against his \$5 that 'A' is true.

These forms of speech should be open in the moment-of-evaluation parameter so that it will make sense to apply a tense if we wish. It is required that t_1 and t_2 denote agents, and that A is a sentence of *LTD*, and furthermore a sentence that contains no free variables. The single quotes remind us that we are working on a concept of speech acts involving *direct* discourse. We suppose that speech acts in *LTD* are always *reported* as in (6), and always *accomplished* by utterances using the indexicals "I" and "you." This would be more difficult to say if we did not have language for the concept of truly uttering. With that language, we can say what is wanted by entering the following.

Assumption 4.2 (Speech-acting is truly uttering) $a_1, a_2, m_c, m_0/h \vDash$ *Speech-act*$(t_1, t_2,$ 'A'$)$ iff $a_1, a_2, m_c, m_0/h \vDash$ *Truly utters*$(t_1, t_2,$ '*Speech-act*(I, you, 'A')').

For example, in *LTD* all promises are made by truly uttering "I promise you ...," and also conversely, in the sense that

Promise(I, you, 'A') and
Truly utters(I, you, '*Promise*(I, you, 'A')')

have exactly the same truth value at every "context-initialized" point,
$a_1, a_2, m_0, m_0/h$.

This constraint is by no means essential or even common to the practice of promising (promises can even be wordless), but it is a harmless supposition for the ideal case, and perhaps makes the discussion easier to follow.

Assumption 4.2 is evidently too obviously circular to be intended as an *analysis* of speech-acting. So what more is there to e.g. promising than truly uttering the words "I promise you"? In the first place, we remind you that although mere uttering is an act, truly uttering an arbitrary sentence, A, is not invariably an act. Nevertheless, in *LTD* truly uttering e.g. "I promise you ..." is indeed always an act. Suppose A is "I will win the Irish Sweepstakes." There is no special problem about choosing to utter A; but you cannot in general choose that A is true. If A has, however, the form "I promise you ...," in *LTD* you can choose not only to utter A, but to utter it truly (you can choose to make a promise by using those words).

Assumption 4.3 (Speech acts are acts) $a_1, a_2, m_c, m_0/h \vDash$ *Speech-act*$(t_1, t_2,$ 'A') iff a_1, a_2, m_c, $m_0/h \vDash [\{t_1, t_2\}$ *stit: Speech-act*$(t_1, t_2,$ 'A')].

Again we choose the deliberative stit. The effect of the choice is that m_0 is the last moment before which the speech act goes into effect, whereas there is in general no first moment.

We may reify speech acts in exactly the same style that we reify utterances.

Definition 4.4 (Reified speech act) A 5-tuple, $\langle a_1, a_2, m_0, H, A\rangle$, *is a speech act of kind Speech-act* iff H is a possible joint choice for $\{a_1, a_2\}$ at m_0 (so that H is a set of histories all of which contain m_0), and for some (or all—it makes no difference) agents a_3 and a_4, we have that a_3, a_4, m_0, $m_0/h \vDash$ *Speech-act*$(a_1{*}, a_2{*},$ 'A') for every h in H. When $\langle a_1, a_2, m_0, H, A\rangle$ is a speech act of any kind, a_1 is defined as its *speaker*, a_2 as its *audience*, m_0 as the *moment* of the speech act, H as the *outcome* of the speech act, and A as its *declarative core*.

We may say, for example, that $\langle a_1, a_2, m_0, H, A\rangle$ is a promise iff a_1, a_2, m_0, $m_0/h \vDash$ *Promise*$(a_1{*}, a_2{*},$ 'A'). Equivalently, we may say that $\langle a_1, a_2, m_0, H, A\rangle$ is a promise iff a_1, a_2, m_0, $m_0/h \vDash$ *Promise*(I, you, 'A'). The declarative core of e.g. a promise will contain no free variables.

What else is common to all speech acts? Much is peculiar to the individual types. We are pushing as far as possible, however, the value of seeing each speech act as creating *norms* that come into play *subsequent* to the speech act, along the lines of section 2. We suggest that the pattern of how and when these norms arise depends (in part) on the semantic content of the declarative core of the speech act. Getting this right requires, we think, double time references. We motivate this by way of an example.

5 Extended example: Themistocles and Eurybiades

We are at some particular moment at the time 4:00 a.m., a moment that we call $m_{4:00}$. At an earlier moment, m_0, whose time was 0:00,

> Themistocles said to Eurybiades, "I promise you that I will choose to fight a sea battle." (7)

So m_0 is the moment of promising, and in direct speech, what Themistocles promised Eurybiades was this:

> I will choose to fight a sea battle. (8)

For analytic convenience, let us render the future tense explicitly as a Prior connective:

$$Will{:}\text{I choose to fight a sea battle,} \tag{9}$$

which we read in English as "It will be true that I choose to fight a sea battle." In addition, we replace "I choose" with "I have just chosen":

$$Will{:}\text{I have just chosen to fight a sea battle.} \tag{10}$$

In English: "It will be true that I have just chosen to fight a sea battle." We shift to "I have just chosen" in order to make the sentence say, without ambiguity, that Themistocles' choice to fight the sea battle is a settled fact.[7]

Formally, we are supposing that

$$a_1, a_2, m_c, m_0/h \vDash Promise(\text{Themistocles, Eurybiades, `}Will{:}\text{I have just chosen to fight a sea battle'}). \tag{11}$$

Here the "declarative core" of the promise is

$$Will{:}\text{I have just chosen to fight a sea battle,} \tag{12}$$

where the context of evaluation is given by speaker = Themistocles, audience = Eurybiades, and moment of context = m_0. Letting T = Themistocles and E = Eurybiades, we may if we wish reify the promise as

$$\langle \text{T, E}, m_0, H, Will{:}\text{I have just chosen to fight a sea battle}\rangle. \quad \text{More briefly: } \langle \text{T, E}, m_0, H, (12)\rangle. \tag{13}$$

5.1 Question of double time references

So what about "double time references" with regard to (11) or (13)? The idea is that in order to understand the nature of the promise (13), we must understand what it means to say that the promise has been "carried out," or, to speak somewhat less intentionally, "satisfied." So what does it mean to say that Themistocles' promise has been satisfied? How can we understand the following?

$$\text{The promise (13) has been satisfied at } m_{4:00}. \tag{14}$$

A natural guess is that we will need two time references. One of the two references will surely be to the 0:00 moment of promising, m_0, and it is equally natural to guess that the other reference will be to a moment at which the satisfaction of promise is under consideration. It turns out, however, that things are not so simple. The other reference is indeed to a later moment, $m_{4:00}$, at which we are considering the "satisfaction

[7] "Have just chosen" is vague, and hence not a proper Prior tense. We shall leave it unanalyzed. One could, however, substitute some precisification such as "within the last two seconds I chose," which would indeed be expressible in standard metric tense logic, and would suit us just as well except for its verbal complexity—and the fact that then we should also need to replace "it will be true that ..." by something like "after at least three seconds have gone by, it will be true that... ."

status" of the promise (13). Easy examples show, however, that there is no simple evaluation of the declarative core (12) itself at such later moments.

5.2 Two wrong answers

Let it be true at $m_{4:00}$ that the promise is satisfied; such is, at $m_{4:00}$, a settled fact. What does this mean? Here are two candidates for using the declarative core (12) that don't work.

At $m_{4:00}$ the promise $\langle$T, E, m_0, H, (12)$\rangle$, is satisfied iff (12) is settled true now, at $m_{4:00}$. In symbols: $\forall h[m_{4:00} \in h \rightarrow \text{T, E}, m_0, m_{4:00}/h \vDash (12)]$. (15)

This is wrong. After all, now, at $m_{4:00}$, what Themistocles promised to make true, namely, (12), is settled false! The point is that when evaluated at $m_{4:00}$, the future tense of (12) would reach forward into times later than that of $m_{4:00}$, long after the sea battle.[8] So let us try again.

At $m_{4:00}$ Themistocles has kept his promise iff (12) is settled true at the moment of promising. In symbols: $\forall h[m_0 \in h \rightarrow \text{T, E}, m_0, m_0/h \vDash (12)]$. (16)

This is also wrong. At the moment of promising, the sentence, (12), was *not* settled true. That is, the declarative core (12) was, at the moment, m_0, of promising, true on some histories through m_0 and not on others—for the choosing still lay in the future—else what was the point of the promise?

5.3 Correct but ungeneralizable answer

In considering whether or not the promise made at m_0 has a "satisfied" status at some future moment, $m_{4:00}$, it seems clear that we do *not* wish to evaluate whether or not the actual future-tensed declarative core, (12), of the promising is settled true. Instead, in this particular and (as it turns out) surprisingly special case, we want to consider only the result of stripping off its future tense:

I have just chosen to fight a sea battle. (17)

When it is settled at $m_{4:00}$ that (17) is true, the promise has been "satisfied" at $m_{4:00}$. We cannot reasonably evaluate the future-tensed declarative core at $m_{4:00}$, but in this case it is reasonable to evaluate the ingredient to which the future tense is applied.

For this example, we had to analyze the declarative core syntactically, and it was easy to do so. It is obvious, however, that there are

[8]We neglect as distractions both the short interval called to mind by "have just" and the possibility that there be *another* sea battle.

other much more complicated examples, perhaps involving connectives and quantification, that would resist such simpleminded syntactic analysis. The specter of these other examples might suggest that a ***uniform*** account of (14), required for an understanding of promise-satisfaction, is beyond hope.

5.4 Right answer: How double time references work

We suggest, however, that a natural account can be found in the doctrine of branching histories and its accompanying semantics. If we pay careful attention to these semantics, we can see exactly how "double time references" work without presuming anything whatever about the various syntactic ingredients of the declarative core of the promise.

At this point it becomes crucial to the analysis that we are following the Prior-Thomason semantics. The fundament of those semantics is this: Because of indeterminism, the truth value of a sentence may depend both on the *moment* of evaluation and also on the *history* of evaluation.[9] So the question is, for which m and which h should we be evaluating the declarative core of the promise at m/h? We need a double time reference. We need to take the *moment* of m/h as the *earlier* moment of promising, m_0. We need to take the *history* of m/h from among those to which the *later* moment, $m_{4:00}$, belongs. You can concentrate on neither m_0 nor $m_{4:00}$ alone. You need them both. This goes not just for promises, but for a wide variety of norms-creating speech acts such as the ones indicated in Example 4.1.

Speaking more generally, the declarative core, A, of a speech act, $\langle a_1, a_2, m_0, H, A\rangle$, has a rich semantic content in terms of its truth value at each moment-history pair, m/h, even when model and context and assignment are held fixed. In principle, the entire semantic content of A is available to us in explaining the contribution of A to the content of the speech act $\langle a_1, a_2, m_0, H, A\rangle$. An essential feature of the double-time-reference account, however, is that speech-acting uses only a small part of this richness; namely, speech-acting uses only the pattern of truth values of A at pairs that feature a single moment, m_0, the moment of speech-acting. The declarative core, A, of the speech act, $\langle a_1, a_2, m_0, H, A\rangle$, contributes nothing but its pattern of truth values at points featuring the various pairs m_0/h, where m_0 is fixed once and for all as the moment of speech-act, and where h ranges over H.

[9] And also of course on the other parameters of truth, the model and the context. That is, the metalinguistic predicate "true on" must be relativized both to a moment and to a history (as well as model and context), so that given model and context, one says "A is true on m/h," where h is one of the histories to which the moment, m, belongs.

It is striking that this be so. One would have thought that one would have to consider the truth value of A at moments *after* the moment of promising in order to get the right "world to words" or "words to world" fit. As we saw in considering (13), however, in looking at these later moments, we did *not* consider the truth value of the declarative core (12) itself. In this special example, we looked instead at the truth value, at later moments, of a syntactic *ingredient* of that declarative core, namely, (17). If, however, we wish to give a uniform account for arbitrary A, the idea of using syntactic decomposition is hopeless: Think only of conditional promises, disjunctive promises, conjunctive promises, quantificational promises, and endless combinations of these. This is the negative lesson.

There is also a positive lesson. By attending to the pattern of values of the declarative core only at the moment of promising (rather than at later moments), we obtain a uniform account that works for arbitrary A as declarative core, regardless of syntactic structure.[10] We may forget the grammatical complexities of A, and consider only its pattern of truth values at points featuring m_0/h, where m_0 is the moment of promising and h is any history in H. This pattern supplies exactly the right semantic content on which to base an account of what the promise promises. The idea is that in analyzing the content of the promise ⟨T, E, m_0, H, (12)⟩, we need use only "(12) is true on m_0/h," where m_0 is the moment of promising, and h is one of the histories in H.

If we keep this in mind, we are led to the following account of what it means to say that Themistocles' promise at m_0 has been satisfied at $m_{4:00}$.

> At $m_{4:00}$ Themistocles' promise ⟨T, E, m_0, H, (12)⟩ made to Eurybiades at m_0 has been satisfied iff at $m_{4:00}$ it is a settled matter that (12) was true at the moment m_0 of promising. In symbols: $\forall h[m_{4:00} \in h \rightarrow \text{T, E}, m_0, m_0/h \models (12)]$. (18)

At $m_{4:00}$ Themistocles has kept his promise iff at $m_{4:00}$ it is settled true that (12) was true *at the moment of promising*, m_0. The key is the "double time reference": The "settled" is evaluated later, at $m_{4:00}$, while what is settled is the truth of (12) at the moment of promising, m_0. The upshot is that you cannot do without "the moment of promising," even though you certainly do not have a name for it.

---

[10]Naturally, in the context of this investigation, we presuppose that A exhibit a structure for which it is illuminating to speak of a rigorous branching-histories semantics. We offer no contribution to other aspects of the theory of promises or other speech acts.

5.5 Verdicts

In sum, we enter the following verdicts.

- Wrong: (15).
- Wrong: (16).
- Right: (18).

In our opinion, these verdicts may be taken to explicate the distinction drawn in Copley 2000 between the doctrine of "bivalence now" that it thinks wrong and the doctrine of "bivalence later" that it endorses.

Here is how the analysis might go in indirect speech, where we can use the $\textit{At-mom}_m$: connective described in Belnap et al. 2001. (The context is presumed to be given by speaker = T, audience = E, and moment of context = m_0; and we shall be using (12) instead of mentioning it.) At $m_{4:00}$ Themistocles' promise is satisfied iff

- Wrong: $\textit{At-mom}_{m_{4:00}}$:*Sett:*(12).
- Wrong: $\textit{At-mom}_{m_0}$:*Sett:*(12).
- Right: $\textit{At-mom}_{m_{4:00}}$:*Sett:*$\textit{At-mom}_{m_0}$:(12).

None of these connectives can be shifted around; and it is essential that there be a "double time reference."

6 Generalization and application

We generalize and simplify the double-time-reference ideas extracted from the Themistocles-Eurybiades example, and apply them to a variety of speech acts.

6.1 The propositional core

We can both put double time references in perspective, and simplify the notation for repeated use, by invoking the Kaplan notion of "content." What this amounts to in this case is freezing the context, and defining a content idea by letting the history parameter vary.[11] Let δ be a speech act (of some kind), $\delta = \langle a_1, a_2, m_0, H, A\rangle$. Then we define

- $H_\delta = \{h\colon h \in H \text{ and } a_1, a_2, m_0, m_0/h \vDash A\}$.

We call H_δ the ***propositional core*** of the speech act δ. Further, we define the following three concepts, which use double time references, but in a fashion hidden in the characterization of δ. Keeping in mind, however, that m_0 is the moment of the speech act, we assume that $m_0 < m_1$ and that $m_1 \in \bigcup H$ (thus m_1 is a moment in the future of possibilities of m_0 that is reachable after the speech act has been accomplished).

- H_δ *is settled true at* m_1 iff $\forall h[m_1 \in h \rightarrow h \in H_\delta]$.

[11] We are not, however, defining the "content" *of the promise.*

- H_δ *is settled false at* m_1 iff $\forall h[m_1 \in h \to h \notin H_\delta]$
- H_δ *is open at* m_1 iff H_δ is neither settled true nor settled false at m_1.

For example, if $\delta = \langle a_1, a_2, m_0, H, A\rangle$ is a promising, then we can transparently define that δ is "satisfied" at m_1 iff H_δ is settled true at m_1.

6.2 Promising

We continue to let $\delta = \langle a_1, a_2, m_0, H, A\rangle$, and we are assuming that δ is a promise. For a_1 to promise is, in part, to commit to a future course of action, or, in more austere terms, to lay on herself a requirement to act in accord with a certain strategy, which we call $s_{a_1,\delta}$, whose field, F, is defined as the set, $\{m_1\colon m_0 < m_1 \text{ and } m_1 \in \bigcup H\}$, of all moments in the future of possibilities of the moment of promising reachable after the act of promising that occurs at m_0. Here by a "strategy," s, for F we mean a partial function defined on some subset of F such that for every m_1 in the domain of s, $s(m_1)$ is a set of histories through m_1 that is closed under choice-equivalence for a_1 at m_1. The particular strategy, $s_{a_1,\delta}$, for a_1 is determined, more or less, by the propositional core, H_δ, of the promise $\delta = \langle a_1, a_2, m_0, H, A\rangle$. Choose any moment m_1 in the future of possibilities of the moment m_0 of promising.

- Suppose H_δ is settled true at m_1. We say that the promise has been *satisfied* at m_1. The strategy $s_{a_1,\delta}$ is undefined at m_1: "There is nothing left for a_1 to do." The strategy $s_{a_2,\delta}$ requires that a_2 act in accord with a strategy that at least gives credit to a_1; and in the case of a contractual promise, the strategy $s_{a_2,\delta}$ should involve what the law of contracts calls "consideration." These elements are not specified in the declarative core itself, but would need separate specification.
- Suppose H_δ is settled false at m_1. We say that the promise has been *infringed.* "There is nothing more that a_1 can do." In this case, $s_{a_1,\delta}$ conditionally requires that a_1 choose in accord with a strategy that compensates a_2, the condition being that a_2 act so as to lay such a requirement on a_1. The strategies $s_{a_1,\delta}$ and $s_{a_2,\delta}$, as well as strategies for other agents, are such as not to prevent a_2 from creating a requirement that he be compensated by a_1. The nature and extent of the portions of these strategies that concern compensation is not given by the declarative core itself, but would need separate specification.
- Suppose H_δ is open at m_1. This case is distinctive of "world-to-words-fitting" speech acts, for it calls upon a_1 to *do* something in the direction of settling true the propositional core. Two cases.

- There is available at m_1 a strategy for a_1 that guarantees the eventual settled truth of H_δ. Then a_1 is required by $s_{a_1,\delta}$ to make a choice at m_1 that is in accord with this guaranteeing strategy. The strategy $s_{a_2,\delta}$ is undefined at m_1.
- No guaranteeing strategy is available for a_1 at m_1. There is, however, bound to be some choice for a_1 at m_1 that favors H_δ; that is, some choice that has nonempty intersection with H_δ, so that on that choice, H_δ remains a possibility. Then a_1 is required by $s_{a_1,\delta}$ to make some such choice. The strategy $s_{a_2,\delta}$ is undefined at m_1.

6.3 Ordering

Ordering is a world-to-words-fitting speech act, structurally similar to promising. The language of "satisfaction" and "infringement" is still apt when H_δ is respectively settled true or settled false. And, crucially, a_2 is required to *do* something at moments subsequent to the moment of ordering at which the propositional core, H_δ, is still open.

6.4 Advising

Advising is also world-to-words. The content of the advice is a strategy for the advisee, a_2, that is structurally like the strategy-content of an order. As Hamblin 1987 says, however, the deontic meaning of the content of advice is very different from ordering. The advisee is *required* to do very little more than consider following the advised strategy, and that only if he has antecedently requested it. The advisee, however, is empowered to require of the advisor that she give reasons why he should follow a strategy forwarding the settled truth of A. Furthermore, the advisee is empowered to require the advisor to compensate, etc., should he get into trouble by following her advice to act in accord with such a strategy.

6.5 Asserting

As before, we continue to let $\delta = \langle a_1, a_2, m_0, H, A\rangle$, and we are assume that δ is an assertion. Assertion has many aspects involving future-oriented norms, of which Brandom 1994 gives the richest account. The assertor, a_1, presumably makes a conditional promise to give reasons for her assertion of A if asked by a_2. The assertee, a_2, is entitled to act on the assertion (but we don't know how to say exactly what that means, though as Brandom says, a_2 is plausibly entitled to assert A to a third party, citing a_1 as authority). If he does so and things go wrong because H_δ turns out settled false, a_2 may require compensation from a_1.

Put in the most austere normative language possible, we may say that an assertion is ***vindicated*** or ***impugned*** at a later moment, m_1, depending on whether at that later moment H_δ is settled true or settled false. At a moment when vindicated or impugned, a_1 may deserve credit or discredit. All of this seems similar to promising, with different words (vindication instead of satisfaction, impugnment instead of infringement). But the double time reference analysis permits a definite and critical structural difference between assertion and promise. It arises in the case of a future moment, m_1, where the propositional core, H_δ, is *open* (neither settled true nor settled false). In the case of a promise, the strategy $s_{a_1,\delta}$ guiding a promisor required that she *do* something in the direction of satisfying A. But in the structurally identical case, when H_δ is as yet open, the strategy guiding an assertor is *undefined.* That is, we suggest, the hallmark of a words-to-world-fitting speech act: Nothing ever needs to be *done* in the direction of settling the propositional core. Strikingly, the difference remains even between "I assert (predict) that I will choose to fight a sea battle" and "I promise that I will choose to fight a sea battle."

6.6 Betting

When $\delta = \langle a_1, a_2, m_0, H, A\rangle$ and δ represents that a_1 bet a_2 her \$3 against his \$5 that 'A' is true, the analysis depending on H_δ is easy. If H_δ is settled true at m_1, then a_2 is required to pay \$5 to a_1 (unless of course the payment has already been made). If H_δ is settled false at m_1, then a_1 is required to pay \$3 to a_2 (unless ...). If H_δ is open at m_1, then nothing is required of anyone. Hence, betting is a words-to-world-fitting speech act.

7 Remarks

There is more than one source of confusion in trying to work through speech acts in an indeterminist setting.

- There is the confusion into which everyone runs when trying to speak cogently about indeterministic phenomena. In my opinion there is no way out of this confusion that does not take into consideration the Prior-Thomason tense-modal semantics that makes the semantic value of many sentences depend—in the logical sense—on how the future unfolds.

- There is the confusion of sorting out the roles of context and of auxiliary parameters in semantic evaluation. For this Kaplan's ideas seem to be essential.

- Further confusion arises from taking seriously that speech acts are choice-based actions, so that ideally the study should require an antecedent theory of agents and their choices. Stit theory seems to help here.
- How should we treat speech-act reports when given in direct discourse? Speech-act reports can certainly be treated from the point of view of indirect discourse, and indeed, as is usual, speech acts can be discussed without any explicit mention of speech-act reports. But these modes of treatment serve to avoid certain confusions rather than to clear them up. To treat direct-discourse reports of speech acts seems to take special care in shifting the context parameters of semantic evaluation.
- Lastly, figuring out how to understand speech acts in terms of their normative consequences for the future seems to require "double time references," which confusingly combine reference to the moment of the speech act with reference to the future moment at which a normative consequence comes under consideration.

We have tried to make a little progress, partly by refusing, as far as possible, to bring in any mental or "intentional" aspects of speech acts, sticking, to a large extent, to strictly causal and normative ideas. We think that by minimizing attention to intentional elements we have made some headway with the rest. Even so, the matter is sufficiently complex as almost to guarantee that at some point or other our treatment has gone wrong. Would that it were otherwise; but, as Alan Ross Anderson used to say, there it is.

Acknowledgements

In addition to M. Weiner whom I mention above, I thank M. Perloff and C. Campbell for a variety of suggestions.

References

Austin, J. L. 1962. *How to do things with words.* Cambridge, MA: Harvard University Press. The William James Lectures, 1955.

Belnap, N. 1991. Backwards and forwards in the modal logic of agency. *Philosophy and phenomenological research* 51:777–807.

Belnap, N. 1996a. Agents in branching time. In *Logic and reality: essays in pure and applied logic, in memory of Arthur Prior*, ed. B. J. Copeland. 239–271. Oxford: Oxford University Press. This article draws heavily on "Before refraining: concepts for agency," 1990, *Erkenntnis* 34:137–169.

Belnap, N. 1996b. Deontic kinematics and austere strategics. In *Logica '96: proceedings of the 10th International Symposium*, ed. P. Kolar T. Childers and V. Svoboda, 21–40.

Belnap, N., and P. Bartha. 1995. Marcus and the problem of nested deontic modalities. In *Modality, morality, and belief: essays in honor of Ruth Barcan Marcus*, ed. W. Sinnott-Armstrong, D. Raffman, and N. Asher. 174–197. Cambridge: Cambridge University Press.

Belnap, N., and M. Green. 1994. Indeterminism and the thin red line. In *Philosophical perspectives*, ed. J. Tomberlin. 365–388. Atascadero, CA: Ridgeview Publishing Company.

Belnap, N., and M. Perloff. 1988. Seeing to it that: a canonical form for agentives. *Theoria* 54:175–199. Corrected version in *Knowledge representation and defeasible reasoning*, ed. H. E. Kyburg, Jr., R. P. Loui, and G. N. Carlson, 1990. 167–190. Dordrecht/Boston/London: Kluwer Academic Publishers.

Belnap, N., and M. Perloff. 1992. The way of the agent. *Studia logica* 51:463–484.

Belnap, N., and M. Perloff. 1993. In the realm of agents. *Annals of mathematics and artificial intelligence* 25–48.

Belnap, N., M. Perloff, and M. Xu. 2001. *Facing the future: Agents and choices in our indeterminist world.* Oxford and New York: Oxford University Press. Forthcoming.

Brandom, B. 1994. *Making it explicit: reasoning, representing, and discursive commitment.* Cambridge, MA: Harvard University Press.

Copley, B. 2000. Conceptualizing the futurate and the future. In *MIT working papers in philosophy and linguistics: The linguistics/philosophy interface*, ed. R. Bhatt, P. Hawley, M. Kackl, and I. Maitra. 45–72.

Hamblin, C. L. 1987. *Imperatives.* Oxford and New York: Basil Blackwell.

Horty, J. 2000. *Agency and deontic logic.* Oxford: Oxford University Press. Forthcoming.

Kaplan, D. 1989. Demonstratives: an essay on the semantics, logic, metaphysics, and epistemology of demonstratives and other indexicals; and Afterthoughts. In *Themes from Kaplan*, ed. J. Almog, J. Perry, and H. Wettstein. 481–563; 565–614. Oxford: Oxford University Press.

Perloff, M. 1991. *Stit* and the language of agency. *Synthese* 86:379–408.

Perloff, M. 1995. Stit and the imperative. *American philosophical quarterly* 32:71–81.

Prior, A. N. 1957. *Time and modality.* Oxford: Oxford University Press.

Prior, A. N. 1967. *Past, Present, and Future.* Oxford: Oxford University Press.

Searle, J. R., and D. Vanderveken. 1985. *Foundations of illocutionary logic.* Cambridge: Cambridge University Press.

Thomason, R. H. 1970. Indeterminist time and truth-value gaps. *Theoria* 36:264–281.

Thomason, R. H. 1984. Combinations of tense and modality. In *Handbook of philosophical logic, vol. II: Extensions of classical logic*, ed. D. Gabbay and F. Guenthner. 135–165. Dordrecht: D. Reidel Publishing Company.

Thomson, J. J. 1990. *The realm of rights.* Cambridge, MA: Harvard University Press.

Xu, M. 1994a. Decidability of deliberative stit theories with multiple agents. In *Temporal Logic, First International Conference, ICTL'94*, ed. D. M. Gabbay and H. J. Ohlbach, 332–348. International Conference of Temporal Logic, Springer-Verlag.

Xu, M. 1994b. Decidability of stit theory with a single agent and refref equivalence. *Studia logica* 53:259–298.

Xu, M. 1994c. Doing and refraining from refraining. *Journal of philosophical logic* 23:621–632.

Xu, M. 1995a. Busy choice sequences, refraining formulas and modalities. *Studia logica* 54:267–301.

Xu, M. 1995b. On the basic logic of stit with a single agent. *Journal of symbolic logic* 60:459–483.

Nuel Belnap
Department of Philosophy
University of Pittsburgh
Pittsburgh, PA 15260
USA
E-mail: belnap@pitt.edu
URL: http://www.pitt.edu/~belnap

4

Homophonic Theory of Truth for Tense Logic

TORBEN BRAÜNER

ABSTRACT. This paper concerns homophonic theories of truth, in the sense of Donald Davidson, for modal and tense logics. Earlier works on this subject have left open the following question: How weak can a modal or tense logic be if it has to permit the existence of a homophonic theory of truth? In this paper we shall show that an answer to the question can be given if attention is restricted to theories of truth which satisfy a very natural further condition.

1 Introduction

Roughly speaking, a semantic theory for a language is called homophonic if the constructions of the language are interpreted in terms of analogous constructions of the metalanguage. For example, the standard truth table semantics for propositional logic is homophonic since $\neg$ is interpreted in terms of negation and $\wedge$ is interpreted in terms of conjunction. This is contrasted by the standard model-theoretic semantics for tense logic where the tense operators P and F are interpreted in terms of quantification; not in terms of tenses.

In this paper we investigate to which extent homophonic theories of truth, in the sense of Donald Davidson, can be constructed for tense logics. The notion of a homophonic theory of truth for tenses was foreshadowed by Arthur N. Prior, the founder of tense logic, as is witnessed by the following quotation from his book *Time and Modality*:

> The function of the operator F, in short, is that of forming a future-tense statement from the corresponding present-tense

Advances in Modal Logic, Volume 3
F. Wolter, H. Wansing, M. de Rijke, and M. Zakharyaschev, eds.

> one, and the future-tense statement is not about the present-tense one, but is about whatever the present-tense statement is about. ... But although the statement 'It will be the case that Professor Carnap is flying to the moon', that is, 'Professor Carnap will be flying to the moon', is not exactly a statement *about* the statement 'Professor Carnap is flying to the moon', we may say that the future-tense statement *is* true if and only if the present-tense statement *will be* true. (Prior 1957, p. 8–9)

However, compared to the total amount of work carried out within the area of theories of truth in general, only little work has been devoted to the problem of constructing theories of truth for modal and tense logics. Notable exceptions are the works of Anil Gupta, Gupta 1978, Christopher Peacocke, Peacocke 1978, and John Wallace, Wallace 1970, Wallace 1975. See also Humberstone 1996, Thomason 1976.

Before introducing Donald Davidson's notion of a theory of truth, we shall consider some related work of Alfred Tarski. In the famous paper Tarski 1956, Tarski showed, for a given formal language (called the object language), how to define in another formal language (called the metalanguage) a 1-place predicate *True* such that for each formula ϕ of the object language, the formula

$$\mathit{True}(\overline{\phi}) \leftrightarrow f(\phi)$$

is provable from the definition. Here $\overline{\phi}$ is a name of, and $f(\phi)$ a translation of, the object language formula ϕ in the metalanguage. Note that $\overline{\phi}$ is a term of, and $f(\phi)$ a formula of, the metalanguage. The biconditionals above are called T-biconditionals. Thus, given a translation, that is, an interpretation, of the object language in the metalanguage, the truth predicate *True* is defined such that any T-biconditional is provable.

In a well-known series of papers, collected in Davidson 1984, Davidson reverses Tarski's priorities: According to Davidson's notion of a theory of truth, starting with the truth predicate *True* as primitive, giving an interpretation of the object language in the metalanguage amounts to specifying a finite number of metalanguage formulas which for any formula ϕ of the object language has as a logical consequence

$$\mathit{True}(\overline{\phi}) \leftrightarrow \psi$$

for some formula ψ of the metalanguage. Again, $\overline{\phi}$ is a name of ϕ. Note that it need not be the case that ψ is a translation of ϕ. But if a translation f is given such that ψ is the translation of ϕ, that is $\psi = f(\phi)$, then the theory of truth is called heterophonic. In this case the biconditionals above are the Tarskian T-biconditionals. If furthermore the object language is included in the metalanguage, and the translation f is the identity function, that is, $\psi = \phi$, then the theory of truth is called homophonic.

Davidson argues that to know such a finite theory of truth for the object language in the metalanguage is a sufficient condition for being able to interpret the object language in the metalanguage. The finitenes requirement is important: According to Davidson, an infinite theory of truth would not be 'learnable'. The Davidsonian notion of a theory of truth can be exposed to various criticisms depending on whether a translation is part of the picture or not. See for example the account given in Sundholm 1986.

As a stepping-stone in our discussion on homophonic theories of truth for tense logics, we shall take a look at the similar problem of constructing theories of truth for modal logics. Gupta, Gupta 1978, Peacocke, Peacocke 1978, and Wallace, Wallace 1970, consider a strategy for constructing, for any modal logic stronger than or equal to S4, a homophonic theory of truth. However, it appears that this strategy does not work for modal logics weaker than S4. They leave open the following question: Which modal logics permit the existence of theories of truth? To quote from Gupta's paper:

> This whole area requires more research. A particularly interesting question is that of a minimal modal logic – how weak can a quantified modal logic be and still meet the Davidsonian adequacy condition? (Gupta 1978, p. 456)

Now, it turns out that the strategy for constructing homophonic theories of truth for modal logics can be adapted to the case with tense logics. The strategy works for tense logics stronger or equal to K_b, the logic of branching time, but it does not work for sufficiently weak tense logics. Again, one can ask the question: What is the minimal condition for the existence of a theory of truth? In this paper we shall show that an answer to the question can be given if attention is restricted to theories of truth which satisfy a very natural further condition. Our result concerns tense logic, but it can easily be simplified to the case with modal logic; thereby an answer is given to the question quoted above. What enables us to answer the question is that we deal with it in model-theoretic, rather than proof-theoretic, terms. This is contrary to what is done in the papers by Gupta, Peacocke, and Wallace.

This paper is structured as follows. In the second section we give a short account of first-order tense logic. In the third section we explain what the difficulty is in giving theories of truth for tense logics and in the fourth section we show that a minimal condition for the existence of a theory of truth can be given if attention is restricted to certain theories of truth. In the fifth section, we make some concluding remarks.

2 A Recap on First-Order Tense Logic

In what follows, we consider the language of first-order tense logic, that is, the language of first-order predicate logic extended with the tense operators P and F. As usual, $H\phi$ and $G\phi$ are abbreviations for respectively $\neg P \neg \phi$ and $\neg F \neg \phi$. To fix terminology, we give a short account of the standard constant domain Kripke semantics for first-order tense logic where constants and function symbols are interpreted rigidly. See Hughes and Cresswell 1968, Hughes and Cresswell 1984. A *frame* is a pair $(T, <)$ where T is a non-empty set and $<$ is a binary relation on T. The elements of T are called *moments* and $<$ is called the *earlier-later* relation. A *model* is a tupel $(T, <, D, V)$ where $(T, <)$ is a frame, D is a non-empty set, and V is a *valuation*, that is, V is a function that

- to each n-place function symbol assigns a function from D^n to D,
- to each pair consisting of an element of T and an n-place predicate letter assigns a subset of D^n.

Constant symbols are treated as 0-place function symbols. Note that V to an element of T and a 0-place predicate letter, that is, a sentential letter, assigns an element of $\{0, 1\}$ where 0 and 1 denotes respectively $\emptyset$ and $\{()\}$. The model $(T, <, D, V)$ is *based* on the frame $(T, <)$. An *assignment* for a model $(T, <, D, V)$ is a function s that to each variable assigns an element of D. Given an assignment s for a model $(T, <, D, V)$, each term is inductively assigned a *value*, that is, an element of D, cf. first-order predicate logic. Given a model $\mathcal{M} = (T, <, D, V)$, the relation $\mathcal{M}, t \models \phi[s]$ is inductively defined in the standard fashion, where t is an element of T, s is an assignment, and ϕ is a first-order tense logical formula. Moreover, $\mathcal{M} \models \phi[s]$ means $\mathcal{M}, t \models \phi[s]$ for all elements t of T and $\mathcal{M} \models \phi$ means $\mathcal{M} \models \phi[s]$ for all assignments s. A formula ϕ is *valid* in a frame $(T, <)$ if and only if $\mathcal{M} \models \phi$ for any model $\mathcal{M}$ that is based on $(T, <)$. A formula ϕ is *valid* in a class of frames $\mathbf{F}$ if and only if ϕ is valid in any frame in $\mathbf{F}$.

We shall frequently make use of propositional versions of the definitions above without commenting it explicitly; in particular, this applies to the notion of a model which in the propositional version is just a triple $(T, <, V)$ where V is a function that to an element of T and a sentential letter assigns an element of $\{0, 1\}$. In what follows, we need the standard result below.

Lemma 2.1 *If ψ is a propositional tense logical formula that is valid in a class of frames* $\mathbf{F}$, *then for any formulas ϕ_1, ..., ϕ_i of first-order tense logic, also the formula $\psi[\phi_1, ..., \phi_i / r_1, ..., r_i]$ is valid in* $\mathbf{F}$.

Proof. A straightforward proof by induction on the structure of ψ. ⊣

As usual, $\psi[\phi_1, ..., \phi_i / r_1, ..., r_i]$ denotes the formula ψ where the formulas ϕ_1, ..., ϕ_i simultaneously have been substituted for occurrences of the sentential letters $r_1, ..., r_i$.

3 Theories of Truth for Tense Logic

As with modal logics, the difficulty in giving theories of truth for tense logics arises in the non-quantificational part. So it suffices to take as object language the language of propositional tense logic. The necessity operator $\Box$ is taken instead of the tense operators P and F in the case of modal logic. For simplicity it is assumed that the object language contains only two sentential letters, say p and q. The metalanguage is constituted by the object language extended with quantifiers and 1-place predicates *Form* and *True*, and furthermore, machinery to form structural descriptive names of the object language formulas: It is assumed that the metalanguage contains, for each symbol e of the object language, a constant symbol $\overline{e}$ naming e, and also, the metalanguage is assumed to contain a binary function symbol $\hat{}$ for concatenation. Then, for each formula ϕ of the object language, a structural descriptive name $\overline{\phi}$ is defined inductively as follows:

$$\begin{aligned} \overline{\psi \wedge \theta} &= \overline{\psi}\,\hat{}\,(\overline{\wedge}\,\hat{}\,\overline{\theta}) \\ \overline{\neg\psi} &= \overline{\neg}\,\hat{}\,\overline{\psi} \\ \overline{P\psi} &= \overline{P}\,\hat{}\,\overline{\psi} \\ \overline{F\psi} &= \overline{F}\,\hat{}\,\overline{\psi} \end{aligned}$$

Thus, $\overline{\phi}$ is formed by concatenating the names of the symbols occurring in ϕ. We ignore parantheses where no confusion can occur.

Now, according to Davidson's notion of a theory of truth, giving a theory of truth for a choosen modal or tense logic amounts to specifying a finite number of metalanguage formulas γ_1, ..., γ_l which satisfies the condition that they have each T-biconditional as a logical consequence, that is, for any formula ϕ of the object language, the metalanguage formula $True(\overline{\phi}) \leftrightarrow \phi$ has to be a logical consequence of $\gamma_1 \ldots \gamma_l$ where the notion of logical consequence employed is the one of the choosen modal or tense logic. The specified formulas typically have the form of recursion clauses; in some cases enclosed by modalities.

In modal and tense logic, one usually considers two different notions of logical consequence that both are semantically characterised:

- A formula ψ is a *local consequence* of ϕ wrt. a class of frames **F** if and only if for any model $\mathcal{M}$ based on a frame in **F**, any moment t, and any assignment s, if $\mathcal{M}, t \models \phi[s]$, then also $\mathcal{M}, t \models \psi[s]$.
- A formula ψ is a *global consequence* of ϕ wrt. a class of frames **F** if and only if for any model $\mathcal{M}$ based on a frame in **F** and any assignment s, if $\mathcal{M} \models \phi[s]$, then also $\mathcal{M} \models \psi[s]$.

Note that ψ is a local consequence of ϕ wrt. **F** if and only if $\phi \rightarrow \psi$ is valid in **F**. Clearly, what is needed is the local consequence relation: We want each T-biconditional to be true at some moment if the formula $\gamma_1 \wedge ... \wedge \gamma_l$ is true at the moment in question. Thus, we end up with the following definition.

Definition 3.1 A finite list of closed metalanguage formulas Γ is a *theory of truth* for a class of frames **F** if and only if for any formula ϕ of the object language, the formula

$$\bigwedge \Gamma \rightarrow (True(\overline{\phi}) \leftrightarrow \phi)$$

is valid in **F**.

As usual, $\bigwedge \Gamma$ denotes the conjunction of the formulas in Γ. We shall later consider further conditions on the theory of truth Γ.

It should be remarked that domains are kept constant since a variable at any moment is supposed to range over the same set, namely the set of strings over the set of object language symbols, and similarly, constant and function symbols are interpreted rigidly since a structural descriptive name $\overline{\phi}$ at any moment is supposed to refer to the same thing, namely the formula ϕ.

Remark 3.2 Care must be taken when the standard notion of a theory of truth is extended to involve modalities. In the literature on theories of truth, one often sees a proof-theoretically characterised notion of logical consequence: A Hilbert system for a logic is given, and a formula ψ is taken to be a logical consequence of a formula θ if and only if ψ is provable from the logical axioms and rules together with θ. So requiring each T-biconditional to be a logical consequence of the formulas Γ then amounts to declaring these formulas to be axioms in the same sense as the logical axioms are axioms and requiring that each T-biconditional is provable. Now, in the case of the the usual Hilbert systems for classical logic, the so called deduction theorem holds: A formula $\theta \rightarrow \psi$ is provable from the logical axioms and rules if and only if ψ is provable from the logical axioms and rules together with θ. Hence, in the case of classical logic one might as well have required that for any object language formula ϕ, the formula $\bigwedge \Gamma \rightarrow (True(\overline{\phi}) \leftrightarrow \phi)$ is provable from the logical axioms and rules, or, in the light of soundness and completeness, valid. But the deduction theorem does not hold for modal and tense logical Hilbert systems in general. See for example Fitting 1993. So in the case of modal and tense logic, giving a Hilbert system and then requiring that each T-biconditional is provable from the logical axioms and rules together with the formulas Γ is not in general equivalent to requiring that each formula $\bigwedge \Gamma \rightarrow (True(\overline{\phi}) \leftrightarrow \phi)$ is provable from the rules and logical axioms alone.

Now, which formulas $\gamma_1, ..., \gamma_l$ are natural candidates to a theory of truth? First, machinery for object language syntax is needed:

$$\begin{array}{l} Form(\overline{p}) \\ Form(\overline{q}) \\ Form(x \mathbin{\hat{}} (\overline{\wedge} \mathbin{\hat{}} y)) \leftrightarrow (Form(x) \wedge Form(y)) \\ Form(\overline{\neg} \mathbin{\hat{}} x) \leftrightarrow Form(x) \\ Form(\overline{P} \mathbin{\hat{}} x) \leftrightarrow Form(x) \\ Form(\overline{F} \mathbin{\hat{}} x) \leftrightarrow Form(x) \end{array}$$

Here and below outermost universal quantifiers are left out, so we really consider the universal closures of the displayed formulas. Concerning the predicate *True*, one option is to take the standard recursion clauses for the classical part extended with clauses for the tense modalities analogous to the clause for negation:

$$\begin{array}{l} True(\overline{p}) \leftrightarrow p \\ True(\overline{q}) \leftrightarrow q \\ (Form(x) \wedge Form(y)) \rightarrow (True(x \mathbin{\hat{}} (\overline{\wedge} \mathbin{\hat{}} y)) \leftrightarrow (True(x) \wedge True(y))) \\ Form(x) \rightarrow (True(\overline{\neg} \mathbin{\hat{}} x) \leftrightarrow \neg True(x)) \\ Form(x) \rightarrow (True(\overline{P} \mathbin{\hat{}} x) \leftrightarrow P\, True(x)) \\ Form(x) \rightarrow (True(\overline{F} \mathbin{\hat{}} x) \leftrightarrow F\, True(x)) \end{array}$$

In the case of modal logic, the natural candidate for the necessity operator $\Box$ is considered instead of the clauses involving P and F.

However, in the case of modal logic such a strategy will not do, as pointed out in Wallace 1970. In fact, this is also the case with tense logic: If Γ_1 denotes the list of formulas constituted by the universal closures of the formulas displayed above together with the formulas for object language syntax, then for example the formula

$$\bigwedge \Gamma_1 \rightarrow (True(\overline{Fp}) \leftrightarrow Fp)$$

is not valid in any frame $(T, <)$ where $<$ is non-empty. The problem is that the recursion clauses do not in general justify replacing the left hand side of a clause by the corresponding right hand side within the context of a tense operator.

In the case of the modal logic S5, such a deficiency can be remedied by enclosing each recursion clause by the necessity operator $\Box$, cf. Wallace 1970. In fact, this is also the case with S4, cf. Gupta 1978. In the case of the logic of linear time, K_l, that is, the tense logic where time is assumed to be transitive, backwards linear, and forwards linear, it turns out that the deficiency above can be remedied by modalising each clause using the modality A which is defined by letting $A\phi$ be an abbreviation for

$$H\phi \wedge \phi \wedge G\phi$$

that is, to be precise, by letting $A\phi$ be an abbreviation for $A[\phi/p]$ where A denotes the formula $Hp \wedge p \wedge Gp$. We might call A an always operator. Hence, in the case of linear time the following clauses for the predicate *True* are taken:

$$A(\mathit{True}(\overline{p}) \leftrightarrow p)$$
$$A(\mathit{True}(\overline{q}) \leftrightarrow q)$$
$$(\mathit{Form}(x) \wedge \mathit{Form}(y)) \rightarrow A(\mathit{True}(x \,\hat{}\, (\overline{\wedge} \,\hat{}\, y)) \leftrightarrow (\mathit{True}(x) \wedge \mathit{True}(y)))$$
$$\mathit{Form}(x) \rightarrow A(\mathit{True}(\overline{\neg} \,\hat{}\, x) \leftrightarrow \neg \mathit{True}(x))$$
$$\mathit{Form}(x) \rightarrow A(\mathit{True}(\overline{P} \,\hat{}\, x) \leftrightarrow P\,\mathit{True}(x))$$
$$\mathit{Form}(x) \rightarrow A(\mathit{True}(\overline{F} \,\hat{}\, x) \leftrightarrow F\,\mathit{True}(x))$$

So if Γ_2 denotes the list of formulas constituted by the universal closures of the formulas displayed above together with the formulas for object language syntax, then Γ_2 is a theory of truth for the class of all linear frames. We shall soon see why this is the case. If K_l is weakened to the logic of branching time, K_b, that is, the logic where time is assumed to be transitive and backwards linear, then the always modality above will not do. But it turns out that if A is defined by letting $A\phi$ be an abbreviation for

$$HG\phi \wedge H\phi \wedge \phi \wedge G\phi$$

then Γ_2 becomes a theory of truth for the class of all branching time frames. Thus, it appears that weakening the logic necessitates strengthning the always modality.

4 A Minimal Condition for Giving a Theory of Truth

Now, how weak can a tense logic be if it has to permit giving an always modality such that the modalised recursion clauses justify replacing the left hand side of a clause by the corresponding right hand side within any context? Or, rather, how weak can a tense logic be if it has to permit the existence of a theory of truth? In what follows, we shall try to give an answer to this question. First an important definition.

Definition 4.1 Let a non-negative integer n be given. Also, let a frame $(T, <)$ be given. Two elements t and u of T are *n-related* if and only if for some non-negative integer $k \leq n$, there exists elements v_0, ..., v_k of T such that $v_0 = t$, $v_k = u$ and for every $i \in \{0, ..., k-1\}$, $v_i < v_{i+1}$ or $v_{i+1} < v_i$. The frame $(T, <)$ is *n-bounded* if and only if any elements t and u of T such that $t <^* u$ are n-related, where $<^*$ is the reflexive, symmetric and transitive closure of $<$. A class of frames $\mathbf{F}$ is *n-bounded* if and only if each frame in $\mathbf{F}$ is n-bounded.

Note that $t <^* u$ if and only if t and u are m-related for some m. So the frame $(T, <)$ is n-bounded if and only if any elements t and u of T that are reachable from each other by moving a finite number of steps in either direction along $<$ are reachable from each other by moving only n or less steps in either direction along $<$.

Remark 4.2 The condition of being n-bounded might appear to be rather strong, but it turns out that most standard tense logics satisfies it for some n. For example, it is straightforward to check that any branching time frame is 2-bounded and that any linear time frame is 1-bounded.

Remark 4.3 In the case of modal logic, an appropriate definition of being n-related can be obtained by strengthening the condition on the elements v_0, ..., v_k of T such that $v_i < v_{i+1}$ has to be the case for *every* $i \in \{0, ..., k-1\}$ and an appropriate definition of being n-bounded can be obtained by considering the reflexive and transitive closure of $<$ rather than the reflexive, transitive, *and symmetric* closure. In that case any transitive frame would be 1-bounded.

With the aim of proving the standard lemma below, we need a convention: For any first-order tense logical formula ϕ, a non-negative integer $depth(\phi)$ is defined as follows.

$$depth(\phi) = \begin{cases} 0 & \text{if } \phi \text{ is an atomic formula} \\ max\{depth(\psi), depth(\theta)\} & \text{if } \phi = \psi \wedge \theta \\ depth(\psi) & \text{if } \phi = \neg\psi \\ depth(\psi) & \text{if } \phi = \forall x\psi \text{ or } \phi = \exists x\psi \\ depth(\psi) + 1 & \text{if } \phi = P\psi \text{ or } \phi = F\psi \end{cases}$$

Now the lemma; it says that the result of evaluating a formula ϕ at a moment t only depends on the result of evaluating predicates occurring in ϕ at moments $depth(\phi)$-related to t.

Lemma 4.4 *Let two models $\mathcal{M} = (T, <, D, V)$ and $\mathcal{M}' = (T, <, D, V')$ be given where $V(f) = V'(f)$ for every function symbol f. For any first-order tense logical formula ϕ, any element t of T, and any assignment s, if $V(u, \sigma) = V'(u, \sigma)$ for every element u of T which is depth(ϕ)-related to t and every predicate letter σ occurring in ϕ, then $\mathcal{M}, t \models \phi[s]$ if and only if $\mathcal{M}', t \models \phi[s]$.*

Proof. Induction on the structure of ϕ. Note that any elements t and u of T are 0-related if and only if $t = u$. ⊣

In what follows, we shall need the convention that for any non-negative integer n, a formula L^n is defined by the stipulations:

$$L^0 = p$$
$$L^{n+1} = HL^n \wedge L^n \wedge GL^n$$

Note that exactly one sentential letter occurs in L^n, namely p. Now a simple lemma concerning L^n.

Lemma 4.5 *Let a model $(T, <, V)$ be given. For any non-negative integer n and any element t of T, $(T, <, V), t \models L^n$ if and only if for every element u of T which is n-related to t, $V(u, p) = 1$.*

Proof. Induction on n. ⊣

The lemma above says that the formula L^n is true at a moment t if and only if the sentential letter p is true at at any moment reachable from t by moving n or less steps in either direction along $<$. Note that if the frame $(T, <)$ is n-bounded, then $(T, <, V), t \models L^n$ if and only if $V(u, p) = 1$ for every element u of T such that $t <^* u$. So we might call L^n an always operator for the class of all n-bounded frames. The history of the L^n operator goes back to the work of Prior, see for example the paper *Tense Logic and the Logic of Earlier and Later* in Prior 1968.

Lemma 4.6 *Let a model $(T, <, V)$ be given and let t be an element of T such that $V(u, o) = V(u, s)$ for every element u of T where $t <^* u$. For every propositional tense logical formula ϕ, $(T, <, V), t \models \phi[o/r] \leftrightarrow \phi[s/r]$.*

Proof. For every propositional tense logical formula ϕ and every element u of T such that $t <^* u$, $(T, <, V), u \models \phi[o/r] \leftrightarrow \phi[s/r]$; this follows by induction on the structure of ϕ. ⊣

Theorem 4.7 *A class of frames* **F** *is n-bounded for some n if and only if* **F** *has a theory of truth Γ with the property that any object language* *model* $(T, <, V)$ *based on a frame in* $\mathbf{F}$ *can be extended to a metalanguage model* $\mathcal{M} = (T, <, D, V)$ *such that* $\mathcal{M} \models \bigwedge\Gamma$.

Proof. Assume that $\mathbf{F}$ is n-bounded. It then follows from Lemma 4.4, Lemma 4.5 and Lemma 4.6 that for any propositional tense logical formula ψ, the formula

$$(L^n(o_1 \leftrightarrow s_1) \wedge ... \wedge L^n(o_i \leftrightarrow s_i)) \rightarrow$$
$$(\psi[o_1, ..., o_i/r_1, ..., r_i] \leftrightarrow \psi[s_1, ..., s_i/r_1, ..., r_i])$$

is valid in $\mathbf{F}$ where o_1, ..., o_i and s_1, ..., s_i are sentential letters. (Recall that $L^n\theta$ is an abbreviation for $L^n[\theta/p]$.) By Lemma 2.1 any formula obtained from the formula displayed above by substituting first-order tense logical formulas for sentential letters, is also valid in $\mathbf{F}$. Now, take as the theory of truth Γ the list of formulas Γ_2 given in the previous section, where A has been replaced by L^n. It then follows that for any propositional tense logical formula ψ and any object language formulas ϕ_1, ..., ϕ_i, the formula

$$\bigwedge\Gamma \rightarrow (\psi[\mathit{True}(\overline{\phi_1}), ..., \mathit{True}(\overline{\phi_i})/r_1, ..., r_i] \leftrightarrow \psi[\phi_1, ..., \phi_i/r_1, ..., r_i])$$

is valid in $\mathbf{F}$. This can be proved by induction on the total number of connectives in the formulas ϕ_1, ..., ϕ_i. (Recall that an object language formula is a formula of propositional tense logic in which only the sentential letters p and q occurs.) Take $i = 1$ and $\psi = r_1$. Having shown that Γ is a theory of truth for $\mathbf{F}$, we just need to show that any object language model $(T, <, V)$ based on a frame in $\mathbf{F}$ can be extended to a metalanguage model $\mathcal{M} = (T, <, D, V)$ such that $\mathcal{M} \models \bigwedge\Gamma$. Let an object language model $(T, <, V)$ be given. Take as D the set of all metalanguage terms and extend V such that the constant symbols and the binary function symbol $\hat{}$ are interpreted syntactically, and moreover, such that

$$V(t, \mathit{Form}) = \{\overline{\phi} \mid \phi \text{ is an object lan. formula}\}$$

and

$$V(t, \mathit{True}) = \{\overline{\phi} \mid \phi \text{ is an object lan. formula where } (T, <, V), t \models \phi\}$$

for any element t of T. It is then straightforward to check that $\mathcal{M} \models \bigwedge\Gamma$. Note that the mapping $\phi \mapsto \overline{\phi}$ is injective.

Conversely, assume that $\mathbf{F}$ has a theory of truth Γ with the mentioned property. Let a frame $(T, <)$ in $\mathbf{F}$ be given, and moreover, let t and u be elements of T such that $t <^* u$. Thus, t and u are m-related for some m. Define an object language valuation V such that $V(w, p) = 1$ for any element w of T. By assumption the object language model $(T, <, V)$ can be extended to a metalanguage model $\mathcal{M} = (T, <, D, V)$ such that

$\mathcal{M} \models \bigwedge \Gamma$. So $\mathcal{M}, t \models \bigwedge \Gamma[s]$ and hence also $\mathcal{M}, t \models \mathit{True}(\overline{L^m}) \leftrightarrow L^m[s]$ where s is an arbitrary assignment. Now, define a valuation V' such that it is exactly as the first-order version of V except that

$$V'(w,p) = \begin{cases} 1 & \text{if } w \text{ is } \mathit{depth}(\bigwedge \Gamma)\text{-related to } t \\ 0 & \text{otherwise} \end{cases}$$

for any element w of T. By Lemma 4.4 $\mathcal{M}', t \models \bigwedge \Gamma[s]$, where $\mathcal{M}' = (T, <, D, V')$ and hence also $\mathcal{M}', t \models \mathit{True}(\overline{L^m}) \leftrightarrow L^m[s]$. But $\mathcal{M}, t \models L^m[s]$ by Lemma 4.5 which implies that $\mathcal{M}, t \models \mathit{True}(\overline{L^m})[s]$. But then obviously $\mathcal{M}', t \models \mathit{True}(\overline{L^m})[s]$ which implies that $\mathcal{M}', t \models L^m[s]$ and this by Lemma 4.5 implies that $V'(u,p) = 1$ wherefore u is $\mathit{depth}(\bigwedge \Gamma)$-related to t. Thus, $\mathbf{F}$ is $\mathit{depth}(\bigwedge \Gamma)$-bounded. $\dashv$

Remark 4.8 The theorem above concerns tense logic, but it can easily be simplified to the case with modal logic. Note that this requires changing the definitions of being n-related and n-bounded, cf. the remark following Definition 4.1. Thereby an answer is given to the question quoted in Section 1.

5 Conclusion

Earlier in this paper we have asked the question: How weak can a tense logic be if it has to permit the existence of a theory of truth? According to Theorem 4.7, n-boundednes is a sufficient condition for the existence of a theory of truth. But is it also necessary? Strictly speaking, according to Definition 3.1 any class of frames has a theory of truth since for any object language formula ϕ, the formula $\bot \rightarrow (\mathit{True}(\overline{\phi}) \leftrightarrow \phi)$ is valid in any frame. But such a trivial theory of truth obtained just by taking $\Gamma = \{\bot\}$ is obviously not acceptable. One might think that it would do just to add the condition that the formula $\neg \bigwedge \Gamma$ must not be valid in the class of frames in question. This is not the case, however, since a theory of truth for any class of frames could then be obtained by appending the formula $G \bot \wedge H \bot$ to list of formulas Γ_1 given in the Section 3. This theory of truth would be unacceptable as the earlier-later relation of any frame would simply be ignored. So stronger conditions have to be imposed on the list of formulas Γ to avoid trivial theories of truth. (In fact, up till now the problem of finding such conditions appears to have been overlooked in the literature on theories of truth for modal and tense logics.)

Now, Theorem 4.7 says that a class of frames is n-bounded for some n if and only if it has a theory of truth Γ which satisfies the condition that any object language model based on a frame in the class can be extended to a metalanguage model where $\bigwedge \Gamma$, and hence any T-biconditional

$True(\overline{\phi}) \leftrightarrow \phi$, is true at any moment. So n-boundednes is a necessary and sufficient condition for the existence of a theory of truth satisfying the mentioned condition. We find this further condition on a theory of truth very natural since it amounts to requiring that the structural descriptive names and the predicates *Form* and *True*, whichever way the world according to an object language model is, can be interpreted such that the value of a structural descriptive name $\overline{\phi}$ at any moment belongs to the extension of the predicate *True* if and only if things are such that the formula ϕ is true at the moment in question.

Acknowledgements

The work presented in this paper was initiated by a discussion with Peter Øhrstrøm and Per Hasle. Thanks to Thomas Müller for comments on an early version. Also thanks for comments to the participants of The 1999 Uppsala Autumn Festival of Philosophical Logic where I gave a talk on the topic of the paper. The work presented here is partially supported by The IT University of Copenhagen.

References

Davidson, D. 1984. *Inquiries into Truth and Interpretation.* Oxford University Press.

Fitting, M. 1993. Basic Modal Logic. In *Handbook of Logic in Artificial Intelligence and Logic Programming, Vol. 1, Logical Foundations*, ed. D. Gabbay et al. 365–448. Oxford University Press, Oxford.

Gupta, A. 1978. Modal Logic and Truth. *Journal of Philosophical Logic* 7:441–472.

Hughes, G. E., and M. J. Cresswell. 1968. *An Introduction to Modal Logic.* Methuen.

Hughes, G. E., and M. J. Cresswell. 1984. *A Companion to Modal Logic.* Methuen.

Humberstone, L. 1996. Homophony, Validity, Modality. In *Logic and Reality: Essays in the Legacy of Arthur Prior*, ed. J. Copeland. 215–236. Oxford University Press/Clarendon Press.

Peacocke, C. 1978. Necessity and Truth Theories. *Journal of Philosophical Logic* 7:473–500.

Prior, A. N. 1957. *Time and Modality.* Clarendon/Oxford University Press.

Prior, A. N. 1968. *Papers on Time and Tense.* Clarendon/Oxford University Press. A revised and extended edition of this book will in 2001 be published by Oxford University Press, edited by P. Hasle, P. Øhrstrøm, T. Braüner, and J. Copeland.

Sundholm, G. 1986. Proof Theory and Meaning. In *Handbook of Philosophical Logic, Vol. III, Alternatives to Classical Logic*, ed. D. Gabbay and F. Guenthner. 471–506. D. Reidel Publishing Company.

Tarski, A. 1956. The Concept of Truth in Formalized Languages. In *Logic, Semantics, Metamathematics.* 152–278. Oxford University Press.

Thomason, R. H. 1976. Necessity, Quotation and Truth: An Indexical Theory. In *Language in Focus*, ed. A. Kasher. 119–138. D. Reidel Publishing Company.

Wallace, J. 1970. On the Frame of Reference. *Synthese* 22:61–94.

Wallace, J. 1975. Nonstandard Theories of Truth. In *The Logic of Grammar*, ed. D. Davidson and G. Harman. 50–60. Dickenson Publishing Company.

Torben Braüner
Department of Computer Science
Roskilde University
P.O. Box 260
DK-4000 Roskilde
Denmark
E-mail: torben@ruc.dk
URL: http://www.dat.ruc.dk/~torben/

5

Weak Necessity on Weak Kleene Matrices

FABRICE CORREIA

ABSTRACT. A possible world semantics for standard modal languages is presented, where the valuation functions are allowed to be partial, the truth–functional connectives are interpreted according to weak Kleene matrices, and the necessity operator is given a "weak" interpretation. Completeness and incompleteness results for some (axiomatic) systems are then established. Extensions of these modal logics in which figure "statability" operators are also examined.

Introduction

Consider a standard modal language L with $\neg$, $\wedge$ and $\Box$ as primitive operators.[1] Suppose now that L is interpreted in *partial* Kripke models, i.e. in triples $< \mathcal{W}, R, V >$ where $\mathcal{W}$ (worlds) is a non–empty set, R (accessibility) is a binary relation on $\mathcal{W}$, and V (valuation) is a (maybe not total) function from $atoms(\mathsf{L}) \times \mathcal{W}$ to $\{0, 1\}$. Then in order to have a full interpretation of L, it suffices to specify truth and falsity clauses for its three primitive operators. Or alternatively, you can specify truth and undefinedness clauses for them.

The most radical view one can imagine about when a complex formula is undefined – and this is the view I shall endorse – is the *Contamination Principle* (CP):

[1]Essentially the same paper was presented at the ESSLLI 2000 students session held in august 2000 at the University of Birmingham (UK). The ESSLLI version did not incorporate the contents of sections 5, 6 and 7 below.

Advances in Modal Logic, Volume 3
F. Wolter, H. Wansing, M. de Rijke, and M. Zakharyaschev, eds.

Truth–value gaps are unconditionally transmitted from simples to complexes – i.e. a formula is undefined provided that at least one of its constituent formulae is.

The CP may seem to be difficult to justify. For e.g. if proposition p is neither true nor false and q is false, then one is tempted to say that the conjunction $p \wedge q$ should be false. However, the intuitive correctness of the Principle depends on what one takes to be the source of truth–value gaps. If the source of gaps is, say, lack of meaning, then the Principle seems to be correct; for instance, if p is meaningless, then arguably so is any conjunction $p \wedge q$.

Assuming the CP forces one to interpret negation and conjunction according to the following three–valued matrices, known as weak Kleene or Bochvar's:

$\neg$	
0	1
1	0
2	2

$\wedge$	0	1	2
0	0	0	2
1	0	1	2
2	2	2	2

where 2 is for 'undefined'. Thus, the modal logics I shall focus on depart from the well investigated modal logics based on the strong Kleene matrices.[2]

As to the truth–clause for necessity, there are two natural possible choices:

- $\Box A$ is true at w iff A is defined at w and A is true at every world accessible from w;
- $\Box A$ is true at w iff A is defined at w and A is true at every world accessible from w at which A is defined.

The first clause defines what one may call *strong necessity*, and the second *weak necessity*. In this paper, I focus on weak necessity, and the main question I will address is simply:

What logical systems does one get when language L is interpreted as previously indicated?

Strong necessity will indirectly enter the scene in section 6, when the basic modal language will be enriched with "statability" operators: as we shall see, strong necessity is definable in terms of weak necessity and any one of these operators.[3]

[2] See (Thijsse 1990), (Thijsse 1992) and (Jaspars and Thijsse 1996).

[3] If we decide not to follow the CP, there appears to be two extra natural truth–clauses for necessity:

- $w \models \Box A$ iff for every v such that wRv and A is defined at v, $v \models \Box A$;
- $w \models \Box A$ iff for every v such that wRv, $v \models \Box A$.

The history of modal logics based on weak Kleene matrices dates back from (Prior 1957), where Prior gives a semantic presentation of his celebrated system Q.[4] As far as I know, the only completeness studies about these logics are to be found in (Bull 1964), (Segerberg 1967) and (Correia *To appear*). Bull axiomatizes Q. Segerberg establishes completeness for a number of modal logics based on weak Kleene matrices. And finally I do the same for some Q–like systems. Prior's semantics for system Q, as well as the semantics presented in (Correia *To appear*) and in the present paper, respects the CP. Segerberg's semantics does not.[5]

Recently, systems for the logic of essentialist claims have been proved sound and complete with respect to partial possible world semantics respecting the CP.[6] My interest for that kind of semantics actually stems from my own work on this topic. In these logics, the modal operators are indexed by expressions (proper names, predicates, class–terms, ...) intended to isolate one or more objects. The intended meaning of e.g. '$\Box_a A$' – where 'a' is a proper name – is that it is true in virtue of the nature of a that A; and the use of predicates or class–terms as indices allows one to express claims to the effect that some given proposition is true in virtue of the natures of several objects taken together, with no restriction on how many these objects might be. It is not the place here to go into details about the logics of essence. Suffices it to say that, as I hope to show elsewhere, the indexed essentialist operators are definable in terms of weak necessity and conditional statability (see section 6 for that notion).

The structure of the paper is the following. Section 1 presents the grammar and the semantics for the basic modal language. Section 2 introduces the family of (axiomatic) systems I am primarily dealing with in this paper. Section 3 presents two canonical constructions. Completeness results are given in section 4, and incompleteness results in section 5. In section 6, the basic modal language is enriched with statability

The first defines another form of weak necessity, the second another form of strong necessity. The existing studies on modal logics based on the strong Kleene matrices all use exclusively this last clause. For that reason, a comparison between these studies and the present investigations on (the first form of) weak necessity appears to be pointless.

[4]Pages 85–86.

[5]In fact, Segerberg's language contains a "truth–operator", T, whose semantics violates the CP: TA is false whenever A is false or has no truth–value. Comparing the present study with Segerberg's paper, the reader will realize that the absence of Segerberg's truth–operator induces very significant differences, in the systems (e.g. Segerberg's systems are closed under a *restricted* version of the rule Modus Ponens, not under full Modus Ponens), as well as in the completeness proofs.

[6]See (Fine 1995), (Correia 2000) and (Fine 2000).

operators and some completeness facts are established. Finally, section 7 is devoted to consequence relations.

1 Grammar and Semantics

Thus, our language is a standard modal language, with $\neg$ (negation), $\wedge$ (conjunction) and $\Box$ (necessity) as primitive operators. Operators $\vee$ (disjunction), $\rightarrow$ (material implication), and $\Diamond$ (possibility) are standardly defined. I shall use A, B, ... (resp. p, q, ...) with or without indices for arbitrary formulae (resp. atoms) of the language. I adopt the following useful definitions:

- at is the set of all atoms;
- $at(A)$ is the set of all atoms in A;
- Where Γ is a set of formulae, $at(\Gamma)$ is the set of all atoms of all formulae in Γ;
- Where X, Y are formulae or sets of formulae, '$X \leq Y$' is used for '$at(X) \subseteq at(Y)$'.

A *frame* is as usual defined as a pair $< \mathcal{W}, R >$, where $\mathcal{W}$ (worlds) is a non–empty set and R (accessibility) is a binary relation on $\mathcal{W}$.

A *valuation* on a set S will be a total function from $at \times S$ to $\{0,1,2\}$.

A *model* is any triple $< \mathcal{W}, R, V >$, with $< \mathcal{W}, R >$ a frame and V a valuation on $\mathcal{W}$. Model $< \mathcal{W}, R, V >$ is said to be *based on* frame $< \mathcal{W}, R >$.

Where $\mathcal{M} = < \mathcal{W}, R, V >$ is a model and w is in $\mathcal{W}$, A is said to be *defined* at w in $\mathcal{M}$ – in short: $d(A, w, \mathcal{M})$ – iff for every atom p in A, $V(p, w) \neq 2$. When no confusion threatens, mention of the model will be omitted. Sometimes '$d(A, B, w)$' will be used for '$d(A, w)$ and $d(B, w)$'.

The truth–predicate $\models$ is defined on every model $< \mathcal{W}, R, V >$ by the following recursion:

- $w \models p$ iff $V(p, w) = 1$;
- $w \models \neg A$ iff $d(A, w)$ and $w \not\models A$;
- $w \models (A \wedge B)$ iff $w \models A$ and $w \models B$;
- $w \models \Box A$ iff $d(A, w)$ and $\forall v \in \mathcal{W}$, if wRv and $d(A, v)$, then $v \models A$.

We have then:

- $d(A, w)$ iff $w \models A$ or $w \models \neg A$;
- $w \models (A \vee B)$ iff $d(A, B, w)$ and ($w \models A$ or $w \models B$);
- $w \models (A \rightarrow B)$ iff $d(A, B, w)$ and ($w \not\models A$ or $w \models B$);
- $w \models \Diamond A$ iff $d(A, w)$ and $\exists v \in \mathcal{W}$ wRv and $v \models A$.

Finally, *validity on a model* is defined as non–falsehood at every world of that model; *validity on a frame* is validity on every model based on that frame; and *validity on a class* of models or frames is validity on every model or frame which belong to that class. A formula schema is said to be valid on a model, or frame, or class of models, or class of frames, when all its instances are.

A word on the definition of validity on a model is required. If validity on a model had been defined as truth at every world of that model, as it is in standard modal logic, then whatever the frame, no formula – and in particular no propositional tautology – would be valid on it; for one can assign 2 to each atom at every world of that frame. According to our definition, every propositional tautology is valid on every model – and hence on every frame. Moreover, given any frame, the non–modal formulae valid on that frame are exactly the propositional tautologies.

2 The Systems

Normal modal system K may be defined as the closure of the set of all propositional tautologies and all instances of schema

K $\quad \Box(A \to B) \to (\Box A \to \Box B)$

under the rules Modus Ponens (A, $A \to B$ / B) and Necessitation (A / $\Box A$). It is not difficult to see that this system is not sound with respect to the class of all frames. In fact, schema **K** is not valid on every frame. Let p, q be distinct atoms, and let $\mathcal{M}_0 = <\{v, w\}, R, V>$ be a model such that R is universal and $V(p, v) = V(q, v) = 1$, $V(p, w) = 2$, $V(q, w) = 0$. Then $\mathcal{M}_0, v \models \Box(p \to q) \wedge \Box p \wedge \neg\Box q$.

However, consider system k, obtained from K by replacing axiom schema **K** by

k $\quad \Box(A \to B) \to (\Box A \to \Box B)$, *provided* $A \leq B$.

Proposition 2.1 *k is sound with respect to the class of all frames.*

Proof. It is easy to verify that every tautology is valid on every frame, as well as **k**, and trivially the rule Necessitation preserves validity on a model (and so, validity on a frame as well). The case of the rule Modus Ponens is not so simple. In fact, Modus Ponens does not preserve validity on a model (p and $p \to q$ are valid on model $\mathcal{M}_0$ mentioned above, but q is not). But it does preserve validity on a frame, as the following argument shows. Let $<\mathcal{W}, R>$ be a frame, and let A and $A \to B$ be valid on it. Let $\mathcal{M} = <\mathcal{W}, R, V>$ be a model. If for every world w of $\mathcal{W}$, B is not defined at w, then B is trivially valid on $\mathcal{M}$. Now suppose that B is defined at some world w. We want to prove that B is true at w in $\mathcal{M}$. Suppose first that every atom in A is in B. Then both A and $A \to B$ are defined at w. So, since both A and $A \to B$ are valid on $< \mathcal{W}, R >$, they both are true at w in $\mathcal{M}$. And by the properties of truth–at–a–world, so is B. Suppose now that some atoms in A are not in B, and let p_1, ..., p_n be these atoms. Consider the model $\mathcal{N} = < \mathcal{W}, R, V' >$ whose valuation V' is defined by:

- $V'(p_i, w) = 1$ for every $i \in \{1, ..., n\}$;
- $V'(p, w') = V(p, w')$ if $w' \neq w$ or p is not one of p_1, ..., p_n.

One can then show (by induction on the length of C) that for every formula C whose atoms are all in B: $\forall w' \in \mathcal{W}, \mathcal{N}, w' \models C$ iff $\mathcal{M}, w' \models C$. As a consequence, $\mathcal{N}, w \models B$ iff $\mathcal{M}, w \models B$. Now by construction, both A and $A \rightarrow B$ are true at w in $\mathcal{N}$ and thus by the properties of truth–at–a–world, so is B. By the previous result, then, B is true at w in $\mathcal{M}$. ⊣

Every system distinct from k to be considered in this paper will be obtained by adding new axioms to k. Of course, since every instance of **k** is an instance of **K**, system k is included in system K; and more generally, each extension of k is included in the corresponding extension of K.

3 Two Canonical Models

In this section, two canonical models for systems extending k are defined, which differ only by their accessibility relation. Each of them will provide us with completeness results we cannot get with the other. In the sequel, S is any system containing k.

Proposition 3.1 *(i)* $\vdash_S \Box A \rightarrow (\Box B \rightarrow \Box(A \wedge B))$; *(ii) If* $\vdash_S A \rightarrow B$ *and* $A \leq B$, *then* $\vdash_S \Box A \rightarrow \Box B$.

Proof. (*i*) Use the fact that $A \rightarrow (B \rightarrow (A \wedge B))$ is a tautology, and **k**. (*ii*) Use Necessitation and **k**. ⊣

The canonical constructions in standard modal logics are based on the notion of maximal consistent sets of formulae. The property of maximality for sets in a canonical model corresponds in the semantics to the fact that every formula is either true or false at every world of every model. In the present context, maximality is too strong. We need a finer notion, that of *r–maximality.* A set Γ of formulae will be said to be *r–maximal* when for every A such that $A \leq \Gamma$, $A \in \Gamma$ or $\neg A \in \Gamma$. And where Γ is a set of formulae and Σ a set of atoms, Γ will be said to be *Σ–maximal* when it is r–maximal and $at(\Gamma) = \Sigma$.

Propositions 3.2, 3.3 and 3.4 below actually hold when S is any system containing classical propositional logic. Their proofs are easy and omitted.

Proposition 3.2 *Let Γ be a consistent set of formulae, and let Σ be a set of atoms such that $\Gamma \leq \Sigma$. Then Γ has an extension which is both S–consistent and Σ–maximal. (Γ is S–consistent iff it is not S–inconsistent. Γ is S–inconsistent iff there are $A_1, \ldots, A_n$ in Γ such that $\vdash_S \neg(A_1 \wedge \ldots \wedge A_n)$.)*

I call $\mathcal{W}^S$ the set of all r–maximal S–consistent sets of formulae. In this section, I shall use w, w', and so on for arbitrary elements of $\mathcal{W}^S$.

Proposition 3.3

1. *If $\vdash_S A$ and $A \leq w$, then $A \in w$;*
2. *If $A \in w$ and $(A \to B) \in w$, then $B \in w$;*
3. *If $\vdash_S (A \to B)$ and $A \in w$ and $B \leq w$, then $B \in w$.*

Proposition 3.4

1. *$\neg A \in w$ iff $A \leq w$ and $A \notin w$;*
2. *$(A \wedge B) \in w$ iff A, $B \in w$.*

I adopt the following definitions:

- $\mu_X(w) = \{B : \Box B \in w \text{ and } B \leq X\}$, for X a formula or a set of formulae;
- $wQ^S w'$ iff for every formula A, if $\Box A \in w$, then $\neg A \notin w'$;
- $wR^S w'$ iff for every formula A, if $A \in w'$, then $\Diamond A \in w$.

Proposition 3.5

1. *$wQ^S w'$ iff $\mu_{w'}(w) \subseteq w'$;*
2. *$wR^S w'$ iff $wQ^S w'$ and $w' \leq w$.*

Proof. 1. (a) From left to right, by the definitions of Q^S and μ, and the fact that w' is r–maximal. (b) For the other direction, suppose $\mu_{w'}(w) \subseteq w'$ and let $\Box A$ be in w. Then either $A \not\leq w'$ or $A \leq w'$. We want to prove that in both cases, $\neg A \notin w'$. If $A \not\leq w'$, then since w' is r–maximal, $\neg A \notin w'$. If on the other hand $A \leq w'$, then $A \in \mu_{w'}(w)$ and so $A \in w'$. Since w' is S–consistent, it follows that $\neg A \notin w'$.
2. (a) Suppose $wR^S w'$. Let $\Box A \in w$. Then by S–consistency, $\neg\Box A \notin w$, and so, $\neg A \notin w'$. Thus, $wQ^S w'$. On the other hand, let p be any atom such that $p \leq w'$. Then by Proposition 3.3 (1), $(p \vee \neg p) \in w'$. So, $\Diamond(p \vee \neg p) \in w$, and consequently $p \leq w$. Thus, $w' \leq w$. (b) Suppose now $wQ^S w'$ and $w' \leq w$. Let A be in w'. Then by Proposition 3.3 (3), $\neg\neg A \in w'$, and given that $wQ^S w'$, $\Box\neg A \notin w$. Now since $w' \leq w$ and $A \leq w'$, $A \leq w$. It follows by r–maximality that $\neg\Box\neg A \in w$, i.e. $\Diamond A \in w$. Thus, $wR^S w'$. ⊣

Proposition 3.6 *If* $\neg\Box A \in w$, *then* $\{\neg A\} \cup \mu_A(w)$ *is* S*–consistent.*

Proof. Suppose $\neg\Box A \in w$. Then

$$\textbf{(a)} \qquad \Box A \leq w.$$

Suppose now that $\{\neg A\} \cup \mu_A(w)$ is S–inconsistent. Then one can find $B_1, \ldots, B_n$ in $\mu_A(w)$ such that $\vdash_S (B_1 \wedge \ldots \wedge B_n) \to A$. Now since $B_1, \ldots, B_n$ are in $\mu_A(w)$, $(B_1 \wedge \ldots \wedge B_n) \leq A$, and so by Proposition 3.1 (*ii*)

$$\textbf{(b)} \qquad \vdash_S \Box(B_1 \wedge \ldots \wedge B_n) \to \Box A.$$

Since $B_1, \ldots, B_n$ are in $\mu_A(w)$, $\Box B_1, \ldots, \Box B_n$ are in w. It follows by Propositions 3.1 (*i*) and 3.3 (3) that

$$\textbf{(c)} \qquad \Box(B_1 \wedge \ldots \wedge B_n) \in w.$$

From (**a**), (**b**), (**c**) and Proposition 3.3 (3), it follows that $\Box A \in w$, and so $\neg\Box A \notin w$. Contradiction. So, $\{\neg A\} \cup \mu_A(w)$ is S–consistent. ⊣

Proposition 3.7

1. $\Box A \in w$ *iff* $A \leq w$ *and for every* w' *such that* $wQ^S w'$, $\neg A \notin w'$;
2. $\Box A \in w$ *iff* $A \leq w$ *and for every* w' *such that* $wR^S w'$, $\neg A \notin w'$.

Proof. From left to right, easy. For the other direction, suppose that $\Box A \notin w$ and $A \leq w$. Then by r–maximality, $\neg\Box A \in w$. So by Proposition 3.6, $\Gamma = \{\neg A\} \cup \mu_A(w)$ is S–consistent. Now, $at(\Gamma) = at(A)$. So by Proposition 3.2, Γ has an $at\{A\}$–maximal S–consistent extension w'. Since $\mu_{w'}(w) = \mu_A(w) \subseteq \Gamma \subseteq w'$, $wQ^S w'$ (Proposition 3.5 (1)). Moreover, since $at(w') = at(A)$ and $A \leq w$, $w' \leq w$. So, $wR^S w'$ (Proposition 3.5 (2)). Finally, $\neg A \in w'$. ⊣

Let V^S be the valuation on $\mathcal{W}^S$ defined by $V^S(p, w) = 0$ iff $\neg p \in w$, $V^S(p, w) = 1$ iff $p \in w$, and $V^S(p, w) = 2$ iff $p, \neg p \notin w$. Our two canonical models are $\mathcal{M}^S =< \mathcal{W}^S, Q^S, V^S >$ and $\mathcal{N}^S =< \mathcal{W}^S, R^S, V^S >$.

Proposition 3.8 $A \in w$ *iff* $\mathcal{M}^S, w \vDash A$
iff $\mathcal{N}^S, w \vDash A$.

Proof. By induction on the length of A, using Propositions 3.4, 3.7 and the fact that for every formula B, $d(B, w, \mathcal{M}^S)$ iff $d(B, w, \mathcal{N}^S)$ iff $B \leq w$. ⊣

Theorem 3.9 *A is a theorem of S iff A is valid on* $\mathcal{M}^S$
iff A is valid on $\mathcal{N}^S$.

Proof. From left to right, by Proposition 3.3 (1). For the other direction, suppose $\nvdash_S A$. Then by Proposition 3.2, $\{\neg A\}$ has an *at*–maximal S–consistent extension. The result follows from Proposition 3.8. $\dashv$

4 Soundness + Completeness

Following a standard terminology, a system is said to be *determined* by a class of models or frames when that system is both sound and complete with respect to it. Proposition 2.1 and Theorem 3.9 establish that:

Theorem 4.1 *k is determined by the class of all frames.*

The aim is now to obtain determination results for systems containing k. I shall use the usual terminology to refer to these systems: kX is k enriched by axiom schema **X**, kXY is k enriched by axiom schemata **X** and **Y**, etc. In this section, I shall focus on systems obtained by enriching k with one or more items from the following classical list:

T $\Box A \rightarrow A$

B $A \rightarrow \Box\Diamond A$

4 $\Box A \rightarrow \Box\Box A$

E $\Diamond A \rightarrow \Box\Diamond A$

If a formula is valid (in our semantics) on a frame, then it is valid *in the sense of normal modal logic* on that frame. Indeed, if A is valid on frame $\mathcal{F}$, then A is valid on every model based on $\mathcal{F}$ whose valuation nowhere gets the value 2, and from this fact one can easily show that A is valid in the normal sense on every Kripke model based on $\mathcal{F}$. As a consequence, given any correspondence of type 'A is valid on a frame iff that frame satisfies condition φ' holding in normal modal logic, half of that condition holds in the present context, namely 'if A is valid on a frame, then that frame satisfies φ'. For each of the four axiom schemata I am dealing with in this section, the other half of correspondence holds as well:

Proposition 4.2

1. **T** *is valid on a frame iff that frame is reflexive;*
2. **B** *is valid on a frame iff that frame is symmetrical;*
3. **4** *is valid on a frame iff that frame is transitive;*
4. **E** *is valid on a frame iff that frame is euclidean.*

Proof. I give the proof for **E** as an illustration. Suppose $< \mathcal{W}, R >$ is euclidean, i.e. such that for all $x, y, z \in \mathcal{W}$ if (xRy and xRz), then yRz. Then let $\mathcal{M} = < \mathcal{W}, R, V >$ be a model, and assume for a *reductio* that for some $w \in \mathcal{W}$, both (i) $\mathcal{M}, w \models \Diamond A$ and (ii) $\mathcal{M}, w \not\models \Box\Diamond A$. By (i), $d(\Box\Diamond A, w, \mathcal{M})$, and so by (ii) we have: (iii) $\mathcal{M}, w \models \Diamond\Box\neg A$. From (i), we know that there is some $w' \in \mathcal{W}$ such that wRw' and $\mathcal{M}, w' \models A$. On the other hand, from (iii), we know that there is some $w'' \in \mathcal{W}$ such that wRw'' and $\mathcal{M}, w'' \models \Box\neg A$. But since $< \mathcal{W}, R >$ is euclidean, $w''Rw'$, and so, $\mathcal{M}, w' \not\models \neg\neg A$, i.e. $\mathcal{M}, w' \not\models A$. Contradiction. So, if (i), then (ii). As a consequence, **E** is valid on $< \mathcal{W}, R >$. ⊣

Things become interesting when one looks at the canonical frames defined in the previous section. In normal modal logic, the (usual) canonical frame of a system S containing any one of **T**, **B**, **4** or **E** satisfies the corresponding condition. As a result, any system obtained from the basic K by adding one or more axiom schemata from the list is determined by the class of all frames satisfying the corresponding conditions. In the present context, things are not so simple. Let us start with a positive fact:

Proposition 4.3

1. *If S contains* **T**, *then both Q^S and R^S are reflexive;*
2. *If S contains* **B**, *then Q^S is symmetrical;*
3. *If S contains* **4**, *then R^S is transitive.*

Proof. The proof for **B** will serve as an illustration. Assume S contains **B**, and let $w, w' \in \mathcal{W}^S$ be such that wQ^Sw'. Suppose now $\neg A \in w$. Then by **B**, $\Box\Diamond\neg A \in w$. So, $\neg\Diamond\neg A \notin w'$, i.e. $\Box A \notin w'$. Thus by contraposition, for every formula A, if $\Box A \in w'$, then $\neg A \notin w$. So $w'Q^Sw$. ⊣

By Propositions 4.2 and 4.3, we have the following

Theorem 4.4 *Each of the systems kT, kB, $k4$, kTB and $kT4$ is determined by the same class of frames as its normal counterpart.*

Unfortunately, one can prove neither that if S contains **B** then R^S is symmetrical, nor that if S contains **4** then Q^S is transitive. Worse, by Proposition 4.6 below, both are *false*. Proposition 4.6 settles the case of axiom schema **E** as well.

Proposition 4.5 *Suppose S is sound with respect to some frame $\mathcal{F}$. Then for every atom p, $\{p\}$, $\{\neg p\}$ and $\{\Box p\}$ are S–consistent. If moreover some worlds of $\mathcal{F}$ are not dead ends (i.e. worlds which are related to no world), then for all atoms p and q, $\{\neg\Box p, \neg\Box q\}$ is S–consistent.*

Proof. Let $< \mathcal{W}, R >$ be a frame with respect to which S is sound, and let p, q be atoms. (i) Let V be a valuation on $\mathcal{W}$ such that $V(p, w) = 1$ for some $w \in \mathcal{W}$. Then $\neg p$ is not valid on $< \mathcal{W}, R, V >$. By soundness, $\{p\}$ is S–consistent. (*ii*) Let V be a valuation on $\mathcal{W}$ such that $V(p, w) = 0$ for some $w \in \mathcal{W}$. Then p is not valid on $< \mathcal{W}, R, V >$. By soundness, $\{\neg p\}$ is S–consistent. (*iii*) Let V be a valuation on $\mathcal{W}$ such that $V(p, w) = 1$ for every $w \in \mathcal{W}$. Then $\neg\Box p$ is not valid on $< \mathcal{W}, R, V >$. By soundness, $\{\Box p\}$ is S–consistent. (*iv*) Suppose that world w in $\mathcal{W}$ is not a dead end, and let V be a valuation on $\mathcal{W}$ such that $V(p, w') = V(q, w') = 0$ for every $w' \in \mathcal{W}$. Then $\neg\Box p$ and $\neg\Box q$ are both true at w. So $\Box p \vee \Box q$ is not valid on $< \mathcal{W}, R, V >$. By soundness, $\{\neg\Box p, \neg\Box q\}$ is S–consistent. ⊣

Proposition 4.6 *Suppose S is sound with respect to some frame $\mathcal{F}$. Then Q^S is neither transitive nor euclidean. If moreover some worlds of $\mathcal{F}$ are not dead ends, then R^S is neither symmetrical nor euclidean.*

Proof. (*1*) Let p, q be two distinct atoms, and assume that S is sound with respect to some frame. By Proposition 4.5, $\{\neg p\}$, $\{q\}$ and $\{\Box p\}$ are S–consistent. Let w be a $\{p\}$–maximal S–consistent extension of $\{\Box p\}$, let w' be a $\{q\}$–maximal S–consistent extension of $\{q\}$, let w'' be a $\{p\}$–maximal S–consistent extension of $\{\neg p\}$. Then trivially, $wQ^S w'$, $w'Q^S w$ and $w'Q^S w''$. But since $\Box p \in w$ and $\neg p \in w''$, it is not the case that $wQ^S w''$. So Q^S is neither transitive nor euclidean. (*2*) Let p, q be two distinct atoms, and assume that S is sound with respect to some frame some of whose worlds are not dead ends. Then by Proposition 4.5, $\{\neg\Box p, \neg\Box q\}$ is S–consistent. Let w be an *at*–maximal S–consistent extension of it. One can then find w' and w'' in $\mathcal{W}^S$ such that $wR^S w'$, $wR^S w''$, $at(w') = \{p\}$ and $at(w'') = \{q\}$ (use Proposition 3.6). Now since w is *at*–maximal, it is not the case that $w'R^S w$. So R^S is not symmetrical. Moreover, it is trivially not the case that $w'R^S w''$. So R^S is not euclidean. ⊣

So, by Theorem 4.4, we have a determination result for every system obtained from k by adding one or more of **T**, **B** and **4**, except $kB4$ and $kTB4$. By Proposition 4.6, completeness for these two systems, cannot be established by means of our canonical constructions. In fact, by Proposition 4.2, both $kB4$ and $kTB4$ are sound with respect to equivalence (i.e. reflexive, symmetrical and transitive) frames, which by reflexivity do not contain dead ends. The situation for systems containing **E** is even more desperate. It should be noted here however that in (Correia *To appear*), system kTE (which is identical to $kTB4$) is proved to be determined by the class of all universal models (i.e. the models defined without accessibility relations, or equivalently, the models whose accessibility relation is universal).

5 Incompleteness

In the previous section, we saw that the usual correspondences for axiom schemata **T**, **B**, **4** and **E** hold in the present context. The problem was just that our two canonical constructions cannot help us to establish determination for some of the systems one can build from these axiom schemata. In this section, we shall see that the usual correspondences do not always hold in our three–valued context; and we shall also see that there are systems extending k which are *incomplete* (i.e. determined by no class of frames), while their normal counterparts are not.

Let us start with a general fact:

Proposition 5.1 *Let* A_1, ..., A_n *be any formulae, and let* B *be any non–modal formula, satisfiable in the sense of propositional calculus. Then* $\neg(\Box A_1 \wedge ... \wedge \Box A_n \wedge B)$ *is valid on a frame only if that frame is reflexive.*

Proof. Let $< \mathcal{W}, R >$ be a non–reflexive frame, and let $w \in \mathcal{W}$ be such that it is not the case that wRw. Let V be any valuation such that:

- V makes B true,
- For every atom p in every A_i, $V(p, w) \neq 2$, and
- For every atom p of the language and every world $w' \neq w$, $V(p, w') = 2$.

Then it is easy to check that $\Box A_1 \wedge ... \wedge \Box A_n \wedge B$ is true at w according to V. ⊣

Now, let us focus on axiom schema

D $\Box A \rightarrow \Diamond A$

In normal modal logic, **D** is valid on a frame $< \mathcal{W}, R >$ iff that frame is serial (i.e. such that for every $x \in \mathcal{W}$, there is a $y \in \mathcal{W}$ such that xRy). But in the present context, this is not the case. Indeed, if **D** is valid on a frame, then so is any formula of type $\neg(\Box A \wedge \Box\neg A \wedge p)$ where p is an atom,[7] and so by the previous Proposition, the frame must be reflexive. But not all serial frames are reflexive. So, here we have a case where the standard correspondence does not hold.

But is there any correspondence fact for **D** at all? Yes. We have the following correspondence on *models*:

(co) **D** is valid on a model iff that model is serial*,

[7]For, of course, in general if a formula A is valid on a frame, then so is $A \vee B$ for every formula B.

where a model $< \mathcal{W}, R, V >$ is serial* iff for all $x \in \mathcal{W}$ and formula A such that $d(A, x, \mathcal{M})$, there is a $y \in \mathcal{W}$ such that $d(A, y, \mathcal{M})$ and xRy. And we also have a correspondence on *frames*:

D is valid on a frame iff that frame is reflexive.

This correspondence can be proved from the fact that a frame is reflexive iff all the models based on it are serial*.

Now, let S be any system extending k containing as a theorem some formula of the type described in Proposition 5.1. From that Proposition and the fact that **T** is valid on every reflexive frame, it follows that if S is determined by some class of frames, then **T** must be a theorem of S. That is, we have the following:

Proposition 5.2 *System kXY... is incomplete if the following two conditions are satisfied:*

1. **T** *is not a theorem of kXY...;*
2. *Some formula of the type described in Proposition 5.1 is a theorem of kXY...*

Given that for every atom p, $\mathbf{D} \to \neg(\Box A \wedge \Box\neg A \wedge p)$ is a theorem of k, we have:

Theorem 5.3 *Every extension of k containing* **D** *but not* **T** *is incomplete.*

Thus, for instance, system kD is incomplete. Indeed, if **T** were a theorem of kD, then it would be a theorem of KD as well (since $kD \subseteq KD$), and we know that this is not the case.

Of course, incompleteness here is *frame*–incompleteness. The frame–incompleteness of a system is no bar to that system's being determined by some class of *models*. Actually, since kD has a canonical model, it is determined by the class consisting of this model only. But we have a more interesting characterization of kD in terms of models. Indeed, since kD has a canonical model, we have by the correspondence (co) that kD is determined by the class of all serial* models.

6 Adding Statability Operators

In this section I examine extensions of the previous logics obtained by adding "statability" operators to the basic modal language. I shall focus on two such operators, Prior's unary necessary statability operator $\mathcal{S}$ and the binary conditional statability operator $\rhd$. Their truth–clauses are the following:

- $w \models \mathcal{S}A$ iff $d(A, w)$ and $\forall v \in \mathcal{W}$ such that wRv, $d(A, v)$;
- $w \models A \rhd B$ iff $d(A, B, w)$ and $\forall v \in \mathcal{W}$ such that wRv and $d(A, v)$, $d(B, v)$.

The first operator is interesting in particular because strong necessity (see the introduction) can be defined as weak necessity plus necessary statability: it is strongly necessary that $A = \mathcal{S}A \wedge \Box A$. The second operator gives us even more expressive power: $\mathcal{S}A$ can be defined as $t \rhd A$, where t is a truth–constant (defined to be true at every world of every model) one may decide to introduce in the language.

I define two systems $k^{\mathcal{S}}$ and $k^{\rhd}$. System $k^{\mathcal{S}}$ is k plus

k$^{\mathcal{S}}$ $(\mathcal{S}p_1 \wedge ... \wedge \mathcal{S}p_n) \wedge \Box(A \to B) \to (\Box A \to \Box B)$, where p_1, ..., p_n are all the atoms of A not in B,

plus the following axioms for $\mathcal{S}$:

NS1 $\mathcal{S}A \to \mathcal{S}p$, where p is any atom of A;

NS2 $(\mathcal{S}p_1 \wedge ... \wedge \mathcal{S}p_n) \to \mathcal{S}A$, where p_1, ..., p_n are all the atoms of A.

System $k^{\rhd}$ is k with axiom **k** replaced by

k$^{\rhd}$ $B \rhd A \wedge \Box(A \to B) \to (\Box A \to \Box B)$,

plus the following axioms for $\rhd$:

CS1 $A \rhd B$, if all atoms in B are in A;

CS2 $(A \rhd B \wedge B \rhd C) \to A \rhd C$;

CS3 $(A \rhd B \wedge A \rhd C) \to A \rhd (B \wedge C)$.

Obviously, $k \subseteq k^{\mathcal{S}}$ and $k \subseteq k^{\rhd}$. I shall use the same notations for extensions of these two systems as for the extensions of **k**.

Both $k^{\mathcal{S}}$ and $k^{\rhd}$ are sound with respect to the class of all frames. Completeness can easily be proved by slight modifications of the canonical constructions of section 3, *but for* $k^{\mathcal{S}}T$ *and* $k^{\rhd}T$: the proof of proposition 6.2 below makes use of axiom **T**. I go through the proofs for $k^{\rhd}T$, leaving the case of $k^{\mathcal{S}}T$ to the reader.

Propositions 3.1–3.4 still hold for S any extension of $k^{\rhd}$, since these systems contain k. We shall also need the following

Proposition 6.1 *Let S be any extension of $k^{\rhd}$. Then*

1. *If $\vdash_S A \to B$, then $\vdash_S B \rhd A \to (\Box A \to \Box B)$;*
2. *If $C_1 \leq A$, ..., $C_n \leq A$, then $\vdash_S (C_1 \rhd B_1 \wedge ... \wedge C_n \rhd B_n) \to A \rhd (B_1 \wedge ... \wedge B_n)$.*

Proof. 1. Use Necessitation and **k**$^{\rhd}$.

2. Use **CS1**, **CS2** and **CS3**. ⊣

The definitions of the accessibility relations and μ change:

- $\mu_X(w) = \{B : \Box B \in w \text{ and } C \rhd B \in w \text{ for some } C \leq X\}$;
- $wQ^S w'$ iff for every formula A,
  - if $\Box A \in w$, then $\neg A \notin w'$;
  - if $A \rhd B \in w$ and $A \leq w'$, then $B \leq w'$;
- $wR^S w'$ iff for every formula A,
  - if $A \in w'$, then $\Diamond A \in w$;
  - if $A \rhd B \in w$ and $A \leq w'$, then $B \leq w'$.

Propositions 3.5–3.7 still hold with the new definitions. In the proof for Proposition 3.5 (1) (right–to–left direction), one has to use the fact that $A \rhd A$ is a theorem. The proof for Proposition 3.6 makes use of Proposition 6.1. In the proof for Proposition 3.7, $at(\Gamma) = \{p : A \rhd p \in w\}$. Proposition 3.8 and Theorem 3.9 can then be proved for S any extension of $k^{\rhd}T$ with the help of the following

Proposition 6.2 *Let S be any extension of $k^{\rhd}T$. Then*

1. *$A \rhd B \in w$ iff $A, B \leq w$ and for every w' such that $wQ^S w'$ and $A \leq w'$, $B \leq w'$;*
2. *$A \rhd B \in w$ iff $A, B \leq w$ and for every w' such that $wR^S w'$ and $A \leq w'$, $B \leq w'$.*

Proof. 1. From left to right, by the definition of Q^S. For the other direction, suppose that $A \rhd B \notin w$ and $A, B \leq w$. Then by r–maximality, $\neg(A \rhd B) \in w$. Now, $\mu_A(w)$ is consistent. For suppose $\mu_A(w)$ is inconsistent. Then one can find B_1, ..., B_n such that (i) $\Box B_1$, ..., $\Box B_n \in w$ and (ii) $\vdash_S \neg(B_1 \wedge ... \wedge B_n)$. It follows that (i) $\Box(B_1 \wedge ... \wedge B_n) \in w$ and (ii) $\Box\neg(B_1 \wedge ... \wedge B_n) \in w$. But by axiom **T**, it follows that w is inconsistent. Thus, $\mu_A(w)$ is consistent. Moreover, $at(\mu_A(w)) = \{p : A \rhd p \in w\}$, and so $at(B) \not\subseteq at(\mu_A(w))$. Let w' be an $at(\mu_A(w))$–maximal consistent extension of $\mu_A(w)$. Then $wQ^S w'$ and $A \leq w'$ but $B \not\leq w'$.

2. From left to right, by the definition of R^S. For the other direction, it suffices to note that in the proof for the previous point, $at(w') = at(\mu_A(w)) \subseteq at(w)$. ⊣

It is easy to see that for S any extension of $k^{\rhd}T$, both Q^S and R^S are reflexive. Thus, $k^{\rhd}T$ is sound and complete with respect to the class of all reflexive frames. The same is true of $k^S T$.

If an extension S of $k^{\rhd}$ contains both modal axiom **4** and

CS4 $\quad A \rhd B \to \Box(A \rhd B)$.

then R^S is transitive. On the other hand, both axioms are valid on every transitive frames. So, $k^{\rhd}T4$ plus **CS4** is sound and complete with respect to the class of all reflexive and transitive frames. The same is true of $k^S T4$ plus the following axiom:

NS3 $\quad \mathcal{S}A \to \Box \mathcal{S}A$.

Completeness results for other extensions of $k^{\triangleright}$ and k^S are hard to establish with our constructions. For instance it is not clear that there is an axiom for $\triangleright$ which, together with axiom **B**, makes the Q–accessibility relation symmetrical. It is worth noting here that in (Correia *To appear*) both $k^{\triangleright}TB4$ plus **CS4** and $k^S TB4$ plus **NS3** are proved sound and complete with respect to the class of all universal frames.

7 Consequence Relations

As we saw, the standard notion of validity on a class of frames is not appropriate in the present context. In contradistinction, there is room here for the standard notion of *semantic consequence* on a class of frames. Let us use the symbol $\Vdash$ for standard semantic consequence. Thus, where $\mathcal{C}$ is a class of frames, Γ a set of formulae, and A a formula, I put:

$\Gamma \Vdash_{\mathcal{C}} A$ iff for every world w of every model based on a frame in $\mathcal{C}$, if $w \vDash \Gamma$ then $w \vDash A$.

The standard notion of consequence is based on the idea that a proposition is a consequence of some propositions when the former *must be true* when the latter are all true. Yet, there is another intuitive notion of consequence: a proposition is a consequence of some propositions when the former *cannot be false* when the latter are all true. Let us use the symbol $\Vvdash$ for this notion of consequence:

$\Gamma \Vvdash_{\mathcal{C}} A$ iff for every world w of every model based on a frame in $\mathcal{C}$, if $w \vDash \Gamma$ then $w \nvDash \neg A$.

It turns out that a formula A is valid on a class $\mathcal{C}$ of frames iff $\varnothing \Vvdash_{\mathcal{C}} A$.

As one can easily see, the deduction theorem holds for the second consequence relation:

$$\Gamma, A \Vvdash_{\mathcal{C}} B \text{ iff } \Gamma \Vvdash_{\mathcal{C}} A \to B.$$

The following Proposition shows that $\Vdash$ is definable in terms of $\Vvdash$:

Proposition 7.1 *Let $\mathcal{C}$ be a class of frames, Γ a set of formulae, and A a formula. Then:*

1. *If Γ is not satisfiable in $\mathcal{C}$, then both $\Gamma \Vdash_{\mathcal{C}} A$ and $\Gamma \Vvdash_{\mathcal{C}} A$;*
2. *If Γ is satisfiable in $\mathcal{C}$, then $\Gamma \Vdash_{\mathcal{C}} A$ iff $\Gamma \Vvdash_{\mathcal{C}} A$ and $A \leq \Gamma$.*

Proof. The first point is trivial. For the second point, the right–to–left direction is almost immediate. For the other direction, suppose that

$\Gamma \Vdash_{\mathcal{C}} A$. Then trivially, $\Gamma \Vvdash_{\mathcal{C}} A$. Remains to prove that $A \leq \Gamma$. Suppose for a *reductio* that $A \nleq \Gamma$, and let p be an atom in A which is foreign to any formula in Γ. Then let $\mathcal{M} = < \mathcal{W}, R, V >$ be a model based on a frame in $\mathcal{C}$ and let w be a world of that model such that $\mathcal{M}, w \models \Gamma$ (such a model–world pair exists since Γ is satisfiable in $\mathcal{C}$). We define a new model $\mathcal{N}$ based on the same frame, and whose valuation V' is identical to V unless $V(p,w) \neq 2$ – in this case, we put $V'(p,w) = 2$. Since p is foreign to Γ, the truth–values of the formulae in Γ do not change, and so we have: $\mathcal{N}, w \models \Gamma$. And of course, we also have: $\mathcal{N}, w \not\models A$. Thus, $\Gamma \nVdash_{\mathcal{C}} A$. But this contradicts the hypothesis. So, $A \leq \Gamma$. ⊣

Further Investigations

They should go in two directions. First, it would be nice to get a proper understanding about how much correspondence and determination facts for propositional extensions of k depart from those we have in normal modal logic, and why. Second, quantificational extensions of k should also be studied, if only because among the logics of essence mentioned in the introduction to this paper, the most natural ones are quantified.

Acknowledgements

This work was supported by the Fond National pour la Recherche Scientifique Suisse, grant No 11–55845.98.

References

Bull, R.A. 1964. An Axiomatization of Prior's Modal Calculus Q. *Notre Dame Journal of Formal Logic* 5:211–214.

Correia, F. 2000. Propositional Logic of Essence. *Journal of Philosophical Logic* 29:295–313.

Correia, F. *To appear.* Adequacy Results for some Priorean Modal Propositional Logics. *Notre Dame Journal of Formal Logic.*

Fine, K. 1995. The Logic of Essence. *Journal of Philosophical Logic* 24:241–273.

Fine, K. 2000. Semantics for the Logic of Essence. *Journal of Philosophical Logic* 29:543–584.

Jaspars, J., and E. Thijsse. 1996. Fundamentals of Partial Modal Logic. In *Partiality, Modality and Nonmonotonicity*, ed. P. Doherty. 111–140. CSLI, Stanford.

Prior, A. N. 1957. *Time and Modality.* Oxford University Press.

Segerberg, K. 1967. Some modal logics based on a three-valued logic. *Theoria* 33:53–71.

Thijsse, E. 1990. Partial Propositional and Modal Logic: the Overall Theory. In *Proceedings of the Seventh Amsterdam Colloquium, Part 2*, ed. M. Stockhof and L. Torenlivet. 555–579. ITLI, Amsterdam.

Thijsse, E. 1992. *Partial Logic and Knowledge Representation.* Eburon, Delft.

Fabrice Correia
Département de Philosophie
Université de Genève
rue de Candolle 2
1211 Genève 4
Switzerland
E-mail: fabrice.correia@lettres.unige.ch

6

Bimodal Logics for Reasoning About Continuous Dynamics

JEN M. DAVOREN AND RAJEEV P. GORÉ

ABSTRACT. We study a propositional bimodal logic consisting of two **S4** modalities $\Box$ and $[a]$, together with the interaction axiom scheme $\langle a\rangle\Box\varphi \rightarrow \Box\langle a\rangle\varphi$. In the intended semantics, the plain $\Box$ is given the McKinsey-Tarski interpretation as the interior operator of a topology, while the labelled $[a]$ is given the standard Kripke semantics using a reflexive and transitive binary relation R_a. The interaction axiom expresses the property that the R_a relation is *lower semi-continuous* with respect to the topology. The class of topological Kripke frames characterised by the logic includes all frames over Euclidean space where R_a is the positive flow relation of a differential equation. We establish the completeness of the axiomatisation with respect to the intended class of topological Kripke frames, and investigate tableau calculi for the logic, although tableau completeness and decidability are still open questions.

1 Introduction

We study a propositional bimodal logic consisting of two **S4** modalities $\Box$ and $[a]$, together with a cross or mix axiom scheme:

$$\langle a\rangle\Box\mathbf{lsc}: \qquad \langle a\rangle\Box\varphi \rightarrow \Box\langle a\rangle\varphi$$

where $\langle a\rangle\varphi \stackrel{\text{def}}{=} \neg[a]\neg\varphi$. This particular bimodal logic is specifically motivated by *hybrid dynamical systems*: systems characterised by interacting discrete and continuous dynamics, which are the subject of a rapidly growing research field at the interface of computer science and control engineering. In the basic case, a hybrid system consists of finitely many differential equations together with reset relations prescribing when the

Advances in Modal Logic, Volume 3
F. Wolter, H. Wansing, M. de Rijke, and M. Zakharyaschev, eds.

system can discretely switch from one differential equation to another. Their engineering applications include air-traffic control, robotics and automated manufacturing (Antsaklis 2000).

Formal reasoning about discrete dynamics is well-studied in the analysis and verification of computer hardware and software using temporal and modal logics; the challenge in the hybrid setting is to also incorporate continuous dynamics into a common logical framework. A crucial move is to consider relations $R_a \subseteq S \times S$ as a generic way of representing continuous as well as discrete dynamics (Davoren and Nerode 2000). A basic relation of interest is the positive *flow relation* or *orbit relation* of a differential equation on a space $S \subseteq \mathbb{R}^n$, where x' is a flow successor of x iff there is a solution curve of the differential equation leading from x to x'. This relation is both reflexive and transitive, hence a *preorder*, and motivates our choice of the **S4** axioms for the labelled modal operator $[a]$, given the standard Kripke semantics. In application to hybrid systems, one works with a **PDL**-like polymodal extension, with modal operators $[a]$ labelled by letters $a \in \Sigma$ in an alphabet of relations Σ, as in (Davoren and Moor 2000).

The purpose of the other **S4** modal operator $\Box$ is to give modal representation to *topological* structure and properties, and thence to notions of *continuity for relations.* The plain box $\Box$ is given the McKinsey-Tarski interpretation as the *interior operator* in an arbitrary *topological space* $(S, \mathcal{T})$, as in (McKinsey 1941), (McKinsey and Tarski 1944) and (Rasiowa and Sikorski 1963). So the denotation set of a formula $\Box\varphi$ is the largest open set in $\mathcal{T}$ contained in the denotation set of φ; recall that a set is open if it is equal to its own interior. The dual $\Diamond$ is interpreted by the topological closure operator. The semantics for our bimodal logic are thus over topological Kripke frames $\mathcal{F} = (S, \mathcal{T}, R_a)$.

The interaction axiom scheme $\langle a\rangle\Box$**lsc** is equivalent to the assertion that the semantic operator $\langle a\rangle$ applied to an open set is always an open set. From the general topology of relations/set-valued maps (Kuratowski 1966), there are two quite distinct topological notions of continuity. As we show in Section 2, an arbitrary relation $R_a \subseteq S \times S$ is *lower semi-continuous* (l.s.c.) with respect to $\mathcal{T}$ exactly when the frame $\mathcal{F} = (S, \mathcal{T}, R_a)$ validates the $\langle a\rangle\Box$**lsc** axiom scheme, and is *upper semi-continuous* (u.s.c.) with respect to $\mathcal{T}$ exactly when $\mathcal{F}$ validates

$$[a]\Box\textbf{usc}: \qquad [a]\Box\varphi \to \Box[a]\varphi$$

which asserts that the operator $[a]$ applied to an open set is an open set. When R_a is a total single-valued map, both notions of semi-continuity reduce to the standard topological notion of continuity for functions. An arbitrary relation $R_a \subseteq S \times S$ is thus called *continuous* if it is both l.s.c.

and u.s.c. ; see Kuratowski (1966), Section 18. As we show in Section 3, the flow relation of a differential equation is always l.s.c. with respect to the standard Euclidean topology on $S \subseteq \mathbb{R}^n$, but it may fail to be u.s.c.

We give the name **LSC** to the logic axiomatised by $\Box$**S4** + $[a]$**S4** + $\langle a \rangle \Box$**lsc**. The Hilbert-style axiomatisation is sound and complete with respect to the class $\mathbb{LSC}$ of topological Kripke frames $(S, \mathcal{T}, R_a)$ for R_a a preorder that is l.s.c. with respect to $\mathcal{T}$, an arbitrary topology. For completeness, the standard canonical Kripke model construction extends to this bimodal configuration, resulting in a canonical frame $\mathcal{F}_{\mathbf{LSC}}$ whose topology $\mathcal{T}_{\preccurlyeq}$ is determined by a preorder $\preccurlyeq \subseteq S \times S$. Topologies in this subclass are variously known as *Alexandroff*, *Kripke* or *cone* topologies, and are in one-one correspondence with preorders. Applying earlier results in (Davoren 1999), we show that the two semi-continuity properties translate in the bi-relational setting to quite familiar "diamond" graph properties; they correspond to the Zig-Zag conditions for $\preccurlyeq$ to be a *bisimulation* with respect to the relation R_a, with l.s.c. mapping to the "Zig" condition, and u.s.c. mapping to the "Zag" condition. A corollary of the completeness proof is that logics such as **LSC** and its fraternal twin **USC** cannot distinguish between frames with Alexandroff topologies and frames with arbitrary topologies.

We attempt to develop a tableau calculus for the logic **LSC**, producing a tableau rule encoding the l.s.c. axiom which uses analytic cuts on a restricted class of analytic super-formulae. This work builds on previous studies by Goré (1991, 1999): monomodal $\Box$**S4.2** is the special case where $[a]$ is identical to $\Box$, so the $\langle a \rangle \Box$**lsc** axiom scheme reduces to the *weak directedness* scheme $\Box$**2** : $\Diamond\Box\varphi \to \Box\Diamond\varphi$. Our tableau calculus is sound with respect to the class of frames but completeness is currently an open question. We nevertheless give details of our current attempt to prove completeness since it may help others to invent additional rules to plug the gap.

Bimodal and polymodal logics *without* any interaction axioms are well-studied as *fusions* of monomodal logics (Wolter 1999) where our base logic $\Box$**S4** + $[a]$**S4** would be written **S4** $\otimes$ **S4**. In particular, by Corollary 2.4 of (Wolter 1999), the base logic is decidable. The source of the difficulty in **LSC** is, of course, the extra interaction axiom.

The remainder of the paper is organised as follows. In Section 2, we give the syntax and topological semantics for the logics, and establish the modal characterisation of the semi-continuity properties. In Section 3, we discuss the motivating applications from the analysis of continuous and hybrid dynamical systems. Section 4 sets out the relationship between topological semantics and the bi-relational Kripke semantics. In Section 5, we give the Hilbert-style axiomatisations for **LSC**, and establish its soundness and completeness. In Section 6 we give our tableau calculus, with a discussion of how the l.s.c. axiom is coded up as a rule, and prove its soundness. Section 7 gives a counter-example to the completeness of our tableau calculus. Section 8 gives an outline of an alternative calculus for **LSC** which is complete and terminating, but whose soundness is open. Section 9 discusses related work on bimodal and polymodal logics, and the concluding Section 10 discusses further work.

2 Syntax and Topological Semantics

Definition 2.1 Let AP $= \{p_0, p_1, ..., q_0, q_1, ...\}$ be a fixed countable set of *atomic propositions.* With $p \in$ AP, the set $\mathcal{L}(\Box, [a])$ of bimodal formulae is recursively generated by the grammar:

$$\varphi \ ::= \ p \mid \neg\varphi \mid \varphi_1 \wedge \varphi_2 \mid \Box\varphi \mid [a]\varphi$$

The other Boolean connectives, logical constants $\top$ and $\bot$, and dual modalities are defined in the standard way. For $\varphi \in \mathcal{L}(\Box, [a])$, the (finite) set $Sf\ \varphi$ of subformulae of φ is defined as usual.

Definition 2.2 A *topological Kripke frame*, or *frame*, for $\mathcal{L}(\Box, [a])$ is a structure $\mathcal{F} = (S, \mathcal{T}, R_a)$ where $S \neq \emptyset$ is the state space, $\mathcal{T} \subseteq \mathcal{P}(S)$ is a topology on S (so $S, \emptyset \in \mathcal{T}$ and $\mathcal{T}$ is closed under finite intersections and arbitrary unions), and $R_a \subseteq S \times S$ is a binary relation on S.

A *valuation* in $\mathcal{F}$ is any function $\xi : \mathrm{AP} \to \mathcal{P}(S)$ assigning a set of states $\xi(p) \subseteq S$ to each atomic $p \in$ AP, and a model over $\mathcal{F}$ is a pair $\mathcal{M} = (\mathcal{F}, \xi)$. For each model $\mathcal{M}$ over $\mathcal{F}$, the denotation set $[\![\varphi]\!]^{\mathcal{M}}$ of a formula $\varphi \in \mathcal{L}(\Box, [a])$ is defined by:

$$\begin{array}{llllll} [\![p]\!]^{\mathcal{M}} & \stackrel{\text{def}}{=} & \xi(p) & & & \\ [\![\neg\varphi]\!]^{\mathcal{M}} & \stackrel{\text{def}}{=} & S - [\![\varphi]\!]^{\mathcal{M}} & [\![\varphi_1 \wedge \varphi_2]\!]^{\mathcal{M}} & \stackrel{\text{def}}{=} & [\![\varphi_1]\!]^{\mathcal{M}} \cap [\![\varphi_2]\!]^{\mathcal{M}} \\ [\![\Box\varphi]\!]^{\mathcal{M}} & \stackrel{\text{def}}{=} & int_{\mathcal{T}}([\![\varphi]\!]^{\mathcal{M}}) & [\![[a]\varphi]\!]^{\mathcal{M}} & \stackrel{\text{def}}{=} & \mathrm{Pre}^{\forall}[R_a]([\![\varphi]\!]^{\mathcal{M}}) \end{array}$$

where the $\forall$-*pre-image operator* $\mathrm{Pre}^{\forall}[R] : \mathcal{P}(S) \to \mathcal{P}(S)$ of R is:

$$\mathrm{Pre}^{\forall}[R](A) \stackrel{\text{def}}{=} \{x \in S \mid (\forall y \in S)[\text{ if } xRy \text{ then } y \in A\,]\}$$

For formulae $\varphi \in \mathcal{L}(\Box, [a])$, and models over a class of frames $\mathbb{F}$:

- φ is *satisfied* at state s in the model $\mathcal{M}$, written $s \Vdash_{\mathcal{M}} \varphi$, if $s \in [\![\varphi]\!]^{\mathcal{M}}$
- φ is *true* in the model $\mathcal{M}$, written $\mathcal{M} \Vdash \varphi$, if $[\![\varphi]\!]^{\mathcal{M}} = S$
- φ is *valid* in the frame $\mathcal{F}$, written $\mathcal{F} \Vdash \varphi$, if $\mathcal{M} \Vdash \varphi$ for *all* models $\mathcal{M} = (\mathcal{F}, \xi)$ over $\mathcal{F}$ and
- φ is $\mathbb{F}$-*valid*, written $\models_{\mathbb{F}} \varphi$, if $\mathcal{F} \Vdash \varphi$ for all frames $\mathcal{F} \in \mathbb{F}$

The universal pre-image operator is the translation of the standard Kripke semantics into operator form, and is a special case of Jónsson and Tarski's operators on a Boolean algebra. It is also Dijkstra's *weakest liberal precondition* operator, and in set-valued analysis, it is known as the *core* operator (Aubin and Frankowska 1990). The semantic operator for $\langle a \rangle$ is the dual $\exists$-pre-image operator:

$$\mathrm{Pre}^{\exists}[R](A) = \{x \in S \mid (\exists y \in S)[\ xRy \ \text{ and } \ y \in A\,]\}$$

On notation, we use letters s, x, y, u, v, w for states/worlds in a frame, and we write $s \Vdash \varphi$ and $[\![\varphi]\!]$ without the $\mathcal{M}$ when the model is clear.

Definition 2.3 A topological Kripke frame $\mathcal{F} = (S,\ \mathcal{T},\ R_a)$ is:

- *relationally preordered* if the relation $R_a \subseteq S \times S$ is reflexive and transitive; i.e. a preorder;
- *lower semi-continuous* if R_a is l.s.c. in $\mathcal{T}$; i.e. for all $A \subseteq S$, if $A \in \mathcal{T}$ then $\mathrm{Pre}^{\exists}[R_a](A) \in \mathcal{T}$;
- *upper semi-continuous* if R_a is u.s.c. in $\mathcal{T}$; i.e., for all $A \subseteq S$, if $A \in \mathcal{T}$ then $\mathrm{Pre}^{\forall}[R_a](A) \in \mathcal{T}$.

Let $\mathbb{TK}$ denote the universal class of all topological Kripke frames, let $\mathbb{TS}4$ denote the class of all relationally preordered frames, let $\mathbb{LSC}$ denote the class of all relationally preordered, lower semi-continuous frames, and let $\mathbb{USC}$ denote the class of all relationally preordered, upper semi-continuous frames. So $(\mathbb{LSC} \cup \mathbb{USC}) \subseteq \mathbb{TS}4 \subseteq \mathbb{TK}$. We now give the modal characterisation of the semi-continuity properties.

Proposition 2.4 *For all frames $\mathcal{F} \in \mathbb{TK}$, the following are equivalent:*

(i) *for all formulae $\varphi \in \mathcal{L}(\Box, [a])$,* $\quad \mathcal{F} \Vdash \langle a \rangle \Box \varphi \to \Box \langle a \rangle \varphi$

(ii) *for all formulae $\varphi \in \mathcal{L}(\Box, [a])$,* $\quad \mathcal{F} \Vdash \langle a \rangle \Box \varphi \leftrightarrow \Box \langle a \rangle \Box \varphi$

(iii) *the relation R_a is l.s.c. with respect to $\mathcal{T}$.*

Proof. Conditions **(i)**, **(ii)** and **(iii)** are respectively equivalent to:

(i)* $\quad \mathrm{Pre}^{\exists}[R_a]\,(int_{\mathcal{T}}\,(A)) \subseteq int_{\mathcal{T}}\,(\mathrm{Pre}^{\exists}[R_a]\,(A)) \;\; \forall A \in \mathcal{P}(S)$

(ii)* $\quad \mathrm{Pre}^{\exists}[R_a]\,(int_{\mathcal{T}}\,(A)) = int_{\mathcal{T}}\,(\mathrm{Pre}^{\exists}[R_a]\,(int_{\mathcal{T}}\,(A))) \;\; \forall A \in \mathcal{P}(S)$

(iii)* $\quad \mathrm{Pre}^{\exists}[R_a]\,(U) = int_{\mathcal{T}}\,(\mathrm{Pre}^{\exists}[R_a]\,(U)) \;\; \forall U \in \mathcal{T}$

Since $U \in \mathcal{T}$ iff $U = int_{\mathcal{T}}\,(U)$ iff $U = int_{\mathcal{T}}\,(A)$ for some $A \in \mathcal{P}(S)$, the equivalence **(ii)*** $\Leftrightarrow$ **(iii)*** is immediate. So it suffices to show **(i)*** $\Rightarrow$ **(iii)*** and **(ii)*** $\Rightarrow$ **(i)***. Since $U = int_{\mathcal{T}}\,(U)$ for all open sets $U \in \mathcal{T}$, we have $\mathrm{Pre}^{\exists}[R_a](U) = \mathrm{Pre}^{\exists}[R_a]\,(int_{\mathcal{T}}\,(U))$. Assuming **(i)***, we get

$$\begin{aligned} int_{\mathcal{T}}\,(\mathrm{Pre}^{\exists}[R_a](U)) &\subseteq \mathrm{Pre}^{\exists}[R_a](U) \\ &= \mathrm{Pre}^{\exists}[R_a]\,(int_{\mathcal{T}}\,(U)) \\ &\subseteq int_{\mathcal{T}}\,(\mathrm{Pre}^{\exists}[R_a]\,(U)) \end{aligned}$$

so **(iii)*** holds. From the inclusion-monotonicity of both operators together with $int_{\mathcal{T}}(A) \subseteq A$, we get:

$$int_{\mathcal{T}}\,(\mathrm{Pre}^{\exists}[R_a]\,(int_{\mathcal{T}}\,(A))) \subseteq int_{\mathcal{T}}\,(\mathrm{Pre}^{\exists}[R_a]\,(A))$$

So **(ii)*** implies **(i)***. $\dashv$

The "3-to-2" box-diamond equivalences for the l.s.c. property have four bi-dual versions:

$$\langle a\rangle\Box\varphi \leftrightarrow \Box\langle a\rangle\Box\varphi \qquad\qquad \Diamond[a]\Diamond\varphi \leftrightarrow [a]\Diamond\varphi$$
$$\Diamond[a]\varphi \leftrightarrow [a]\Diamond[a]\varphi \qquad\qquad \langle a\rangle\Box\langle a\rangle\varphi \leftrightarrow \Box\langle a\rangle\varphi$$

Note that, while the upper two equivalences only require $[a]$ be **K**, the lower pair of equivalences require $[a]$ be **S4**.

The proof of Proposition 2.4 only uses the inclusion-monotonicity of $\mathrm{Pre}^{\exists}[R_a]$, so a uniform substitution of $\mathrm{Pre}^{\forall}$ for $\mathrm{Pre}^{\exists}$ in Proposition 2.4 gives the u.s.c. characterisation, with the characteristic formulae all of the form box-box or diamond-diamond.

Proposition 2.5 *For frames $\mathcal{F} \in \mathbb{TK}$, the following are equivalent:*

(i) *for all formulae $\varphi \in \mathcal{L}(\Box, [a])$, $\mathcal{F} \Vdash [a]\Box\varphi \to \Box[a]\varphi$*
(ii) *for all formulae $\varphi \in \mathcal{L}(\Box, [a])$, $\mathcal{F} \Vdash [a]\Box\varphi \leftrightarrow \Box[a]\Box\varphi$*
(iii) *the relation R_a is u.s.c. with respect to $\mathcal{T}$.*

3 Applications to Continuous and Hybrid Systems

Consider a state space $S \subseteq \mathbb{R}^n$ equipped with the standard Euclidean metric topology $\mathcal{T}_{\mathbb{R}}$, and consider a differential equation $\frac{d}{dt}x(t) = F(x(t))$ on S. Assuming standard conditions on the vector field $F : S \to \mathbb{R}^n$ such as Lipschitz continuity, there is a unique solution $\gamma_x : T \to S$ from each initial condition $x = \gamma_x(0) \in S$, for some time interval $T \subseteq \mathbb{R}$ including 0. Assume further that solutions exist for the whole non-negative time axis $\mathbb{R}^+ = [0, \infty)$, so they can be aggregated as the *semi-flow* $\Phi : S \times \mathbb{R}^+ \to S$ of F, which is a continuous function on $S \times \mathbb{R}^+$ satisfying the *flow laws*: $\Phi(x, 0) = x$ and $\Phi(x, t+s) = \Phi(\Phi(x,t), s)$ and $\frac{\partial}{\partial t}\Phi(x,t) = F(\Phi(x,t))$.

Definition 3.1 Given a semiflow $\Phi : S \times \mathbb{R}^+ \to S$, the *positive orbit relation* or *flow relation* $f \subseteq S \times S$ of Φ is given by:

$$x \xrightarrow{f} x' \quad \text{iff} \quad (\exists t \in \mathbb{R}^+)\ x' = \Phi(x, t)$$

In words, x' is an f-successor of x iff there is a solution curve of the differential equation leading from x to x'. We show that the resulting frame $\mathcal{F} = (S, \mathcal{T}_{\mathbb{R}}, f)$ lies in $\mathbb{LSC}$. The reflexivity and transitivity of f is immediate from the flow laws. Since time in $\mathbb{R}^+$ is linearly ordered, f is also *weakly connected*, giving the modal logic **S4.3**, but we ignore linearity here. Flow relations of reasonable classes of *differential inclusions* from set-valued vector fields also obey the basic **S4** conditions.

Proposition 3.2 *The positive orbit relation of a semi-flow is l.s.c.*

Proof. We appeal to an alternative formulation of the l.s.c. property for relations on metric spaces from (Aubin and Frankowska 1990): $R \subseteq S \times S$ is l.s.c. iff for all $x \in \text{dom}(R)$, if $(x_n)_{n\in\omega}$ is any sequence converging to x and $x\,R\,y$, then there exists a sequence $(y_n)_{n\in\omega}$ converging to y with $x_n\,R\,y_n$ for all $n \in \omega$. Let $(x_n)_{n\in\omega}$ be a sequence converging to x and suppose $x \xrightarrow{f} y$. So $y = \Phi(x, t_0)$ for some $t_0 \in \mathbb{R}^+$. Since Φ is continuous on its product domain, it is continuous in both arguments separately. Thus the map $\Phi_{t_0} := \Phi(-, t_0) : S \to S$ is continuous, and thus preserves converging sequences. Hence the sequence $(y_n)_{n\in\omega}$ defined by $y_n = \Phi_{t_0}(x_n)$ must converge to y, and $x_n \xrightarrow{f} y_n$ for all $n \in \omega$. ⊣

In contrast, the u.s.c. property is possessed by some but not all flow relations. As a concrete example, consider the piecewise-linear differential equation over $S = \mathbb{R}$ given by $\frac{dx}{dt} = -x$ if $x \leq 1$ and $\frac{dx}{dt} = x - 2$ if $x > 1$. The semi-flow is explicitly defined by:

$$\Phi(x,t) = \begin{cases} xe^{-t} & \text{if } x \leq 1 \text{ and } t \in \mathbb{R}^+ \\ e^{\ln\left(\frac{1}{2-x}\right)-t} & \text{if } 1 < x < 2 \text{ and } t \geq \ln\left(\frac{1}{2-x}\right) \\ (x-2)e^t + 2 & \text{if } (\ 1 < x < 2 \text{ and } 0 \leq t < \ln\left(\frac{1}{2-x}\right)\ ) \\ & \text{or } (\ x \geq 2 \text{ and } t \in \mathbb{R}^+\ ) \end{cases}$$

Then consider the open interval $A = (-\infty, 3)$. For the relation $R_a = f$, the box or universal pre-image $[a]A$ is the ***flow-invariant subset*** of A, namely the points in A *all* of whose flow successors are in A; this is the set $(-\infty, 2]$, which is closed and not open[1], hence f is not u.s.c. The diamond or existential pre-image $\langle a\rangle A$ is the set of points which have *some* flow successor in A. For $A = (-\infty, 3)$ in the example, this is just A itself.

Kripke models of hybrid systems typically have a state space of the form $S = Q \times X$ where Q is a finite set and $X \subseteq \mathbb{R}^n$. Such systems may either flow according to the differential equation associated with a

[1]This example is due to Thomas Moor, Dept. Systems Engineering, RSISE, ANU.

discrete state $q \in Q$, or they may switch discrete states with an accompanying reset of the real-valued variables (possibly the identity map). Hybrid trajectories consist of a sequence of segments of solution curves of differential equations, with endpoints of successive segments related by discrete reset relations. Each discrete state $q \in Q$ also has associated with it a set $Inv_q \subseteq X$, known as the *mode invariant*, and to be a segment of a hybrid trajectory, a solution curve of the state q differential equation is required to lie entirely inside the set Inv_q. In order to reason about hybrid trajectories in relational terms, using alternating compositions of continuous flow relations and discrete reset relations, we actually need to work with *restricted* flow relations.

Given a semi-flow $\Phi : S \times \mathbb{R}^+ \to S$ and any set $A \subseteq S \subseteq \mathbb{R}^n$, define the relation $e(A, \Phi) \subseteq S \times S$ of *restricted evolution* by:

$$x \overset{e(A,\Phi)}{\longrightarrow} x' \quad \overset{\text{def}}{\Leftrightarrow} \quad (\exists t \in \mathbb{R}^+)[x' = \Phi(x,t) \wedge (\forall s \in [0,t])\, \Phi(x,s) \in A]$$

The (unrestricted) flow or orbit relation $f = f(\Phi)$ is then the special case given by $f = e(S, \Phi)$.

Restricted evolution relations $e(A, \Phi)$ are transitive, weakly connected, and *quasi-reflexive*, in the sense that they are reflexive on their domain A. The axiom schemata $[a]\mathbf{T_Q} : ([a]\varphi \wedge \langle a \rangle \top) \to \varphi$ characterises quasi-reflexivity; note that the domain of the relation R_a is the denotation set of the formula $\langle a \rangle \top$.

Various topological and dynamic conditions on the domain A suffice to ensure that a relation $e(A, \Phi)$ is l.s.c. in the standard Euclidean topology, with quite distinct conditions sufficing for the u.s.c. property. For example, suppose the domain A is *f-convex*, in the sense that if $x_1, x_2 \in A$ and $x_1 \overset{f}{\longrightarrow} x \overset{f}{\longrightarrow} x_2$ then $x \in A$. Letting p_A be an atomic proposition denoting the set A, and $\langle f \rangle$ and $[f]$ be the modalities for f, this property is captured modally by the formula $p_A \to [f](\langle f \rangle p_A \to p_A)$, and is equivalent to the relational equality $e = f \cap (A \times A)$, where $e = e(A, \Phi)$. In such a case, the modal operators for e are definable in terms of those for f. Under the hypothesis that A is f-convex, e is l.s.c. iff A is an open set. If additionally, f is known to be u.s.c. , then e is u.s.c. iff A is a closed set. As developed in (Davoren and Nerode 2000), a rich array of properties of hybrid trajectories can be encoded as formulae of **PDL**, the modal μ-calculus $\mathbf{L}\mu$, and extensions obtained by adding $\Box$**S4** topological modalities.

In addition to the standard Euclidean topology on $S \subseteq \mathbb{R}^n$, work in hybrid systems also leads one to consider *finite* topologies generated from finite covers of S, which are necessarily Alexandroff. Operationally, finite topologies arise from (set-valued) *analog-to-digital* conversion maps, and

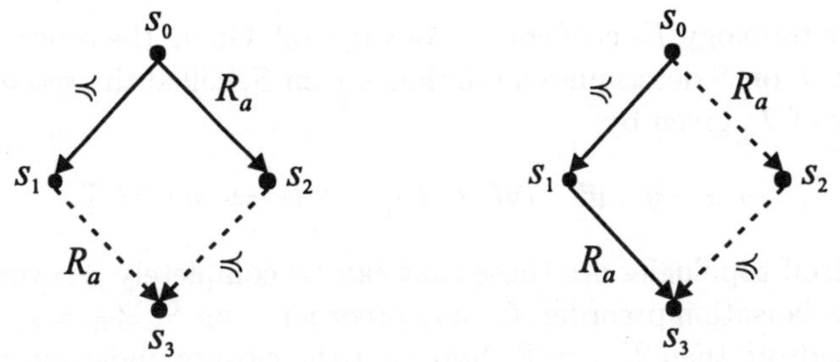

(a) "Lower" diamond property (b) "Upper" diamond property

FIGURE 1 Bi-relational forms of semi-continuity properties

from finite discretisations of S used for computational approximation.

Modal logic has been used quite successfully in (Davoren and Moor 2000) to give a framework and methodology for the design and synthesis of switching controllers for hybrid systems. In that work, the base logic is a polymodal fusion of **K** modalities, which is further extended with axioms for restricted evolution relations, and for *metric tolerance relations* $B_\varepsilon \subseteq S \times S$ defined by: $x\, B_\varepsilon\, y$ iff $d(x, y) < \varepsilon$, where d is a metric on S. The latter relations are captured by the modal logic **KTB** (for reflexivity and symmetry). Note, however, that (Davoren and Moor 2000) does not yet make use of any semi-continuity interactions axioms – although continuity arguments are used in proofs in natural language mathematics. Thus the question of what happens when we *do* add the lower semi-continuity axiom to the basic bi-modal $\Box$**S4**+$[a]$**S4** logic is of considerable interest.

4 Bi-relational Kripke Semantics

We now discuss how the relational Kripke semantics for **S4** are a special case of the McKinsey-Tarski topological semantics; see (Davoren 1998, Davoren 1999). We then characterise the semi-continuity properties in bi-relational frames as "diamond" properties, as illustrated in Figure 1.

Given a preorder $\preccurlyeq$ on S, the *Alexandroff topology* $\mathcal{T}_\preccurlyeq$ on S determined by $\preccurlyeq$ is defined by taking $U \in \mathcal{T}_\preccurlyeq$ iff U is *up-closed* with respect to $\preccurlyeq$, *i.e.* if $x \in U$ and $x \preccurlyeq y$ then $y \in U$. Dually, the closed sets under the topology $\mathcal{T}_\preccurlyeq$ are exactly the down-closed sets with respect to $\preccurlyeq$, *i.e.* if $y \in C$ and $x \preccurlyeq y$ then $x \in C$. The topology $\mathcal{T}_\preccurlyeq$ is closed under arbitrary *intersections* as well as unions, and for all sets $A \subseteq S$,

$$int_{\mathcal{T}_\preccurlyeq}(A) = \mathrm{Pre}^\forall[\preccurlyeq](A) \qquad \text{and} \qquad cl_{\mathcal{T}_\preccurlyeq}(A) = \mathrm{Pre}^\exists[\preccurlyeq](A)$$

More generally, a topology $\mathcal{T}$ on S is called *Alexandroff* if for every point $x \in S$ there is a *smallest open set* containing x. In particular, every *finite* topology on a (arbitrary) set S is Alexandroff. For a preorder $\preccurlyeq$

on S, the topology $\mathcal{T}_{\preccurlyeq}$ is of course Alexandroff. Going the other way, any topology $\mathcal{T}$ on S determines a relation $\preccurlyeq_{\mathcal{T}}$ on S, called the *specialisation preorder* of $\mathcal{T}$, given by:

$$x \preccurlyeq_{\mathcal{T}} y \quad \text{iff} \quad (\forall U \in \mathcal{T})[\ x \in U \ \Rightarrow\ y \in U\ ]$$

Alexandroff topologies are those that can be completely recovered from their specialisation preorder: for any preorder $\preccurlyeq$ on S, $\preccurlyeq_{\mathcal{T}_{\preccurlyeq}} = \preccurlyeq$, and if $\mathcal{T}$ is Alexandroff, then $\mathcal{T}_{\preccurlyeq_{\mathcal{T}}} = \mathcal{T}$. Note that the closure under intersections means that Alexandroff topologies occupy a far corner in the space of all topologies, and look pathological from the viewpoint of general topology: the only Alexandroff topology that is Hausdorff is the discrete topology.

For a class $\mathbb{F}$ of topological Kripke frames, we let $\mathbb{AF}$ denote the subclass of frames $(S, \mathcal{T}, R_a)$ in $\mathbb{F}$ where $\mathcal{T}$ is Alexandroff. We also refer to these as *bi-relational* frames, and write them as $(S, \preccurlyeq, R_a)$.

Proposition 4.1 *Let $\mathcal{F} = (S, \mathcal{T}_{\preccurlyeq}, R_a)$ be any frame with an Alexandroff topology, with $\preccurlyeq$ the corresponding preorder. Then:*

- *The relation R_a is l.s.c. with respect to $\mathcal{T}_{\preccurlyeq}$ iff R_a satisfies the "lower" diamond property w.r.t. $\preccurlyeq$: for all $s_0, s_1, s_2 \in S$*

$$s_0 \preccurlyeq s_1 \text{ and } s_0\, R_a\, s_2 \quad \Rightarrow \quad (\exists s_3 \in S)[\ \ s_1\, R_a\, s_3 \text{ and } s_2 \preccurlyeq s_3\ ]$$

- *The relation R_a is u.s.c. with respect to $\mathcal{T}_{\preccurlyeq}$ iff R_a satisfies the "upper" (or "sideways") diamond property w.r.t. $\preccurlyeq$: for all $s_0, s_1, s_3 \in S$*

$$s_0 \preccurlyeq s_1 \text{ and } s_1\, R_a\, s_3 \quad \Rightarrow \quad (\exists s_2 \in S)[\ \ s_0\, R_a\, s_2 \text{ and } s_2 \preccurlyeq s_3\ ]$$

Proof. Apply Proposition 2.2 of (Davoren 1999). ⊣

The "lower" and "upper" diamond properties are illustrated in Figure 1, parts (a) and (b) respectively, and correspond to the Zig-Zag conditions for $\preccurlyeq$ to be a *bisimulation* with respect to R_a (Davoren 1999).

5 Hilbert-style Axiomatisation

Our Hilbert-style proof system $\mathcal{H}LSC$ for the logic **LSC** contains the tautologies of classical propositional logic, the rule of *modus ponens*, and the axiom schemata and inference rules below:

$\Box\mathbf{K}\wedge$:	$\Box(\varphi_1 \wedge \varphi_2) \leftrightarrow (\Box\varphi_1 \wedge \Box\varphi_2)$	$\Box\mathbf{K}\top$:	$\Box\top$
$\Box\mathbf{T}$:	$\Box\varphi \to \varphi$	$\Box\mathbf{4}$:	$\Box\varphi \to \Box\Box\varphi$
$[a]\mathbf{K}\wedge$:	$[a](\varphi_1 \wedge \varphi_2) \leftrightarrow ([a]\varphi_1 \wedge [a]\varphi_2)$	$[a]\mathbf{K}\top$:	$[a]\top$
$[a]\mathbf{T}$:	$[a]\varphi \to \varphi$	$[a]\mathbf{4}$:	$[a]\varphi \to [a][a]\varphi$

$$\langle a\rangle\Box\mathbf{lsc} : \quad \langle a\rangle\Box\varphi \to \Box\langle a\rangle\varphi$$

$\Box$-*monotonicity:* $\dfrac{\varphi \to \psi}{\Box\varphi \to \Box\psi}$ $\qquad$ $[a]$-*monotonicity:* $\dfrac{\varphi \to \psi}{[a]\varphi \to [a]\psi}$

We write $\vdash_{\mathcal{H}LSC} \varphi$ if the formula $\varphi \in \mathcal{L}(\Box, [a])$ is derivable in $\mathcal{H}LSC$. A set $\Gamma \subseteq \mathcal{L}(\Box, [a]$ of formulae is *$\mathcal{H}LSC$-consistent* if it is not the case that $\vdash_{\mathcal{H}LSC} (\varphi_1 \wedge \cdots \wedge \varphi_n) \to \bot$ for any finite subset $\{\varphi_1, \ldots, \varphi_n\} \subseteq \Gamma$.

The $\Box\mathbf{S4}$ axiom schemes and the $\Box$-monotonicity rule as presented correspond precisely to Kuratowski's (1966) axiomatisation of the topological interior operator (McKinsey 1941, McKinsey and Tarski 1944), (Rasiowa and Sikorski 1963). Note that the usual box-necessitation rules can be derived from the monotonicity rules using the box-$\mathbf{K}\top$ axioms. Likewise, the $[a]\mathbf{S4}$ axiom schemata capture the properties of the operator $\mathrm{Pre}^{\forall}[R_a]$ for preordered relations R_a. The verification that $\mathcal{F} \Vdash \langle a\rangle\Box\varphi \to \Box\langle a\rangle\varphi$ for all frames $\mathcal{F} \in \mathbb{LSC}$ is given in Proposition 2.4. This establishes soundness. For completeness of the axiomatisation, the standard canonical model construction using maximal consistent sets of formulae suffices, as independently discovered by Fischer-Servi (1981).

Proposition 5.1 (Soundness and completeness of $\mathcal{H}LSC$)
For all $\varphi \in \mathcal{L}(\Box, [a])$, $\models_{\mathbf{LSC}} \varphi$ iff $\vdash_{\mathcal{H}LSC} \varphi$.

Proof. For completeness, define an Alexandroff frame $\mathcal{F}_{\mathbf{LSC}}$ where:

$$\begin{array}{rcl} S & := & \{\, s \subseteq \mathcal{L}(\Box, [a]) \mid s \text{ is maximal } \mathcal{H}LSC\text{-consistent} \,\} \\ s_1 \preccurlyeq s_2 & \text{iff} & (\forall \varphi \in \mathcal{L}(\Box, [a]))\ [\ \Box\varphi \in s_1 \ \Rightarrow\ \varphi \in s_2\ ] \\ & \text{iff} & (\forall \varphi \in \mathcal{L}(\Box, [a]))\ [\ \varphi \in s_2 \ \Rightarrow\ \Diamond\varphi \in s_1\ ] \\ s_1\, R_a\, s_2 & \text{iff} & (\forall \varphi \in \mathcal{L}(\Box, [a]))\ [\ [a]\varphi \in s_1 \ \Rightarrow\ \varphi \in s_2\ ] \\ & \text{iff} & (\forall \varphi \in \mathcal{L}(\Box, [a]))\ [\ \varphi \in s_2 \ \Rightarrow\ \langle a\rangle\varphi \in s_1\ ] \end{array}$$

and define the canonical valuation $\xi_0 : \mathrm{AP} \to \mathcal{P}(S)$ as usual by $\xi_0(p) := \{s \in S \mid p \in s\}$. Let $\mathcal{M}_{\mathbf{LSC}} = (\mathcal{F}_{\mathbf{LSC}}, \xi_0)$ denote the canonical model. By the standard arguments, the axioms for $\Box\mathbf{S4}$ and $[a]\mathbf{S4}$ entail that the relations $\preccurlyeq$ and R_a are both preorders, hence $\mathcal{F}_{\mathbf{LSC}} \in \mathbb{ATS}4$.

To establish that $\mathcal{F}_{\mathbf{LSC}}$ is l.s.c. with respect to $\mathcal{T}_{\preccurlyeq}$, and so $\mathcal{F}_{\mathbf{LSC}} \in \mathbb{LSC}$, we need to verify the l.s.c. diamond property: if $s_0 \preccurlyeq s_1$ and

$s_0 \, R_a \, s_2$, then there exists a state s_3 such that $s_1 \, R_a \, s_3$ and $s_2 \preccurlyeq s_3$. So suppose $s_0 \preccurlyeq s_1$ and $s_0 \, R_a \, s_2$. Then consider the set of formulae

$$c = \{\varphi \in \mathcal{L}(\Box, [a]) \mid [a]\varphi \in s_1 \text{ or } \Box\varphi \in s_2\}$$

If c is non-empty and $\mathcal{H}LSC$-consistent, then there exists a maximal $\mathcal{H}LSC$-consistent set $s_3 \supseteq c$, and by construction, $s_1 \, R_a \, s_3$ and $s_2 \preccurlyeq s_3$.

The non-emptiness of c is immediate, since $[a]\top \in s$ and $\Box\top \in s$ for all s. To prove that c is $\mathcal{H}LSC$-consistent, suppose, for a contradiction, that it is not. Then there is a finite subset $c_0 = \{\varphi_1, ..., \varphi_n, \psi_1, ..., \psi_m\}$ of c, such that $\vdash_{\mathcal{H}LSC} (\varphi_1 \wedge \cdots \wedge \varphi_n \wedge \psi_1 \wedge \cdots \wedge \psi_m) \to \bot$, where each $[a]\varphi_i \in s_1$ and each $\Box\psi_j \in s_2$. Let $\varphi = \varphi_1 \wedge \cdots \wedge \varphi_n$ and likewise $\psi = \psi_1 \wedge \cdots \wedge \psi_m$, hence $\vdash_{\mathcal{H}LSC} \psi \to \neg\varphi$. Since maximal consistent sets s are closed under conjunction, and boxes distribute over conjunction, we have $[a]\varphi \in s_1$ and $\Box\psi \in s_2$. Now $\Box\psi \in s_2$ and $s_0 \, R_a \, s_2$ imply $\langle a\rangle\Box\psi \in s_0$, so by the $\langle a\rangle\Box$**lsc** axiom $\langle a\rangle\Box\psi \to \Box\langle a\rangle\psi$, we then have $\Box\langle a\rangle\psi \in s_0$. By the monotonicity of both diamonds and boxes, $\vdash_{\mathcal{H}LSC} \Box\langle a\rangle\psi \to \Box\langle a\rangle\neg\varphi$, hence $\Box\langle a\rangle\neg\varphi \in s_0$. Then since $s_0 \preccurlyeq s_1$, we have $\langle a\rangle\neg\varphi \in s_1$, and hence $\neg[a]\varphi \in s_1$. But this contradicts $[a]\varphi \in s_1$, so c must be $\mathcal{H}LSC$-consistent.

The usual induction on formulae establishes the "Truth Lemma": $s \in [\![\varphi]\!]^{\mathcal{M}_{\mathbf{LSC}}}$ iff $\varphi \in s$ for all maximal $\mathcal{H}LSC$-consistent sets $s \subseteq \mathcal{L}(\Box, [a])$ and all formulae $\varphi \in \mathcal{L}(\Box, [a])$. Then for any φ, if it is not the case that $\vdash_{\mathcal{H}LSC} \varphi$, then there is some maximal $\mathcal{H}LSC$-consistent set s such that $\varphi \notin s$, hence $\mathcal{M}_{\mathbf{LSC}} \nVdash \varphi$. $\dashv$

Corollary 5.2 *For all* $\varphi \in \mathcal{L}(\Box, [a])$, $\models_{\mathbb{LSC}} \varphi$ *iff* $\models_{\mathbb{ALSC}} \varphi$.

Proof. In the non-trivial direction, suppose $\models_{\mathbb{ALSC}} \varphi$. Then in particular, $\mathcal{M}_{\mathbf{LSC}} \Vdash \varphi$ in the canonical model, hence $\vdash_{\mathcal{H}LSC} \varphi$. Then since $\mathcal{H}LSC$ is sound with respect to the whole class $\mathbb{LSC}$, we have $\models_{\mathbb{LSC}} \varphi$. $\dashv$

The canonical model construction can be applied uniformly to the family of bimodal logics obtained by axiomatising the relational modality $[a]$ by any consistent normal canonical monomodal logic, and adding one or both of the two semi-continuity axioms; in particular to the natural twin logic **USC** given by $\Box$**S4**$+[a]$**S4**$+[a]\Box$**usc** (Coulthard 2000). Consequently, the (finitary) bimodal language cannot distinguish between validity in the class of all appropriate frames, with arbitrary topologies, and validity in the class of appropriate frames restricted to Alexandroff topologies. Hence we can restrict attention to the corresponding class of bi-relational frames without any loss.

It remains on open question as to whether standard filtration quotient techniques (see, for example, Chapter 4 of (Goldblatt 1992)) can be successfully extended to the canonical model $\mathcal{M}_{\mathbf{LSC}}$ to prove the finite model property for the logic **LSC**, or analogously for **USC**, or whether filtration breaks down for these logics. What is not clear is how to choose a finite filtration set Γ adequate to push through the semi-continuity properties. As an alternative method of attack we have investigated a tableau calculus for the logic **LSC** as reported next.

6 The Tableau Calculus $\mathcal{T}LSC$

Our tableau calculus $\mathcal{T}LSC$ for $\mathbb{LSC}$ is based upon the tableau calculus for the logic **S4.2** from (Goré 1991). The logic **S4.2** is the monomodal logic obtained from $\mathbb{LSC}$ if $R_a = \preccurlyeq = R$. All the rules for $\mathcal{T}LSC$ are given in Figure 2 where $X, Y \subseteq \mathcal{L}(\Box, [a])$ are finite sets of formulae and $X; \varphi$ means $X \cup \{\varphi\}$: see (Goré 1999) for an introduction to modal tableau calculi. The subset of these rules consisting of the top three rows gives another calculus called $\mathcal{T}(\Box S4 + [a]S4)$, discussed later.

A $\mathcal{T}LSC$-tableau for a finite set of formulae X is an upside down tree with root X, such that each node in the tree is obtained from its parent node by an application of a rule of $\mathcal{T}LSC$. A tableau branch is *closed* if the final rule application on this branch is $(\bot)$, otherwise the branch is *open*. A tableau is open if some branch is open, otherwise it is closed. If there is a closed $\mathcal{T}LSC$-tableau for the finite set $X \cup \{\neg\varphi\}$, we write $X \vdash_{\mathcal{T}LSC} \varphi$. In particular, when X is empty we write $\vdash_{\mathcal{T}LSC} \varphi$.

Following (Goré 1999), all rules are *static* except $(K4\neg\Box)$ and $(K4\neg[a])$ which are *transitional* since they correspond to the creation of successor worlds. The (LSC) rules create bigger formulae from smaller formulae. To keep this process from repeating itself *ad infinitum*, the new formula is marked with a $\star$ and the (LSC) rules are not applicable to such starred formulae. All other rules must treat starred formulae as if they were unstarred. Also, if the formula to be added by an (LSC) rule is already present without a star, then that rule is not applied. This stops a formula from appearing both starred and unstarred in a node.

As usual, each tableau rule can be read downwards as: if the set of formulae above the horizontal line is $\mathbb{LSC}$-satisfiable, then so is at least one of the sets below the horizontal line.

Proposition 6.1 (Soundness of $\mathcal{T}LSC$)
For all formulae $\varphi \in \mathcal{L}(\Box, [a])$, *if* $\vdash_{\mathcal{T}LSC} \varphi$ *then* $\models_{\mathbb{LSC}} \varphi$.

Proof. All rules except the (LSC) rules are standard **S4** rules, so consider the (LSC) rules. The premiss of $(LSC\neg\Box)$ is that $X; \neg\Box\varphi$ is **LSC**-satisfiable. Thus there is an $\mathbb{LSC}$-model with a world w_0 such that $w_0 \Vdash X$ and $w_0 \Vdash \neg\Box\varphi$. Now if $w_0 \Vdash [a]\neg\Box\varphi$ then we are done, since

$$(\bot)\ \frac{X;\varphi;\neg\varphi}{\bot} \quad (\wedge)\ \frac{X;\varphi\wedge\psi}{X;\varphi;\psi} \quad (\neg\wedge)\ \frac{X;\neg(\varphi\wedge\psi)}{X;\neg\varphi \mid X;\neg\psi} \quad (\neg\neg)\ \frac{X;\neg\neg\varphi}{X;\varphi}$$

$$(KT\Box)\ \frac{X;\Box\varphi}{X;\Box\varphi;\varphi} \qquad (KT[a])\ \frac{X;[a]\varphi}{X;[a]\varphi;\varphi}$$

$$(K4\neg\Box)\ \frac{X;\Box Y;\neg\Box\varphi}{\Box Y;Y;\neg\varphi} \qquad (K4\neg[a])\ \frac{X;[a]Y;\neg[a]\varphi}{[a]Y;Y;\neg\varphi}$$

$$(cut\neg\Box)\ \frac{X;\neg\Box\varphi}{X;\neg\Box\varphi;\varphi \mid X;\neg\Box\varphi;\neg\varphi} \quad (cut\neg[a])\ \frac{X;\neg[a]\varphi}{X;\neg[a]\varphi;\varphi \mid X;\neg[a]\varphi;\neg\varphi}$$

$$(LSC\neg\Box)\ \frac{X;\neg\Box\varphi}{X;\neg\Box\varphi;\Box(\neg[a]\neg\Box\varphi)^\star \mid X;\neg\Box\varphi;[a]\neg\Box\varphi}\ \neg\Box\varphi \text{ unstarred}$$

$$(LSC\neg[a])\ \frac{X;\neg[a]\varphi}{X;\neg[a]\varphi;[a](\neg\Box\neg[a]\varphi)^\star \mid X;\neg[a]\varphi;\Box\neg[a]\varphi}\ \neg[a]\varphi \text{ unstarred}$$

FIGURE 2 Tableau Calculi $\mathcal{T}(\Box S4 + [a]S4)$ and $\mathcal{T}LSC$

then w_0 satisfies the right child of $(LSC\neg\Box)$. So suppose $w_0 \not\Vdash [a]\neg\Box\varphi$, that is, $w_0 \Vdash \langle a\rangle\Box\varphi$. Then by Proposition 2.4, $w_0 \Vdash \Box\langle a\rangle\Box\varphi$, or equivalently, $w_0 \Vdash \Box\neg[a]\neg\Box\varphi$. Hence w_0 satisfies the left child of $(LSC\neg\Box)$. The soundness of $(LSC\neg[a])$ is proved analogously. $\dashv$

The two branches of $(LSC\neg\Box)$ are mutually exclusive since the right child contains $[a]\neg\Box\varphi$, and an application of $(KT\Box)$ to the left child brings $(\neg[a]\neg\Box\varphi)^\star$ into the left child. If this formula were unstarred then the $(LSC\neg[a])$ rule would be applicable to it, sending the system into an infinite loop. Thus $(LSC\neg\Box)$ encodes a cut on the *larger* formula $[a]\neg\Box\varphi$, for some subformula $\neg\Box\varphi$. The fact that the left child contains $\Box(\neg[a]\neg\Box\varphi)^\star$ rather than just $(\neg[a]\neg\Box\varphi)^\star$ encodes the axiom $\langle a\rangle\Box\varphi \leftrightarrow \Box\langle a\rangle\Box\varphi$, but does not lead to an infinite loop since no rule creates bigger formulae from $\Box$-formulae. The $(LSC\neg[a])$ rule encodes an analogous cut rule and axiom. This "analytic super-formula property" is the key to termination of proof search (Goré 1999).

7 A counter-example to completeness of $\mathcal{T}LSC$

In previous versions of this paper, we conjectured that $\mathcal{T}LSC$ was complete, but the following counter-example to completeness was recently found by Nicolette Bonnette.

Example 7.1 The set $X = \{p_1, p_2, p_3, \neg p_1 \vee \neg p_2 \vee \neg p_3\}$ is classically unsatisfiable and every non-empty proper subset of X is classically satisfiable. Consider the formula

$$\varphi_0 \quad := \quad \Box[a]p_1 \wedge \Diamond[a]p_2 \wedge [a]\Box p_3 \wedge \langle a\rangle\Box(\neg p_1 \vee \neg p_2 \vee \neg p_3)$$

Proposition 7.2 *The formulae φ_0 is not $\mathbb{LSC}$-satisfiable.*

Proof. For a contradiction suppose that there is an $\mathbb{LSC}$-model with a world s_0 such that $s_0 \Vdash \varphi_0$. Then there must exist two worlds s_1 and s_2 such that $s_0 \preccurlyeq s_1$ and $s_0 R_a s_2$, $s_1 \Vdash [a]p_1, [a]p_2$ and $s_2 \Vdash \Box p_3, \Box(\neg p_1 \vee \neg p_2 \vee \neg p_3)$. By the l.s.c. condition there must exist a world s_3 such that $s_1 R_a s_3$ and $s_2 \preccurlyeq s_3$. Then we must have $s_3 \Vdash p_1, p_2, p_3, (\neg p_1 \vee \neg p_2 \vee \neg p_3)$, which we know to be impossible. $\dashv$

However, as a detailed attempt will show, there is no closed $\mathcal{T}LSC$-tableau for the set $\{\varphi_0\}$.

We are currently trying to pinpoint exactly why completeness fails using the model-graph techniques outlined in (Goré 1999) so we can invent new rules to plug the gap. The l.s.c. condition is unusual because of its non-local effects: we require a global view of the counter-model in order to detect and repair incomplete half-diamonds. A similar situation occurs in tense logics where events in the future can effect worlds in the past, and vice-versa. We believe that this is the cause of the incompleteness of our purely local tableau calculus $\mathcal{T}LSC$.

An alternative approach is to use a complete and terminating explicit algorithm, and to prove its soundness, as sketched next.

8 An Alternative Tableau Calculus $\mathcal{T}lscSAT$

For a class of bi-relational frames $\mathbb{B}$, let $\mathbb{B}^{\mathrm{fr}}$ denote the subclass of frames $(S, \preccurlyeq, R_a)$ in $\mathbb{B}$ such that S is finite and the frame consists of a finite rooted tree of nodes: the nodes are finite R_a-clusters and $\preccurlyeq$-clusters when the underlying relations are transitive (Goré 1999).

Let $\mathcal{T}(\Box S4 + [a]S4)$ consist of the rules $(\bot)$, $(\wedge)$, $(\neg\wedge)$, $(\neg\neg)$, $(KT\Box)$, $(KT[a])$, $(K4\neg\Box)$, and $(K4\neg[a])$ from Figure 2. Thus, $\mathcal{T}(\Box S4 + [a]S4)$ is the union of two standard calculi for **S4** (Goré 1999), one for each relation R_a and $\preccurlyeq$.

Proposition 8.1 *For all $\varphi \in \mathcal{L}(\Box, [a])$, the following are equivalent:* **(i)** $\models_{\mathbb{TS}4} \varphi$ **(ii)** $\models_{\mathbb{ATS}4} \varphi$ **(iii)** $\models_{\mathbb{ATS}4^{fr}} \varphi$ **(iv)** $\vdash_{\mathcal{H}(\Box S4+[a]S4)} \varphi$ **(v)** $\vdash_{\mathcal{T}\Box S4+[a]S4} \varphi$

Given a finite set of formulae X, the systematic tableau procedure (Goré 1999) attempts to construct a closed $\mathcal{T}(\Box S4 + [a]S4)$-tableau for X. If it succeeds then by soundness, the set X is not $\mathbb{TS}4$-satisfiable. If it fails, then no $\mathcal{T}(\Box S4 + [a]S4)$-tableau for X closes so the systematic procedure constructs a (finite) $\mathbb{ATS}4^{fr}$-model $\mathcal{M}$ whose root satisfies X. In such a representation, the number of actual edges is minimised by including only the essential edges, and implicitly making the edges "transitive". Thus $\mathcal{M}$ is really a data structure (graph), but there is no explicit "database" of tuples representing the relations.

The tableau calculus $\mathcal{T}lscSAT$ is a "hybrid" calculus that uses the tableau calculus $\mathcal{T}(\Box S4 + [a]S4)$ to handle the $\Box$**S4** + $[a]$**S4** aspects of our logic, and uses an additional global data-structure to explicitly track and repair the failures of the l.s.c. condition. A similar technique is used in (Fariñas del Cerro and Gasquet 1999)[2].

In our hybrid method, we start by applying the systematic $\mathcal{T}(\Box S4 + [a]S4)$-tableau construction to $\{\neg\varphi\}$. If this procedure finds a closed $\mathcal{T}(\Box S4 + [a]S4)$-tableau, then we know that φ is $\mathbb{TS}4$-valid, and hence φ is also $\mathbb{ALSC}$-valid. Otherwise, we have at least one finite $\mathbb{ATS}4^{fr}$-model $\mathcal{M}_1$ whose root satisfies $\neg\varphi$. In constructing $\mathcal{M}_1$, the systematic method chooses either the left or right disjunct when applying the $(\neg\wedge)$ rule. But the other disjunct may also lead to a different finite $\mathbb{ATS}4^{fr}$-model $\mathcal{M}_2$ for $\{\neg\varphi\}$. In general, there will be some finite number k of such $\mathbb{ATS}4^{fr}$-models $\mathcal{M}_1, \cdots, \mathcal{M}_k$ for $\{\neg\varphi\}$. We want to extend one of them to an $\mathbb{ALSC}^{fr}$-model $\mathcal{M}_o$, the root node of which still satisfies $\neg\varphi$. To do this we have to ensure that it satisfies the l.s.c. condition.

Let $(s_0 \preccurlyeq s_1, s_0 R_a s_2)$ stand for three worlds s_0, s_1 and s_2 in $\mathcal{M}_1$ such that $s_0 \preccurlyeq s_1$ and $s_0 R_a s_2$ both hold, but such that there is no x for which both $s_1 R_a x$ and $s_2 \preccurlyeq x$ hold. The l.s.c. condition demands that every such "incomplete lsc-half-diamond" be completed. To create such a world we start a new systematic $\mathcal{T}(\Box S4 + [a]S4)$-tableau construction for the set $x := \{[a]\psi \mid [a]\psi \in s_1\} \cup \{\Box\psi \mid \Box\psi \in s_2\}$. If this procedure finds a closed $\mathcal{T}(\Box S4 + [a]S4)$-tableau then the current model $\mathcal{M}_1$ cannot be extended into an $\mathbb{LSC}$-model so we try $\mathcal{M}_2$, and so on. If the systematic procedure finds no closed $\mathcal{T}(\Box S4 + [a]S4)$-tableau, it will return an $\mathbb{ATS}4^{fr}$-model $\mathcal{M}_x$ with a root node r containing x. We can safely add

[2]The claim on page 317 of (Fariñas del Cerro and Gasquet 1999) that "a decidability result for confluence with transitivity was an open problem" is erroneous as Segerberg proved the finite model property for **K4.2** in his doctoral dissertation and (Goré 1991) contains a decision procedure for **S4.2**.

the "transitive" edges $s_1 R_a r$ and $s_2 \preccurlyeq r$ as desired, thereby appending $\mathcal{M}_x$ to $\mathcal{M}_1$. If the new model contains no further incomplete lsc-half-diamonds we have produced an $\mathbb{ALSC}^{\text{fr}}$-model which falsifies φ at its root node. Otherwise we must complete all remaining lsc-half-diamonds in the new model. Eventually, we will either report that each original $\mathbb{ATS}4^{\text{fr}}$-model for $\{\neg\varphi\}$ contained at least one lsc-half-diamond which could not be completed, or until we find some $\mathbb{ATS}4^{\text{fr}}$-model for $\{\neg\varphi\}$ in which we can complete all incomplete lsc-half-diamonds.

The procedure $\mathcal{T}lscSAT$ is complete and terminates, but we have been unable to prove its soundness. The main problem is that a node x inserted when completing a half-diamond may appear already higher up in the current counter-model under construction, thereby creating an infinite path. A standard trick is to insert the two required edges so they loop back to the previous occurrence of x, thereby maintaining finiteness at the cost of two loops in the current counter-model. Although this trick can be made to work in the mono-modal setting (Fariñas del Cerro and Gasquet 1999), it is by no means clear how to treat such loops within an inductive argument in the bimodal setting.

9 Related Work on Bimodal and Polymodal Logics

While topological semantics for **S4**-like modalities have been used in a number of recent studies of bimodal logics, our work is the first to directly address continuity properties of *relations* in a bimodal framework. Earlier work by Davoren (1998), and independent work by Kremer and Mints (1997), uses the topological semantics for $\Box$**S4** together with the scheme $[a]$**F** : $[a]\varphi \leftrightarrow \langle a\rangle\varphi$, which characterises $R_a = g$ as a *total function* $g : S \to S$. The resulting bimodal logic is adequate to formalise the continuity of functions, via the equivalent schemata $\langle a\rangle\Box$**lsc** or $[a]\Box$**usc**, plus other topological properties such as being an open map or being a homeomorphism. The thesis (Davoren 1998) proves the decidability of the bimodal logic $\Box$**S4** + $[a]$**KF** + $\langle a\rangle\Box$**usc** by giving a sound and complete tableau proof system, and taking a finite quotient of the infinite systematic tableau constructed for completeness. That thesis also begins an investigation of polymodal logics built on $\Box$**S4** + **DPDL** (i.e. *deterministic* **PDL**) with the atomic actions $a \in \Sigma$ interpreted by continuous total functions. While the l.s.c. property is inherited under each of the regular expression operations of relational composition, finite unions and Kleene-star, inheritance of the u.s.c. property fails for the Kleene-star (Davoren and Nerode 2000). As noted in an unpublished manuscript by Kremer, following (Kremer and Mints 1997), there are continuous functions whose Kleene-star is not u.s.c.

Dabrowski, Moss and Parikh (1996), and Heinemann (1998, 2000), examine bimodal logics for expressing elementary reasoning about points and sets in general topology, motivated by applications to reasoning about knowledge in multi-agent systems. Their semantics are given over *subset frames* $(S, \mathcal{O})$ where $\mathcal{O}$ is an arbitrary family of non-empty subsets of S, so for a topology $\mathcal{T}$ on S, the family $\mathcal{O} = \mathcal{T} - \{\emptyset\}$ gives a special case. In (Dabrowski et al. 1996, Heinemann 2000) the logics are built on the bimodal base $\Box$**S4**+K**S5**, where the two box modalities $\Box$ and K are interpreted by different aspects of the relationship between points $s \in S$ and sets $U \in \mathcal{O}$, and are further interconnected by the $K\Box$**usc** axiom scheme; these logics also have *persistence* axioms for literals: $p \to \Box p$ and $\neg p \to \Box\neg p$ for atomic propositions p. When $\mathcal{O}$ comes from a topology, the topological interior operator is recovered by their compound modality $\Diamond K$. The papers (Dabrowski et al. 1996, Heinemann 2000) also extend the canonical Kripke model construction to their bimodal setting, with the two modalities realised by a preorder and an equivalence relation respectively; (Dabrowski et al. 1996) further establishes the decidability of their bimodal logic by taking a finite filtration of the birelational canonical Kripke model. In (Heinemann 1998), the **S4** box is weakened to that from a partial function, to give a "next-time" operator.

Fischer-Servi (1981, 1989) studies bimodal logics where two relationally interpreted modalities are subject to interaction axioms of the same form as those we study. These logics have an axiomatisation equivalent to $\Box$**L** + $[a]$**S4** + $\langle a\rangle\Box$**lsc** + $[a]\Box$**usc**, where **L** is any one of **KT**, **S4**, **S4.1** and **S4.2**. The equivalence can be seen by mapping the box operators L_1 and L_2 of (Fischer-Servi 1981) to our $[a]$ and $\Box$ respectively, and using the equivalence between $\Diamond[a]\varphi \to [a]\Diamond\varphi$ and $\langle a\rangle\Box$**lsc**. Fischer-Servi interprets both modalities using relations, varies the axioms for $\Box$ and fixes $[a]$ as an **S4**-modality, whereas we fix $\Box$ as a topologically interpreted **S4**-modality and vary the axioms for $[a]$. The notion of "finite axiomatizability" used in (Fischer-Servi 1989) is *without* the necessitation rules, hence Fischer-Servi's negative results do not directly apply to questions of decidability of our logics.

10 Further Work and Conclusions

We have developed a simple and elegant logical framework in which to reason about continuous dynamics, using the very familiar ingredients of two **S4** modalities together with an interaction axiom. As demonstrated, the interaction is non-trivial, and its origins lie in a natural mathematical phenomenon, giving solid motivation for the resulting logic **LSC**.

We have given a tableau calculus $\mathcal{T}LSC$ which is sound but not complete for **LSC**. We have outlined a procedure $\mathcal{T}lscSAT$ which is complete and terminating for **LSC**, but for which soundness is still open. We have also considered various filtration methods, all to no avail. It is clear that the problems of completeness of $\mathcal{T}LSC$, the soundness of $\mathcal{T}lscSAT$ and the solution to the filtration method are all three vertices of the same problem: if we can solve one, then we can solve them all.

A wide spectrum of interesting bimodal and polymodal logics can be obtained by keeping the topological $\Box$**S4** fixed and varying the axioms for the relational modalities $[a]$ and $\langle a \rangle$. For example, one may strengthen to $[a]$**S4.3** to cover the weak connectedness of flow relations, or consider polymodal variants to cover discrete reset relations. Further investigation is required of logics such as $\Box\mathbf{S4} + \mathbf{PDL}$ and $\Box\mathbf{S4} + \mathbf{L}\mu$, combining the topological modalities with infinitary constructs such as the Kleene-star or more general fixed-point quantification, and the effect on them of adding one or more semi-continuity axioms.

Acknowledgements

Thanks to Nicolette Bonnette, Marcus Kracht, Maarten Marx, Ulrike Sattler and Stephan Tobies for discussions on the decidability of **LSC**, and to Thomas Moor for discussions on logics for hybrid systems.

References

Antsaklis, P.J. (ed.). 2000. *Proceedings of the IEEE.* July. Special Issue on Hybrid Systems: Theory and Applications.

Aubin, J-P., and H. Frankowska. 1990. *Set-Valued Analysis.* Boston: Birkhäuser.

Coulthard, V. 2000. A Study of Topological Polymodal Logics. Technical report. CSL, RSISE, ANU, November. BSc Honours thesis.

Dabrowski, A., L.S. Moss, and R. Parikh. 1996. Topological Reasoning and the Logic of Knowledge. *Annals of Pure and Applied Logic* 78:73–110.

Davoren, J.M. 1998. *Modal Logics for Continuous Dynamics.* Doctoral dissertation, Dept. of Mathematics, Cornell University.

Davoren, J.M. 1999. Topologies, Continuity and Bisimulations. *Theoretical Informatics and Applications* 33:357–381.

Davoren, J.M., and T. Moor. 2000. Logic-based design and synthesis of controllers for hybrid systems. Technical report. Dept. Sys. Eng. , RSISE, ANU, July 2000. `http://csl.anu.edu.au/~davoren/`.

Davoren, J.M., and A. Nerode. 2000. Logics for Hybrid Systems. *Proc. of the IEEE* 88:985–1010.

Fariñas del Cerro, L., and O. Gasquet. 1999. Tableaux Based Decision Procedures for Modal Logics of Confluence and Density. *Fundamenta Informaticae* 40:317–333.

Fischer-Servi, G. 1981. Teoremi di Completezza per Calcoli Bimodali. *Riv. Mat. Univ. Parma* 4(7):347–350.

Fischer-Servi, G. 1989. Finite Axiomatizability in Bimodal Calculi. *Riv. Mat. Univ. Parma* 4(15):271–287.

Goldblatt, R. 1992. *Logics of Time and Computation.* Stanford: CSLI.

Goré, R. 1991. Semi-analytic tableaux for propositional normal modal logics with applications to nonmonotonicity. *Logique et Analyse* 133-134:73–103. (Printed in 1994).

Goré, R. 1999. Tableaux Methods for Modal and Temporal Logics. In *Handbook of Tableaux Methods*, ed. M. D'Agostino *et al.* 297–396. Kluwer.

Heinemann, B. 1998. Topological Nexttime Logic. In *Advances in Modal Logic, Volume 1*, ed. M. Kracht *et al.* 99–113. CSLI.

Heinemann, B. 2000. Generalizing the Modal and Temporal Logic of Linear Time. In *8th International Conference on Algebraic Methodology and Software Technology (AMAST'2000)*, ed. T. Rus *et al.* Berlin. Springer-Verlag.

Kremer, P., and G. Mints. 1997. Dynamic Topological Logic. *Bulletin of Symbolic Logic* 3:371–372.

Kuratowski, K. 1966. *Topology.* New York: Academic Press.

McKinsey, J.C.C. 1941. A solution to the decision problem for the Lewis systems S2 and S4, with an application to topology. *Journal of Symbolic Logic* 6:117–141.

McKinsey, J.C.C., and A. Tarski. 1944. The Algebra of Topology. *Annals of Mathematics* 45:141–191.

Rasiowa, H., and R. Sikorski. 1963. *The Mathematics of Metamathematics.* Warsaw: PWN.

Wolter, Frank. 1999. Decision Problems for Combined (Modal) Logics. Technical report. `http://www.informatik.uni-leipzig.de/~wolter/`: Institüt für Informatik, Univesität Leipzig, September.

J.M. Davoren
Computer Sciences Laboratory
Research School of Information Sciences and Engineering
The Australian National University
Canberra ACT 0200 AUSTRALIA
E-mail: `j.m.davoren@anu.edu.au`
URL: `http://csl.anu.edu.au/~davoren`

R.P. Goré
Computer Sciences Laboratory
Research School of Information Sciences and Engineering
and
Department of Computer Science
Faculty of Engineering and Information Technology
The Australian National University
Canberra ACT 0200 AUSTRALIA
E-mail: rpg@arp.anu.edu.au
URL: http://csl.anu.edu.au/~rpg

7

Modal Logics with Weak Forms of Recursion: PSPACE Specimens

STÉPHANE DEMRI

ABSTRACT. We analyze the computational complexity of extensions of the multimodal version of the standard modal logic K by finite addition of axiom schemes that can be read as the production rules of a formal grammar. By using proof-theoretical means, we show that every right linear grammar logic has a satisfiability problem in deterministic exponential time and we exhibit countably infinite classes of right linear grammar logics that contain weak forms of recursion for which the satisfiability problem can be solved in polynomial space.

1 Introduction

In order to explain the algorithmic properties of many modal logics, a possible approach consists in studying very expressive decidable logical theories in which can be easily embedded the modal logics. The modal μ-calculus (see e.g. (Kozen 1983)), the guarded fixed point logic μLGF (Grädel and Walukiewicz 1999) and the monadic second-order theory of two successors (Rabin 1969) are good representatives of such theories. Weaker logical theories such as the Propositional Dynamic Logic (see e.g. (Pratt 1979, Fischer and Ladner 1979)) and the guarded fragment (Andreka et al. 1998) are also serious candidates although strictly less expressive, respectively. Moreover, since both the μ-calculus and the guarded fixed point logic μLGF with bounded arity have an **EXPTIME**-complete satisfiability problem (Grädel and Walukiewicz 1999), those logical formalisms inherit the algorithmic properties of weak extensions of the modal logic K (by adding the universal modal connective for example) that are already of the same complexity level. By

Advances in Modal Logic, Volume 3
F. Wolter, H. Wansing, M. de Rijke, and M. Zakharyaschev, eds.

contrast, many standard modal logics are known to be in **PSPACE** (Ladner 1977, Halpern and Moses 1992) and this algorithmic property is not reflected by the analysis of most identified fragments of second order logic. As far as we know, a rare exception can be found in (Marx 1997) (see also (Lutz et al. 1999, Marx et al. 2000b)), where a **PSPACE** first-order fragment is defined capturing multimodal logics with K, T and B modal connectives plus inclusion, to quote a few examples of logics. This situation is all the more surprising since many modal logics with weak forms of recursion (K4, S4, ...) are also in **PSPACE**. So, what is the adequate first-order/second-order fragment that is responsible for the algorithmic behavior of **PSPACE** modal logics? This question could become meaningful for optimizing the efficiency of mechanical reasoning for such logics and thus avoiding a blind translation into a rich decidable logical theory, possibly algorithmically more expensive in the worst-case. Although we have no answer for this question, this motivates the developments made in this paper. We study a countably infinite class of multimodal logics that can be embedded uniformly into first-order logic with the relational translation. The target fragment is not known to belong to identified decidable fragments but we exhibit modal logics with weak forms of recursion that are in **PSPACE**.

In order to understand the **PSPACE** modal logics with weak forms of recursion, a class of multimodal logics that are worth investigating are the grammar logics defined in (Fariñas del Cerro and Penttonen 1988) that are closely related to formal grammars. With each production rule $i_1 \cdot \ldots \cdot i_n \to j_1 \cdot \ldots \cdot j_{n'}$ in the grammar is associated a reduction principle $[i_1]\ldots[i_n]\mathrm{p} \Rightarrow [j_1]\ldots[j_{n'}]\mathrm{p}$ (see e.g. (van Benthem 1976)) which is a particular form of Sahlqvist formula (Sahlqvist 1975). Observe that the logical view on grammar presented in (Kracht 2000) differs from the approach in (Fariñas del Cerro and Penttonen 1988). In the present paper, we mainly study the extensions of the multimodal logic K_m with m independent K modal connectives by finite addition of axiom schemes of the above form such that the associated finite set of production rules forms a right linear formal grammar. The right linear grammar logics contain a weak form of recursion although different from the one in the logics defined in (Halpern and Reif 1983) (e.g., we do not assume any determinism). For instance, consider the multimodal logic $\mathcal{L} = K_5 + [1]\mathrm{p} \Rightarrow [3][2]\mathrm{p}, [2]\mathrm{p} \Rightarrow [4][1]\mathrm{p}, [2]\mathrm{p} \Rightarrow [5]\mathrm{p}$. Each modal connective of $\mathcal{L}$ corresponds to a PDL modal connective (see e.g. (Demri 2000)). For example, [2] in $\mathcal{L}$ corresponds to the PDL modal connective $[(\mathsf{c}_4;\mathsf{c}_3)^*;(\mathsf{c}_2 \cup (\mathsf{c}_4;\mathsf{c}_1) \cup \mathsf{c}_5)]$. Though the PDL equivalent of [2] contains the star operator, in the paper we show that $\mathcal{L}$-satisfiability is in **PSPACE** (Theorem 8.1(1)). By contrast, the bimodal logic $K_2 + [1]\mathrm{p} \Rightarrow [1][2]\mathrm{p}$ is

EXPTIME-complete (see Theorem 7.5). More generally, we wonder which grammar logics are in **PSPACE**. In the paper, by proof-theoretical means we characterize a class of right linear grammar logics that are in **PSPACE**.

A standard way to find **PSPACE** upper bound for modal logics consists in designing sound and complete tableaux-like calculi (see e.g. (Kripke 1963, Ladner 1977, Halpern and Moses 1992, Basin et al. 1997, Massacci 1998, Baader and Sattler 2000, Marx et al. 2000a)) augmented with adequate strategies, most of the time depth-first visit of the proof tree with a controlled amount of contractions. One can however distinguish the works that establish **PSPACE** upper bounds but not necessarily the tightest ones (see (Ladner 1977, Halpern and Moses 1992)) from the works that improve the space function by reducing the exponents of the polynomials (see e.g. (Huldelmaier 1993, Hudelmaier 1996)). This is also sometime a matter of natural chronology as determining the decidability status of a logic may precede its computational complexity characterization. The present work belongs rather to the first category since we wish to establish **PSPACE** complexity upper bounds by proof-theoretical means in a uniform way. Although we know that improvements are possible in many cases, we rather concentrate on the gain of generality and uniformity. Moreover, we want to refine the borderline between **EXPTIME**-hard right linear logics and **PSPACE** right linear logics in order to partially answer to the following question inspired from (Vardi 1997, Grädel 1999): why so many (multi)modal logics with weak forms of recursion are in **PSPACE**?

For any right linear grammar logic, we shall define an additive sequent calculi that is proved to be sound and complete. In the spirit of (Ohnishi and Matsumoto 1957), the calculi use neither labels nor a generalized form of sequents. Other kinds of sequent-style calculi for these logics already exist in the literature, see e.g. (Kracht 1996, Szałas 1996, Basin et al. 1998, Baldoni 1998) and we believe that the present formulation of the calculi is quite adequate to find complexity upper bounds mainly because of our treatment of contraction. Then, we show that given a right linear grammar $\mathcal{G}$ and a modal formula ϕ, deciding whether the formula is satisfiable in the extension of K_m with axiom schemes from $\mathcal{G}$ can be done in deterministic exponential-time in the size of $\mathcal{G}$ and ϕ. We refer to this problem as the *general satisfiability problem* for right linear grammar logics. The complexity upper bound is established by using a standard loop checking method dual to the one in (Pratt 1979). An extension is also presented for the global logical consequence problems. This improves upper bounds from (Baldoni 1998, Baldoni et al. 1998). We also easily show that the general satisfiability problem for right linear grammar logics is **EXPTIME**-hard by exhibiting a decidable countably infinite class of **EXPTIME**-hard right linear grammar logics. Further classes of **EXPTIME**-hard grammar logics can be found in the companion paper (Demri 2000).

In the second part of the paper, we propose a characterization of **PSPACE** decision procedures from the sequent calculi that allows us to show uniformly that all the right linear grammar logics from some identified countably infinite classes of logics are in **PSPACE**. We have indeed found decidable sufficient syntactic properties of the right linear grammars that guarantee that the generated logics from the grammars are in **PSPACE**. All the complexity upper bounds established in the paper are obtained by analyzing proofs in the sequent calculi and thus this follows the proof-theoretic alternative described in Section 8 in (van Benthem 2000) to explain the algorithmic behavior of modal logics.

For instance, we are able to show that given a bimodal extension of K_2 obtained from K_2 by additing axiom schemes from either a left linear or a right linear grammar, deciding whether the satisfiability problem of the logic is in **PSPACE** can be done in linear-time in the size of the grammar. Although the right linear grammars generate the same class of languages as the left linear grammars, this correspondence is not relevant at the level of grammar logics.

2 Logics

Given the set PRP $= \{\mathrm{p}_i : i \in \mathbb{N}\}$ of propositional variables, the set FORM of modal formulae is defined as the smallest set such that PRP $\subseteq$ FORM and, if $\phi, \psi \in$ FORM, then $\phi \wedge \psi \in$ FORM, $\neg\phi \in$ FORM and for $i \geq 1$, $[i]\phi \in$ FORM. For $m \geq 1$, we write L_m to denote the restriction of the modal language to the modal connectives in $\{[i] : i \in \{1, \ldots, m\}\}$. Standard abbreviations include $\vee$, $\Rightarrow$, $\langle i \rangle$. A **necessity formula** is a formula of the form $[i]\phi$ for some $i \geq 1$. The set $\mathrm{sub}(\phi)$ of **subformulae** of the formula ϕ is defined in the standard way. The **modal depth** of an occurrence of a formula ψ in ϕ is the number of occurrences of modal connectives that dominate ψ in ϕ. We write $\mathrm{md}(\phi)$ to denote the maximal modal depth of the subformulae of ϕ. An occurrence of the subformula ψ in ϕ is **positive** [resp. **negative**] $\overset{\text{def}}{\Leftrightarrow}$ it is in the scope of an even [resp. odd] number of negations. The **possibility weight** [resp. **necessity weight**] of a formula ϕ, denoted $\mathrm{pw}(\phi)$ [resp. $\mathrm{nw}(\phi)$], is the number of occurrences of subformulae of the form $[i]\psi$ with negative [resp. positive] polarity. The notation Γ, ϕ, where Γ is a finite multi-set of formulae and ϕ is a formula, designates a multi-set which is the union of Γ with the singleton multi-set containing only ϕ. Let f be a map

$f : \mathrm{FORM} \to \mathbb{N}$. If $\Gamma = \phi_1, \ldots, \phi_n$ is a finite multi-set of formulae, by $f^+(\Gamma)$ we mean the natural number $f(\phi_1) + \ldots + f(\phi_n)$. Similarly, for $i \in \{1, \ldots, m\}$, for any multi-set $\Gamma = \phi_1, \ldots, \phi_n$ of formulae, we write $[i]\Gamma$ [resp. $\neg\Gamma$] to denote $[i]\phi_1, \ldots, [i]\phi_n$ [resp. $\neg\phi_1, \ldots, \neg\phi_n$]. We also write $Set(\Gamma)$ to denote the set of formulae occurring in Γ and $\phi \in \Gamma$ for $\phi \in Set(\Gamma)$.

For any $\mathtt{L}_m$-formula ϕ, we write $\mathrm{r}(\phi)$ to denote the **rank** of ϕ; that is, the number of occurrences of members of $\mathrm{PRP} \cup \{\neg, \wedge\} \cup \{[i] : 1 \leq i \leq m\}$. For example $\mathrm{r}(\mathrm{p} \wedge (\mathrm{q} \wedge \neg\mathrm{p})) = 6$. Under reasonable hypothesis, the length of an $\mathtt{L}_m$-formula ϕ, noted $|\phi|$, is in $\mathcal{O}(\mathrm{r}(\phi) \times (log\ \mathrm{r}(\phi) + log\ m))$. As usual in complexity theory, the extra logarithmic factor is due to the fact that we need an index of size $log\ \mathrm{r}(\phi)$ for the different propositional variables.

An $\mathtt{L}_m$**-frame** is a structure $\mathcal{F} = \langle W, R_1, \ldots, R_m \rangle$ such that W is a nonempty set and for $i \in \{1, \ldots, m\}$, R_i is a binary relation on W. An $\mathtt{L}_m$**-model** is a structure $\mathcal{M} = \langle W, R_1, \ldots, R_m, V \rangle$ such that $\langle W, R_1, \ldots, R_m \rangle$ is an $\mathtt{L}_m$-frame and V is a valuation $V : \mathrm{PRP} \to \mathcal{P}(W)$. The standard definition of the satisfiability relation $\models$ is omitted here (see e.g. (Blackburn et al. 2001)). An $\mathtt{L}_m$-formula ϕ is said to be **true** in the $\mathtt{L}_m$-model $\mathcal{M}$ (written $\mathcal{M} \models \phi$) $\stackrel{\text{def}}{\Leftrightarrow}$ for all $x \in W$, $\mathcal{M}, x \models \phi$. An $\mathtt{L}_m$-formula ϕ is said to be **true** in the $\mathtt{L}_m$-frame $\mathcal{F}$ (written $\mathcal{F} \models \phi$) $\stackrel{\text{def}}{\Leftrightarrow}$ ϕ is true in all the $\mathtt{L}_m$-models based on $\mathcal{F}$.

In this paper, a **modal logic** $\mathcal{L}$ is understood as a pair $\langle \mathtt{L}_m, \mathcal{S} \rangle$ where $\mathtt{L}_m$ is a modal language with m modal connectives and $\mathcal{S}$ is a nonempty class of $\mathtt{L}_m$-frames. The class $\mathcal{S}$ is usually defined in terms of properties that the relations in the frames of $\mathcal{S}$ are supposed to satisfy. An $\mathtt{L}_m$-formula is said to be $\mathcal{L}$**-satisfiable** $\stackrel{\text{def}}{\Leftrightarrow}$ there is an $\mathtt{L}_m$-model based on some $\mathcal{F} \in \mathcal{S}$ and $x \in W$ such that $\mathcal{M}, x \models \phi$. An $\mathtt{L}_m$-formula is said to be $\mathcal{L}$**-valid** $\stackrel{\text{def}}{\Leftrightarrow}$ for all the $\mathcal{L}$-models $\mathcal{M}$ based on some frame in $\mathcal{S}$, ϕ is true in $\mathcal{M}$. $\mathcal{L}$-satisfiability and $\mathcal{L}$-validity can be easily extended to finite sets of formulae understood as conjunctions.

3 Grammar Logics

For any alphabet Σ (finite set of symbols), we write Σ^* [resp. Σ^+] to denote the set of [resp. nonempty] finite strings built over elements of Σ. ϵ denotes the empty string and $u_1 \cdot u_2$ denotes the concatenation of two strings. For any finite string u, we write $|u|$ to denote its length. For any $u \in \Sigma^*$, we write u^k to denote the string composed of k copies of u. By convention, $u^0 = \epsilon$.

A **(formal) grammar** $\mathcal{G}$ is a quadruple $\mathcal{G} = \langle N, \Sigma, P, S \rangle$ such that N and Σ are disjoint finite sets of **nonterminal symbols** and **terminal**

symbols, respectively (in the paper we allow Σ empty). P is a finite set of **production rules**, each production rule is of the form $u \rightarrow v$ such that $u \in (N \cup \Sigma)^* N (N \cup \Sigma)^*$ and $v \in (N \cup \Sigma)^*$. Finally, $S \in N$ is a special symbol called the **start symbol** (see e.g. (Hopcroft and Ullman 1979)). For the grammar $\mathcal{G}$, the size of $\mathcal{G}$, denoted $|\mathcal{G}|$, is

$$|\mathcal{G}| \stackrel{\text{def}}{=} (card(N) + card(\Sigma) + \Sigma_{u \rightarrow v \in P}(|u \cdot v| + 1)) \times log(card(N) + card(\Sigma))$$

Let $\Rightarrow_{\mathcal{G}}$ be the direct derivation relation defined as the subset of $(N \cup \Sigma)^* \times (N \cup \Sigma)^*$ such that $u \Rightarrow_{\mathcal{G}} v \stackrel{\text{def}}{\Leftrightarrow}$ there is a production rule $u' \rightarrow v' \in P$ such that $u = u_1 \cdot u' \cdot u_2$, $v = u_1 \cdot v' \cdot u_2$, $u_1, u_2 \in (N \cup \Sigma)^*$. Let $\Rightarrow_{\mathcal{G}}^*$ be the reflexive and transitive closure of $\Rightarrow_{\mathcal{G}}$. For $i \in (N \cup \Sigma)$, we write $\mathtt{L}_i(\mathcal{G})$ to denote the set of strings $\{u \in \Sigma^* : i \Rightarrow_{\mathcal{G}}^* u\}$. For instance, for $i \in \Sigma$, $\mathtt{L}_i(\mathcal{G}) = \{i\}$. A grammar $\mathcal{G}$ is said to be **strongly finite** [resp. **finite**] $\stackrel{\text{def}}{\Leftrightarrow}$ for $i \in N$, $\{u \in (N \cup \Sigma)^* : i \Rightarrow_{\mathcal{G}}^* u\}$ is finite [resp. $\mathtt{L}_i(\mathcal{G})$ is finite]. It is possible that for some $i \in N$, $\mathtt{L}_i(\mathcal{G})$ is empty although $\mathcal{G}$ is not strongly finite.

In the rest of the paper we assume that each grammar $\langle N, \Sigma, P, S \rangle$ satisfies $N = \{1, \ldots, k\}$ for some $k \geq 1$, $\Sigma = \{k+1, \ldots, m\}$ for some $k \leq m$ (we allow Σ to be empty) and $S = 1$.

Let $\mathcal{G} = \langle N, \Sigma, P, S \rangle$ be a grammar and $\rightsquigarrow$ be the binary relation in $N \times (N \cup \Sigma)^*$ such that $i \rightsquigarrow u \stackrel{\text{def}}{\Leftrightarrow}$ either there is $j \rightarrow u \in P$ such that $i \Rightarrow_{\mathcal{G}}^* j$ or $u \in N$ and $i \Rightarrow_{\mathcal{G}}^* u$. If $\mathcal{G}$ is right [resp. left] linear, then by using the technique for eliminating the unit production rules (Hopcroft and Ullman 1979), one can compute in polynomial-time in $|\mathcal{G}|$ a right [resp. left] linear grammar $\mathcal{G}' = \langle N', \Sigma', P', S' \rangle$ such that $N = N'$, $\Sigma = \Sigma'$, $S = S'$, $P \subseteq P'$ and for $\langle i, u \rangle \in (N \cup \Sigma)^*$, $i \rightsquigarrow u$ in $\mathcal{G}$ iff $i \rightarrow u \in P'$. For $i \in N$, we have $\{u \in (N \cup \Sigma)^* : i \Rightarrow_{\mathcal{G}}^* u\} = \{u \in (N \cup \Sigma)^* : i \Rightarrow_{\mathcal{G}'}^* u\}$. In the case when $\mathcal{G}$ is either left linear or right linear (called **regular** in the sequel), $\rightsquigarrow$ can be computed in polynomial-time in $|\mathcal{G}|$ and for all $i \in N$, $\sum_{i \rightsquigarrow u} |u| \leq |\mathcal{G}|$. Although $\mathcal{G}$ and $\mathcal{G}'$ generate the same language ($\mathtt{L}_1(\mathcal{G}) = \mathtt{L}_1(\mathcal{G}')$), in the sequel we do not assume that the grammars are necessarily of the form of $\mathcal{G}'$. Indeed, grammars generating the same language, may engender different grammar logics. The binary relation $\rightsquigarrow$ is used in Section 4 to define sequent calculi.

Let $\mathcal{G}$ be a grammar and $\mathcal{S}$ be a class of $\mathtt{L}_m$-frames. We write $\mathcal{S}^{\mathcal{G}}$ to denote the subset of $\mathcal{S}$ such that for any $\mathcal{F} = \langle W, R_1, \ldots, R_m \rangle \in \mathcal{S}$, $\mathcal{F} \in \mathcal{S}^{\mathcal{G}} \stackrel{\text{def}}{\Leftrightarrow}$ for any production rule $i_1 \ldots i_{k'} \rightarrow j_1 \ldots j_{k''}$ in $\mathcal{G}$, $R_{j_1} \circ \ldots \circ R_{j_{k''}} \subseteq R_{i_1} \circ \ldots \circ R_{i_{k'}}$. For the logic $\mathcal{L}_m = \langle \mathtt{L}_m, \mathcal{S}_m \rangle$ where $\mathcal{S}_m$ is the class of all the $\mathtt{L}_m$-frames, we write $\mathcal{L}_m^{\mathcal{G}}$ to denote the logic $\langle \mathtt{L}_m, \mathcal{S}_m^{\mathcal{G}} \rangle$. $\mathcal{L}_m^{\mathcal{G}}$ is said to be a **grammar logic** (Fariñas del Cerro and Penttonen 1988).

For any string $u = i_1 \cdot \ldots \cdot i_n$ in $\{1, \ldots, m\}^*$, we write R_u to denote $R_{i_1} \circ \ldots \circ R_{i_n}$. When $u = \epsilon$, $R_u \stackrel{\text{def}}{=} \{\langle x, x\rangle : x \in W\}$. Moreover, we write $[u]\phi$ to denote the L_m-formula $[i_1] \ldots [i_n]\phi$ where $u = i_1 \cdot \ldots \cdot i_n$. If $u = \epsilon$, then $[u]\phi$ is simply ϕ.

Theorem 3.1 *Let $\mathcal{G} = \langle N, \Sigma, P, S\rangle$. For $u, v \in (N \cup \Sigma)^*$, (I) $u \Rightarrow^*_{\mathcal{G}} v$ iff (II) $[u]\mathrm{p} \Rightarrow [v]\mathrm{p}$ is $\mathcal{L}^{\mathcal{G}}_m$-valid iff (III) for all $\mathcal{L}^{\mathcal{G}}_m$-models $R_v \subseteq R_u$.*

The equivalence between (II) and (III) is a classical correspondence result in modal logic theory (see e.g. (van Benthem 1984)). (I) implies (II) can be proved by induction on the length of the derivation whereas (II) implies (I) can be shown by using part of the proof of Theorem 3 in (Chagrov and Shehtman 1994). In order to study the grammar logic $\mathcal{L}^{\mathcal{G}}_m$, what is essential is the value of the set P of production rules whereas once P is fixed, the value of the start symbol S and the distribution of the terminal and nonterminal symbols are immaterial for $\mathcal{L}^{\mathcal{G}}_m$-satisfiability. Hence, semi-Thue rewriting systems are also appropriate to define grammar logics.

The **general satisfiability problem** GSP(REG) [resp. GSP(LIN), GSP(RLIN_f), GSP(RLIN)] for regular grammar [resp. linear grammar, finite right linear grammar, right linear grammar] logics is defined as follows:

- Inputs: a regular [resp. right linear, finite right linear, linear] grammar $\mathcal{G}$ and an L_m-formula ϕ;
- Question: Is ϕ $\mathcal{L}^{\mathcal{G}}_m$-satisfiable?

The above general satisfiability problems can be viewed as syntactic variants of satisfiability problems for fragments of the well-known description logic $\mathcal{ALC}$ augmented with role value maps (see details in (Demri 2000)).

It is known that the multimodal logic K_m, $m \geq 1$, has a **PSPACE**-complete satisfiability problem (see e.g. (Halpern and Moses 1992)). Adding a regular set of modal axioms preserves the **PSPACE** complexity lower bound.

Theorem 3.2 *Let $\mathcal{G}$ be either a regular grammar or a context-free grammar with a nonempty set of terminal symbols. Then, $\mathcal{L}^{\mathcal{G}}_m$-satisfiability is* **PSPACE**-*hard.*

A natural proof consists in reducing satisfiability for either the modal logic K or the modal logic T into $\mathcal{L}^{\mathcal{G}}_m$-satisfiability. The only difficulty in the proof is to show that for any binary relation R there is an $\mathcal{L}^{\mathcal{G}}_m$-frame with $R_m = R$. Additionally, GSP(LIN) is undecidable. This can

$$\Gamma, \phi \vdash \Delta, \phi \text{ (initial sequents)} \quad \frac{\Gamma \vdash \Delta, \phi}{\Gamma, \neg\phi \vdash \Delta} (\neg \vdash) \quad \frac{\Gamma, \phi \vdash \Delta}{\Gamma \vdash \Delta, \neg\phi} (\vdash \neg)$$

$$\frac{\Gamma, \phi_1, \phi_2 \vdash \Delta}{\Gamma, \phi_1 \wedge \phi_2 \vdash \Delta} (\wedge \vdash) \quad \frac{\Gamma \vdash \Delta, \phi_1 \quad \Gamma \vdash \Delta, \phi_2}{\Gamma \vdash \Delta, \phi_1 \wedge \phi_2} (\vdash \wedge)$$

FIGURE 1 **Initial sequents and standard rules for propositional connectives**

be proved by reducing the problem of empty intersection between linear languages into GSP(LIN) (see e.g. (Rozenberg and Salomaa 1994)) by using either prefixed tableaux calculi (Baldoni et al. 1998) or the equational characterization of context-free languages (Demri 2000).

4 Sequent calculi

4.1 Definitions

Let $\mathcal{G}$ be a right linear grammar. The basic syntactic objects in the calculi are **sequents.** A **sequent** is an expression of the form $\Gamma \vdash \Delta$ where Γ and Δ are finite multi-sets of formulae, Γ is the **antecedent** and Δ the **succedent**. We write SEQ_m to denote the class of all sequents built over the modal language L_m. The additive sequent calculus $\mathrm{G}\mathcal{L}_m^{\mathcal{G}}$ for the logic $\mathcal{L}_m^{\mathcal{G}}$ contains the rules from Figure 1 for the propositional fragment of $\mathcal{L}_m^{\mathcal{G}}$. The rules are read upwards and other rules depending of $\mathcal{G}$ are presented below.

The left-hand side introduction rule $[i]$ is defined below only if $i \Rightarrow_{\mathcal{G}}^* \epsilon$:

$$\frac{\Gamma, [i]\phi, \phi \vdash \Delta}{\Gamma, [i]\phi \vdash \Delta} ([i] \vdash)$$

The right-hand side introduction rule for $[i]$ is defined as follows. For $i \in \{1, \dots, k\}$, let $\rightsquigarrow (i) = \{u_{i,1}, \dots, u_{i,l_i}\}$ where given a binary relation R on U and $x \in U$, $R(x) \stackrel{\text{def}}{=} \{y \in U : \langle x, y \rangle \in R\}$. For $j \in \{1, \dots, m\}$ and for $i \in \{1, \dots, k\}$, let $start_j^i = \{\langle i, k' \rangle : u_{i,k'} = j \cdot v_{i,k'}\}$. Let $i \in \{k+1, \dots, m\}$. The $(\vdash [i])$ rule is defined as follows:

$$\frac{\bigcup_{\langle 1,k' \rangle \in start_i^1} [v_{1,k'}]\Gamma_1, \dots, \bigcup_{\langle k,k' \rangle \in start_i^k} [v_{k,k'}]\Gamma_k, \Gamma_i \vdash \phi}{\Gamma', [1]\Gamma_1, \dots, [k]\Gamma_k, [i]\Gamma_i \vdash [i]\phi, \Delta} (\vdash [i])$$

Moreover, we assume that in Γ', there is no formula of the form $[j]\psi$ for some $j \in \{1, \dots, k\} \cup \{i\}$. It remains to define for $i \in \{1, \dots, k\}$, the $(\vdash [i])$ rule:

$$\frac{\Gamma_1', \dots, \Gamma_k' \vdash \phi}{\Gamma', [1]\Gamma_1, \dots, [k]\Gamma_k \vdash [i]\phi, \Delta} (\vdash [i])$$

where for $j \in \{1, \ldots, k\}$, if $j \rightsquigarrow i$, then $\Gamma'_j \stackrel{\text{def}}{=} \Gamma_j$, otherwise $\Gamma'_j \stackrel{\text{def}}{=} \emptyset$ (empty multi-set). Moreover, we assume that in Γ', there is no formula of the form $[j]\psi$ for some $j \in \{1, \ldots, k\}$.

For $i \in \{1, \ldots, m\}$ we abbreviate the $(\vdash [i])$ rule by

$$\frac{\Gamma^{[i]} \vdash \phi}{\Gamma \vdash [i]\phi, \Delta} \ (\vdash [i])$$

where the appropriate definition of $\Gamma^{[i]}$ from Γ is immediate from the above cases.

The $([i] \vdash)$ rules and the $(\vdash [i])$ rules are defined from the strings u satisfying $i \rightsquigarrow u$ and not only from the ones satisfying $i \to u \in P$. This is the price we may have to pay in the $(\vdash [i])$ rules since we want to introduce a $[i]$-formula at the right-hand side and to take into account the grammatical properties of the logic simultaneously.

An implicit contraction is operated in the $([i] \vdash)$-rule (if $i \Rightarrow^*_{\mathcal{G}} \epsilon$) whereas implicit contractions can be found also in the applications of the $(\vdash [i])$-rule but this depends on the structure of $\mathcal{G}$ (see e.g. Section 8 for calculi with no implicit contractions). In order to get the **PSPACE** upper bounds, our main task is to control contraction and this requires a careful analysis.

Example 4.1 Let $\mathcal{G}^{rl} = \langle \{1,2\}, \{3,4\}, \{1 \to 3{\cdot}3{\cdot}1, 1 \to 2, 1 \to 4{\cdot}2, 2 \to 4 \cdot 4 \cdot 4 \cdot 2\}, 1 \rangle$ be a right linear grammar. The $(\vdash [4])$ rule is defined as follows:

$$\frac{[2]\Gamma_1, [4][4][2](\Gamma_1, \Gamma_2), \Gamma_4 \vdash \phi}{\Gamma', [1]\Gamma_1, [2]\Gamma_2, [4]\Gamma_4 \vdash [4]\phi, \Delta} \ (\vdash [4])$$

Observe the duplication of Γ_1.

As is usual, a **proof** Π in $\mathrm{G}\mathcal{L}^{\mathcal{G}}_m$ is a tree whose nodes are labelled by sequents satisfying the following conditions: the topmost sequents of Π are initial sequents and every sequent of Π, except the lowest one is an upper sequent of an inference whose lower sequent is also in Π. A sequent $\Gamma \vdash \Delta$ is **provable** in $\mathrm{G}\mathcal{L}^{\mathcal{G}}_m \stackrel{\text{def}}{\Leftrightarrow}$ there is a proof where the lowest sequent is $\Gamma \vdash \Delta$. A formula ϕ is **provable in** $\mathrm{G}\mathcal{L}^{\mathcal{G}}_m \stackrel{\text{def}}{\Leftrightarrow}$ the sequent $\emptyset \vdash \phi$ (also noted $\vdash \phi$) is provable in $\mathrm{G}\mathcal{L}^{\mathcal{G}}_m$. A sequent $\Gamma \vdash \Delta$ is **consistent** $\stackrel{\text{def}}{\Leftrightarrow}$ $\Gamma \vdash \Delta$ is not provable in $\mathrm{G}\mathcal{L}^{\mathcal{G}}_m$.

A sequent $\Gamma \vdash \Delta$ is **downward saturated** $\stackrel{\text{def}}{\Leftrightarrow}$ for $\phi \in \Gamma$ and for $\psi \in \Delta$:

- $\phi = \phi_1 \wedge \phi_2$ implies $\phi_1, \phi_2 \in \Gamma$; $\phi = [i]\phi_1$ and $i \Rightarrow^*_{\mathcal{G}} \epsilon$ imply $\phi_1 \in \Gamma$;
- $\phi = \neg\phi_1$ implies $\phi_1 \in \Delta$; $\psi = \neg\psi_1$ implies $\psi_1 \in \Gamma$;
- $\psi = \psi_1 \wedge \psi_2$ implies either $\psi_1 \in \Delta$ or $\psi_2 \in \Delta$.

A **derivation** is defined as a proof except that the topmost sequents are not necessarily initial sequents. The derivations are supposed to grow upwards. If Π is a derivation of $\Gamma \vdash \Delta$ we write $\sigma = \Gamma_0 \vdash \Delta_0 \prec (r_1)\Gamma_1 \vdash \Delta_1 \prec (r_2) \ldots \prec (r_{n-1})\Gamma_n \vdash \Delta_n$ to denote the fact that there is an initial segment σ of a branch in Π from the root $\Gamma_0 \vdash \Delta_0 = \Gamma \vdash \Delta$ such that for $i \in \{0, \ldots, n-1\}$, $\Gamma_{i+1} \vdash \Delta_{i+1}$ is one of the premisses of the inference of the rule (r_{i+1}) with conclusion $\Gamma_i \vdash \Delta_i$. We omit to write the r_i's when they are of no use. We write $Ant(\sigma)$ [resp. $Suc(\sigma)$] to denote the antecedent set $\bigcup_{0 \leq i \leq n} Set(\Gamma_i)$ [resp. the succedent set $\bigcup_{0 \leq i \leq n} Set(\Delta_i)$]. We write $last(\sigma)$ [resp. $first(\sigma)$] to denote $\Gamma_n \vdash \Delta_n$ [resp. $\Gamma_0 \vdash \Delta_0$]. The sequence σ is said to be **local** $\overset{\text{def}}{\Leftrightarrow}$ no r_i is the right-hand side introduction rule $(\vdash [j])$ for some $j \in \{1, \ldots, m\}$. The sequence σ is said to be **consistent** $\overset{\text{def}}{\Leftrightarrow}$ all the $\Gamma_i \vdash \Delta_i$ are consistent. The sequence σ is said to be **maximal** $\overset{\text{def}}{\Leftrightarrow}$ σ is local and $Ant(\sigma) \vdash Suc(\sigma)$ is downward saturated. The maximal and consistent sequences play the role of downward saturated sets in the standard terminology for tableaux (see e.g. (Goré 1999) for further details and historical notes). This complication is due to the fact that we consider multi-sets instead of sets in the sequents. The reward is that we can more easily control contraction and this shall be helpful to get **PSPACE** complexity upper bounds.

4.2 Properties

Let ϕ be a formula. The closure of ϕ with respect to $\mathcal{G}$ is the smallest set $\mathrm{cl}_{\mathcal{G}}(\phi)$ of formulae such that $\mathrm{cl}_{\mathcal{G}}(\phi)$ is closed under subformulae, $\mathrm{sub}(\phi) \subseteq \mathrm{cl}_{\mathcal{G}}(\phi)$ and if $i \rightsquigarrow u$ and $[i]\psi \in \mathrm{cl}_{\mathcal{G}}(\phi)$, then $[u]\psi \in \mathrm{cl}_{\mathcal{G}}(\phi)$. One can prove that $card(\mathrm{cl}_{\mathcal{G}}(\phi))$ is bounded by $|\mathcal{G}| \times \mathrm{r}(\phi)$. We write $\mathrm{SEQ}(\phi)$ to denote the set of sequents $\Gamma \vdash \Delta$ such that $\bigcup_{\psi \in \Gamma, \Delta} \mathrm{cl}_{\mathcal{G}}(\psi) \subseteq \mathrm{cl}_{\mathcal{G}}(\phi)$.

Lemma 4.2 *Let* $\Gamma \vdash \Delta$ *be a sequent. Then, every formula occurring in a derivation of* $\Gamma \vdash \Delta$ *belongs to* $\bigcup_{\psi \in \Gamma, \Delta} \mathrm{cl}_{\mathcal{G}}(\psi)$.

The (easy) proof is by induction on the depth of the proof tree. Following for instance the terminology from (Goré 1999), $\mathrm{G}\mathcal{L}_m^{\mathcal{G}}$ has therefore the analytical superformula property and obviously $\mathrm{G}\mathcal{L}_m^{\mathcal{G}}$ does not have necessarily the subformula property.

Lemma 4.3 $\Gamma \vdash \Delta$ *is a provable sequent in* $\mathrm{G}\mathcal{L}_m^{\mathcal{G}}$ *iff* $\Gamma \vdash \Delta$ *has a proof in* $\mathrm{G}\mathcal{L}_m^{\mathcal{G}}$ *such that all the initial sequents are of the form* $\Gamma', \mathrm{p} \vdash \mathrm{p}, \Delta'$ *where* p *is a propositional variable.*

The proof of Lemma 4.3 is standard. It is sufficient to show in the induction step that for every initial sequent $\Gamma', \phi \vdash \phi, \Delta'$ with $\mathrm{r}(\phi) \geq 2$, there is a proof of $\Gamma', \phi \vdash \phi, \Delta'$ such that all the initial sequents are of the form $\Gamma', \psi \vdash \psi, \Delta'$ with $\mathrm{r}(\psi) < \mathrm{r}(\phi)$.

A rule is **invertible** $\overset{\text{def}}{\Leftrightarrow}$ for every inference of the rule, the conclusion has a proof iff the premises have proofs.

Lemma 4.4 *The rules* $(\wedge \vdash)$, $(\vdash \wedge)$, $(\neg \vdash)$, $(\vdash \neg)$ *and* $([i] \vdash)$ *if* $i \Rightarrow^*_{\mathcal{G}} \epsilon$ *are invertible.*

The proof of Lemma 4.4 uses Lemma 4.3 and is not difficult to show. For instance, invertibility of $([i] \vdash)$ if $i \Rightarrow^*_{\mathcal{G}} \epsilon$, is immediate.

Lemma 4.5 *Let* $\Gamma \vdash \Delta$ *be a consistent sequent. Then, there is a maximal and consistent sequence* $\sigma = \Gamma_0 \vdash \Delta_0 \prec \Gamma_1 \vdash \Delta_1 \prec \ldots \prec \Gamma_n \vdash \Delta_n$ *with* $\Gamma_0 \vdash \Delta_0 = \Gamma \vdash \Delta$.

Proof. Since $\Gamma \vdash \Delta$ is consistent, we know that no proof of $\Gamma \vdash \Delta$ exists and there is no propositional variable occurring in both Γ and Δ. Let $\Gamma_0 \vdash \Delta_0 = \Gamma \vdash \Delta$. If $\Gamma_0 \vdash \Delta_0$ is downward saturated, then we are done. Now suppose that $\sigma_i = \Gamma_0 \vdash \Delta_0 \prec \Gamma_1 \vdash \Delta_1 \prec \ldots \prec \Gamma_i \vdash \Delta_i$ is a local sequence, $Ant(\sigma_i) \vdash Suc(\sigma_i)$ is not downward saturated and each $\Gamma_j \vdash \Delta_j, 0 \leq j \leq i$, is consistent. For instance, suppose that $\phi_1 \wedge \phi_2 \in Ant(\sigma_i)$ and $\phi_1, \phi_2 \notin Ant(\sigma_i)$. Hence, $\phi_1 \wedge \phi_2 \in \Gamma_i$ since otherwise $\phi_1, \phi_2 \in \Gamma_{i'}$ for some $i' < i$. Apply the $(\wedge \vdash)$-rule to an occurrence of $\phi_1 \wedge \phi_2$ in $\Gamma_i \vdash \Delta_i$ and let $\Gamma_{i+1} \vdash \Delta_{i+1}$ be $(\Gamma_i \setminus \{\phi_1 \wedge \phi_2\}), \phi_1, \phi_2 \vdash \Delta_i$. Since the $(\wedge \vdash)$-rule is invertible (see Lemma 4.4), $\Gamma_{i+1} \vdash \Delta_{i+1}$ is also consistent. If $\Gamma_0 \vdash \Delta_0 = \Gamma \vdash \Delta$ and $Ant(\sigma_i) \vdash Suc(\sigma_i)$ is not downward saturated for some other reason, we use a similar reasoning with obvious adaptations (we may have to choose between two branches). Lemma 4.2 guarantees termination and the length of σ can be bounded by $card(\mathrm{cl}_{\mathcal{G}}(\Gamma, \Delta))$. $\dashv$

4.3 Completeness

Following (Rautenberg 1983) (see also (Goré 1999)), we introduce the central notion of model graph. We prove completeness using the well-known technique due to Schütte (see e.g. (Schütte 1967, Takeuti 1975)).

Definition 4.6 A **model graph** for some sequent $\Gamma \vdash \Delta$ is an L_m-frame of the form $\langle W, R_{1+m}, \ldots, R_{2\times m}\rangle$ such that W is a countable set of maximal and consistent sequences such that

1. for $\sigma \in W$, $Ant(\sigma) \cup Suc(\sigma) \subseteq \bigcup_{\psi \in \Gamma, \Delta} \mathrm{cl}_{\mathcal{G}}(\psi)$;
2. there is σ_0 such that $Set(\Gamma) \subseteq Ant(\sigma_0)$ and $Set(\Delta) \subseteq Suc(\sigma_0)$;
3. for $\sigma \in W$, if $[i]\phi \in Suc(\sigma)$ for some $i \in \{1, \ldots, m\}$, then there is $\sigma' \in W$ such that $\sigma R_{i+m} \sigma'$ and $\phi \in Suc(\sigma')$;

4. for $\sigma, \sigma' \in W$, if $\sigma R_{i+m} \sigma'$ and $j \rightsquigarrow i \cdot u$ for some $\langle i, j \rangle \in \{1, \ldots, m\} \times \{1, \ldots, k\}$, and $[j]\phi \in Ant(\sigma)$, then $[u]\phi \in Ant(\sigma')$;
5. for $\sigma, \sigma' \in W$, if $\sigma R_{i+m} \sigma'$ for some $i \in \{k+1, \ldots, m\}$, and $[i]\phi \in Ant(\sigma)$, then $\phi \in Ant(\sigma')$.

A direct consequence of Definition 4.6(4) is that Definition 4.6(5) holds true even for $i \in \{1, \ldots, k\}$. The cornerstone of the completeness proof is the following result.

Theorem 4.7 *If there is a model graph for $\Gamma \vdash \Delta$, then $Set(\Gamma) \cup Set(\neg\Delta)$ is $\mathcal{L}^{\mathcal{G}}_m$-satisfiable.*

Proof. (sketch) For $i \in \{1, \ldots, k\}$, there is $l_i \geq 0$ such that $i \rightarrow u_{i,1}, \ldots, i \rightarrow u_{i,l_i}$ are the only production rules in P having i as left-hand side. Let W be a countable non-empty set and $R_{m+1}, \ldots, R_{2\times m}$ be binary relations on W. Let $f : \mathcal{P}(W \times W)^m \rightarrow \mathcal{P}(W \times W)^m$ be the map such that

$$f(R_1, \ldots, R_m) \stackrel{\text{def}}{=}$$

$$\langle (R_{1+m} \cup \bigcup_{1 \leq j \leq l_1} R_{u^\star_{1,j}}), \ldots, (R_{k+m} \cup \bigcup_{1 \leq j \leq l_k} R_{u^\star_{k,j}}), R_{k+1+m}, \ldots, R_{2\times m} \rangle$$

where $u^\star_{i,j}$ is a string obtained from $u_{i,j}$ by replacing $i \in \{k+1, \ldots, m\}$ by $i+m$. Let $\leq$ be the binary relation on $\mathcal{P}(W \times W)^m$ defined as follows: $\langle R_1, \ldots, R_m \rangle \leq \langle R'_1, \ldots, R'_m \rangle \stackrel{\text{def}}{\Leftrightarrow}$ for all $i \in \{1, \ldots, m\}$, $R_i \subseteq R'_i$. The structure $\langle \mathcal{P}(W \times W)^m, \leq \rangle$ is a complete lattice and f is continuous and order-preserving. By Kleene's Theorem, the least fixed point of f exists and is equal to

$$\mu(f) = \bigcup_{i \in \mathbb{N}} f^i(\emptyset, \ldots, \emptyset) \stackrel{\text{def}}{=} \langle \mathcal{R}_1, \ldots, \mathcal{R}_m \rangle.$$

By construction, $\langle W, \mathcal{R}_1, \ldots, \mathcal{R}_m \rangle$ is an $\mathcal{L}^{\mathcal{G}}_m$-frame such that for $i \in \{1, \ldots, m\}$, $R_{i+m} \subseteq \mathcal{R}_i$ and $\mathcal{R}_i \subseteq (R_{m+1} \cup \ldots \cup R_{2\times m})^*$.

Let $\langle W, R_{1+m}, \ldots, R_{2\times m} \rangle$ be a model graph for $\Gamma \vdash \Delta$. We define an $\mathcal{L}^{\mathcal{G}}_m$-model $\mathcal{M} = \langle W, \mathcal{R}_1, \ldots, \mathcal{R}_m, V \rangle$ as follows: $\langle \mathcal{R}_1, \ldots, \mathcal{R}_m \rangle$ is the least fixed point of f defined with $R_{1+m}, \ldots, R_{2\times m}$ and for any propositional variable p, $V(\mathrm{p}) \stackrel{\text{def}}{=} \{\sigma \in W : \mathrm{p} \in Ant(\sigma)\}$.

By induction on formulae we can show that for all $\sigma \in W$, for $\phi \in Ant(\sigma) \cup Suc(\sigma)$, if $\phi \in Ant(\sigma)$, then $\mathcal{M}, \sigma \models \phi$ otherwise $\mathcal{M}, \sigma \not\models \phi$. Moreover, there is $\sigma_0 \in W$ such that $Set(\Gamma) \subseteq Ant(\sigma_0)$ and $Set(\Delta) \subseteq Suc(\sigma_0)$. Consequently, $\mathcal{M}, \sigma_0 \models Set(\Gamma) \cup Set(\neg\Delta)$. $\dashv$

Theorem 4.8 *For any sequent $\Gamma \vdash \Delta$, $(\bigwedge_{\phi \in \Gamma} \phi) \Rightarrow (\bigvee_{\phi \in \Delta} \phi)$ is $\mathcal{L}^{\mathcal{G}}_m$-valid iff the sequent $\Gamma \vdash \Delta$ is provable in $\mathrm{G}\mathcal{L}^{\mathcal{G}}_m$.*

Proof. The soundness proof is standard by using an induction on the depth of the proof tree. In order to prove completeness, we assume that $(\bigwedge_{\phi\in\Gamma}\phi) \Rightarrow (\bigvee_{\phi\in\Delta}\phi)$ is $\mathcal{L}^{\mathcal{G}}_m$-valid and suppose that $\Gamma \vdash \Delta$ is consistent. The first step is to create a maximal and consistent sequence σ_0 starting with $\Gamma \vdash \Delta$ (see Lemma 4.5). Since σ_0 is consistent, $last(\sigma_0)$ is consistent. We use this fact to construct an $\mathtt{L}_m$-frame whose infinite limit will be a graph model. This is a meta-level construction for which we can visit all derivations for $last(\sigma_0)$, choosing nodes at will, since all such derivations cannot be completed as proofs. We use the successor relations R_i, $1 \leq i \leq m$ while building this frame. By Theorem 4.7, $(\bigwedge_{\phi\in\Gamma}\phi)\wedge(\bigwedge_{\phi\in\Delta}\neg\phi)$ is $\mathcal{L}^{\mathcal{G}}_m$-satisfiable which will lead to a contradiction.

Let us show how to build the model graph. If no $[i]\phi$ occurs in the succedent part of $last(\sigma_0)$, then the structure $\langle\{\sigma_0\},\emptyset,\ldots,\emptyset\rangle$ is a model graph for $\Gamma \vdash \Delta$. Otherwise, for $i \in \{1,\ldots,m\}$, let $\psi_i^1,\ldots,\psi_i^{s_i}$ be all the formulae such that $[i]\psi_i^j$ occurs in the succedent part of $last(\sigma_0)$. Let Γ' be the antecedent part of $last(\sigma_0)$. Since $last(\sigma_0)$ is consistent, for $i \in \{1,\ldots,m\}$, for $j \in \{1,\ldots,s_i\}$, $\Gamma'^{[i]} \vdash \psi_i^j$ is also consistent. For $i \in \{1,\ldots,m\}$, for $j \in \{1,\ldots,s_i\}$, create a maximal and consistent sequence $\sigma_{i,j}$ starting with $\Gamma'^{[i]} \vdash \psi_i^j$ and put $\sigma_0 R_i \sigma_{i,j}$. The sequences $\sigma_{i,j}$ are said to be of level 1. $\sum_{i=1}^m s_i$ is bounded by $|\mathcal{G}| \times (\sum_{\psi\in\Gamma}\mathrm{r}(\psi) + \sum_{\psi\in\Delta}\mathrm{r}(\psi))$ and this shall hold true at any level. Continue to create the nodes of further levels in a similar way. Either this procedure can go forever (but the infinite limit frame is a model graph for $\Gamma \vdash \Delta$) or the procedure stops after a finite amount of time (the number of level is finite as well as the resulting model graph). ⊣

5 GSP(RLIN) is EXPTIME-complete

No single rule ($\vdash [i]$) for some $i \in \{1,\ldots,m\}$ in $\mathrm{G}\mathcal{L}^{\mathcal{G}}_m$ captures all the properties of the relation R_i in the $\mathcal{L}^{\mathcal{G}}_m$-frames unlike the combination of all the introduction rules for necessity formulae. As in the *single steps calculi* in (Massacci 1994) (see also (Goré 1999, Massacci 2000)), the closure property of the $\mathcal{L}^{\mathcal{G}}_m$-frames (see the proof of Theorem 4.7) is encoded step by step and this is the key point to characterize the complexity of GSP(RLIN). That is why, we can improve the complexity upper bound of GSP(RLIN) from (Baldoni 1998, Baldoni et al. 1998).

Theorem 5.1 *GSP(RLIN) is in* **EXPTIME**.

Proof. We use a technique that is dual to the method in (Pratt 1979) that shows that PDL satisfiability is in **EXPTIME**. By Theorem 4.8, ϕ is $\mathcal{L}^{\mathcal{G}}_m$-satisfiable iff $\neg\phi$ is not provable in $\mathrm{G}\mathcal{L}^{\mathcal{G}}_m$ and therefore we concentrate on validity instead of satisfiability since our procedure is deterministic. Actually, we use a variant of $\mathrm{G}\mathcal{L}_m^{\mathcal{G}}$, namely $\mathrm{SETG}\mathcal{L}_m^{\mathcal{G}}$, where the sequents are pairs of finite sets (instead of finite multi-sets). A contraction is explicitly operated for the rules $(\vdash \wedge)$, $(\wedge \vdash)$, $(\neg \vdash)$ and $(\vdash \neg)$. Following the developments of Section 4, one can show that $X \vdash Y$ is provable in $\mathrm{SETG}\mathcal{L}_m^{\mathcal{G}}$ iff the formula $(\bigwedge_{\phi \in X} \phi) \Rightarrow (\bigvee_{\phi \in Y} \phi)$ is $\mathcal{L}_m^{\mathcal{G}}$-valid. Let ϕ be a formula for which we want to know whether ϕ is $\mathcal{L}_m^{\mathcal{G}}$-valid. The rules of the proof system $\mathrm{SETG}\mathcal{L}_m^{\mathcal{G}}$ can be computed in polynomial-time in $|\mathcal{G}|$ since the binary relation $\rightsquigarrow$ can be computed in polynomial-time in $|\mathcal{G}|$ and all the rules are of size in $\mathcal{O}(|\mathcal{G}|)$. Their applicability can be checked in polynomial-time in $|\mathcal{G}|$ and in the size of the premiss and conclusion sequents. The cost of the computation of $\mathrm{SETG}\mathcal{L}_m^{\mathcal{G}}$ is relevant here since $\mathcal{G}$ is part of the inputs.

The cardinality of the set $\mathrm{cl}_{\mathcal{G}}(\phi)$ is at most $|\mathcal{G}| \times \mathrm{r}(\phi)$. Let $\mathrm{SETSEQ}(\phi)$ be the finite set of sequents $X \vdash Y$ for $X, Y \subseteq \mathrm{cl}_{\mathcal{G}}(\phi)$. $\mathrm{SETSEQ}(\phi)$ is obviously a subset of the countably infinite set $\mathrm{SEQ}(\phi)$ where only sets can occur in the sequents of $\mathrm{SETSEQ}(\phi)$. The cardinality of $\mathrm{SETSEQ}(\phi)$ is bounded by $2^{|\mathcal{G}| \times \mathrm{r}(\phi)+1}$ and the size of each $X \vdash Y$ is at most $2 \times (|\mathcal{G}| \times |\phi|)^2$. Obviously, $\vdash \phi$ belongs to $\mathrm{SETSEQ}(\phi)$. We define a sequence of sets $Z_1 \subseteq Z_2 \subseteq Z_3 \subseteq \ldots$ included in $\mathrm{SETSEQ}(\phi)$. Z_1 is defined as the set of sequents $X \vdash Y$ from $\mathrm{SETSEQ}(\phi)$ such that $X \cap Y \neq \emptyset$. Now suppose that Z_i is defined and let us define Z_{i+1}. For $X \vdash Y \in \mathrm{SETSEQ}(\phi)$, $X \vdash Y \in Z_{i+1} \overset{\text{def}}{\Leftrightarrow}$ either (C1) $X \vdash Y \in Z_i$ or (C2) there are $X_1 \vdash Y_1, X_l \vdash Y_l \in Z_i$ such that

$$\frac{X_1 \vdash Y_1 \quad \ldots \quad X_l \vdash Y_l}{X \vdash Y}\ (r)$$

is a correct inference of the rule (r) in $\mathrm{SETG}\mathcal{L}_m^{\mathcal{G}}$. The index l takes the value either 1 or 2 according to the form of the rule (r). Since $card(X) + card(Y) \leq 2 \times |\mathcal{G}| \times \mathrm{r}(\phi)$, each formula in $X \vdash Y$ can be principal in only one rule, checking the condition (C2) can be done in exponential time in $|\mathcal{G}| + |\phi|$. If $\vdash \phi \in Z_{i+1}$, we stop and return 'yes' ϕ is $\mathcal{L}_m^{\mathcal{G}}$-valid. Otherwise, we continue the construction. Since $Z_i \subseteq Z_{i+1}$ and $\mathrm{SETSEQ}(\phi)$ has at most $2^{|\mathcal{G}| \times \mathrm{r}(\phi)+1}$ elements, this construction terminates after at most exponentially many stages. Computing Z_{i+1} can be done in deterministic exponential time in $|\mathcal{G}| + |\phi|$. Hence, the whole construction can be done in deterministic exponential time in $|\mathcal{G}| + |\phi|$. Whenever $Z_i = Z_{i+1}$ and $\vdash \phi \notin Z_i$, we return 'no', ϕ is not $\mathcal{L}_m^{\mathcal{G}}$-valid. The procedure can be shown to be correct. $\dashv$

The procedure in the proof of Theorem 5.1 is more suited for proving theoretical results than for being used in applications. It can be viewed as the addition of a highly inefficient loop checking to the calculi $\mathrm{G}\mathcal{L}_m^{\mathcal{G}}$

(see e.g. (Ladner 1977, Fitting 1983, Cerrito and Cialdea Mayer 1997, Heuerding 1998) for related matters). Besides the **EXPTIME** upper bound is sharp enough.

Theorem 5.2 *For $m \geq 2$, there is a decidable countably infinite set of right linear grammars such that $\mathcal{L}_m^{\mathcal{G}}$-satisfiability is* **EXPTIME**-*hard.*

Proof. (sketch) Let $\mathcal{G}_i$, $i \geq 1$, be the right linear grammar

$$\langle \{1\}, \{2, \ldots, m\}, \{1 \rightarrow \epsilon, 1 \rightarrow 2^i \cdot 1\}, 1 \rangle.$$

For any two different prime numbers n, n', we have $\mathtt{L}_1(\mathcal{G}_n) \neq \mathtt{L}_1(\mathcal{G}_{n'})$. By Theorem 3.1, this guarantees that we have defined a countably infinite set of essentially different right linear grammar logics. Let $\mathtt{L}(\Box)$ be the standard modal language for the modal logic K. Let K-GSAT be the set of standard modal formulae ϕ such that there is a Kripke model $\mathcal{M} = \langle W, R, V \rangle$ satisfying $\mathcal{M} \models \phi$. The global satisfiability problem K-GSAT is known to be **EXPTIME**-hard (Chen and Lin 1994) (see also (Hemaspaandra 1996)). Let ϕ be a formula of $\mathtt{L}(\Box)$. One can show that ϕ belongs to K-GSAT iff $\bigwedge_{0 \leq \alpha \leq i-1} [2^\alpha \cdot 1]\phi'$ is $\mathcal{L}_m^{\mathcal{G}_i}$-satisfiable where ϕ' is obtained from ϕ by replacing every occurrence of $\Box$ by $[2]$. ⊣

Theorem 5.3 *GSP(RLIN) is* **EXPTIME**-*complete.*

There is also a natural log-space transformation from GSP(RLIN) into satisfiability for the μ-calculus with multiple fixed points (see e.g. (Park 1981, Streett and Emerson 1989, Grädel et al. 2000). By way of example, the translation $t([2]\phi)$ of $[2]\phi$ for the right linear grammar logic $\mathcal{L}$ from Section 1 is

$$\nu X_2(X_1, X_2, X_3).\langle [1]X_3 \wedge [3]X_2, [2]X_3 \wedge [4]X_1 \wedge [5]X_3, t(\phi) \rangle$$

where X_3 is used as a renaming variable. The principle behind this example can be generalized to any modal connective of a right linear grammar logic. However, by using a polynomial-time transformation into satisfiability for PDL with finite automata one can show that GSP(REG), the extension of GSP(RLIN) with left linear grammars, is **EXPTIME**-complete (Demri 2000).

The general global logical consequence problem for right linear grammar logics GGLC(RLIN) takes as inputs a right linear grammar $\mathcal{G}$ and two $\mathtt{L}_m$-formulae ϕ, ψ and checks whether for all $\mathcal{L}_m^{\mathcal{G}}$-models $\mathcal{M}$, $\mathcal{M} \models \phi$ implies $\mathcal{M} \models \psi$. For any right linear grammar $\mathcal{G}$, GLC($\mathcal{L}_m^{\mathcal{G}}$) denotes the problem obtained from GGLC(RLIN) by fixing the grammar to $\mathcal{G}$. The calculus G$\mathcal{L}_m^{\mathcal{G}}$ can be extended in order to deal with GGLC(RLIN). The sequents are of the form $\Gamma \vdash_\phi \Delta$ for the L_m-formula ϕ. $G\mathcal{L}^{\mathcal{G}}_m$ is extended by writing $\vdash_\phi$ instead of $\vdash$. However, one rule is added:

$$\frac{\Gamma, \phi \vdash_\phi \Delta}{\Gamma \vdash_\phi \Delta}\ glc_\phi$$

glc_ϕ is an obvious adaptation of existing rules for capturing global logical consequence (see e.g. (Fitting 1983, Heuerding 1998, Massacci 2000)). One can show that for any sequent $\Gamma \vdash_\phi \Delta$, the formula $\langle \phi, (\bigwedge_{\psi \in \Gamma} \psi) \Rightarrow (\bigvee_{\psi \in \Delta} \psi) \rangle \in \mathrm{GLC}(\mathcal{L}^{\mathcal{G}}_m)$ iff the sequent $\Gamma \vdash_\phi \Delta$ is derivable in $G\mathcal{L}^{\mathcal{G}}_m + glc_\phi$. Similarly, one can show that GGLC(RLIN) is in **EXPTIME** by adapting the proof of Theorem 5.1.

6 A sufficient condition for PSPACE decision procedures

The completeness proof of Theorem 4.8 and the proof of Lemma 4.5 induce a depth-first systematic procedure to determine whether a sequent $\Gamma \vdash \Delta$ is provable in $G\mathcal{L}^{\mathcal{G}}_m$ or not. In the proof of Lemma 4.5, in order to obtain a maximal sequence σ starting from a given sequent in a derivation $\vdash \phi$, essential backtracking points are introduced when the rule $(\vdash \wedge)$ needs to be applied. Moreover, the length of such a sequence σ is bounded by $|\mathcal{G}| \times \mathrm{r}(\phi)$. By Lemma 4.2 and since $card(\mathrm{cl}_{\mathcal{G}}(\phi))$ is bounded by $|\mathcal{G}| \times \mathrm{r}(\phi)$, the number of backtracking points is bounded by $|\mathcal{G}| \times \mathrm{r}(\phi)$. Similarly, in the proof of Theorem 4.8, other backtracking points are introduced when the rule $(\vdash [i])$ for some $i \in \{1, \ldots, m\}$ needs to be applied. The number of such backtracking points is also bounded by $|\mathcal{G}| \times \mathrm{r}(\phi)$. So one can use a bit-string of length $|\mathcal{G}| \times \mathrm{r}(\phi)$ to remember which choices have been already tried. Consequently, in order to show that a given right linear grammar logic $\mathcal{L}^{\mathcal{G}}_m$ has a polynomial space satisfiability problem (or equivalently a polynomial space validity problem), it is sufficient to show that there is a polynomial $p(.)$ such that in the depth-first systematic procedure for proving $\vdash \phi$, the $(\vdash [i])$ rules are applied on a branch at most $p(|\phi|)$ times (see also the proof of Lemma 6.1). This means that in the proof of Theorem 4.8 the number of levels is finite and is bounded by a polynomial in the size of the input sequent $\Gamma \vdash \Delta$.

This observation is not really surprising, but all the point now is to refine the above statement in order to be able to prove the polynomial space upper bounds for countably many right linear grammar logics.

Let $\langle S, \ll \rangle$ be a well-founded set, $meas : \mathrm{SEQ}_m \rightarrow S$ be a map and $p_{card}(.)$, $p_{length}(.)$ be polynomials such that for any L_m-formula ϕ,

(C3) for $\Gamma_1 \vdash \Delta_1, \ldots, \Gamma_n \vdash \Delta_n \in \mathrm{SEQ}(\phi)$, $meas(\Gamma_n \vdash \Delta_n) \ll \ldots \ll$

$meas(\Gamma_1 \vdash \Delta_1) \ll meas(\vdash \phi)$ implies $n < p_{card}(|\phi|)$;

(C4) for any inference

$$\frac{\Gamma_1 \vdash \Delta_1 \quad \ldots \quad \Gamma_l \vdash \Delta_l}{\Gamma \vdash \Delta} \ (r)$$

(a) with (r) different from $(\vdash [i])$ for all $i \in \{1, \ldots, m\}$ ($l \in \{1, 2\}$ according to the form of the rule (r));

(b) $\Gamma_1 \vdash \Delta_1, \ldots, \Gamma_l \vdash \Delta_l, \Gamma \vdash \Delta \in \mathrm{SEQ}(\phi)$;

we have for $i \in \{1, l\}$, either $meas(\Gamma_i \vdash \Delta_i) \ll meas(\Gamma \vdash \Delta)$ or $meas(\Gamma_i \vdash \Delta_i) = meas(\Gamma \vdash \Delta)$;

(C5) for any sequence of sequents in $\mathrm{SEQ}(\phi)$,

$$\sigma_0 \prec (\vdash [i_1])\Gamma_1 \vdash \Delta_1 \prec \sigma_1 \prec (\vdash [i_2]) \ldots \sigma_{n-1} \prec (\vdash [i_n])\Gamma_n \vdash \Delta_n$$

such that for $i \in \{0, \ldots, n-1\}$, σ_i is maximal and $n > p_{length}(|\phi|)$, we have $meas(\Gamma_n \vdash \Delta_n) \ll meas(first(\sigma_0))$.

The condition (C3) holds true when $\{meas(\Gamma \vdash \Delta) : \Gamma \vdash \Delta \in \mathrm{SEQ}(\phi)\}$ is finite since $\langle S, \ll \rangle$ is a well-founded set. For example, this is the case with $meas(\Gamma \vdash \Delta) = \mathrm{md}(\bigwedge_{\psi \in \Gamma, \Delta} \psi)$.

Lemma 6.1 *Let $\mathcal{G}$ be a right linear grammar. If there exist a well-founded set $\langle S, \ll \rangle$, a map meas : $\mathrm{SEQ}_m \rightarrow S$ and polynomials $p_{card}(.)$, $p_{length}(.)$ satisfying the conditions (C3), (C4) and (C5), for any L_m-formula ϕ, then $\mathcal{L}_m^{\mathcal{G}}$-satisfiability is in* **PSPACE**.

Proof. (sketch) The proof is by an easy verification by using the depth-first systematic procedure induced by the proofs of Lemma 4.5 and Theorem 4.8 in order to check whether $\vdash \phi$ is provable in $\mathrm{G}\mathcal{L}_m^{\mathcal{G}}$. ⊣

The above method is not new (see e.g. (Ladner 1977, Fitting 1983, Massacci 1998)) but it allows us to use a uniform depth-first procedure from the proof of Theorem 4.8 and from the proof of Lemma 4.5.

Let $\mathcal{G}^{rl}$ be the right linear grammar defined in Example 4.1. Because of the duplication of Γ_1 in the application of the $(\vdash [4])$-rule, none of the obvious measures work to prove the polynomial space upper bound. Obviously, $\mathcal{G}^{rl}$ is finite since $\mathrm{L}_1(\mathcal{G}^{rl})$ and $\mathrm{L}_2(\mathcal{G}^{rl})$ are empty.

7 Finiteness implies PSPACE

This section is devoted to show that for any finite right linear grammar $\mathcal{G}$, the $\mathcal{L}_m^{\mathcal{G}}$-satisfiability problem is in **PSPACE** (forthcoming Theorem 7.4 is even a bit more general). We need to introduce a few definitions. Let $\mathcal{G}$ be a finite right linear grammar. The set N of non terminal symbols can be partionned into three sets N_1, N_2 and N_3 such that for $i \in N$,

- $i \in N_1 \overset{\mathrm{def}}{\Leftrightarrow} \mathrm{L}_i(\mathcal{G}) = \emptyset$;

- $i \in N_2 \stackrel{\text{def}}{\Leftrightarrow} i \notin N_1$ and for some $j \in N_1$ and $u \in \Sigma^*$, $i \Rightarrow_{\mathcal{G}}^* u \cdot j$;
- $i \in N_3 \stackrel{\text{def}}{\Leftrightarrow} i \notin N_1 \cup N_2$.

N_1, N_2 and N_3 can be computed in polynomial-time in $|\mathcal{G}|$. A $N_2 \to N_1$-**derivation** is a sequence of the form

$$i_0 \Rightarrow_{\mathcal{G}} u_1 \cdot i_1 \Rightarrow_{\mathcal{G}} \ldots \Rightarrow_{\mathcal{G}} u_1 \ldots u_{n-1} \cdot i_{n-1} \Rightarrow_{\mathcal{G}} u_1 \ldots u_n \cdot i_n$$

such that $n \geq 1$, $\{i_0, \ldots, i_{n-1}\} \subseteq N_2$ and $i_n \in N_1$. The main properties of the partition $\{N_1, N_2, N_3\}$ are the following.

Lemma 7.1 *Let $\mathcal{G}$ be a finite right linear grammar and $\{N_1, N_2, N_3\}$ be the partition on N defined above. Then,*

(I) if $i \Rightarrow_{\mathcal{G}}^* u$ with $i \in N_1$, then $u \in \Sigma^* \cdot N_1$;

(II) if $i_0 \Rightarrow_{\mathcal{G}} u_1 \cdot i_1 \Rightarrow_{\mathcal{G}} \ldots \Rightarrow_{\mathcal{G}} u_1 \ldots u_n \cdot i_n$ is a $N_2 \to N_1$-derivation, then

1. for $j \in \{0, \ldots, n-2\}$, $i_j \neq i_{j+1}$;

2. $n \leq card(N_2) - 1$; $|u_1 \cdot \ldots \cdot u_n| \leq |\mathcal{G}|$;

(III) if $i \Rightarrow_{\mathcal{G}}^* u$ with $i \in N_3$, then $u \in (\Sigma^* \cdot N_3) \cup \Sigma^*$;

(IV) if $i \Rightarrow_{\mathcal{G}}^* u$ with $i \in N_2 \cup N_3$ and $u \in \Sigma^*$, then $|u| \leq |\mathcal{G}|$.

For $i \in N \cup \Sigma$, we define $w_{\mathcal{G}}(i)$, the **weight** of i in $\mathcal{G}$, as follows:

- for $i \in N_1 \cup \Sigma$, $w_{\mathcal{G}}(i) \stackrel{\text{def}}{=} 1$;
- for $i \in N_3$, $w_{\mathcal{G}}(i) \stackrel{\text{def}}{=} max\{1 + |u| : u \in \mathrm{L}_i(\mathcal{G})\}$;
- for $i \in N_2$,

$$w_{\mathcal{G}}(i) \stackrel{\text{def}}{=} max(\{1 + |u| : u \in \mathrm{L}_i(\mathcal{G})\} \cup \{1 + |u_1 \cdot \ldots \cdot u_n| : i \Rightarrow_{\mathcal{G}} u_1 i_1 \Rightarrow_{\mathcal{G}} \ldots \Rightarrow_{\mathcal{G}} u_1 \ldots u_n i_n \ N_2 \to N_1\text{−derivation}\}).$$

Observe that for $i \in N \cup \Sigma$, $w_{\mathcal{G}}(i) \leq |\mathcal{G}|$. Lemma 7.2 contains the main properties of the map $w_{\mathcal{G}}(.)$.

Lemma 7.2 *Let $\mathcal{G}$ be a finite right linear grammar. Then,*

(I) if $i \rightsquigarrow u \cdot i'$ with $i \in N_3$ and $u \in \Sigma^+$, then $i' \in N_3$ and $|u| - 1 + w_{\mathcal{G}}(i') < w_{\mathcal{G}}(i)$;

(II) if $i \rightsquigarrow u$ with $i \in N_3$ and $u \in \Sigma^+$, then $|u| < w_{\mathcal{G}}(i)$;

(III) if $i \rightsquigarrow u$ with $i \in N_2$ and $u \in \Sigma^+$, then $|u| < max\{1 + |v| : v \in \mathrm{L}_i(\mathcal{G})\} \leq w_{\mathcal{G}}(i)$;

(IV) if $i \rightsquigarrow u \cdot i'$ with $i \in N_2$, $i' \in N_3$ and $u \in \Sigma^+$, then $|u| - 1 + w_{\mathcal{G}}(i') < w_{\mathcal{G}}(i)$;

(V) if $i \rightsquigarrow u \cdot i'$ with $i, i' \in N_2$ and $u \in \Sigma^+$, then $|u| - 1 + w_{\mathcal{G}}(i') < w_{\mathcal{G}}(i)$.

Proof. By way of example, we show (V). Assume $i \rightsquigarrow u \cdot i'$ for some $i, i' \in N_2$ and $u \in \Sigma^+$. It is obvious that

$$max\{|v| + 1 : v \in \mathrm{L}_{i'}(\mathcal{G})\} + |u| - 1 < max\{|v| + 1 : v \in \mathrm{L}_i(\mathcal{G})\}.$$

It remains to show that

$$max\{1 + |u_1 \cdot \ldots \cdot u_n| : i' \Rightarrow_{\mathcal{G}} u_1 i_1 \Rightarrow_{\mathcal{G}} \ldots \Rightarrow_{\mathcal{G}} u_1 \ldots u_n i_n\} + |u| - 1$$
$$< max\{1 + |u_1 \cdot \ldots \cdot u_n| : i \Rightarrow_{\mathcal{G}} u_1 i_1 \Rightarrow_{\mathcal{G}} \ldots \Rightarrow_{\mathcal{G}} u_1 \ldots u_n i_n\}.$$

Since $i \rightsquigarrow u \cdot i'$, there are non terminal symbols $i_0, \ldots, i_\alpha$, $\alpha \geq 0$, such that $i_0 = i$ and $i_0 \Rightarrow_{\mathcal{G}} i_1 \Rightarrow_{\mathcal{G}} \ldots \Rightarrow_{\mathcal{G}} i_\alpha \Rightarrow_{\mathcal{G}} u \cdot i'$. In the case all the non terminal symbols $i_0, \ldots, i_\alpha$ belongs to N_2 we can easily conclude the proof since

$$max\{1 + |u_1 \cdot \ldots \cdot u_n| : i \Rightarrow_{\mathcal{G}} u_1 i_1 \Rightarrow_{\mathcal{G}} \ldots \Rightarrow_{\mathcal{G}} u_1 \ldots u_n i_n\}$$
$$\geq |u| + max\{1 + |u_1 \cdot \ldots \cdot u_n| : i' \Rightarrow_{\mathcal{G}} u_1 i_1 \Rightarrow_{\mathcal{G}} \ldots \Rightarrow_{\mathcal{G}} u_1 \ldots u_n i_n\}.$$

Suppose $i_j \in N_3$ for some $j \in \{1, \ldots, \alpha\}$. If $i_j \in N_3$, then there is some $j' \in N_1$ such that $i_j \Rightarrow^*_{\mathcal{G}} v \cdot j'$, which leads to a contradiction by definition of the partition $\{N_1, N_2, N_3\}$. Suppose $i_j \in N_1$ for some $j \in \{1, \ldots, \alpha\}$. Hence $i_j \Rightarrow^*_{\mathcal{G}} u \cdot i'$ and $u \cdot i' \notin \Sigma^* \cdot N_1$ which is in contradiction with Lemma 7.1(I). ⊣

We define the maps pw^a : FORM $\rightarrow \mathbb{N}$ and pw^s : FORM $\rightarrow \mathbb{N}$ as follows ("a" stands for antecedent and "s" for succedent). Let ϕ be an L_m-formula. Let $[i]\psi$ be the occurrence of a formula occurring negatively [resp. positively] in ϕ. To be precise, one should define the notion of occurrence (as a finite sequence of natural numbers for instance). For the sake of simplicity, this is omitted here. In order to define $\mathrm{pw}^a(\phi)$ [resp. $\mathrm{pw}^s(\phi)$], we define an auxiliary value $\mathrm{pw}^a(\phi, [i]\psi)$. Let $[i_1]\psi_1, \ldots, [i_n]\psi_n$, be the positive [resp. negative] occurrences of necessity formulae of ϕ such that the very occurrence of $[i]\psi$ is a subformula of each $[i_j]\phi_{i_j}$ (if any). In the case when some element of N_1 is in $\{i_1, \ldots, i_n\}$, $\mathrm{pw}^a(\phi, [i]\psi) \stackrel{\text{def}}{=} 0$ [resp. $\mathrm{pw}^s(\phi, [i]\psi) \stackrel{\text{def}}{=} 0$], otherwise $\mathrm{pw}^a(\phi, [i]\psi) \stackrel{\text{def}}{=} 1 + \sum_{j=1}^{n} w_{\mathcal{G}}(i_j)$ [resp. $\mathrm{pw}^s(\phi, [i]\psi) \stackrel{\text{def}}{=} 1 + \sum_{j=1}^{n} w_{\mathcal{G}}(i_j)$]. We are now in a position to define $\mathrm{pw}^a(\phi)$ [resp. $\mathrm{pw}^s(\phi)$]. $\mathrm{pw}^a(\phi) \stackrel{\text{def}}{=} \sum \mathrm{pw}^a(\phi, [i]\psi)$ [resp. $\mathrm{pw}^s(\phi) \stackrel{\text{def}}{=} \sum \mathrm{pw}^s(\phi, [i]\psi)$] where the sum is on the set of negative [resp. positive] occurrences of necessity formula in ϕ. For each $\psi \in \mathrm{cl}_{\mathcal{G}}(\phi)$, the number of negative occurrences of necessity formula is bounded by $|\mathcal{G}| \times \mathrm{r}(\phi)$ as well as the number of positive occurrences of necessity formula. So for $\psi \in \mathrm{cl}_{\mathcal{G}}(\phi)$,

$$max(\mathrm{pw}^a(\psi), \mathrm{pw}^s(\psi)) \leq |\mathcal{G}|^2 \times \mathrm{r}(\phi)^2 \times (|\mathcal{G}| + 1).$$

Theorem 7.3 *Let $\mathcal{G}$ be a finite right linear grammar logic. Then, $\mathcal{L}_m^{\mathcal{G}}$-satisfiability is* **PSPACE**-*complete.*

Deciding whether $\mathcal{G}$ is a finite right linear grammar can be done in polynomial time in $|\mathcal{G}|$.

Proof. (sketch) Let us define S, $meas$, p_{card} and p_{length}. We shall write $\mathrm{md}_N(\Gamma \vdash \Delta)$ to denote the maximal nesting of modal connectives $[i]$ with $i \in N$ in $\bigwedge_{\psi \in \Gamma, \Delta} \psi$. Observe that for $\Gamma \vdash \Delta \in \mathrm{SEQ}(\phi)$, $\mathrm{md}_N(\Gamma \vdash \Delta) \leq \mathrm{r}(\phi)$. We write $\mathrm{pw}'(\Gamma \vdash \Delta)$ to denote

$$\mathrm{pw}'(\Gamma \vdash \Delta) \stackrel{\text{def}}{=} max(\{\mathrm{pw}^a(\psi) : \psi \in \Gamma\} \cup \{\mathrm{pw}^s(\psi) : \psi \in \Delta\}).$$

- $\langle S, \ll \rangle \stackrel{\text{def}}{=} \langle \mathbb{N}^2, < \rangle$ where $<$ is the standard lexicographical ordering on $\mathbb{N}^2$ extending the standard $<$ on $\mathbb{N}$;
- $meas(\Gamma \vdash \Delta) \stackrel{\text{def}}{=} \langle \mathrm{md}_N(\Gamma \vdash \Delta), \mathrm{pw}'(\Gamma \vdash \Delta) \rangle$;
- $p_{card}(x) = 1 + |\mathcal{G}|^2 \times (|\mathcal{G}| + 1) \times x^3$; $p_{length}(x) = 1$.

Let ϕ be an L-formula. Condition (C3) is satisfied because

$$card(\{meas(\Gamma \vdash \Delta) : \Gamma \vdash \Delta \in \mathrm{SEQ}(\phi)\}) \leq |\mathcal{G}|^2 \times (|\mathcal{G}| + 1) \times |\phi|^3.$$

The condition (C4) is obviously satisfied. In order to check the condition (C5), by way of example consider an inference of a $(\vdash [i])$ rule for some $i \in \Sigma$ (see notations in Section 4.1). Since for $j \in N$ and $u \in (N \cup \Sigma)^*$, $j \rightsquigarrow u$ implies $u \in \Sigma^* \cup \Sigma^* \cdot N$, for $j \in \{1, \ldots, k\}$, $\mathrm{md}_N([j]\Gamma_j) \geq \mathrm{md}_N(\bigcup_{\langle j,k' \rangle \in start_i^j} [v_{j,k'}]\Gamma_j)$. Additionnally, $\mathrm{md}_N([i]\phi) = \mathrm{md}_N(\phi)$, which guarantees that $\mathrm{md}_N(.)$ does not strictly increase (when reading the rules upwards).

In order to show that $\mathrm{pw}'(.)$ strictly decreases, it is sufficient to sec that $\mathrm{pw}^s([i]\phi) = 1 + \mathrm{pw}^s(\psi)$ and to check that

- for $j \in N_1$, $max(\{\mathrm{pw}^a(\psi) : \psi \in [j]\Gamma_j\}) = max(\{\mathrm{pw}^a(\psi) : \psi \in \bigcup_{\langle j,k' \rangle \in start_i^j} [v_{j,k'}]\Gamma_j\}) = 0$ by Lemma 7.1(I);
- for $j \in N_2$ [resp. $j \in N_3$],

$$max(\{\mathrm{pw}^a(\psi) : \psi \in [j]\Gamma_j\}) \geq$$
$$max(\{\mathrm{pw}^a(\psi) : \psi \in \bigcup_{\langle j,k' \rangle \in start_i^j} [v_{j,k'}]\Gamma_j\})$$

  by Lemma 7.2(III,IV,V) [resp. by Lemma 7.2(I,II)].

⊣

Theorem 7.4 below is slightly stronger than Theorem 7.3.

Theorem 7.4 *GSP(RLIN$_f$) is in* **PSPACE**.

The proof of Theorem 7.4 follows the line of the proof of Theorem 7.3 by observing that for any sequent $\Gamma \vdash \Delta \in \mathrm{SEQ}(\phi)$, $\mathrm{md}_N(\Gamma \vdash \Delta) \leq \mathrm{r}(\phi)$ and $\mathrm{pw}'(\Gamma \vdash \Delta) \leq |\mathcal{G}|^2 \times (|\mathcal{G}| + 1) \times |\phi|^2$. Moreover, from the space analysis in Section 6, we can decide whether $\vdash \phi$ is provable in $\mathrm{G}\mathcal{L}_m^{\mathcal{G}}$ in polynomial space in $|\mathcal{G}| + |\phi|$.

It seems difficult to extend Theorem 7.4 to a larger class of finite context-free grammars. Indeed, for the finite left linear grammar $\mathcal{G}^{ll} = \langle \{1\}, \{2\}, \{1 \to 1 \cdot 2\}, 1 \rangle$, $\mathcal{L}_2^{\mathcal{G}^{ll}}$-satisfiability is already **EXPTIME**-hard (see Theorem 7.5(1) below). By contrast, for any strongly finite context-free grammar $\mathcal{G}$, $\mathcal{L}_m^{\mathcal{G}}$-satisfiability is in **PSPACE**. This can be shown by translation into PDL without Kleene star. By analogy to Theorem 7.4, it is open whether the general satisfiability problem for strongly finite context-free grammar logics is in **PSPACE**.

By using Theorem 7.3 and (Demri 2000), one can characterize the complexity of all the bimodal regular grammar logics.

Theorem 7.5 *Let $\mathcal{L}_2^{\mathcal{G}}$ be a bimodal regular grammar logic, that is $N \cup \Sigma = \{1, 2\}$ ($m = 2$) and $S = 1$.*

1. *If $1 \to 1 \cdot 2^i$ is a production rule of $\mathcal{G}$ for some $i \geq 1$, then, $\mathcal{L}_2^{\mathcal{G}}$-satisfiability is* **EXPTIME**-*complete.*
2. *If $\mathcal{G}$ is right linear and finite [resp. infinite], then $\mathcal{L}_2^{\mathcal{G}}$-satisfiability is* **PSPACE**-*complete [resp.* **EXPTIME**-*complete].*

Theorem 7.5 exhausts all the possibilities of bimodal regular grammar logics. As a corollary, given a bimodal regular grammar logic $\mathcal{L}_2^{\mathcal{G}}$, deciding whether $\mathcal{L}_2^{\mathcal{G}}$ is **PSPACE**-complete can be done in linear time in $|\mathcal{G}|$. **EXPTIME**-hardness is roughly due to the fact that the concerned bimodal logics contain a modal connective [2] and another one [1] that is a variant of its reflexive and transitive closure (noted [2*]), see also (Spaan 1993, Sattler 1996, Castilho et al. 1999) for logics with a similar attribute. For instance, if $\mathcal{M}, x \models \langle 1 \rangle \phi \wedge [1]\psi$ for some $\mathcal{L}_2^{\mathcal{G}^{ll}}$-model $\mathcal{M}$, then there is $y \in R_1(x)$ such that $\mathcal{M}, y \models \phi$ and for all $z \in R_2^*(y)$, $\mathcal{M}, z \models \psi$, in symbols $\mathcal{M}, y \models \phi \wedge [2^*]\psi$. These are typically the formulae of that form that are responsible for the **EXPTIME**-hardness of PDL (Fischer and Ladner 1979, Spaan 1993). However, not every variant [1] of [2*] leads to **EXPTIME**-hardness. Consider the right linear grammar $\mathcal{G} = \langle \{1\}, \{2\}, \{1 \to 2 \cdot 1\}, 1 \rangle$ that is closely related to $\mathcal{G}^{ll}$. Although [2] [resp. [1]] can be viewed as the PDL modal connective $[\mathsf{c}_2]$ [resp. $[\mathsf{c}_2^*; \mathsf{c}_1]$], $\mathcal{L}_2^{\mathcal{G}}$ cannot isolate the modal connective $[\mathsf{c}_2^*]$ (or equivalently [2*]) because in $[\mathsf{c}_2^*; \mathsf{c}_1]$, there is always a last step that is a c_1 transition. Indeed, we have shown (see Theorem 7.3) that $\mathcal{L}_2^{\mathcal{G}}$-satisfiability is in **PSPACE**. By Theorem 3.2 we conclude that $\mathcal{L}_2^{\mathcal{G}}$-satisfiability is **PSPACE**-complete.

8 Infinity and PSPACE

Theorem 7.3 roughly states that finiteness for right linear grammars implies an **PSPACE** complexity upper bound for the corresponding grammar logics. By contrast, we can show that infinity does not imply **EXPTIME**-hardness. The class of right linear grammar logics introduced below contains for $m \geq 2$, countably infinite **PSPACE** logics $\mathcal{L}_m^{\mathcal{G}}$ with $N \cup \Sigma = \{1, \ldots, m\}$.

Theorem 8.1 *Let $\mathcal{G} = \langle N, \Sigma, P, S \rangle$ be a right linear grammar such that for $i \in N$, (1) $i \rightsquigarrow u$ implies $|u| \geq 1$ and (2) $i \rightsquigarrow j \cdot u$ and $i \rightsquigarrow j \cdot u'$ for some $j \in \Sigma$ imply $u = u'$. Then, $\mathcal{L}_m^{\mathcal{G}}$-satisfiability is* ***PSPACE****-complete.*

If $\mathcal{G}$ is viewed as a finite automaton $\mathcal{A}$, the assumption (2) in Theorem 8.1 can be interpreted as a requirement on the determinism of $\mathcal{A}$.

Proof. (sketch) Let us define S, $meas$, p_{card} and p_{length}:

- $\langle S, \ll \rangle \stackrel{\text{def}}{=} \langle \mathbb{N}, < \rangle$; $p_{card}(x) = x + 1$; $p_{length}(x) = 1$;
- for any sequent $\Gamma \vdash \Delta$, $meas(\Gamma \vdash \Delta) = \mathrm{pw}^+(\Gamma) + \mathrm{nw}^+(\Delta)$ (see Section 2 for the definitions of $\mathrm{pw}^+(.)$ and $\mathrm{nw}^+(.)$).

The assumptions of Lemma 6.1 can be shown to hold. ⊣

For any grammar $\mathcal{G}$ satisfying the assumptions of Theorem 8.1, the proof system $\mathrm{G}\mathcal{L}_m^{\mathcal{G}}$ involves no rule with implicit contraction. For $m \geq 5$, let $\mathcal{G} = \langle \{1, 2, 3\}, \{4, \ldots, m\}, P, 1 \rangle$ be the infinite right linear grammar with P defined as the union of $\{1 \rightarrow 4^m\}$ with

$$\bigcup_{i \in \{5, \ldots, m\}} \{1 \rightarrow i^{(2^m)} \cdot 2, 2 \rightarrow (i \cdot (i+1) \cdot \ldots \cdot m)^i \cdot 3, 3 \rightarrow i^i \cdot (i+1)^{i+1} \cdot \ldots \cdot m^m 1\}$$

The grammar $\mathcal{G}$ is not finite but by Theorem 8.1, $\mathcal{L}_m^{\mathcal{G}}$-satisfiability is in **PSPACE**.

In Theorem 8.1, it is unlikely that one can significantly relax the condition $|u| \geq 1$ by allowing $|u| = 0$. By Theorem 7.5(2), $\mathcal{L}_2^{\mathcal{G}}$-satisfiability is **EXPTIME**-hard for the right linear grammar $\mathcal{G} = \langle \{1\} \{2\} \{1 \rightarrow \epsilon, 1 \rightarrow 2 \cdot 1\} 1 \rangle$.

9 An open problem

By a proof-theoretical analysis, we have designed polynomial space decision procedures in a uniform framework for countably infinite right linear grammar logics. Understanding the complexity/decidability status of context-free grammar logics by proof-theoretical means seems to be a challenge worth being attacked in order to further characterize the complexity of modal logics. More precisely, let $\mathcal{G}$ be a context-free grammar such that for $i \in N$, $\{u \in (\Sigma \cup N)^* : i \Rightarrow_{\mathcal{G}}^* u\}$ is a regular language. In (Demri 2000) it is shown that $\mathcal{L}_m^{\mathcal{G}}$-satisfiability can be polynomially reduced to PDL satisfiability by replacing any occurrence of $[i]$ by $[\pi_i]$ where π_i is a regular expression (program) generating precisely the language $\{u \in (\Sigma \cup N)^* : i \Rightarrow_{\mathcal{G}}^* u\}$. Hence, $\mathcal{L}_m^{\mathcal{G}}$-satisfiability is in **EXPTIME** and showing that $\mathcal{L}_m^{\mathcal{G}}$-satisfiability is in **PSPACE** is equivalent to showing that PDL satisfiability restricted to the program expressions $\pi_1, \ldots, \pi_m$ (if $\Sigma \cup N = \{1, \ldots, m\}$) is in **PSPACE**. Is there a natural class of **PSPACE** context-free grammar logics? K4, S4 should preferably fall into this hypothetical class.

Acknowledgements

I wish to express my gratitude to the anonymous referee for useful suggestions and remarks, to G. de Giacomo and E. Grädel for pointing me to works related to modal μ-calculus with simultaneous fixed points, to the participants of AiML-ICTL 2000 for their feedback on this work and to the AiML organizers for their kind invitation.

References

Andreka, H., I. Nemeti, and J. van Benthem. 1998. Modal languages and Bounded Fragments of Predicate Logic. *JPL* 27(3):217–274.

Baader, F., and U. Sattler. 2000. Tableau algorithms for Description Logics. In *TABLEAUX-9*, 1–18. LNAI 1847, Springer.

Baldoni, M. 1998. *Normal Multimodal Logics: Automated Deduction and Logic Programming.* Doctoral dissertation, Università degli Studi di Torino.

Baldoni, M., L. Giordano, and A. Martelli. 1998. A Tableau Calculus for Multimodal Logics and Some (Un)Decidability results. In *TABLEAUX-7*, 44–59. LNAI 1397, Springer.

Basin, D., S. Matthews, and L. Vigano. 1997. A new method for bounding the complexity of modal logics. In *Computational Logic and Proof Theory*, 89–102. LNCS 1289, Springer.

Basin, D., S. Matthews, and L. Viganò. 1998. Natural Deduction for Non-Classical Logics. *Studia Logica* 60(1):119–160.

Blackburn, P., M. de Rijke, and Y. Venema. 2001. *Modal Logic.* Cambridge University Press. to appear.

Castilho, M., O. Gasquet, and A. Herzig. 1999. Formalizing action and change in modal logic I: the frame problem. *JLC* 9(5):701–735.

Cerrito, S., and M. Cialdea Mayer. 1997. A Polynomial Translation of S4 into T and Contraction-free Tableaux for S4. *Logic Journal of the IGPL* 5(2):287–300.

Chagrov, A., and V. Shehtman. 1994. Algorithmic aspects of propositional tense logics. In *CSL-8*, 442–455. LNCS 933, Springer.

Chen, C., and I. Lin. 1994. The complexity of propositional modal theories and the complexity of consistency of propositional modal theories. In *LFCS-3*, 69–80. LNCS 813, Springer.

Demri, S. 2000. The complexity of regularity in grammar logics and related modal logics. Submitted.

Fariñas del Cerro, L., and M. Penttonen. 1988. Grammar logics. *Logique et Analyse* 123–134.

Fischer, M., and R. Ladner. 1979. Propositional Dynamic Logic of Regular Programs. *Journal of Computer and System Sciences* 18:194–211.

Fitting, M. 1983. *Proof methods for modal and intuitionistic logics.* D. Reidel Publishing Co.

Goré, R. 1999. Tableaux methods for modal and temporal logics. In *Handbook of Tableaux Methods*, ed. M. d'Agostino, D. Gabbay, R. Hähnle, and J. Posegga, 297–396. Kluwer.

Grädel, E. 1999. Why are modal logics so robustly decidable? *Bulletin of the EATCS* 68:90–103.

Grädel, E., C. Hirsch, and M. Otto. 2000. Back and forth between guarded and modal logics. In *LICS'2000*, 217–228.

Grädel, E., and I. Walukiewicz. 1999. Guarded Fixed Point Logic. In *LICS'99*, 45–54.

Halpern, J., and Y. Moses. 1992. A guide to completeness and complexity for modal logics of knowledge and belief. *AI* 54:319–379.

Halpern, J., and H. Reif. 1983. The propositional dynamic logic of deterministic, well-structured programs. *TCS* 27:127–165.

Hemaspaandra, E. 1996. The price of Universality. *NDJFL* 37(2):173–203.

Heuerding, A. 1998. *Sequent Calculi for Proof Search in Some Modal Logics.* Doctoral dissertation, University of Bern.

Hopcroft, J., and J. Ullman. 1979. *Introduction to automata theory, languages, and computation.* Addison-Wesley Publishing Company, Reading, M.A.

Hudelmaier, J. 1996. Improved decision procedures for the modal logics K, T and S4. In *CSL'95*, 320–334. LNCS 1092, Springer.

Huldelmaier, J. 1993. An O(n log n)-space decision procedure for intuitionistic propositional logic. *JLC* 3(1):63–75.

Kozen, D. 1983. Results on the Propositional μ-calculus. *TCS* 27:333–354.

Kracht, M. 1996. Power and Weakness of the modal display calculus. In *Proof theory of modal logic*, ed. H. Wansing, 93–121. Kluwer.

Kracht, M. 2000. Logic and Syntax - A Personal Perspective. In *Advances in Modal Logics, Vol. 2*, ed. M. Zakharyaschev, K. Segerberg, M. de Rijke, and H. Wansing, 355–384. CSLI Publications.

Kripke, S. 1963. Semantical analysis of modal logic I: normal modal propositional calculi. *Zeitschrift für Mathematik Logik und Grundlagen der Mathematik* 9:67–96.

Ladner, R. 1977. The computational complexity of provability in systems of modal propositional logic. *SIAM Journal of Computing* 6(3):467–480.

Lutz, C., U. Sattler, and S. Tobies. 1999. A suggestion for an n-ary description logic. In *International Workshop on Description Logics, Sweden*, 81–85.

Marx, M. 1997. Complexity of Modal Logics of Relations. Technical report. ILLC.

Marx, M., S. Mikulas, and M. Reynolds. 2000a. The Mosaic Method for Temporal Logics. In *TABLEAUX-9*, 324–340. LNAI 1847, Springer.

Marx, M., S. Mikulas, and S. Schlobach. 2000b. Labelled Deduction for the guarded fragment. In *Labelled Deduction*, ed. D. Basin, M. D'Agostino, D. Gabbay, and S. Matthews, 195–214. Kluwer.

Massacci, F. 1994. Strongly analytic tableaux for normal modal logics. In *CADE-12*, 723–737. LNAI 814, Springer.

Massacci, F. 1998. *Efficient Approximate Deduction and an Application to Computer Security*. Doctoral dissertation, Università Degli Studi Di Roma "'La Sapienza".

Massacci, F. 2000. Single Steps Tableaux for Modal Logics. *JAR* 24(3):319–364.

Ohnishi, M., and K. Matsumoto. 1957. Gentzen Method in modal calculi. *Osaka Mathematical Journal* 9:113–130.

Park, D. 1981. Concurrency and Automata on infinite sequences. In *5th GI Conference on Theoretical Computer Science*, 167–183. LNCS 104, Springer.

Pratt, V. 1979. Models of program logics. In *Proceedings of the 20th IEEE Symposium on Foundations of Computer Science*, 115–122.

Rabin, M. 1969. Decidability of second-order theories and automata on infinite trees. *Transactions of the American Mathematical Society* 41:1–35.

Rautenberg, W. 1983. Modal Tableau Calculi and Interpolation. *JPL* 12:403–423.

Rozenberg, G., and A. Salomaa. 1994. *Cornerstones of Undecidability*. Prentice Hall.

Sahlqvist, H. 1975. Completeness and correspondence in the first and second order semantics for modal logics. In *3rd Scandinavian Logic Symposium, 1973*, 110–143. North Holland.

Sattler, U. 1996. A concept language extended with different kinds of transitive roles. In *20. Deutsche Jahrestagung für Künstliche Intelligenz*. LNM 1137, Springer.

Schütte, K. 1967. *Proof theory*. Springer-Verlag.

Spaan, E. 1993. *Complexity of Modal Logics*. Doctoral dissertation, ILLC, Amsterdam University.

Streett, R., and A. Emerson. 1989. An automata theoretic decision procedure for the propositional μ-calculus. *Information and Computation* 81:249–264.

Szałas, A. 1996. On natural deduction in first-order fixpoint logics. *Fundamenta Informaticae* 26(1):81–94.

Takeuti, G. 1975. *Proof Theory.* North-Holland Publishing Company.

van Benthem, J. 1976. Modal Reduction principles. *JSL* 2:301–312.

van Benthem, J. 1984. Correspondence Theory. In *Handbook of Philosophical Logic, Volume II*, ed. D. Gabbay and F. Günthner, 167–247. Reidel, Dordrecht.

van Benthem, J. 2000. Modal Logic in Two Gestalts. In *Advances in Modal Logics, Vol. 2*, ed. M. Zakharyaschev, K. Segerberg, M. de Rijke, and H. Wansing, 91–118. CSLI Publications.

Vardi, M. 1997. Why is modal logic so robustly decidable? In *DIMACS Series in Discrete Mathematics and Theoretical Computer Science 31, American Mathematical Society*, 149–183.

Stéphane Demri
Lab. Spécification et Vérification
ENS de Cachan & CNRS UMR 8643
61 Av. Pdt. Wilson
94235 Cachan Cedex, France
email: `demri@lsv.ens-cachan.fr`
URL: `http://www.lsv.ens-cachan.fr/~demri/`
On leave from Lab. LEIBNIZ, Grenoble.

8

Towards a Many-Dimensional Modal Logic for Semantic Processing

TIM FERNANDO

ABSTRACT. Notions of context for natural language interpretation are factored in terms of three processes: translation, entailment and attunement. The processes are linked by accessibility relations of the kind studied in many-dimensional modal logic, modulo complications from constraints between translation and entailment (violations in which may trigger re-attunement) and from refinement and underspecification.

1 Introduction

It is familiar practice in semantics to encode context as a sequence, selected components of which can be varied while others are held constant. The indexical *now* was, for instance, brought into temporal logic in Kamp 1971 via an early form of *multi-dimensional modal logic* (e.g. Marx and Venema 1997). And to treat presupposition, Stalnaker 1978 injected content into context, feeding the development of "dynamic" approaches to semantics such as *Discourse Representation Theory* (DRT). The present work is motivated largely by what is identified in Kamp 1990 as "the principal challenge to DRT" — viz. "the exact formulation of the construction algorithm" which takes syntactic analyses of English sentences and returns logical forms. Our emphasis, however, is not so much on any particular kind of syntactic analysis or (target) logical form as it is on the incrementality and non-determinism of some such algorithm. That incrementality and non-determinism are analyzed by bringing out contextual parameters that condition processing.

Advances in Modal Logic, Volume 3
F. Wolter, H. Wansing, M. de Rijke, and M. Zakharyaschev, eds.

1.1 Background: translation and entailment

To fix notation, let E be the set of inputs to the algorithm and Φ be the set of outputs. Without making any assumptions about E or Φ, let us form the set

$$\mathit{Stage} \;\subseteq\; \bigcup_{n\geq 0}(E^n \times \Phi^n)$$

of pairs $(\vec{e},\vec{\varphi})$ of input and output sequences. These i/o pairs can be construed as stages of a *translation* process insofar as every pair

$$(e_1 e_2 \cdots e_n, \varphi_1\varphi_2\cdots\varphi_n) \in \mathit{Stage}$$

can be traced to the initial pair (ϵ,ϵ) (of empty input and output) by the sequence of transitions

$$(\epsilon,\epsilon) \stackrel{e_1}{\to} (e_1,\varphi_1) \stackrel{e_2}{\to} (e_1e_2,\varphi_1\varphi_2) \stackrel{e_3}{\to} \cdots \stackrel{e_n}{\to} (e_1e_2\cdots e_n,\varphi_1\varphi_2\cdots\varphi_n) \qquad (1)$$

where for every $e \in E$,

$$\stackrel{e}{\to} \;=\; \bigcup_{(\vec{e},\vec{\varphi})\in \mathit{Stage}} \{((\vec{e},\vec{\varphi}),(\vec{e}e,\vec{\varphi}\varphi)) \;:\; (\vec{e}e,\vec{\varphi}\varphi)\in \mathit{Stage}\}\,.$$

The existence of the transition sequence (1) can be simplified to the requirement that *Stage* be ***prefix-closed*** — i.e., whenever $(\vec{e}e,\vec{\varphi}\varphi) \in \mathit{Stage}$, $(\vec{e},\vec{\varphi}) \in \mathit{Stage}$. The transitions $\stackrel{e}{\to}$ are worth isolating, however, to examine context, the semantic component of which is given by a notion of entailment

$$\vdash \;\subseteq\; \Phi^* \times \Phi$$

(which presumably is the point behind passing from E to Φ). Together, a translation-entailment pair $(\mathit{Stage},\vdash)$ provides an interpretation of a modal language $\mathcal{L}_\circ$ consisting of formulas A generated according to

$$A \quad ::= \quad \varphi \mid \langle e\rangle A \mid \neg A \mid A\wedge A$$

via a "forcing" predicate $\Vdash_\circ \subseteq \mathit{Stage}\times\mathcal{L}_\circ$ defined by

$$\begin{array}{rcl}
(\vec{e},\vec{\varphi}) \Vdash_\circ \varphi & \text{iff} & \vec{\varphi}\vdash\varphi \\
(\vec{e},\vec{\varphi}) \Vdash_\circ \langle e\rangle A & \text{iff} & (\vec{e}e,\vec{\varphi}\varphi) \Vdash_\circ A \text{ for some } \varphi \\
(\vec{e},\vec{\varphi}) \Vdash_\circ \neg A & \text{iff} & \text{not } (\vec{e},\vec{\varphi}) \Vdash_\circ A \\
(\vec{e},\vec{\varphi}) \Vdash_\circ A\wedge B & \text{iff} & (\vec{e},\vec{\varphi}) \Vdash_\circ A \text{ and } (\vec{e},\vec{\varphi}) \Vdash_\circ B
\end{array}$$

(with $[e]A$, $A\vee B$, $A\supset B$ and $A\equiv B$ understood as abbreviations of $\neg\langle e\rangle\neg A$, $\neg(\neg A\wedge\neg B)$, $\neg(A\wedge\neg B)$ and $(A\supset B)\wedge(B\supset A)$, respectively).[1]

[1] The multi-modal character of $\mathcal{L}_\circ$ is reminiscent of dynamic logic. For applications to DRT, the difference can be put very briefly as follows: quantified dynamic logic pertains to the model-theoretic interpretation of DRT's logical forms (called DRSs), $\mathcal{L}_\circ$ to the construction algorithm. (More in Fernando 1999a.)

This is just Kripke semantics, with accessibility relations $\stackrel{e}{\rightarrow}$ determined by *Stage*, and valuation by $\vdash$. Accordingly, let us henceforth refer to pairs $(Stage, \vdash)$ as $\mathcal{L}_\circ$*-models*. A measure of the ambiguity of e at $(\vec{e}, \vec{\varphi})$ is given by $\mathcal{L}_\circ$-formulas A, if any, such that

$$(\vec{e}, \vec{\varphi}) \quad \Vdash_\circ \quad \langle e \rangle A \wedge \langle e \rangle \neg A \,,$$

discerning any difference between logical forms φ and φ' such that $(\vec{e}e, \vec{\varphi}\varphi)$ and $(\vec{e}e, \vec{\varphi}\varphi') \in Stage$. Quantifying over all stages, let us call an $\mathcal{L}_\circ$-formula A $(Stage, \vdash)$*-valid* if for every $(\vec{e}, \vec{\varphi}) \in Stage$, $(\vec{e}, \vec{\varphi}) \Vdash A$. For example, suppose E includes the following English sentences e_1 and e_2.

e_1 : Doris informed Chloe she aced the exam.
e_2 : In fact, she was the only Texan who passed.

Ambiguous though e_1 and e_2 may be, to say that the *she*'s in $e_1 e_2$ co-vary in the $\mathcal{L}_\circ$-model $(Stage, \vdash)$ is to require that the $\mathcal{L}_\circ$-formula

$$[e_1][e_2] \quad (\dot{\forall} x \dot{\in} \text{Texan})(\text{passed}(x) \mathbin{\dot{\supset}} \text{aced}(x))$$

be $(Stage, \vdash)$-valid, where the dotted connectives are connectives in Φ (as opposed to $\mathcal{L}_\circ$).

Before worrying about multiple outputs, we might check that an input e at a stage $(\vec{e}, \vec{\varphi})$ has any output at all, which can be expressed as

$$(\vec{e}, \vec{\varphi}) \quad \Vdash_\circ \quad \langle e \rangle \top \qquad [\text{ iff } (\vec{e}, \vec{\varphi}) \in dom(\stackrel{e}{\rightarrow}) \,]$$

where $\top$ is an $\mathcal{L}_\circ$-tautology (valid in every $\mathcal{L}_\circ$-model). In accordance with Karttunen 1974, presupposition can be equated with pre-condition for translation,[2] and presupposition satisfaction in Heim 1983 modified slightly to

$$(\vec{e}, \vec{\varphi}) \textit{ satisfies the presuppositions of } e \quad \text{iff} \quad (\vec{e}, \vec{\varphi}) \Vdash_\circ \langle e \rangle \top \,.$$

Consider, for example,

$\hat{e}$: The king of Buganda is bald.

That "ê presupposes there is a king of Buganda" can then be stated as

(2) $\langle \hat{e} \rangle \top \;\equiv\;$ '∃KoB'

[2]The term presupposition has been applied to so many things that any account of it is likely to be contentious. What is at stake in the present paper is the claim that failures in translation are interesting, as are the adjustments (to $\mathcal{L}_\circ$-models) made before, during or after translation. The present shift in presupposition satisfaction from the realm of semantics (Heim 1983) to the construction algorithm is compatible not only with the ***presupposition-as-anaphora*** view of van der Sandt 1992 but also with the ***familiarity condition*** of Heim 1982.

where '∃KoB' $\in \Phi$ asserts there is a king of Buganda. Moreover, the *negation test* becomes the $\mathcal{L}_\circ$-formula

$$(3) \qquad \langle \dot{\neg} e \rangle \top \;\equiv\; \langle e \rangle \top$$

(where $\dot{\neg}$ is a connective in E, not $\mathcal{L}_\circ$), while the presupposition projection behavior of E-conjunction $\dot{\wedge}$ and E-implication $\dot{\supset}$ (accounted for in Heim 1983, going back at least to Karttunen) translate in $\mathcal{L}_\circ$ to the sequentiality conditions

$$\langle e \dot{\wedge} e' \rangle \top \;\equiv\; \langle e \rangle \langle e' \rangle \top$$
$$\langle e \dot{\supset} e' \rangle \top \;\equiv\; \langle e \rangle \langle e' \rangle \top \,.$$

Turning to the logic of $\mathcal{L}_\circ$, Fernando 1999a establishes the following. Call an entailment relation $\vdash$ *$\mathcal{L}_\circ$-transparent* if for all $\varphi_1, \ldots, \varphi_n, \varphi \in \Phi$, we have $\varphi_1 \cdots \varphi_n \vdash \varphi$ iff the $\mathcal{L}_\circ$-formula $(\varphi_1 \wedge \cdots \wedge \varphi_n) \supset \varphi$ is $(Stage, \vdash)$-valid, for every $Stage \subseteq \bigcup_{n \geq 0} (E^n \times \Phi^n)$.

Proposition. *$\vdash$ is $\mathcal{L}_\circ$-transparent iff $\vdash$ verifies*

$$\textit{(Weak)} \;\; \frac{\vec{\varphi} \vdash \varphi}{\psi \vec{\varphi} \vdash \varphi} \;\; \frac{\vec{\varphi} \vdash \varphi}{\vec{\varphi} \psi \vdash \varphi} \qquad \textit{(Ref)} \;\; \frac{}{\varphi \vdash \varphi}$$

$$\textit{(Cut)} \;\; \frac{\vec{\varphi} \vdash \psi \qquad \psi \vec{\varphi} \vdash \varphi}{\vec{\varphi} \vdash \varphi} \,.$$

As formulated above, (Weak), (Ref) and (Cut) imply contraction and exchange

$$\text{(Con)} \;\; \frac{\vec{\varphi} \varphi \varphi \vdash \psi}{\vec{\varphi} \varphi \vdash \psi} \qquad \text{(Exc)} \;\; \frac{\vec{\varphi} \varphi \psi \vec{\psi} \vdash \varphi'}{\vec{\varphi} \psi \varphi \vec{\psi} \vdash \varphi'}$$

(as (Exc) is implicit in doubling (Weak), and (Con) in halving the conclusion of (Cut) from $\vec{\varphi}\vec{\varphi} \vdash \varphi$ to $\vec{\varphi} \vdash \varphi$). Next, let us add Φ-conjunction and Φ-implication (distinct from $\mathcal{L}_\circ$-connectives) to (Weak), (Ref) and (Cut) in the assumption

(a$\vdash$) (Weak), (Ref) and (Cut) hold in $\vdash$, and there are binary connectives $\dot{\wedge}$ and $\dot{\supset}$ in Φ s.t. for all $\varphi, \psi, \varphi' \in \Phi$ and $\vec{\varphi} \in \Phi^*$,

$$(\varphi \dot{\wedge} \psi)\vec{\varphi} \vdash \varphi' \quad \text{iff} \quad \varphi \psi \vec{\varphi} \vdash \varphi'$$
$$\vec{\varphi} \varphi \vdash \psi \quad \text{iff} \quad \vec{\varphi} \vdash \varphi \dot{\supset} \psi \,.$$

As for *Stage*, let us embed the set Φ of logical forms in the set E of inputs to translation, with the understanding that *Stage* keeps the logical forms fixed, and contains the pair (ϵ, ϵ) — i.e., (a∘) $\Phi \subset E$, $(\epsilon, \epsilon) \in \mathit{Stage}$ and for every $(e_1 \cdots e_n, \varphi_1 \cdots \varphi_n) \in \mathit{Stage}$,

$$(\forall i \in \{1, \ldots, n\}) \quad \text{if } e_i \in \Phi \text{ then } e_i = \varphi_i$$
$$(\forall \varphi \in \Phi) \quad (e_1 \cdots e_n\varphi, \varphi_1 \cdots \varphi_n\varphi) \in \mathit{Stage} \ .$$

The assumption (a∘) says that every logical form φ in Φ can be obtained unambiguously and without risk of presupposition failure by choosing an appropriate input, conveniently named φ. This re-use of φ is perhaps confusing, but otherwise harmless. Now, applying $\dot{\wedge}$ to combine any finite number of logical forms entailed at a certain stage into one, and $\dot{\supset}$ to anticipate logical forms entailed at a next stage, we get

Theorem. *The set of $\mathcal{L}$-formulas valid in all $\mathcal{L}_\circ$-models $(\mathit{Stage}, \vdash)$ satisfying (a$\vdash$) and (a∘) is decidable, and consists of the theorems of normal modal logic, supplemented with the axiom schemes*

(A1)	$\varphi \dot{\supset} \varphi$	*(A4)*	$\varphi \supset [e]\varphi$
(A2)	$\varphi \wedge \psi \equiv \varphi \dot{\wedge} \psi$	*(A5)*	$\langle\varphi\rangle\psi \equiv \varphi \dot{\supset} \psi$
(A3)	$\varphi \dot{\supset} \psi \supset \varphi \supset \psi$	*(A6)*	$\langle\varphi\rangle A \supset [\varphi]A$

For a fixed entailment relation $\vdash$ satisfying (a$\vdash$), it suffices to strengthen (A1)-(A3) to the scheme

$$\varphi_1 \wedge \cdots \wedge \varphi_n \quad \supset \quad \varphi$$

for all $\varphi_1, \ldots, \varphi_n, \varphi$ such that $\varphi_1 \cdots \varphi_n \vdash \varphi$.

Observe that adding $E = \Phi$ to (a∘) and (a$\vdash$) would validate the commuting squares

$$\langle e\rangle\langle e'\rangle A \quad \equiv \quad \langle e'\rangle\langle e\rangle A$$

in Marx and Venema 1997. Our analysis of ambiguity and presupposition, however, depends on inputs E to *Stage* going beyond the outputs Φ — or more to the point: on *Stage* being unrulier than a function (let alone a bijection).

1.2 Plan of present paper

Identifying the presupposition of e with $\langle e\rangle\top$ raises the question: what if $\langle e\rangle\top$ fails? Take, for instance, $\hat{e}$ above, uttered say, in isolation — i.e. at the initial stage (ϵ, ϵ) of some $\mathcal{L}_\circ$-model $(\mathit{Stage}, \vdash)$. If $\epsilon \vdash$ '∃KoB' then $\hat{e}$'s presupposition is met. But if not — and note that we can hardly expect before hearing $\hat{e}$ to have chosen $\vdash$ such that $\epsilon \vdash$ '∃KoB' — then according to (2), $\hat{e} \notin \mathit{dom}(\mathit{Stage})$. How then can we translate $\hat{e}$? If interpretation is to proceed (and many have argued it may), we have no choice but to leap out of $(\mathit{Stage}, \vdash)$, and into some other $\mathcal{L}_\circ$-model $(\mathit{Stage}', \vdash')$ where '∃KoB' is taken for granted:

$$\epsilon \vdash' \text{'∃KoB'} \quad \text{whence (2) implies} \quad \hat{e} \in \mathit{dom}(\mathit{Stage}').$$

Such leaps, called accommodation in the presupposition literature, can be quite tricky, subject, as they are, to various constraints such as consistency and plausibility which may or may not be amenable to formalization. Without settling which leaps are legitimate, we can nonetheless survey the possibilities by expanding $\mathcal{L}_\circ$ to a language $\mathcal{L}$, the models of which are formed from families of $\mathcal{L}_\circ$-models (Fernando 1999b). Indexing a family of $\mathcal{L}_\circ$-models by a set I, the idea is to expose the heretofore implicit contextual parameter $\alpha \in I$ determining an $\mathcal{L}_\circ$-model $(\mathit{Stage}_\alpha, \vdash_\alpha)$. Passing from $\mathcal{L}_\circ$-stages $(\vec{e}, \vec{\varphi})$ to $\mathcal{L}$-stages $(\vec{e}, \vec{\varphi})_\alpha$, where α keeps track of the $\mathcal{L}_\circ$-model to which $(\vec{e}, \vec{\varphi})$ is understood to belong, the present paper explores the problem of factoring an $\mathcal{L}$-transition

$$(4) \qquad (\vec{e}, \vec{\varphi})_\alpha \overset{e}{\rightsquigarrow} (\vec{e}e, \vec{\psi}\psi)_\beta$$

that changes the index α and previous translations $\vec{\varphi}$ (to β and $\vec{\psi}$, respectively) in the course of translating e to ψ. An example of (4) is considered in section 2, motivating investigations in sections 3 and 4 of the question: can we treat the three changes (to α and to $\vec{\varphi}$, plus the translation of e) separately? Section 3 provides a partial positive reply, centering around merged translation sets. Complications involving entailment are taken up in section 4, refining (as well as underspecifying) the notion of logical form.

2 An example: *attunement*

Recalling §1.1 above, consider the negation $\dot{\neg}\hat{e}$ of $\hat{e}$

$\dot{\neg}\hat{e}$: The king of Buganda is not bald.

Combining the negation test (3) with (2) yields

$$(5) \qquad \langle \dot{\neg}\hat{e} \rangle \top \ \equiv \ \text{`}\exists\text{KoB'} .$$

Assuming (for the sake of the argument) that $\dot{\neg}\hat{e}$ is uttered at stage (ϵ, ϵ) of an $\mathcal{L}_\circ$-model $(\mathit{Stage}_0, \vdash_0)$ for which $\epsilon \not\vdash_0$ '∃KoB' (and $\dot{\neg}\hat{e} \notin \mathit{dom}(\mathit{Stage}_0)$), a minimal adjustment of background assumptions leads to $(\mathit{Stage}_1, \vdash_1)$ where

$$(\vec{e}, \vec{\varphi}) \in \mathit{Stage}_1 \quad \text{iff} \quad (\text{`}\exists\text{KoB'}\vec{e}, \text{`}\exists\text{KoB'}\vec{\varphi}) \in \mathit{Stage}_0$$
$$\vec{\varphi} \vdash_1 \varphi \quad \text{iff} \quad \text{`}\exists\text{KoB'}\vec{\varphi} \vdash_0 \varphi$$

for all $\vec{\varphi} \in \Phi^*$, $\varphi \in \Phi$ and $\vec{e} \in E^*$. The adjustment would be no different were $\hat{e}$ uttered instead of $\dot{\neg}\hat{e}$. But suppose we were to follow $\dot{\neg}\hat{e}$ by

$\dot{\neg}$'∃KoB' : Buganda has no king.

Or, for that matter, what if $\epsilon \vdash_0 \dot{\neg}$'∃KoB'? Under such circumstances, uttering $\dot{\neg}\hat{e}$ at $(\epsilon, \epsilon)_0$ might be odd, but surely not outright contradictory the way $\hat{e}$ would? Perhaps (5) is too strong and ought to be weakened to

$$(6) \qquad \neg\dot{\neg}\text{'}\exists\text{KoB'} \supset ((\langle\dot{\neg}\hat{e}\rangle\top \equiv \text{'}\exists\text{KoB'})$$

where the $\mathcal{L}_\circ$-formula $\neg\dot{\neg}$'∃KoB' expresses the (syntactic) consistency of '∃KoB'. Then, assuming $\epsilon \vdash_0 \dot{\neg}$'∃KoB', $\dot{\neg}\hat{e}$ need not trigger a leap between $\mathcal{L}_\circ$-models. But if (as is eminently plausible) neither $\epsilon \vdash_0$ '∃KoB' nor $\epsilon \vdash_0 \dot{\neg}$'∃KoB', then a leap to $(Stage_1, \vdash_1)$ would run afoul of a subsequent utterance of $\dot{\neg}$'∃KoB'. For (6), we could leap instead to an $\mathcal{L}_\circ$-model $(Stage_2, \vdash_2)$ that rejects '∃KoB'

$$(\vec{e}, \vec{\varphi}) \in Stage_2 \quad \text{iff} \quad (\dot{\neg}\text{'}\exists\text{KoB'}\vec{e}, \dot{\neg}\text{'}\exists\text{KoB'}\vec{\varphi}) \in Stage_0$$

$$\vec{\varphi} \vdash_2 \varphi \quad \text{iff} \quad \dot{\neg}\text{'}\exists\text{KoB'}\vec{\varphi} \vdash_0 \varphi \,.$$

However, if after $\dot{\neg}\hat{e}$, we are told that the king of Buganda has curls (rather than $\dot{\neg}$'∃KoB'), then we get a contradiction in $(Stage_2, \vdash_2)$, from which we would have been immune in $(Stage_1, \vdash_1)$. Inasmuch as there is no telling what may come after the input $\dot{\neg}\hat{e}$, no correct choice can be made at $(\epsilon, \epsilon)_0$ between $(Stage_1, \vdash_1)$ and $(Stage_2, \vdash_2)$.[3]

2.1 Leaps triggered by pre- and post-conditions

Notice that if one of the logical forms in Φ is contradictory, call it $\bot$, then assumption $(a_\circ)$ from §1.1 above provides stages forcing $\bot$ — e.g. $(\bot, \bot)$. Garbage in, garbage out. That said, we can, with $\mathcal{L}_\circ$, discriminate between stages that force $\bot$ and those that do not (i.e. those forcing $\neg\bot$). These consistency considerations can trigger leaps between $\mathcal{L}_\circ$-models before or after logical forms are constructed. It is noteworthy that Gazdar 1979 lumps presupposition with Gricean implicature, conceived as a pragmatic adjustment made after logical forms have been constructed. Whatever picture of presupposition one adopts, there are many reasons (including implicatures) for adding background information α to the explicit discourse context $(\vec{e}, \vec{\varphi})$. The generality of logical forms (e.g. McCarthy and S.Buvač 1997) is brought out by keeping α separate from the translation $\vec{\varphi}$ of an input $\vec{e}$ (rather than packing it into $\varphi^{\vec{\alpha}}$) and then applying $\vec{\varphi}$ to different α's. The process of choosing α I propose to label ***attunement***. It is clearly related to the two other processes of entailment and translation, both of which take on a monotonic character in §1.1. The expansion from $\mathcal{L}_\circ$ to $\mathcal{L}$ allows non-monotonicity to enter through transitions

$$(\vec{e}, \vec{\varphi})_\alpha \overset{-}{\to} (\vec{e}, \vec{\varphi})_\beta \qquad \text{and} \qquad (\vec{e}, \vec{\varphi})_\alpha \overset{\diamond}{\to} (\vec{e}, \vec{\psi})_\alpha$$

[3] This is not to deny that there might be a preference (which can be expressed through modal operators; e.g. Fernando 1999b) for the presupposition '∃KoB', but only to affirm that, as a default, it is subject to retraction (through another leap).

that change the indices and translations respectively. Section 3 explores how to factor $\mathcal{L}$-transitions $\overset{e}{\rightsquigarrow}$ in (4) between $\overset{-}{\rightarrow}$, $\overset{\diamond}{\rightarrow}$ and $\overset{e}{\rightarrow}$.

2.2 From denials to underspecification

Rather than leaping to conclusions that are defeasible, we might try to avoid non-monotonicity by minimizing any choices made, postponing these until the necessary contextual elements become available. Returning to $\hat{e}$, for example, van der Sandt 1991 argues that $\dot{\neg}\hat{e}$ can be used not only as an assertion, in which case (5) holds, but also as a *denial*, in which case it is not an assertion and not subject to (5). Assuming this argument were correct, the challenge becomes: what if we are told "the king of Buganda is not bald" but are unable to tell if the utterance is an assertion or a denial? Matters would be simple were we to restrict our inputs to representations for which the question "am I an assertion or a denial?" is answered. The problem is that inputs often come in the form of *underspecified representations*, and just how much of the underspecification we should try to resolve is tricky. Whether or not we should resolve only what would support *monotonic semantic interpretation* (Alshawi and Crouch 1992), it is natural to leave some underspecification, suggesting transitions

$$(\vec{e}, \vec{\varphi})_\alpha \rightsquigarrow (\vec{\varphi}, \vec{\psi})_\beta \tag{7}$$

that feed outputs $\vec{\varphi}$ of α as inputs to β (to resolve some underspecification left over after α). While (7) drops the label e on $\overset{e}{\rightsquigarrow}$ in (4), it also changes $\vec{e}$, making it no more obvious to factor than (4). We shall see in section 4.

3 $\mathcal{L}$-models and merged translations

Consider a family of $\mathcal{L}_\circ$-models $(Stage_\alpha, \vdash_\alpha)_{\alpha\in I}$ indexed by some set I. To allow the notion of logical form to vary with α, we will *not* assume that there is a single set Φ of logical forms on which every entailment relation $\vdash_\alpha$ is defined. However, let us assume (as in $(a_\circ)$ in §1.1) that all logical forms belong to E,[4] and combine the different $\mathcal{L}_\circ$-stages in a set

$$\bigcup_{\alpha\in I}\{(\vec{e},\vec{\varphi})_\alpha : (\vec{e},\vec{\varphi}) \in Stage_\alpha\} \subseteq \bigcup_{n\geq 0}(E^n \times E^n \times I)$$

[4]Imposing $(a_\circ)$ for $\Phi = E$ would, as previously noted, trivialize translation. A more interesting adaptation of $(a_\circ)$ to $\mathcal{L}$ would for each $\alpha \in I$, use some subset of E for Φ.

where we write $(\vec{e}, \vec{\varphi})_\alpha$ instead of $(\vec{e}, \vec{\varphi}, \alpha)$. So, without mentioning $\mathcal{L}_\circ$, we can define an *$\mathcal{L}$-model* to be a pair $(STAGE, \{\vdash_\alpha\}_{\alpha \in I})$ such that

$$STAGE \subseteq \bigcup_{n \geq 0} (E^n \times E^n \times I) \quad \text{and for each } \alpha \in I, \ \vdash_\alpha \subseteq E^* \times E \,.$$

Having obliterated in $\mathcal{L}$ the distinction (in $\mathcal{L}_\circ$) between E and Φ, we will, nevertheless, continue to write $(\vec{e}, \vec{\varphi})$ instead of $(\vec{e}, \vec{e'})$, and sometimes φ instead of e. For $\mathcal{L}$ to extend $\mathcal{L}_\circ$ conservatively, let us agree that

$$\begin{array}{rcl}
(\vec{e}, \vec{\varphi})_\alpha \Vdash \varphi & \text{iff} & \vec{\varphi} \vdash_\alpha \varphi \\
(\vec{e}, \vec{\varphi})_\alpha \Vdash \langle e \rangle A & \text{iff} & (\vec{e}e, \vec{\varphi}\varphi)_\alpha \Vdash A \text{ for some } \varphi \\
(\vec{e}, \vec{\varphi})_\alpha \Vdash \neg A & \text{iff} & \text{not } (\vec{e}, \vec{\varphi})_\alpha \Vdash A \\
(\vec{e}, \vec{\varphi})_\alpha \Vdash A \wedge B & \text{iff} & (\vec{e}, \vec{\varphi})_\alpha \Vdash A \text{ and } (\vec{e}, \vec{\varphi})_\alpha \Vdash B
\end{array}$$

for all $(\vec{e}, \vec{\varphi})_\alpha \in STAGE$ and $\varphi, e \in E$. Now, the point in expanding $\mathcal{L}_\circ$ to $\mathcal{L}$ is to allow changes in $\alpha \in I$, the simplest of which is given by

$$(\vec{e}, \vec{\varphi})_\alpha \Vdash \langle - \rangle A \quad \text{iff} \quad (\vec{e}, \vec{\varphi})_\beta \Vdash A \text{ for some } \beta$$

Such a modality is, along with $\Diamond$ where

$$(\vec{e}, \vec{\varphi})_\alpha \Vdash \Diamond A \quad \text{iff} \quad (\vec{e}, \vec{\psi})_\alpha \Vdash A \text{ for some } \vec{\psi} \,,$$

familiar from multi-dimensional modal logic, with accessibility relations $\stackrel{-}{\rightarrow}, \stackrel{\Diamond}{\rightarrow} \subseteq STAGE \times STAGE$ that restrict changes to the index and translations respectively

$$\begin{array}{lcl}
(\vec{e}, \vec{\varphi})_\alpha \stackrel{-}{\rightarrow} (\vec{e'}, \vec{\varphi'})_{\alpha'} & \text{iff} & \vec{e} = \vec{e'} \text{ and } \vec{\varphi} = \vec{\varphi'} \\
(\vec{e}, \vec{\varphi})_\alpha \stackrel{\Diamond}{\rightarrow} (\vec{e'}, \vec{\varphi'})_{\alpha'} & \text{iff} & \vec{e} = \vec{e'} \text{ and } \alpha = \alpha' \,.
\end{array}$$

3.1 Factoring via merged translations

To factor an $\mathcal{L}$-transition of the form (4) through the relations $\stackrel{e}{\rightarrow}, \stackrel{-}{\rightarrow}$ and $\stackrel{\Diamond}{\rightarrow}$, let us consider the assumption

(a_+) for every pair $\alpha, \beta \in I$, there is an index $\gamma \in I$ such that

$$\text{prefix-closure}(Stage_\alpha \cup Stage_\beta) \ \subseteq \ Stage_\gamma$$

where prefix-closure(S) is (by definition)

$$\{(\vec{e}, \vec{\varphi}) \ : \ (\vec{e}\vec{e'}, \vec{\varphi}\vec{\varphi'}) \in S \text{ for some } \vec{e'}, \vec{\varphi'} \text{ of the same length}\} \,.$$

Note that (a_+) holds vacuously if there is just one index in I, the stage set of which is prefix-closed (a natural assumption for $\mathcal{L}_\circ$). We get more than one index in I once we differentiate in $\mathcal{L}$ between different contextual assumptions α and β. From here, it is a small step to forming disjunctions of these assumptions — suggesting that I be closed under a binary function + such that for all $\alpha, \beta \in I$, prefix-closure($Stage_\alpha \cup Stage_\beta$) = $Stage_{\alpha+\beta}$. In fact, (a_+) is weaker than this requirement, as (a_+) can also be met in one fell swoop by a prefix-closed "monster" index $\star$ that is maximally underspecified/uncommitted/ignorant/indeterminate in that $Stage_\star = \bigcup_{\alpha \in I} Stage_\alpha$. At any rate, ($a_+$) allows us to factor $(\vec{e}, \vec{\varphi})_\alpha \overset{e}{\rightsquigarrow} (\vec{e}e, \vec{\psi}\psi)_\beta$, as

$$(\vec{e}, \vec{\varphi})_\alpha \;\bar{\to}\; (\vec{e}, \vec{\varphi})_\gamma \overset{\Diamond}{\to} (\vec{e}, \vec{\psi})_\gamma \overset{e}{\to} (\vec{e}e, \vec{\psi}\psi)_\gamma \;\bar{\to}\; (\vec{e}e, \vec{\psi}\psi)_\beta \;,$$

with an index γ mediating the transformation of α into β.

More generally, given $e_1 \cdots e_n \in E^*$, let $\langle\!\langle e_1 \cdots e_n \rangle\!\rangle$ be the regular language consisting of the string $\langle e_1 \rangle \cdots \langle e_n \rangle$ and all other strings obtained from it by inserting any number of occurrences of $\langle - \rangle$ or $\Diamond$. That is,

$$\langle\!\langle e_1 \cdots e_n \rangle\!\rangle \quad = \quad L \cdot \langle e_1 \rangle \cdot L \cdots L \cdot \langle e_n \rangle \cdot L$$

where L is $\{\langle - \rangle, \Diamond\}^*$. The closure condition (a_+) on I validates the $\mathcal{L}$-scheme

(+1) $\quad \sigma A \supset \langle - \rangle \Diamond \langle e_1 \rangle \langle e_2 \rangle \cdots \langle e_n \rangle \langle - \rangle A \quad$ for $\sigma \in \langle\!\langle e_1 \cdots e_n \rangle\!\rangle$, $A \in \mathcal{L}$

and, if, moreover, $Stage_\alpha$ is prefix-closed for every $\alpha \in I$,

(+2) $\quad \sigma A \supset \langle - \rangle \Diamond \langle - \rangle \langle e_1 \rangle \langle e_2 \rangle \cdots \langle e_n \rangle A \quad$ for $\sigma \in \langle\!\langle e_1 \cdots e_n \rangle\!\rangle$, $A \in \mathcal{L}$.

Factoring $(\vec{e}, \vec{\varphi})_\alpha \overset{e}{\rightsquigarrow} (\vec{e}e, \vec{\psi}\psi)_\beta$ as

$$(\vec{e}, \vec{\varphi})_\alpha \quad \overset{e}{\to} \quad (\vec{e}e, \vec{\varphi}\varphi)_\alpha \quad \bar{\to} \quad (\vec{e}e, \vec{\varphi}\varphi)_\gamma \quad \cdots \quad (\vec{e}e, \vec{\psi}\psi)_\beta$$

might be blocked by presupposition failure of e at $(\vec{e}, \vec{\varphi})_\alpha$

3.2 Further axioms

But if the presupposition $\langle e \rangle \top$ of e holds, $\overset{e}{\to}$ commutes with $\overset{\Diamond}{\to}$ and with $\bar{\to}$

$$\begin{array}{lll} (e\Diamond) & \langle e \rangle \top & \supset \; (\langle e \rangle \Diamond A \equiv \Diamond \langle e \rangle A) \\ (e-) & \langle e \rangle \top & \supset \; (\langle e \rangle \langle - \rangle A \equiv \langle - \rangle \langle e \rangle A) \end{array}$$

provided, for every α, $Stage_\alpha$ is prefix-closed. The closure condition (a_+) yields

($\mathcal{E}-$) $\quad \langle - \rangle A \wedge \langle - \rangle B \supset \langle - \rangle (A \wedge \Diamond B) \quad$ for A and $B \in \mathcal{E}$

where the set $\mathcal{E}$ of "existential" $\mathcal{L}$-formulas A is generated by

$$A \quad ::= \quad \varphi \mid \neg\varphi \mid \langle e \rangle A \mid \langle - \rangle A \mid \Diamond A \mid A \wedge A \mid A \vee A$$

(recalling that $A \vee B$ abbreviates $\neg(\neg A \wedge \neg B)$). Beyond (+1), (+2), ($e\Diamond$), ($e-$) and ($\mathcal{E}-$), we have the S5 axioms for $\langle - \rangle$ and for $\Diamond$. But we lose symmetry for combinations $\langle - \rangle \Diamond$ etc because $\bar{\to}$ may fail to commute with $\overset{\Diamond}{\to}$.

4 Processing layers of representations

To factor the transition $(\vec{e}, \vec{\varphi})_\alpha \rightsquigarrow (\vec{\varphi}, \vec{\psi})_\beta$, let us cheat a bit, and rewrite it as

$$(\vec{e}, \vec{\varphi})_\alpha \quad \overset{;}{\to} \quad (\vec{e}, \vec{\psi})_{\alpha;\beta}$$

where $Stage_{\alpha;\beta} = \{(\vec{e}, \vec{\psi}) : (\exists \vec{\varphi})\ (\vec{e}, \vec{\varphi}) \in Stage_\alpha \ \&\ (\vec{\varphi}, \vec{\psi}) \in Stage_\beta\}$. By passing to $\overset{;}{\to}$, we preserve not only the inputs $\vec{e}$, but also the intuition that $\vec{\varphi}$ is the underspecified expression obtained from translating $\vec{e}$ using α, while $\vec{\psi}$ is its refinement from β. Now, we can factor $(\vec{e}, \vec{\varphi})_\alpha \overset{;}{\to} (\vec{e}, \vec{\psi})_{\alpha;\beta}$ as

$$(\vec{e}, \vec{\varphi})_\alpha \ \bar{\to}\ (\vec{e}, \vec{\varphi})_\gamma \ \overset{\Diamond}{\to}\ (\vec{e}, \vec{\psi})_\gamma \ \bar{\to}\ (\vec{e}, \vec{\psi})_{\alpha;\beta} \ ,$$

appealing to the closure condition (a_+) on I for a γ such that

$$\text{prefix-closure}(Stage_\alpha \cup Stage_{\alpha;\beta}) \ \subseteq\ Stage_\gamma \ .$$

4.1 Monotonic interpretation (adding translation to entailment)

The index $\alpha; \beta$ need *not* be introduced above through a global closure condition on I taking ; to be a total binary function on I. Instead, we might, as in category theory, conceive of ; as a partial operation that is defined only when certain conditions are met (matching domains and co-domains, etc).[5] With this in mind, define $\overset{;}{\to} \subseteq STAGE \times STAGE$ by

$$(\vec{e}, \vec{\varphi})_\alpha \ \overset{;}{\to}\ (\vec{e'}, \vec{\varphi'})_{\alpha'} \quad \text{iff} \quad \vec{e} = \vec{e'} \text{ and for some } \beta \text{ s.t. } (\alpha, \beta) \in dom(;), \ \alpha' = \alpha; \beta \text{ and } (\vec{\varphi}, \vec{\varphi'})_\beta \in STAGE \ ,$$

the modal operator for which let us denote $\langle ; \rangle$. Beyond the $\mathcal{L}$-scheme

$(; - \Diamond) \qquad \langle ; \rangle A \supset \langle - \rangle \Diamond \langle - \rangle A \qquad \text{for } A \in \mathcal{L}$

(recording our factorization of $\overset{;}{\to}$ above), we might also stipulate

(;mon) $\qquad \varphi \supset [;]\varphi \qquad \text{for } \varphi \in E$

(where $[;]$ is $\neg\langle ; \rangle\neg$) which unwinds to the constraint that for all $(\alpha, \beta) \in dom(;)$,

$$\text{whenever } (\vec{e}, \vec{\varphi})_\alpha \overset{;}{\to} (\vec{\varphi}, \vec{\psi})_\beta \text{ and } \vec{\varphi} \vdash_\alpha \varphi, \ \ \vec{\psi} \vdash_\beta \varphi.$$

[5]The same conditions presumably for $(\vec{e}, \vec{\varphi})_\alpha \rightsquigarrow (\vec{\varphi}, \vec{\psi})_\beta$.

Thus, to interpret underspecified representations of α monotonically (i.e. in accordance with (;mon)), it suffices, assuming $Stage_\beta$ is prefix-closed for every $\beta \in I$, that

$$(8) \qquad \varphi_1 \cdots \varphi_n \vdash_\alpha \varphi \quad \text{iff} \quad (\epsilon, \epsilon)_\alpha \Vdash [;][\varphi_1] \cdots [\varphi_n]\varphi$$

for all $\varphi_1, \ldots, \varphi_n, \varphi \in E$. Given definitions of $\vdash_\beta$ for every β such that $(\alpha, \beta) \in dom(;)$, (8) can serve as a recipe for cooking up $\vdash_\alpha$.

4.2 Conclusion

In closing, let us turn to the obvious question: what about an axiomatization of $\mathcal{L}$, extending say that in §1.1? Short of establishing a completeness theorem (which I believe is within reach), I hope to have outlined what $\mathcal{L}$ ought to look like for such a theorem to be interesting. A distinctive feature of $\mathcal{L}$ is its representationalism, eschewing reductions of expressions in E to sets of possible worlds. In this connection, one further construct might be considered for $\mathcal{L}$ — viz. atomic $\mathcal{L}$-formulas of the form $\underline{\varphi}$, for $\varphi \in E$, marking explicit (as opposed to merely entailed) information inasmuch as

$$(\vec{e}, \vec{\varphi})_\alpha \Vdash \underline{\varphi} \quad \text{iff} \quad \varphi \text{ is listed among } \vec{\varphi} \,.$$

Alternatively, $\underline{\varphi}$ might be reconceptualized as $\underline{\Diamond}\varphi$ for some modal operator $\underline{\Diamond}$ expressing a transition from α to an index whose notion of entailment is look-up.

References

Alshawi, H., and R. Crouch. 1992. Monotonic semantic interpretation. In *Proc. 30th Annual Meeting of the Association for Computational Linguistics.*

Fernando, T. 1999a. A modal logic for non-deterministic discourse processing. *Journal of Logic, Language and Information* 8(4).

Fernando, T. 1999b. Non-monotonicity from constructing semantic representations. In *Proc. 12th Amsterdam Colloquium.* ILLC, University of Amsterdam.

Gazdar, G. 1979. *Pragmatics: Implicature, Presupposition and Logical Form.* Academic Press, New York.

Heim, I. 1982. *The Semantics of Definite and Indefinite Noun Phrases.* Dissertation, University of Massachusetts, Amherst (Garland Press, New York, 1988).

Heim, I. 1983. On the projection problem for presuppositions. In *Proc. West Coast Conference on Formal Linguistics.* Stanford Linguistics Association.

Kamp, H. 1971. Formal properties of 'now'. *Theoria* 37.

Kamp, H. 1990. Prolegomena to a structural account of belief and other attitudes. In *Propositional Attitudes.* CSLI Lecture Notes Number 20, Stanford.

Karttunen, L. 1974. Presupposition and linguistic context. *Theoretical Linguistics* 1:181–194.

Marx, M., and Y. Venema. 1997. *Multi-Dimensional Modal Logic.* Applied Logic Series Number 4. Kluwer Academic Publishers, Dordrecht.

McCarthy, J., and S.Buvač. 1997. Formalizing context. In *Computing Natural Language.* CSLI, Stanford.

Stalnaker, R. 1978. Assertion. In *Syntax and Semantics.* Academic Press, New York.

van der Sandt, R. 1991. Denial. In *Chicago Linguistic Society 27: Parasession on Negation.*

van der Sandt, R. 1992. Presupposition projection as anaphora resolution. *Journal of Semantics* 9(4).

Tim Fernando
Computer Science Department
Trinity College
Dublin 2, Ireland
E-mail: Tim.Fernando@tcd.ie
URL: http://www.cs.tcd.ie/Tim.Fernando/

9

Algorithmic Correspondence Theory for Substructural Categorial Logic

MARCELO FINGER

ABSTRACT. Substructural categorial connectives can be treated as modalities. Such binary connectives have a possible worlds semantics based on ternary accessibility relations. This modal treatment allows one to explore *categorial correspondence theory*, in analogy to the usual correspondence theory for modal logics. Its aim is to find a first-order restriction over ternary frames corresponding to a categorial sequent.

This paper proposes an algorithmic method that deals with categorial correspondence theory. It proposes a proof theoretical method based on *SLaKE-tableaux* that produces a second-order formula corresponding to a given categorial sequent. When the SLaKE-tableau is finite, a valuation for the propositional atoms is obtained from the tableau; a first-order formula equivalent to the given sequent over ternary frames is thus computed. For infinite tableaux, no first-order formula is computed.

1 Introduction

Since Kripke (Kripke 1963) has proposed a possible worlds semantics for modal logics, it has been noted that the presence of certain modal axioms impose specific restrictions on the binary accessibility relation of Kripke frames. For example, it is well known that the axiom $\Box p \to p$ is true at any frame $(W, <)$ whose binary accessibility relation is reflexive, $\forall x(x < x)$; conversely, models of any system that has such a formula as a theorem must have a reflexive accessibility relation.

In fact, any modal axiom can be translated to a second-order formula. Some of these second-order formulas are equivalent to first-order

Advances in Modal Logic, Volume 3
F. Wolter, H. Wansing, M. de Rijke, and M. Zakharyaschev, eds.

formulas over Kripke frames. At the heart of *modal correspondence theory* lies the identification of which axioms correspond to some first-order restriction over Kripke frames (van Benthem 1984). A special class of modal axioms, known as the Sahlqvist formulas, is guaranteed to generate first-order restrictions over Kripke frames, and such restrictions can be obtained algorithmically (Sahlqvist 1975).

The work of Routley and Meyer (1973) has shown that relevance logics can be treated as a modal logic with a Kripke-style semantics. This was later extended to other logics in the family of substructural logics (Došen 1993, Restall 2000), and in her PhD thesis, Natasha Kurtonina (1994) presented such semantics for a fragment of substructural logics, known as categorial logics, without any structural pressuposition. In analogy to traditional modal logics, categorial logics have three binary connectives and their semantics is based on a *ternary*[1] accessibility relation $R \subseteq W^3$. The connectives generally found in categorial logics (Moortgat 1997, Carpenter 1997) are here[2] represented as $\bullet$ (called *product*, or *fusion*, or *multiplicative conjunction*), / (*slash* or *right-implication*) and \ (*backslash* or *left-implication*).

Categorial logics are just a fragment of substructural logics. In a sequent presentation of substructural logics, it is also known that the presence of certain structural rules correspond to derivability of certain sequents. For example, any system that allows for the structural rule of associativity derives $A/B \vdash (A/C)/(B/C)$.

With a Kripke-style semantics for categorial logics, a semantic connection between derivable sequents and semantic restrictions could be investigated. The idea of a correspondence theory for substructural logics was proposed by Roorda (1991). Kurtonina (1994) later developed several methods to obtain first-order restrictions over ternary Kripke frames corresponding to categorial formulas; each method deals with a different fragment of categorial logics. One of these methods involved translations of formulas into a suitable fragment; another method was more in the modal logic tradition, generating a second-order formula and a valuation of second-order variables that yields a first-order equivalent formula.

This work pursues further the topic of correspondence theory for categorial logics started by Roorda (1991) and Kurtonina (1994). Our approach is algorithmic, pursuing a unifying method for all categorial formulas. We propose a tableaux-based method, called *SLaKE*[3]*-tableaux*,

[1] In general, an n-ary connective is defined in terms of an $(n+1)$-ary accessibility relation.

[2] These connectives are also found in the literature as $\otimes$, $\rightarrow$and $\leftarrow$.

[3] **S**ubstructural **La**belled **KE**-tableaux

to construct a first-order formula corresponding to a categorial sequent. SLaKE-tableaux are based on KE-tableaux (D'Agostino 1992) over T- and F-signed formulas. Furthermore, to deal with its substructural side, formulas are labelled following the Labelled Deductive System discipline (Gabbay 1996, D'Agostino and Gabbay 1994, Broda et al. 1996a). We show that if a SLaKE-tableau is deterministic and finitely saturates (ie, no branches can be further expanded) a first-order formula corresponding to the input sequent can be computed.

The presentation of this tableau-based method for correspondence theory proceeds as follows. Section 2 provides the background for ternary frames and KE-tableaux. This enables the definition of SLaKE-tableaux in Section 3 and the algorithm for constructing a correspondence formula for categorial sequents. We then show several examples of how to construct such a formula in Section 4, followed by a demonstration of the method's correctness in Section 5. We conclude in Section 6 listing the work that still has to be done in substructural correspondence theory.

2 Background

2.1 Ternary Frames

Ternary frames are relational structures that allow us to view the connectives of (substructural) categorial logic as binary modalities. In our case, we will deal with the fragment containing the usual categorial binary connectives, / (slash), \ (backslash) and • (product). The • connective is assumed to be left-associative, that is, $A \bullet B \bullet C \equiv (A \bullet B) \bullet C$. A sequent calculus presentation for such fragment is shown in Figure 1. The antecedents of sequents are binary trees, where $\Gamma[A, B]$ indicates that (A, B) is a subtree of Γ; antecedents are also left associative, that is, $\Gamma, \Delta, \Sigma \equiv (\Gamma, \Delta), \Sigma$. Consequents of sequents are simple formulas. Note that no structural rules are being admitted a priori, and therefore we are in the *non-associative Lambek Calculus* Lambek 1958.

On the semantic side, we define a *ternary frame* as a pair $\mathfrak{F} = (W, R)$, where W is a non-empty set of *possible worlds* and R is a ternary *accessibility relation.* A *ternary model* $\mathcal{M} = (W, R, V)$ consists of a ternary frame plus a *valuation* $V : \mathcal{P} \rightarrow 2^W$, mapping propositional letters to sets of worlds.

The semantic interpretation of categorial formulas in the $\{/, \backslash, \bullet\}$-fragment over ternary frames is:

$$
\begin{array}{lll}
\mathfrak{F}, V, a \models p & \text{iff} & a \in V(p) \\
\mathfrak{F}, V, a \models A/B & \text{iff} & \forall b \forall c (Rcab \wedge \mathfrak{F}, V, b \models B \Rightarrow \mathfrak{F}, V, c \models A) \\
\mathfrak{F}, V, a \models B \backslash A & \text{iff} & \forall b \forall c (Rcba \wedge \mathfrak{F}, V, b \models B \Rightarrow \mathfrak{F}, V, c \models A) \\
\mathfrak{F}, V, a \models B \bullet A & \text{iff} & \exists b \exists c (Rabc \wedge \mathfrak{F}, V, b \models A \wedge \mathfrak{F}, V, c \models B)
\end{array}
$$

$$\frac{}{A \vdash A}\ (Axiom)$$

$$\frac{\Gamma[A] \vdash C \quad \Delta \vdash B}{\Gamma[A/B, \Delta] \vdash C}\ (/ \vdash) \qquad \frac{\Gamma, B \vdash A}{\Gamma \vdash A/B}\ (\vdash /)$$

$$\frac{\Gamma[A] \vdash C \quad \Delta \vdash B}{\Gamma[\Delta, B\backslash A] \vdash C}\ (\backslash \vdash) \qquad \frac{B, \Gamma \vdash A}{\Gamma \vdash B\backslash A}\ (\vdash \backslash)$$

$$\frac{\Gamma[A, B] \vdash C}{\Gamma[A \bullet B] \vdash C}\ (\bullet \vdash) \qquad \frac{\Gamma \vdash A \quad \Delta \vdash B}{\Gamma, \Delta \vdash A \bullet B}\ (\vdash \bullet)$$

FIGURE 1 A sequent calculus presentation of the non-associative Lambek Calculus.

We also use the abbreviations:

$$Rab(cd) =_{def} \exists x(Rabx \wedge Rxcd)$$
$$Ra(bc)d =_{def} \exists y(Rayd \wedge Rybc)$$

Such abbreviation associates to the left: $Ra(bcd)e =_{def} Ra((bc)d)e$, etc. As usual, we write $\mathfrak{F}, a \models A$ when $\mathfrak{F}, V, a \models A$ for any valuation V; if a is omitted, this means that the condition holds for any possible world.

A sequent $A_1, \ldots A_n \vdash C$ holds at a world $a \in W$ in ternary frame $\mathfrak{F}$ (notation: $\mathfrak{F}, a \models (A_1, \ldots A_n \vdash C)$) iff $\mathfrak{F}, V, a \models A_1 \bullet \ldots \bullet A_n$ implies $\mathfrak{F}, V, a \models C$. This is equivalent to saying that if for some $a_1, \ldots, a_n$, $Ra(a_1 \ldots a_{n-1})a_n$ and, for $1 \leq i \leq n$, $\mathfrak{F}, V, a_i \models A_i$, this implies $\mathfrak{F}, V, a \models C$.

Note that $\mathfrak{F}$ can be seen as a first-order model for a language over R and $P_1, \ldots, P_n$, where each P_i is a predicate symbol corresponding to the propositional letter p_i. The notation $(\forall P)\varphi$ indicates the universal closure of all second-order variables in φ, and $(\exists P)\varphi$ the existential one. Following the modal logic tradition, the *standard translation* of a categorial formula A into second-order logic formula is $(\forall P)\forall a ST_a(A)$, where:

$$\begin{aligned} ST_a(p_i) &= P_i(a) \\ ST_a(A \bullet B) &= \exists bc(Rabc \wedge ST_b(A) \wedge ST_c(B)) \\ ST_a(B/A) &= \forall bc(Rcab \wedge ST_b(A) \Rightarrow ST_c(B)) \\ ST_a(A\backslash B) &= \forall bc(Rcba \wedge ST_b(A) \Rightarrow ST_c(B)) \end{aligned}$$

The second-order quantification is over the relevant predicate symbols and reflects all relevant valuations in a frame. So every categorial (modal) sequent $A \vdash C$ corresponds to a second-order formula such that:

$$\mathfrak{F} \models (A \vdash C) \textit{ iff } \mathfrak{F} \models (\forall P)\forall a\, (ST_a(A) \Rightarrow ST_a(C))$$

The crucial point of correspondence theory is to know when such a second-order formula defines a *first-order* frame property, that is, if there is a first-order formula ψ such that

$$\mathfrak{F} \models (A \vdash C) \textit{ iff } \mathfrak{F} \models \psi$$

It is the computation of such a property, when it exists, that we investigate next by means of KE- and SLaKE-tableaux.

2.2 KE Tableaux

D'Agostino has shown in (D'Agostino 1992) that analytic tableaux, in the style proposed by Smullyan (Smullyan 1968), cannot polynomially simulate truth tables and in some cases perform exponentially worse than them. To avoid such problems in a principled way, KE-tableaux were introduced.[4]

As usual, KE-tableaux deal with *signed formulas*, where each formula is signed with a T (truth) or an F (falsity). The signed formulas $T\ A$ and $F\ A$ are called *opposite formulas*. A tableau branch that contains a pair of opposite formulas is *closed*. A theorem is proved by refutation, trying to close all branches of the tableau. Each connective has a pair of expansion rules, that decompose a signed formula into smaller signed formulas.

Unlike analytic tableaux, all decomposition rules in a KE-tableaux are linear. Some of the rules are single premised, while others take two premises. For instance, we present here the KE-rules for classical implication ($\Rightarrow$): a two-premised rule for positively signed formulas ($T\ B \Rightarrow A$) with a single conclusion, and one single premised rule for negatively signed ones ($F\ B \Rightarrow A$) with a double conclusion:

$$\frac{\begin{array}{c}T\ B \Rightarrow A\\ T\ B\end{array}}{T\ A} \qquad \frac{F\ B \Rightarrow A}{\begin{array}{c}T\ B\\ F\ A\end{array}}$$

Similarly, a pair of linear rules can be given to any of the classical boolean connectives. The only branching rule in a KE-tableau is the *Principle of Bivalence* (PB), that states that a formula can be either true or *not true*[5]:

[4]No one seems to know what KE stands for; apparently K stands for "klassisch", used in analogy to Gentzen's LK system; E may stand for "elimination" (of what?). KE was supposed to be just an initial working name, but somehow it stuck.

[5]Note that this is different from *true or false* — the principle of excluded middle — which is not valid for all substructural logics.

$$\overline{T\,A \qquad F\,A}$$

Usually, the branching formula A is chosen to generate a second premise to some linear rule, so it is always a subformula of some existing formula in the tableau. As a consequence, KE-tableaux branch fewer times than a normal analytic tableau, and it can p-simulate a truth table.

3 Substructural Labelled KE Tableaux

Substructural Labelled KE (SLaKE) tableaux are the main proof theoretical equipment we use in the generation of correspondence formulas for substructural sequents.

The use of KE tableaux for substructural logics have been proposed in (D'Agostino and Gabbay 1994), by attaching a label to the signed formula, as in $T\ A : a$. We use this idea, but without following the labelling discipline developed in (D'Agostino and Gabbay 1994, Broda and Finger 1995, Broda et al. 1996b). Instead, we simply add a new label at each new node of the tableau. Formally, let $\mathcal{L}$ be a countable set of labels, let A be a categorial formula; then for every $a \in \mathcal{L}$, the formulas $T\ A : a$ and $F\ A : a$ are *signed labelled formulas.*

Furthermore, each finite SLaKE-tableau is associated with a *correspondence formula.* The original sequent is associated with an initial tableau and with a second-order formula. Each of the tableau linear expansion rules is associated with an expansion of the correspondence formula of the form $\sharp_i := \varphi(R, A_1, \ldots, A_n, \sharp_{i+1})$, where R is the ternary accessibility relation, $A_1, \ldots, A_n$ are the formulas generated in the expansion, and $\sharp$ is the "substitution place" for next expansion and can be read simply as *truth.* The tableau rules for SLaKE-tableaux are illustrated in Figure 2.

In each linear rule in Figure 2, the formulas above the horizontal line are the *premises* of the rule, and those below it are the *conclusions* of the rule. There are one-premised and two-premised rules, but each rule has exactly one premise that is a *compound* formula, which is called the *main premise*; other premises are called *auxiliary.* Two-premised rules are $\forall$-rules and one-premised rules are $\exists$-rules. If either of the conclusions of an $\exists$-rule is present on the current branch, it is not added again with a new label. $\forall$-rules always generate a new conclusion.

The last rule in Figure 2 is the Principle of Bivalence (PB) branching rule. It is only applied for a formula A following the *branching heuristics*:

> PB is used for a formula A that serves as an auxiliary premise for a $\forall$-rule; PB is only applied in a branch when no other linear expansion is possible.

SLaKE Expansion	*Formula Expansion*	*Rxyz*
$T\ A/B : a$ $T\ B : b$ ——— $T\ A : c$ (new c)	$\sharp_i := \forall c(Rcab \Rightarrow (ST_c(A) \wedge \sharp_{i+1}))$	$Rcab$
$F\ A/B : a$ ——— $T\ B : b$ (new b) $F\ A : c$ (new c)	$\sharp_i := \exists b \exists c(Rcab \wedge ST_b(B) \wedge \neg ST_c(A) \wedge \sharp_{i+1})$	$Rcab$
$T\ B\backslash A : a$ $T\ B : b$ ——— $T\ A : c$ (new c)	$\sharp_i := \forall c(Rcba \Rightarrow (ST_c(A) \wedge \sharp_{i+1}))$	$Rcba$
$F\ B\backslash A : a$ ——— $T\ B : b$ (new b) $F\ A : c$ (new c)	$\sharp_i := \exists b \exists c(Rcba \wedge ST_b(B) \wedge \neg ST_c(A) \wedge \sharp_{i+1})$	$Rcba$
$T\ A \bullet B : a$ ——— $T\ A : b$ (new b) $T\ B : c$ (new c)	$\sharp_i := \exists b \exists c(Rabc \wedge ST_b(A) \wedge ST_c(B) \wedge \sharp_{i+1})$	$Rabc$
$F\ A \bullet B : a$ $T\ A : b$ ——— $F\ B : c$ (new c)	$\sharp_i := \forall c(Rabc \Rightarrow (\neg ST_c(B) \wedge \sharp_{i+1}))$	$Rabc$
——— $T\ A : x \quad F\ A : x$	$\sharp_i := \forall x((ST_x(A) \wedge \sharp^1_{i+1}) \vee (\neg ST_x(A) \wedge \sharp^2_{i+1}))$	—

FIGURE 2 SLaKE rules

The main premises that trigger the application of PB for A are: $F\ A \bullet B$, $T\ A\backslash B$ and $T\ B/A$. The corresponding $\forall$-rule will be applicable on the $T\ A$ branch. This heuristics guarantees that only subformulas of the original sequent will be introduced by PB. It introduces two "substitution places" in the correspondence formula, $\sharp^1_{i+1}$ and $\sharp^2_{i+1}$, one for each new branch. A branch that can still be expanded is called *active*. Each active branch in a SLaKE tableau always has exactly one substitution place.

The importance of substitution places is that they guarantee that each formula introduced in the correspondence formula will "see the correct context", that is, it will be in the scope of the correct quantifiers.

A sequent of the form $A_1, \dots, A_n \vdash C$ is transformed into the initial SLaKE-tableau:

$$\begin{array}{c} T\ A_1 : a_1 \\ \vdots \\ T\ A_n : a_n \\ F\ C : a \end{array}$$

Since the tableau is a refutation method, this induces the correspondence formula:

$$\begin{aligned} \neg\varphi = \neg\exists a a_1 \ldots a_n[\quad & ST_{a_1}(A_1) \wedge \ldots \wedge ST_{a_n}(A_n) \wedge \neg ST_a(C) \\ & \wedge Ra(a_1 \ldots a_{n-1})a_n \wedge \sharp] \end{aligned}$$

A single premised sequent $A \vdash C$ generates the initial tableau containing $T\ A : a$ and $F\ C : a$, with the initial correspondence formula $\neg\varphi_1(\sharp_1) = \neg\exists a(ST_a(A) \wedge \neg ST_a(C) \wedge \sharp_1)$. We could extend the method for sequents with empty antecedents, but we do not pursue this topic here.

The aim of the SLaKE-tableau construction is *not* to close every tableau branch, but to expand each tableau branch until no more expansions are possible. Each expansion step will also give us a new version of the correspondence formula. If we can finitely expand all tableau branches, a valuation for the atomic formula is constructed, so that we obtain a first-order formula by substituting in the final formula the evaluated values. It is also possible that there will be some infinite branches (something that would not happen in simple propositional tableaux), in which case the above method is not applicable.

A *SLaKE-saturated set* Γ is a set of labelled signed formulas and of $Rxyz$ formulas such that, with respect to the rules of Figure 2:

(a) If the premise of a rule is in Γ, each of its consequence is in Γ for some label, and the $Rxyz$ formula from the $Rxyz$-column is also added to Γ.

(b) For each compound formula in Γ that is a main premise of a $\forall$-rule, there must be in Γ either an auxiliary premise or the opposite of it.

The expansion of a tableau aims at constructing branches that are SLaKE-saturated sets. Item (a) corresponds to normal branch expansion. The fact that the consequence of a $\forall$-rule is always added to Γ with a new label rises the possibility of having infinite SLaKE-saturated sets. Item (b) guides the branching heuristics. At the end of a finite expansion, a correspondence formula will be built from a suitable valuation. Let us show it through examples.

Example 3.1 Consider the sequent $p/q \vdash q\backslash p$. Its associated SLaKE tableau is:

1.	$T\ p/q : a$		
2.	$F\ q\backslash p : a$		$\neg\varphi = \neg\exists a(ST_a(p/q) \wedge \neg ST_a(q\backslash p) \wedge \sharp_1)$
3.	$T\ q : b$	from 2	
4.	$F\ p : c$	from 2	$\sharp_1 := \exists b \exists c(Rcba \wedge Q(b) \wedge \neg P(c) \wedge \sharp_2)$
5.	$T\ p : d$	from 1, 3	$\sharp_2 := \forall d(Rdab \Rightarrow P(d) \wedge \sharp_3)$

Initially, we expand line 2, and simultaneously, using the semantics of $q\backslash p$, we expand the correspondence formula substituting $\sharp_1$ into it. We then use lines 1 and 3 for another expansion, obtaining $\sharp_2$. At this point, the tableau is saturated and closed (remember our main goal is not to close a tableau, but to saturate every branch of it). So we make $\sharp_3 := \top$. We have thus built the second-order formula:

$$\begin{aligned}\neg\varphi = \neg\exists a(\quad & ST_a(p/q) \wedge \neg ST_a(q\backslash p) \wedge \\ & \exists b \exists c(Rcba \wedge Q(b) \wedge \neg P(c) \wedge \forall d(Rdab \Rightarrow P(d))))\end{aligned}$$

The formula $\neg\varphi$ is equivalent to the original sequent, so a suitable valuation has to be constructed to obtain a first-order formula. Such a valuation is implicitly built in the following way. If $T\ r : x$ occurs in a branch, then $P(x)$ must hold iff x is different from all worlds y such that $F\ r : y$ occurs in the same branch above it. Similarly, if $F\ r : x$ occurs in a branch, then $\neg P(x)$ must hold iff x is different from all worlds y such that the $T\ r : y$ occurs in the same branch above it. By applying this rule to the tableau above, we get the implicit *canonical valuations* of P and Q:

$$Q(b) \leftrightarrow \top$$
$$(\neg P(c) \leftrightarrow \top) \wedge (P(d) \leftrightarrow d \neq c)$$

Saturation guarantees that the compound formulas in $\neg\varphi$ can be substituted by $\top$. By substituting each second-order variable in $\neg\varphi$ by their canonical valuation, we obtain the first-order formula

$$\neg\exists a \exists b \exists c(\top \wedge \top \wedge Rcba \wedge \forall d(Rdab \Rightarrow d \neq c))$$

which is equivalent to $\forall a \forall b \forall c(Rcba \Rightarrow Rcab)$, meaning that R must have the *commutativity* property for its second and third positions. It is easy to verify that any model satisfying this property also satisfies the original sequent, and vice-versa. □

We will prove the correctness of this method in Section 5. First, let us present the method in detail.

A valuation V', defined over the set of possible worlds W', is said to *extend* V over the set $W \subset W'$ if the two valuations agree on the truth of all atoms at all worlds in W.

The expansion of a tableau is the stepwise construction of a *counter-model* for the input sequent. Each step generates:

- a valuation V_{i+1} extending V_i over the set of labels in the tableau at step i; and
- a second-order formula $\neg\varphi_{i+1}(\sharp_{i+1})$ that is equivalent to $\neg\varphi_i(\sharp_i)$.

Definition 3.2 Let $\mathcal{X} \in \{T, F\}$; define $\bar{\mathcal{X}}$ such that $\bar{T} = F$ and $\bar{F} = T$. For each SLaKE atomic formula $\mathcal{X}\, p : x$ in the tableau, define:

$$O(\mathcal{X}\, p : x) = \{y \mid \bar{\mathcal{X}}\, p : y \text{ occurs "above" } \mathcal{X}\, p : x\}$$

For every atomic p, define the *canonical valuation* of $P(x)$ implicitly as:

$$\bigwedge_{\mathcal{X} p:x} ((P(x) \leftrightarrow \bigwedge_{y \in O(Tp:x)} x \neq y) \wedge (\neg P(x) \leftrightarrow \bigwedge_{y \in O(Fp:x)} x \neq y))$$

The rationale of the canonical valuation is simply that an atom cannot be both true and false at the same world. This is the *minimal* condition that any model must verify. To obtain a first-order formula we need to substitute the *canonical* valuation into $\neg\varphi$. The canonical valuation is defined such that no scope violation is possible, for the substitution introduces no free variables. If the tableau branches, each sub-branch is developed independently and is not affected the labels introduced at other branches. This has the effect of restricting the scope of quantifiers in the construction of the correspondence formula.

If the tableau saturates, such a valuation satisfies all compound formulas (Lemma 5.3). The first-order formula $\neg\psi$ thus obtained is obviously implied by $\neg\varphi$.

Let $(\exists P)\varphi$ be the second order existential closure of φ. If the tableau saturates in finitely many steps, then a countermodel $\mathfrak{F}^*$ must satisfy $\mathfrak{F}^* \models (\exists P)\varphi$. That is, the class of models that validate the initial sequent does not contain $\mathfrak{F}^*$. If there are only finitely many possible refutations of the initial sequent, we obtain finitely many (say m) second-order formulas $\neg\varphi_j$, one for each refutation, that when substituted by the correspondent canonical valuation each generate $\neg\psi_j$ $(1 \leq j \leq m)$, all implied by $\neg\varphi$.

Furthermore, due to saturation, any possible refutation must satisfy one of the φ_j, so a countermodel to the input sequent satisfies the second-order formula $(\exists P) \bigvee \varphi_j$, which is equivalent to $(\exists P)\varphi$. But, with the constructed counter valuations we have a witness for that existential second-order quantification, so each φ_j implies its correspondent ψ_j. And therefore $\neg\psi_j$ implies $\neg\varphi_j$ and $\bigwedge_{j=1}^{m} \neg\psi_j$ implies (and thus is equivalent to) $\neg\varphi_j$.

As a consequence, if there is a single way to finitely refute a sequent, the first-order formula generated $\neg\psi$ is equivalent to the second-order formula $\neg\varphi$, which is equivalent to the validity of the input sequent. This motivates the following definition.

A SLaKE-tableau is *deterministic* if at every point of its extension there is only one applicable rule. From what has just been explained above, finitely-saturated deterministic tableaux generate a first-order formula equivalent to its input sequent. If the tableau is finite and non-deterministic, one has to generate all the possible expansions, and for each one compute its associated formula $\neg\psi_i$; the first-order formula equivalent to the validity of the input sequent is the conjunction of all such formulas.

We now present the full algorithm for computing the correspondence formula.

Algorithm 3.1
Input: a sequent $A_1, \ldots, A_n \vdash C$.
Output: its first-order correspondence formula, if there are only finite refutations.

1. Initialize the tableau for $T\ A_1 : a_1, \ldots, A_n : a_n$ and $F\ C : a$, with initial correspondence formula $\neg\varphi(\sharp)$.
2. Repeat while the tableau is not saturated nor an infinite branch has been detected.
   (a) If there is an applicable rule, expand the tableau and the correspondence formula according to the rules in Figure 2.
   (b) If there are no linear rules applicable but the tableau is not saturated, choose a complex formula over which to apply the branching rule PB and continue expanding both branches.
3. If the tableau has an infinite branch, stop.
4. Otherwise, a second-order formula $\neg\varphi$ was generated. Construct the canonical valuations for every atom p.
5. If the tableau is deterministic, output the first-order formula $\neg\psi$ obtained by substituting all compound formulas by $\top$ and all $P(x)$ by its canonical valuation at the time of introduction. Otherwise, repeat items 2, 3 and 4 for each of the possible refutations, each generating $\neg\psi_j$; output $\bigwedge \neg\psi_j$.

Next we see a few more examples.

4 Examples

Example 4.1 [Finite Deterministic Non-Branching Tableau]

Consider the sequent $p/q \vdash (p/r)/(q/r)$; such sequent is not in the format of the "Sahlqvist-van Benthem" Theorem in (Kurtonina 1994) and had to be dealt with by means of a translation method. A SLaKE-tableau constructed for it looks like:

1.	$T\ p/q : a$		$\neg\varphi = \neg\exists a(ST_a(p/q) \wedge$
2.	$F\ (p/r)/(q/r) : a$		$\neg ST_a((p/r)/(q/r)) \wedge \sharp_1)$
3.	$T\ (q/r) : b$	from 2	$\sharp_1 := \exists b \exists c(Rcab \wedge$
4.	$F\ (p/r) : c$	from 2	$ST_b(q/r) \wedge \neg ST_c(p/r) \wedge \sharp_2)$
5.	$T\ r : d$	from 4	
6.	$F\ p : e$	from 4	$\sharp_2 := \exists d \exists e(Recd \wedge R(d) \wedge \neg P(e) \wedge \sharp_3)$
7.	$T\ q : f$	from 3,5	$\sharp_3 := \forall f(Rfbd \Rightarrow (Q(f) \wedge \sharp_4))$
8.	$T\ p : g$	from 1,7	$\sharp_4 := \forall g(Rgaf \Rightarrow (P(g) \wedge \sharp_5))$

The *canonical valuation* for $P(w)$ in $\neg\varphi$ is:

$$(P(g) \leftrightarrow g \neq e) \wedge (\neg P(e) \leftrightarrow \top)$$
$$Q(f) \leftrightarrow \top$$
$$R(d) \leftrightarrow \top$$

By substituting the canonical valuation to $\neg\varphi$, which also makes all compound formulas true, we obtain:

$$\neg\exists a \exists b \exists c(Rcab \wedge \exists d \exists e(Recd \wedge \forall f(Rfbd \Rightarrow \forall g(Rgaf \Rightarrow g \neq e)))$$

which is equivalent to

$$\forall a \forall b \forall d \forall e(\quad Re(ab)d \Rightarrow Rea(bd) \quad).$$

That is, the sequent $A/B \vdash (A/C)/(B/C)$ imposes a restriction of *left-associativity* to the second and third positions of the ternary relation R. Note that it is well know from substructural logics that the sequent $A/B \vdash (A/C)/(B/C)$ is provable whenever the structural rule of associativity is accepted (Došen 1993). This shows a remarkable connection between proof-theoretical properties and ternary frame semantics, which holds for other axioms too.

Example 4.2 [Finite Deterministic Branching Tableau]

Now let us see an example with branching. For that, consider Peirce's Axiom $(p\backslash q)\backslash p \vdash p$. When the tableau is initialized, no linear rules are applicable, so the branching heuristics is applied to $p\backslash q$, corresponding to the valid disjunctive statement $\forall b(ST_b(p\backslash q) \vee \neg ST_b(p\backslash q))$:

$$\begin{array}{cc} \multicolumn{2}{c}{1.\ T\ (p\backslash q)\backslash p : a} \\ \multicolumn{2}{c}{2.\ \ F\ p : a} \\ \hline 3(i).\ T\ p\backslash q : b & 3(ii).\ F\ p\backslash q : b \\ \hline 4(i).\ T\ p : c & 4(ii).\ T\ p : e \\ \cline{1-1} 5(i).\ T\ q : d & 5(ii).\ F\ q : f \end{array}$$

which has the corresponding second-order expansion:

$$\begin{array}{ll} \multicolumn{2}{c}{\neg\varphi = \neg\exists a(ST_a((p\backslash q)\backslash p) \wedge \neg P(a) \wedge \sharp_1)} \\ \hline \multicolumn{2}{c}{\sharp_1 := \forall b((ST_b(p\backslash q) \wedge \sharp_2^i) \vee (\neg ST_b(p\backslash q) \wedge \sharp_2^{ii})))} \\ \hline \sharp_2^i = \forall c(Rcba \Rightarrow (P(c) \wedge \sharp_3^i)) & \\ \cline{1-1} \sharp_3^i = \forall d(Rdcb \Rightarrow (Q(d) \wedge \sharp_4^i)) & \sharp_2^{ii} = \exists e \exists f(Rfeb \wedge P(e) \wedge \neg Q(f) \wedge \sharp_3^{ii}) \end{array}$$

The branches are independently developed. The left branch will cause substitutions in $\sharp_2^i$ and the right branch will cause substitutions in $\sharp_2^{ii}$. Since the substitution occurs in a negative context, the branching will impose a conjunction of constraints, each of which can be computed separately. New labels could be repeated on both branches, since there will never be a quantifier scope confusion, but for clarity reasons we will always use variables new to the entire tableau.

If we were looking for a closed tableau, we could stop the development of the left branch at $4(i)$; however, our aim here is to obtain a *SLaKE-saturated* set at each branch, so we proceed to obtain $5(i)$. The correspondence formula obtained after both branches are saturated and negation is pushed inside is:

$$\begin{array}{ll} \forall a \exists b[& \neg(ST_a((p\backslash q)\backslash p) \wedge \neg P(a) \wedge ST_b(p\backslash q) \wedge \\ & \forall c(Rcba \Rightarrow (P(c) \wedge \forall d(Rdcb \Rightarrow Q(d))))) \ \ ] \wedge \\ \forall a \exists b[& \neg(ST_a((p\backslash q)\backslash p) \wedge \neg P(a) \wedge \neg ST_b(p\backslash q) \wedge \\ & \exists e \exists f(Rfeb \wedge P(e) \wedge \neg Q(f))) \ \ ] \end{array}$$

The first conjunct corresponds to the left branch, and the second conjunct corresponds to the right one. Since each branch is developed independently, the canonical valuation is developed as before, that is:

$$\begin{array}{c} (\neg P(a) \leftrightarrow \top) \wedge (P(c) \leftrightarrow c \neq a) \wedge (P(e) \leftrightarrow e \neq a) \\ (Q(d) \leftrightarrow \top) \wedge (\neg Q(f) \leftrightarrow \top) \end{array}$$

The first-order correspondence formula thus obtained is:

$$\begin{array}{c} \forall a \exists b[\neg(\forall c(Rcba \Rightarrow (c \neq a \wedge \forall d(Rdcb \Rightarrow \top)))) \\ \wedge \\ \neg(\exists e \exists f(Rfeb \wedge e \neq a \wedge \top))] \end{array}$$

which is equivalent to

$$\forall a \exists b(Raba \wedge \forall ef(Rfeb \Rightarrow e = a))$$

Kurtonina (Kurtonina 1994) has shown that this condition plus associativity (Example 4.1) and commutativity (Example 3.1) makes the ternary model collapse, that is, they imply $\forall abc(Rabc \Rightarrow a = b = c)$.

Example 4.3 [Finite Non-Deterministic Tableau]

All SLaKE-tableaux seen so far have been *deterministic* in the sense that, at every step, there was a single expansion rule to be applied to a branch. Consider now a sequent for which we have more than one applicable expansion at the initial step:

(1) $$A/B, B, A/C, C \vdash A$$

whose initial expansion has two choices: $T\ A/B : a$ with $T\ B : b$, or $T\ A/C : c$ with $T\ C : d$. If the first pair is chosen first, the correspondence formula obtained is

(2) $$\forall abcde(\ Re(abc)d \Rightarrow (Reab \vee (Recd \wedge \exists fRfab))\ ),$$

and if the second pair is chosen first, a different correspondence formula is obtained:

(3) $$\forall abcde(\ Re(abc)d \Rightarrow (Recd \vee (Reab \wedge \exists gRgcd))\ ).$$

Since both formulas constrain the frame, we take the *conjunction* of (2) and (3), thus obtaining

(4) $$\forall abcde(\ Re(abc)d \Rightarrow (Reab \vee Recd)\ ).$$

That is, with non-deterministic SLaKE-tableaux, we have to take the conjunction of the correspondence formulas generated by all possible tableaux. Since we have exhausted *all possible refutations*, this means that a frame validates (1) iff it satisfies (4).

4.1 Infinite Branches

It is not always the case that a tableau branch can be finitely saturated. In those cases we cannot apply the method above, so we do not get a first-order formula. We present next two categorial formulas that generate infinite tableaux. It is known from (Kurtonina 1994) that these formulas have no correspondent first-order formula, so it is quite reassuring that our method does not produce one.

For example, consider the tableau for $A/A \vdash A\backslash A$:

1.	$T\ A/A : a$	
2.	$F\ A\backslash A : a$	
3.	$T\ A : b$	from 2
4.	$F\ A : c$	from 2
5.	$T\ A : d$	from 1 and 3
6.	$T\ A : e$	from 1 and 5
7.	$T\ A : f$	from 1 and 6
	$\vdots$	

Another example is $(A\backslash A)\backslash A \vdash A$:

1. $T\ (A\backslash A)\backslash A : a$
2. $F\ A : a$

$3i.$	$T\ A\backslash A : b$	$3ii.$	$F\ A\backslash A : b$
$4i.$	$T\ A : c$	$4ii.$	$T\ A : e$
$5i.$	$T\ A : d$ from $3i$ and $4i$	$5i.$	$F\ A : f$
$6i.$	$T\ A : e$ from $3i$ and $5i$		
	$\vdots$		

Note that the infinite branches in the tableau above are characterized by cycles of T-signed formulas $T\ A/A$ or $T\ A\backslash A$ and $T\ A$. Such cycles can be of larger size and involve more complex formulas, but as long as a single element of the cycle is derived a second time (with a new label), an infinite branch is guaranteed, for this element will then be regenerated infinitely often.

5 Correctness of the Method

We now prove the correctness of the method developed in this work. We first note the following property of SLaKE expansions.

Lemma 5.1 *Suppose a branch expansion step has transformed* $\neg\varphi_n(\sharp_n)$ *into* $\neg\varphi_{n+1}(\sharp_{n+1})$*. Then* $\neg\varphi_n(\sharp_n)$ *and* $\neg\varphi_{n+1}(\sharp_{n+1})$ *are logically equivalent over the class of all frames.*

Proof. Clearly, $\varphi_{n+1}(\sharp_{n+1}) \models \varphi_n(\sharp_n)$ over the class of all frames. For the other direction, consider first linear expansions. Note that each $\sharp$-expansion in Figure 2 is a logical consequence of the semantic definition of the rule premises. Since the semantic definition of the premises always occur positively in $\varphi_n(\sharp_n)$, it is the case that $\varphi_n(\sharp_n) \models \varphi_{n+1}(\sharp_{n+1})$. Also note that the expansion for PB is logically valid.

So $\varphi_{n+1}(\sharp_{n+1})$ and $\varphi_n(\sharp_n)$ are equivalent over the class of all frames, and so is their negation. $\dashv$

As a consequence, a simple induction on the expansion of $\neg\varphi$ gives us the following.

Lemma 5.2 *If $\neg\varphi$ is the second-order formula computed by the SLaKE-expansion of $A_1, \ldots, A_n \vdash C$, then for every ternary frame $\mathfrak{F}$:*

$$\mathfrak{F} \models A_1, \ldots, A_n \vdash C \text{ iff } \mathfrak{F} \models \neg\varphi$$

So $\neg\varphi$ is equivalent to the input sequent. Any valuation of the second-order variables in $\neg\varphi$ will generate a first-order formula implied by the original sequent. All we need now is to show that the canonical valuation of Definition 3.2 really satisfies all formulas in a branch.

For that, we consider a frame $\mathfrak{F}$, a SLaKE-saturated set Γ and a canonical valuation built on Γ. Consider a mapping of the labels in Γ into the worlds of $\mathfrak{F}$ that respects the restrictions of the canonical valuation (ie the labels that V forces to be different are not mapped into the same world); we say that $\mathfrak{F}, V \models T\ A : a$ iff $\mathfrak{F}, V, a \models A$ and $\mathfrak{F}, V \models F\ A : a$ iff $\mathfrak{F}, V, a \not\models A$.

Lemma 5.3 *Let Γ be a SLaKE saturated set and let $\mathfrak{F}$ be a frame that satisfies all the R-formulas in Γ. Let V be the canonical valuation built from Γ. Then for all signed formula $S \in \Gamma$, $\mathfrak{F}, V \models S$.*

Proof. We prove the lemma by induction on the structure of the formulas in Γ. The canonical valuation satisfies all the atomic formulas, taking care only that no atomic formula is true and false in the same world.

For the compound formulas, the single premises of $\exists$-rules replace a compound formula by its definition, so if the conclusions are satisfied, so is the compound premise. For the main premise of a $\forall$-rule, saturation guarantees that every possible conclusion is added to Γ, so if all such conclusions are satisfied so is the main premise. As a consequence, the frame $\mathfrak{F}$ also satisfies the compound main premise. $\dashv$

By putting together Lemmas 5.2 and 5.3 we obtain the following.

Theorem 5.4 *If $\mathcal{T}$ is a finite saturated SLaKE-tableau for the sequent $A_1, \ldots, A_n \vdash C$ computing a first-order formula $\neg\psi$ then*

$$\mathfrak{F} \models (A_1, \ldots, A_n \vdash C) \Rightarrow \mathfrak{F} \models \neg\psi.$$

For the converse, we first consider deterministic tableaux.

Theorem 5.5 *Let $\mathcal{T}$ be a finite, saturated and deterministic SLaKE-tableau for sequent $A_1, \ldots, A_n \vdash C$ computing a first-order formula $\neg\psi$. Then*

$$\mathfrak{F} \models (A_1, \ldots, A_n \vdash C) \Leftarrow \mathfrak{F} \models \neg\psi.$$

Proof. The construction of $\mathcal{T}$ is a refutation for the input sequent, such that every $\mathfrak{F}$ with $\mathfrak{F} \not\models (A_1, \ldots A_n \vdash C)$ must satisfy the existential closure of the second-order formula φ, that is, for every frame $\mathfrak{F}$ refuting the original sequent, $\mathfrak{F} \models (\exists P)\varphi$.

Since $\mathcal{T}$ is deterministic and saturated, it is associated to a SLaKE saturated set Γ. Any refutation frame $\mathfrak{F}$ must satisfy all the atomic formulas and the R-formulas in Γ. The canonical valuation built from it is such that it satisfies all the atomic formulas.

By Lemma 5.3, $\mathfrak{F}$ with the canonical valuation satisfies all the formulas in $\mathcal{T}$, that is $\mathfrak{F}, V \models \varphi$. So when we apply the canonical valuation to φ obtaining ψ, it follows that:

$$\mathfrak{F} \not\models (A_1, \ldots, A_n \vdash C) \Rightarrow \mathfrak{F} \models \psi,$$

which is equivalent to the desired result. ⊣

To eliminate the restriction on deterministic tableaux, we have to consider all possible refutations. This yields the final result.

Theorem 5.6 *Given a sequent $A_1, \ldots, A_n \vdash C$ and all possible finite SLaKE-tableaux refuting it, $\mathcal{T}_1, \ldots, \mathcal{T}_m$ generating, respectively, first-order formulas $\neg\psi_1, \ldots, \neg\psi_m$. Then*

$$\mathfrak{F} \models (A_1, \ldots, A_n \vdash C) \Longleftrightarrow \mathfrak{F} \models \bigwedge_{j=1}^{m} \neg\psi_j.$$

Proof. ($\Rightarrow$) This is Theorem 5.4.

($\Leftarrow$) Suppose there is a frame $\mathfrak{F}$ that refutes the input sequent, such that there is a valuation V and a point a such that $\mathfrak{F}, V, a \not\models (A_1, \ldots, A_n \vdash C)$. The tableaux $\mathcal{T}_1, \ldots, \mathcal{T}_m$ are all the possible refutations of the input sequent, when we consider the second-order formulas $\varphi_1, \ldots, \varphi_m$ generated by each refutation, we get $\mathfrak{F}, a \models \bigvee_{j=1}^{m} (\exists P)\varphi_j$.

Without loss of generality, suppose $\mathfrak{F}$ satisfies one such disjunct, φ_j: $\mathfrak{F} \models (\exists P)\varphi_j$. We claim that there is a *finite* valuation V_f that satisfies φ_j built in the following way:

$$V_f(p) = \{a | T\ p : a \text{ belongs to } \mathcal{T}_j\}$$

It is clear that $V_f(p)$ is finite for each p, for there are only finitely many formulas in T_j.

By induction on the construction of $\mathcal{T}_j$, we show that the satisfaction of φ_j depends on finitely many possible worlds. The base case is the initial tableau, which clearly relies on finitely many possible worlds.

If the expansion of $\mathcal{T}_j$ was done by a single-premised $\exists$-rule, then only two new possible worlds have to be considered, and by induction hypothesis the number of worlds remains finite.

If the expansion of $\mathcal{T}_j$ was done by PB, then no new world needs to be considered (and one extra variable is introduced), so the induction hypothesis guarantees that the number of worlds remains finite.

If the expansion of $\mathcal{T}_j$ was done by a double-premised $\forall$-rule, we need to examine each case, so consider, for example, a main formula of the form $T\ A/B : a$; to satisfy such formula, we need to enforce that:

$$\forall bc(Rcab \wedge T\ B : b \Rightarrow T\ A : c)$$

There may be infinitely many valid formulas of the form $Rcab$, but by induction hypothesis, only finitely many worlds b where the truth of B has been enforced, so the truth of A will be enforced only at finitely many worlds c. [6] The proof is analogous for the other main premises of $\forall$-rules. We have proved our claim.

At the end of the tableau construction, all atomic formulas in the tableau are satisfied by V_f. We have thus shown a *finite model property* for the sequents that have a finitely saturated tableaux. This means that if we have opposite signed formulas of the form $T\ p : x$ and $F\ p : y$, we only have to impose $x \neq y$ for finitely many points. Such a procedure gives us the same substitution as the canonical valuation; it follows that $\mathfrak{F}, V_f, a \models \psi_j$.

We have thus shown:

$$\mathfrak{F} \not\models (A_1, \ldots, A_n \vdash C) \Rightarrow \mathfrak{F} \models \bigvee_{j=1}^{m} \psi_j,$$

which is equivalent to the desired result. $\dashv$

6 Conclusion and Further Work

We have proposed an original method to construct a first-order correspondence formula for substructural categorial formulas. This construction is based on the novel SLaKE-tableaux. Our method unifies under an algorithmic approach the generation of first-order correspondence formulas initially treated by (Kurtonina 1994).

[6] Note that such an argument fails if we had in our language a formula that is true at all worlds; luckily, we do not, not even A/A, as can be verified by its semantics.

We have the following conjecture:

Conjecture 6.1 *A sequent has a first-order equivalent iff it has a finite SLaKE-tableaux.*

The *if*-part has already been proved, but for the *only if*-part all we can say for the moment is that our method does not apply to infinite tableaux. To prove the conjecture, we have to show that it is not possible to have a sequent with an infinite SLaKE-tableau but with a first-order equivalent.

Acknowledgements

The author would like to thank Wamberto Vasconcelos and Angela Weiss for helpful comments on the paper. This work was partly supported by the Brazilian Research Council (CNPq), grant PQ 300597/95-5.

References

Broda, K., and M. Finger. 1995. KE-tableaux for a Fragment of Linear Logic. In *Proceedings of the 4th International Workshop on Analytic Tableaux and Related Methods.* Koblenz, May.

Broda, K., M. Finger, and A. Russo. 1996a. LDS-Natural Deduction for Substructural Logics. In *3rd Workshop on Logic, Language, Information and Computation (WoLLIC96)*, 15–18. Abstract only.

Broda, Krysia, Marcello D'Agostino, and Alessandra Russo. 1996b. Transformation methods in LDS. In *Logic, Language and Reasoning. An Essay in Honor of Dav Gabbay.* Kluwer Academic Publishers.

Carpenter, Bob. 1997. *Type-Logical Semantics.* MIT Press.

D'Agostino, M., and D. Gabbay. 1994. A Generalization of Analytic Deduction via Labelled Tableaux, part I: Basic Substructural Logics. *Journal of Automated Reasoning* 13:243–281.

D'Agostino, Marcello. 1992. Are Tableaux an Improvement on Truth-Tables? — Cut-free Proofs and Bivalence. *Journal of Logic, Language and Information* 1:235–252.

Došen, K. 1993. A Historical Introduction to Substructural Logics. In *Substructural Logics*, ed. P. Schroeder-Heister and K. Došen. 1–31. Oxford University Press.

Gabbay, Dov M. 1996. *Labelled Deductive Systems.* Oxford University Press.

Kripke, Saul. 1963. Semantical Analysis of Modal Logic I. *Zeitschrift für Mathematische Logik und Grundlagen der Mathematik.* 9(67–96).

Kurtonina, Natasha. 1994. *Frames and Labels — A Modal Analysis of Categorial Inference.* Doctoral dissertation, Research Institute for Language and Speech (OTS), Utrecht, and Institute of Logic, Language and Information (ILLC), Amsterdam.

Lambek, J. 1958. The Mathematics of Sentence Structure. *American Mathematical Monthly* 65:154–169.

Moortgat, M. 1997. Categorial Type Logics. In *Handbook of Logic and Language*, ed. J. Van Benthem and A. ter Meulen. 93–178. Elsevier North-Holland/The MIT Press.

Restall, G. 2000. *An Introduction to Substructural Logics*. Routledge.

Roorda, Dirk. 1991. *Resource Logics: Proof-theoretical Investigations*. Doctoral dissertation, Institute of Logic, Language and Information (ILLC), University of Amsterdam.

Routley, Richard, and Robert K. Meyer. 1973. The Semantics of Entailment. In *Truth, Syntax and Modality*, ed. Hugues Leblanc. Studies in Logic and the Foundations of Mathematics, Vol. 68, 199–243. North-Holland.

Sahlqvist, H.1975. Completeness and Correspondence in the First- and Second-order Semantics for Modal Logic. In *Proceedings of the 3rd Scandinavian Logic Symposium*.

Smullyan, Raymond M. 1968. *First-Order Logic*. Springer-Verlag.

van Benthem, J. 1984. Correspondence Theory. In *Handbook of Philosophical Logic*. 167–248. D. Reidel Publishing Company.

Marcelo Finger
Department of Computer Science
Institute of Mathematics and Statistics
University of São Paulo
Rua do Matão, 1010
05508-900, São Paulo, SP
Brazil
E-mail: `mfinger@ime.usp.br`
URL: `http://www.ime.usp.br/~mfinger`

10

Dealing with the Ramification Problem in Extended Propositional Dynamic Logic

NORMAN FOO AND DONGMO ZHANG

ABSTRACT. This paper presents a solution to the ramification problem in the presence of nondeterministic effect and nondeterministic choice of actions based on the extended propositional dynamic logic. We first show that the causation-based approach can be re-coded in the extended dynamic logic. Then by taking advantage of the extra expressive power of dynamic logic for compound actions and programs, any complex ramifications involved in an automatic system can be expressed and specified. Specially, Thielscher's and Shanahan's benchmark examples can be more effectively elaborated.

1 Introduction

The study of reasoning about action and change has been a focus of AI for over twenty years. Several formalisms of action have been proposed and most fundamental problems in this area, such as the frame problem, ramification problem, qualification problem and so on, have been widely investigated with varying degrees of success. Recently, the research has addressed applications in more pragmatic areas of AI, e.g., in planning, cognitive robots and intelligent agents ([Giacomo et al. 1998] [Levesque et al. 1997] [Levesque and Reiter 1998] [Prendinger and Schurz 1996] [Shapiro et al. 1998]). However, these applications require action theories that can express compound actions, which are necessary to represent complex plans, robot controllers, automatic systems and agent-

Advances in Modal Logic, Volume 3
F. Wolter, H. Wansing, M. de Rijke, and M. Zakharyaschev, eds.

oriented programs.[1] [Levesque et al. 1997] introduced an approach to reasoning about compound actions and programs in the situation calculus. As remarked in [Levesque et al. 1997], however, compound actions were not introduced as the formal objects of the language of the situation calculus. The difficulty is that such actions could include some conditions which are hard to be embedded into a first-order term (c.f. [Levesque et al. 1997]).

Dynamic logic is a logic of programs and a formalism for actions ([Harel 1979] [Goldblatt 1987]). It is suitable for expressing compound actions and programs. Therefore dynamic logic can serve as the action theory for the aforementioned applications. The question is *whether it is capable of not only providing solutions as good as other formalisms, but also resolving problems that they could not.*

Reiter's monotonic solution to the frame problem ([Reiter 1991]) has been shown in [Giacomo and Lenzerini 1995] and [Prendinger and Schurz 1996] to be easily simulated in dynamic logic. Special inference mechanisms have been introduced for reasoning with action descriptions ([Goranko and Passy 1992] [Prendinger and Schurz 1996] [Castilho et al. 1999] [Zhang and Foo 2000]). Syntactical extensions on classical propositional dynamic logic have been made to represent concurrence and non-execution of actions by [Giacomo and Lenzerini 1995] and to express direct effects of actions and causal propagations by [Zhang and Foo 2000]. This paper will show further that the causation-based solution to the ramification problem can also be re-coded in dynamic logic. However, a stronger motivation of this paper is to show that a general solution to the ramification problem can be obtained by taking advantage of dynamic logic in the extra expressive power of compound actions and programs.

The *ramification problem* is the problem of determining the indirect effects (or ramifications) of actions with incomplete action description. As one of the fundamental challenges in reasoning about action, the ramification problem has been intensively investigated in the last ten years ([Lin 1995] [McCain and Turner 1995] [Sandewall 1996] [Thielscher 1997] [Giunchiglia et al. 1997] [Kakas and Miller 1997] [Denecker et al. 1998] [Shanahan 1999]). Such researches were mostly derived by so-called benchmark examples, such as Lin's suitcase ([Lin 1995]), Thielscher's circuits and their variants ([Thielscher 1997] [Denecker et al. 1998] [Shanahan 1999]). In this paper we propose an approach of programming automatic systems in order to deal with complex ramifications in a dynamic

[1] A compound action is an action generated from primitive actions by the program connectives $;, \cup, ?, *$.

system. We check our approach with the main benchmark examples in this area. More significantly, our solution to the ramification problem shows that dynamic logic has its unique superiority in reasoning about dynamic systems.

Our solution will be built on an extended system of propositional dynamic logic, called *EPDL*, introduced in [Zhang and Foo 2000]. *EPDL* enables a unified representation of action laws and causal laws, and a unified treatment of reasoning about direct and indirect effects of actions in a succinct axiomatized formal system.

2 Preliminaries of *EPDL*

In this section, we summarize some basic facts of the extended propositional dynamic logic (*EPDL*) which will be used in this paper. See [Zhang and Foo 2000] for more details.

A language $\mathcal{L}_{EDPL}$ of *EPDL* consists of a set **Flu** of fluent symbols and a set $\mathbf{Act}_P$ of primitive action symbols. For the purpose of this paper, we shall only consider finite languages of *EPDL*, which means both **Flu** and $\mathbf{Act}_P$ are finite.

The syntax for actions is the same as that for programs in *PDL*. The only new syntax introduced are formulas of the form $[\varphi]A$, called *propositional causation*, representing indirect effects of actions.[2] For example, $[\neg alive]\neg walking$ means "$\neg alive$ causes $\neg walking$".

Proposition ($\varphi \in \mathbf{Pro}$), formula ($A \in \mathbf{Fma}$) and action ($\alpha \in \mathbf{Act}$) are defined by the following BNF rules:

$\varphi ::= f \mid \neg\varphi \mid \varphi_1 \to \varphi_2$
$A ::= f \mid \neg A \mid A_1 \to A_2 \mid [\alpha]A \mid [\varphi]A$
$\alpha ::= a \mid \alpha_1;\alpha_2 \mid \alpha_1 \cup \alpha_2 \mid \alpha^* \mid A?$
where $f \in \mathbf{Flu}$ and $a \in \mathbf{Act}_P$.

The following two abbreviations will be useful:

$\langle[\alpha]\rangle A =_{df} \langle\alpha\rangle\top \wedge [\alpha]A$, meaning "$\alpha$ must cause A";
$\prec\alpha\succ A =_{def} \langle\alpha\rangle\top \to \langle\alpha\rangle A$, meaning "$\alpha$ may cause A if it is feasible."

The semantics of *EPDL* is the standard *PDL* semantics plus the interpretation of propositional causation. A model for $\mathcal{L}_{EPDL}$ is a structure $M = (W, \mathcal{R}, V)$, where $\mathcal{R} = \{R_\alpha : \alpha \in \mathbf{Act}\} \cup \{R_\varphi : \varphi \in \mathbf{Pro}\}$. The components of W, $\{R_\alpha : \alpha \in \mathbf{Act}\}$ and V are the same as those of *PDL* models. For any propositional formula φ, R_φ is also a binary relation on W. The satisfaction relation $M \models_s A$ is defined as usual:

[2]The approach of using a proposition as a modality has been used by [Groeneveld 1995] to deal with changes in knowledge. Besides the motivation, however, the semantics and deduction system are totally different from ours.

- $M \models_w f$ iff $f \in V(f)$, for any $f \in \mathbf{Flu}$;
- $M \models_w A_1 \to A_2$ iff $M \models_w A_1$ implies $M \models_w A_2$;
- $M \models_w [\gamma]A$ iff for all $w' \in W$, $wR_\gamma w'$ implies $M \models_{w'} A$, where $\gamma \in \mathbf{Act_P} \cup \mathbf{Pro}$.

As in any modal logic, $A \in \mathbf{Fma}$ is valid in M, written as $M \models A$, if $M \models_w A$ for all $w \in W$. "$\models A$" means "A is valid in any model".

It is easy to see that if $\models_{PL}$ denotes the satisfiability relation of the classical propositional logic, then for any $\varphi \in \mathbf{Pro}$, $\models_{PL} \varphi$ if and only if $\models \varphi$.

The conditions for *standard models* of $\mathcal{L}_{EPDL}$ consist of the ones for program connectives: $R_{\alpha;\beta} = R_\alpha \circ R_\beta$; $R_{\alpha\cup\beta} = R_\alpha \cup R_\beta$; $R_{\alpha^*} = R_\alpha^*$; $R_{A?} = \{(w,w) : M \models_w A\}$. The extra conditions for propositional causation are as follows:

- If $M \models_w \varphi$, then $(w,w) \in R_\varphi$.
- If $\models \varphi_1 \to \varphi_2$, then $R_{\varphi_1} \subseteq R_{\varphi_2}$.

We can easily see that the semantics for propositional causation is some sort of weakening of the one for test action. An example given in [Zhang and Foo 2000] illustrated they can be different. In fact, test action only concerns the current state of a system while a causal relation is sensitive to both the history and the future evolution of the system because it is action-relevant.

The axiomatic deductive system of $EPDL$ consists of the following axiom schemes and inference rules:

- Axiom schemes:
  - all tautologies of the propositional calculus.
  - all axioms for compound programs.[3]
  - EK axiom: $[\gamma](A \to B) \to ([\gamma]A \to [\gamma]B)$
  - CW axiom: $[\varphi]A \to [\varphi?]A$
- Inference rules:
  - MP: From A and $A \to B$, infer B.
  - EN: From A, infer $[\gamma]A$.
  - LC: From $\varphi_1 \to \varphi_2$, infer $[\varphi_2]A \to [\varphi_1]A$.

  where $\varphi, \varphi_1, \varphi_2 \in \mathbf{Pro}$, $A \in \mathbf{Fma}$, $\alpha \in \mathbf{Act}$ and $\gamma \in \mathbf{Pro} \cup \mathbf{Act}$.

A formula A is a theorem, denoted by $\vdash A$, if it can derived from the axiom system.

A special case of $EPDL$ is when the set of primitive action symbols is empty, $EPDL$ will degenerate into a multi-modal logic, denoted by

[3]See Goldblatt 1987 for details.

$EPDL^-$, which concentrates on the propositional causation. The properties of $EPDL^-$ in causal reasoning have been investigated in [Zhang and Foo 2000].

3 Action Description and its Normal Forms

$EPDL$ provides a formal language to describe behaviors and internal relations of a dynamic system. Those sentences which describe the generic effects of actions, domain constraints and causal ramifications are generally called *action description.* In this paper, an action description of a dynamic system is a finite set of $EPDL$ formulas.

3.1 Reasoning with action description

Let Σ be an action description. A formula A is a Σ-*theorem*, written by $\vdash^{\Sigma} A$, if it belongs to the least set of formulas which contains all the theorems of $EPDL$, all the elements of Σ, and is closed under MP and EN.

If Γ is a set of formulas, then $\Gamma \vdash^{\Sigma} A$ means there exists $A_1, \cdots$, $A_n \in \Gamma$ such that $\vdash^{\Sigma} (A_1 \wedge \cdots \wedge A_n) \to A$.

An action description Σ is *uniformly consistent* if $\not\vdash^{\Sigma} \bot$. Γ is Σ-*consistent* if $\Gamma \not\vdash^{\Sigma} \bot$.

We call a formula of the form $\varphi \wedge l \to [a]l$ a *strong frame axiom*, a formula of the form $\varphi \wedge l \to \prec a \succ l$ a *weak frame axiom* and a formula of the form $\varphi \to \langle a \rangle \top$ a *qualification axiom*, where φ is a proposition, a is a primitive action and l is a literal.

Definition 3.1 Let Σ be uniformly consistent and Γ be Σ-consistent. For a given formula A, if a set Δ of (strong or weak) frame axioms and qualification axioms satisfies:

1. Γ is $\Sigma \cup \Delta$-consistent;
2. $\Gamma \vdash^{\Sigma \cup \Delta} A$,

then A is said to be Σ-*provable from* Γ *with* Δ, denoted by $\Gamma \vdash^{\Sigma}_{\Delta} A$. The members of Δ are called *supplementary axioms.*

Roughly speaking, $\Gamma \vdash^{\Sigma}_{\Delta} A$ means that A can be derived from Γ under the action description Σ with the additional assumption Δ of frame axioms and qualification axioms. We shows in a separate paper that $\Gamma \vdash^{\Sigma}_{\Delta} A$ if and only if there is a set Δ' of supplementary axioms which only contain the symbols occurred in Γ, Σ and A (note that some slight extra assumption is required). Moreover, if Σ is in normal form (see the following subsection), all the action symbols in Δ' can only occur in Γ and A. Therefore the frame problem can be mitigated and even avoided if we postpone listing the frame axioms until a query occurs. We call this solution to the frame problem *lazy-formalization.*

3.2 Normal action descriptions

We say a formula in $\mathcal{L}_{EPDL}$ is in *normal form* if it is written in the following forms:

- $[\varphi]\psi$ (causal law)
- $\varphi \rightarrow [a]l$ (deterministic action law)
- $\varphi \rightarrow \prec a \succ l$ (nondeterministic action law)
- $\varphi \rightarrow \langle a \rangle \top$ (qualification law)

  where, $\varphi \in \mathbf{Pro}$, $a \in \mathbf{Act}_P$ and l is a fluent literal.

An action description Σ is *normal* if each formula in Σ is in normal form. We denote by Σ_a the set of all the action laws about action a in Σ, and by Σ_i the set of causal laws in Σ.

Even though the normal form is restricted (in fact it can not express disjunctive direct effects of actions), it is still expressive. For instance, causal laws can express any "causal law" in the sense of McCain and Turner ([McCain and Turner 1995]). Both of "action law" and "causal relationship" in sense of Thielscher ([Thielscher 1997]) can be represented by normal forms. It is not hard to prove that precondition axioms and successor state axioms in propositional language in terms of Reiter's situation calculus [Reiter 1991] can be translated into normal form. Additionally, any sentence in action language $\mathcal{A}$ ([Gelfond and Lifschitz 1998]) can also be expressed in normal form.

From now on, we will only consider normal action descriptions.

Theorem 3.2 [Zhang and Foo 2000] *Let* Σ *be normal and* Σ_i *be the set of causal laws in* Σ*. Then* $\vdash^{\Sigma_i} [\varphi]\psi$ *iff* $D(\Sigma_i) \cup \bigcup_{\varphi \vdash \chi} \{\lambda : [\chi]\lambda \in \Sigma_i\} \vdash \psi$, *where* $D(\Sigma_i) = \{\varphi \rightarrow \psi : [\varphi]\psi \in \Sigma_i\}$.

This property is important because it gives a sufficient and necessary condition for causal propagation which tells us what causal laws can or cannot be derived from a set of given causal laws. It is easy to see that the provability about $\vdash^{\Sigma_i} [\varphi]\psi$ can be manipulated in $EPDL^-$ because there is no action involved.

To show how to reason with action description, we examine a simple example.

Example 3.3 Consider the circuit shown by Figure 1.

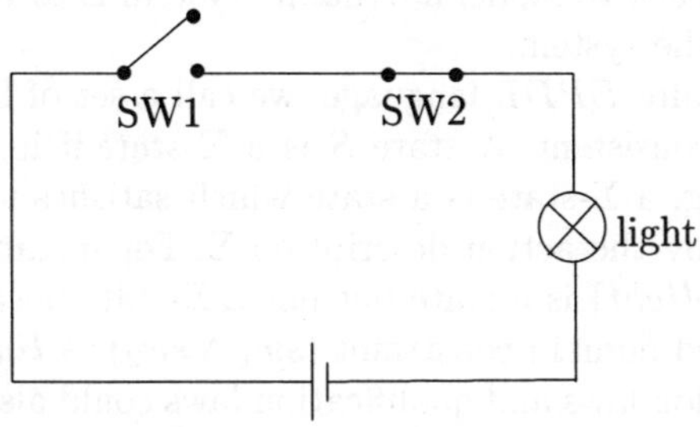

Figure 1: Thielscher's two switches circuit

Let Σ be the action description of this circuit:

$$\Sigma = \left\{ \begin{array}{l} [sw_1 \wedge sw_2]light \\ [\neg sw_1 \vee \neg sw_2]\neg light \\ \neg sw_i \rightarrow [Toggle_i]sw_i \\ sw_i \rightarrow [Toggle_i]\neg sw_i \\ \langle Toggle_i \rangle \top \\ i = 1, 2 \end{array} \right\}$$

It is not hard to prove the following inference relations in $EPDL$:

1. $\vdash^{\Sigma} (sw_1 \wedge sw_2) \leftrightarrow light$
2. $sw_1 \vdash^{\Sigma} [Toggle_1]\neg light$
3. $\neg sw_1 \wedge \neg sw_2 \vdash^{\Sigma}_{\Delta} [Toggle_1; Toggle_2]light$, where $\Delta = \{\neg sw_2 \rightarrow [Toggle_1]\neg sw_2,\ sw_1 \rightarrow [Toggle_2]sw_1\}$.
4. $\neg light \vdash^{\Sigma}_{\Delta} [Toggle_1 \cup Toggle_2 \cup (Toggle_1; Toggle_2)]light$, where $\Delta = \{sw_2 \rightarrow [Toggle_1]sw_2,\ \neg sw_2 \rightarrow [Toggle_1]\neg sw_2,\ sw_1 \rightarrow [Toggle_2]sw_1\}$.
5. $\vdash^{\Sigma_i} [\neg sw_1]\neg light$.
6. $\not\vdash^{\Sigma_i} [sw_2 \wedge \neg light]\neg sw_1$. where Σ_i is the set of causal laws in Σ.

Item 1 shows that the causal laws in the action description ($[sw_1 \wedge sw_2]light$ and $[\neg sw_1 \vee \neg sw_2]\neg light$) imply domain constraints. Item 2, 3 and 4 show how effects of actions are propagated through causal laws. Among them, the inference relations 3 and 4 require frame axioms which did not list in the action description. Item 5 and 6 show which causal laws can be or can not be derived from the given causal laws. Specially, item 6 says that sw_2 and $\neg light$ can never cause $\neg sw_1$ even though $\vdash^{\Sigma}$ $(sw_2 \wedge \neg light) \rightarrow \neg sw_1$. This illustrates the difference between causation and logical implication.

4 State Changes under Actions

A standard approach to model a dynamic system is to reveal the evolution of states of the system.

For a given finite $EPDL$ language, we call a set of literals a *state* if it is maximally consistent. A state S is a Σ-*state* if it is Σ-consistent. Roughly speaking, a Σ-state is a state which satisfies the domain constraints implied by the action description Σ. For instance, in Example 3.3, $\{sw_1, sw_2, \neg light\}$ is a state but not a Σ-state because it does not satisfy the implied domain constraint $(sw_1 \wedge sw_2) \leftrightarrow light$.

Note that action laws and qualification laws could also imply domain constraints. For instance, $\{f_1 \rightarrow [a]f_3, f_2 \rightarrow [a]\neg f_3, \langle a\rangle\top\}$ implies $\neg f_1 \vee \neg f_2$.

Suppose that Σ is a normal action description. Σ_a is the set of action laws in Σ. We call a set E of literals a *direct effect* of action a under state S if it is a maximal set of literals which satisfies $S \vdash^{\Sigma_a} \langle a\rangle E$ (Note that a direct effect of an action means here a set of direct effects of the action. We could have several direct effect sets for one action due to nondeterministic effects of actions). If additionally $S \vdash^{\Sigma_a} [a]E$, then E is called *the deterministic direct effect.* Therefore the condition for deterministic case is $S \vdash^{\Sigma_a} \langle[a]\rangle E$.

For the circuit in Example 3.3, the (deterministic) direct effect of the action $Toggle_1$ under state $\{\neg sw_1, sw_2, \neg light\}$ is $\{sw_1\}$.

4.1 Optional next states

Now we consider how a dynamic system changes its states under an action. Let S_0 be a Σ-state, called an *initial state.* A Σ-state S is an *optional next state of* S_0 *under action* α if there is a set Δ of frame axioms such that $S_0 \vdash^{\Sigma}_{\Delta} \langle\alpha\rangle S$. If $S_0 \vdash^{\Sigma}_{\Delta} \langle[\alpha]\rangle S$, S is called *an optional deterministic next state of* S_0.

Example 4.1 Consider the action description in Example 3.3 again. Let the initial state $S_0 = \{\neg sw_1, \neg sw_2, \neg light\}$. Then $S_1 = \{sw_1, \neg sw_2, \neg light\}$ is an optional deterministic next state of S_0 under $Toggle_1$ since $S_0 \vdash^{\Sigma}_{\Delta} \langle[Toggle_1]\rangle(sw_1 \wedge \neg sw_2 \wedge \neg light)$. Note that it is the unique optional deterministic next state of S_0.

It is not hard to verify that there are two optional deterministic next states of S_1 under action $Toggle_2$:

1. $S_2' = \{sw_1, sw_2, light\}$, where supplementary axioms $\Delta = \{sw_1 \rightarrow [Toggle_2]sw_1\}$;
2. $S_2'' = \{\neg sw_1, sw_2, \neg light\}$, where supplementary axioms $\Delta = \{\neg light \rightarrow [Toggle_2]\neg light\}$.

Note that the second one is undesirable.

Incidentally, we can also verify that there are two optional deterministic next states of S_0 under action $Toggle_1 \cup Toggle_2$:

$S' = \{sw_1, \neg sw_2, \neg light\}$

$S'' = \{\neg sw_1, sw_2, \neg light\}$.

Both the optional next states are expected.

Example 4.2 Consider a variant of the Yale Shooting Problem. Suppose that the action description is:

$$\Sigma = \left\{ \begin{array}{l} \neg loaded \rightarrow [Load]loaded \\ loaded \rightarrow \prec Spin \succ \neg loaded \\ loaded \rightarrow [Shoot]\neg loaded \\ loaded \rightarrow [Shoot]\neg alive \\ \langle Load\rangle\top, \langle Spin\rangle\top, \langle Shoot\rangle\top \end{array} \right\}.$$

It is not hard to verify that the following states are the optional next states[4] of $S_0 = \{\neg loaded, alive\}$ under action $Load; Spin; Shoot$:

$S' = \{\neg loaded, \neg alive\}$

$S'' = \{\neg loaded, alive\}$.

Both the optional next states are expected.

These examples show us the three possible resources of the multiplicity of optional next states:

- optional uses of frame axioms (supplementary axioms Δ);
- nondeterministic choice of actions($\cup$ and $*$);
- nondeterministic effects of actions ($\langle a \rangle$).

The multiplicity due to the last two is expected. However, the first one may appear contrived, so is perhaps avoidable. The problem lies in the incompleteness of action description. How to avoid this kind of multiplicity is exactly the *ramification problem*, which has attracted considerable attention of the KR community in the last ten years. The extant solutions to the problem can be grouped into two categories: *minimization-based* and *causation-based* approaches. The second one seems to be more popular nowadays because it de-emphasizes meta-hypotheses and can take advantage of causal information as well as domain constraints. The extra requirements for this approach are that it should provide a special syntax for expressing causal information and some inference mechanism for modeling causal propagation. The *EPDL* has been equipped with these facilities. In the next subsection we show how to use them to deal with the ramification problem.

[4]but not deterministic next state.

4.2 Possible next states

The key meta-hypothesis of causation-based approach is that any change in a dynamic system is either a *direct* or *indirect* effect of an action. In our terminology, suppose that S_0 is the initial state, S is one of the optional next state of S_0 under action a. If we denote $U = S_0 \cap S$ as the unchanged part of S and E as a direct effect of a, then the rest of changed part I of S, i.e., $I = S \backslash (E \cup U)$, should be the indirect effects of the action a (See Figure 2).

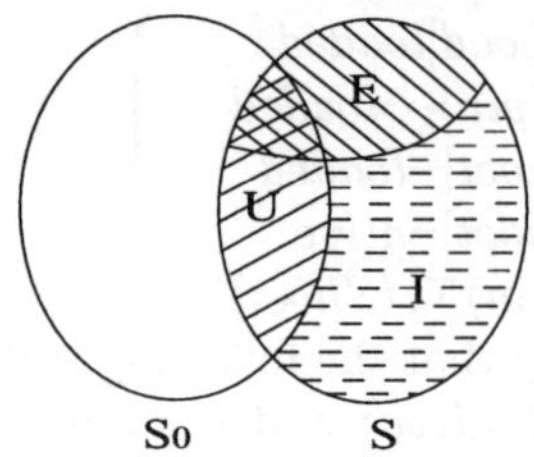

Figure 2: Causation-based approach to the ramification problem

According to the characteristics of causal propagation, the indirect effects of actions are the propagation of the direct effects through the causal laws. Therefore it seems to be the case that $\vdash^{\Sigma_i} [E]I$, i.e., the indirect effects I should be caused by the direct effects E under the action laws Σ_i. However, this condition ignores the effects of frame axioms (making something unchanged is also a kind of effects of actions). As we permit postponing the listing of frame axioms in action descriptions (see Section 3), not all the effects of frame axioms are included in the direct effects of the action. Therefore a reasonable condition for the indirect effects of action should be $\vdash^{\Sigma_i} [E \cup U]I$. This leads to the following definition.

Definition 4.3 Let Σ be a normal action description and S_0 a Σ-state. Let E be a direct effect of a primitive action a under S_0. Then an optional next state S of S_0 under a is called a ***possible next state of*** S_0 ***under*** a if $E \subseteq S$ and $\vdash^{\Sigma_i} [E \cup U]I$, where $U = S \cap S_0$ and $I = S \backslash (E \cup U)$.

A possible next state is *deterministic* if it is an optional deterministic next state.

As we have mentioned before, the provability of $\vdash^{\Sigma_i}$ can be checked in $EPDL^-$. This means that if the direct effects of an action has been given, $EPDL^-$ is enough to determine how effects of actions are propagated by causal laws.

Example 4.4 Continuation of Example 4.1. We have know that $S_2' = \{sw_1, sw_2, light\}$ is an optional next state of $S_1 = \{sw_1, \neg sw_2, \neg light\}$. It is easy to verify that S_2' is also a possible deterministic next state of S_1 under action $Toggle_2$. S_2'' is not because $\not\vdash^{\Sigma_i} [sw_2 \wedge \neg light]\ \neg sw_1$, where $E = \{sw_2\}$, $U = \{\neg light\}$ and $I = \{\neg sw_1\}$.

We haven't finished the definition of possible next state because we haven't considered compound actions.

For any action α, we define S to be *a possible next state of* S_0 *under action* α inductively as follows:

1. If α is a primitive action, then S is a possible next state of S_0 under α.
2. If $\alpha = A?$, then $S = S_0$ and $S_0 \vdash^{\Sigma} A$.
3. If $\alpha = \alpha_1; \alpha_2$, then there exists a possible next state S_1 of S_0 under action α_1 such that S is a possible next state of S_1 under action α_2.
4. If $\alpha = \alpha_1 \cup \alpha_2$, then S is a possible next state of S_0 under action α_1 or α_2.
5. If $\alpha = \alpha_1^*$, then $S = S_0$ or there exists a possible next state S_1 of S_0 under action α_1^* such that S is a possible next state of S_1 under action α_1.

For the deterministic case, the definition is similar. Note that a possible deterministic next state is not necessarily unique because of non-deterministic choice of actions.

If S is the unique possible next state of S_0 under action α, we will call it *the next state of* S_0 *under action* α.

If we only consider the deterministic case, take E as initially given and replace $\vdash^{\Sigma_i}$ by $\vdash_C$, this definition will coincide with the definition of possible next state in [McCain and Turner 1995]. Therefore McCain and Turner's solution to the ramification problem can be re-coded in the extended dynamic logic.

The following two propositions are straightforward by the above definition and Theorem 3.2, respectively.

Proposition 4.5 *If all the direct effects of action a are empty, then* S_0 *is the unique possible next state of* S_0 *under a.*

Proposition 4.6 *Any possible next state* S *of* S_0 *under action a satisfies the fix-point property:* $S = \{l : E \cup (S_0 \cap S) \cup D(\Sigma_i) \vdash_C l\}$, *where* $D(\Sigma_i) = \{\varphi \to \psi : [\varphi]\psi \in \Sigma_i\}$, $C = \{\varphi \Rightarrow \psi : [\varphi]\psi \in \Sigma_i\}$ *and the natations,* $\Rightarrow$ *and* $\vdash_C$, *take the meaning in* [McCain and Turner 1995].

The first proposition shows the essential difference between McCain and Turner's causal theory in [McCain and Turner 1997] and our approach; their system allow "self-caused" effects whereas ours does not. The second one shows the relationship between McCain and Turner's earlier causal theory in [McCain and Turner 1995] and our approach. Note that the set of the possible next states of S_0 under action a in our framework could be different from the set $Res^4_{\Sigma_i}(E, S_0)$ because there are slight differences between the inference power of $\vdash^{\Sigma_i}$ and $\vdash_C$.

5 Modeling Automatic Systems

Let's consider the circuit of a flash lamp (see Figure 3), which is a variation of Thielscher's light-detector circuit [Thielscher 1997], a well-known benchmark for solutions to the ramification problem.

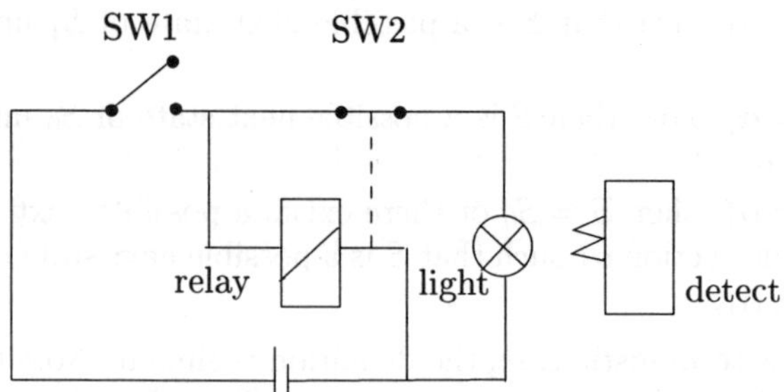

Figure 3: A circuit for flash lamp

Initially, the switch 1 is open and switch 2 is closed. When the switch 1 is toggled, the relay is triggered and then it opens the switch 2. So the light makes a flash. The detector can detect the flash if any.

First, let us specify this device by the following action description:

$$\Sigma_1 = \left\{ \begin{array}{ll} [sw_1 \wedge sw_2]light & [light]detect \\ [\neg sw_1 \vee \neg sw_2]\neg light & \neg sw_i \rightarrow [Toggle_i]sw_i \\ [sw_1]relay & sw_i \rightarrow [Toggle_i]\neg sw_i \\ [\neg sw_1]\neg relay & \langle Toggle_i \rangle \top \\ [relay]\neg sw_2 & i = 1, 2 \end{array} \right\}$$

Suppose that the initial state is $S_0 = \{\neg sw_1,\ sw_2,\ \neg relay,\ \neg light,\ \neg detect\}$. It is easy to verify that the only possible next state under action $Toggle_1$ is $S' = \{sw_1,\ \neg sw_2,\ relay,\ \neg light,\ \neg detect\}$. Note that the detector has not detected the flash of the light.

The problem here should not lie in the circuit because it can be realized in practice. So it should be the action description. Notice that although time is not explicitly expressed in the *EPDL*, the inference mechanism of *EPDL* actually implies that effect propagation is instantaneous — there is no delay in the procedure of propagation from cause propositions to effect propositions. Consequently, as far as this example is concerned, there is no time gap between the effect propagation from sw_1 to $light$ with sw_2 and the one from sw_1 to $\neg light$ through $relay$ and $\neg sw_2$. Therefore the detector has no time to react to the flash of the light. In fact, this device works only when there is a delay between the relay being activated and it actuating other things. [5]

5.1 State change of an automatic system

To specify this device correctly, we should explicitly represent the actions of the relay. We consider the following two actions of a relay: *Activate* and *Release*. The former action activate the relay in the result of opening the switch 2:

$$sw_2 \rightarrow [Activate]\neg sw_2$$

The qualification law for this action is that the switch 1 is on:

$$sw_1 \rightarrow \langle Activate\rangle\top$$

The later action releases the relay in the result of closing the switch 2:

$$\neg sw_2 \rightarrow [Release]sw_2$$

The qualification law for this action is that the switch 1 is off:

$$\neg sw_1 \rightarrow \langle Release\rangle\top$$

These four action laws replace the causal laws $[sw_1]relay$, $[\neg sw_1]\neg relay$ and $[relay]\neg sw_2$ in Σ_1. Let Σ_2 be the resulting action description, that is,

$$\Sigma_2 = \left\{ \begin{array}{ll} [sw_1 \wedge sw_2]light & sw_2 \rightarrow [Activate]\neg sw_2 \\ [\neg sw_1 \vee \neg sw_2]\neg light & \neg sw_2 \rightarrow [Release]sw_2 \\ [light]detect & \langle Toggle_i\rangle\top \\ \neg sw_i \rightarrow [Toggle_i]sw_i & sw_1 \rightarrow \langle Activate\rangle\top \\ sw_i \rightarrow [Toggle_i]\neg sw_i & \neg sw_1 \rightarrow \langle Release\rangle\top \\ i = 1,2 & \end{array} \right\}$$

Based on the action description Σ_2, it is easy to see that the next state of $S_0 = \{\neg sw_1, sw_2, \neg light, \neg detect\}$ under action $Toggle_1$ is $S_1 = \{sw_1, sw_2, light, detect\}$.

The next state of S_1 under the action *Activate* is $S_2 = \{sw_1, \neg sw_2, \neg light, detect\}$.

So the detector has detected a flash.

---

[5] The related discussions about delays of causal propagations can be also found in [Kakas and Miller 1997] [Denecker et al. 1998] [Thielscher 1998].

Now the new action description correctly reflects the evolution of state changes of the circuit. A bit pity is that the current approach can not capture the peculiarity of action *Activate* and *Release*: they run *automatically* whenever their preconditions are satisfied. In computing state changes, no distinction is made between such automatic changes and those due to external inputs like *Toggle*. In fact, the behavior of this circuit is a typical feature of an *automatic system* : after being activated by some external actions, it conducts a set of internal actions automatically, such as a computer, robot, or more visibly, a domino game. One way to specify such systems is programming them.

5.2 Programming an automatic system

In an automatic system, we can classify actions into two sorts: *endogenous actions* and *exogenous actions*.[6] The former are those that can run automatically as long as the trigger conditions hold while the later originate from outside. For instance, in the above example, *Activate* and *Release* are endogenous actions, whereas $Toggle_1$ and $Toggle_2$ are exogenous ones. We denote the set of the exogenous primitive actions of an automatic system as $\mathbf{Act}_{Ex}$, the set of the endogenous primitive actions as $\mathbf{Act}_{En}$.

To specify effects of actions in an automatic system, it is not sufficient to only consider the exogenous actions because each exogenous action could trigger some endogenous actions which could produce further effects that trigger yet more endogenous actions. More concretely, suppose that S_0 is the initial state of an automatic system and now we perform an exogenous action a. The final state of this system after executing a is not necessarily the next state of S_0 under a because the effects of action a may trigger some other actions and the state of the system could evolve with the running of these actions. So, the final state should be the next state of S_0 under some action: "$a; \alpha$", where α is a (possibly complex) action, called the *tail* of a, which runs automatically after a.

We call a (possible) next state of S_0 under action $a; \alpha$ a *(possible) final* state of a. The states between S_0 and the final state are called *intermediate* states.

It is easy to see that the key problem of modelling effects of actions in an automatic system is to find the tails of exogenous actions. The following examples show that tails mainly depend on the structure of automatic systems rather than the associated exogenous actions.

[6] A similiar separation was discussed in [Levesque et al. 1997].

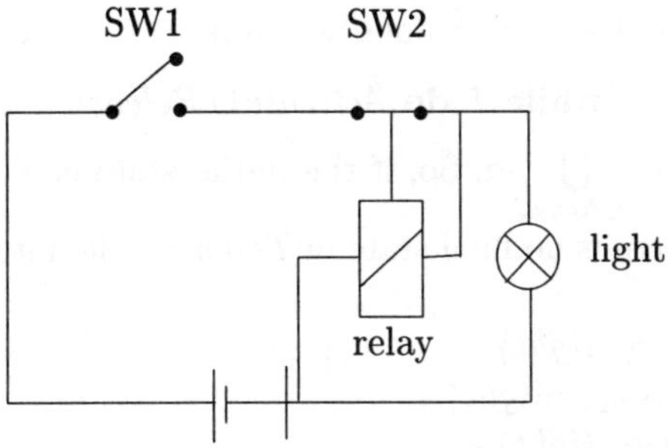

Figure 4: A circuit for Christmas light

Example 5.1 (*Running once*) Consider the circuit shown in Figure 3 again. The tail of both exogenous actions $Toggle_1$ and $Toggle_2$ can be $Activate \cup Release$. So the *final state* of $Toggle_1$ from $S_0 = \{\neg sw_1, sw_2, \neg light, \neg detect\}$ is the *next state* of $Toggle_1; (Activate \cup Release)$, i.e.,
$S = \{sw_1, \neg sw_2, \neg light, detect\}$.
The *intermediate state* is
$S_1 = \{sw_1, sw_2, light, detect\}$.
In more detail, starting from S_0, the system proceeds to S_1 after $Toggle_1$ is conducted. Then one of *Activate* and *Release* is triggerd, in fact, only *Activate* can be triggered in the current state because sw_1 is true. Finally the system goes for the state $S = \{sw_1, \neg sw_2, \neg light, detect\}$, as desired.

In [Thielscher 1997], two possible final states of action $Toggle_1$ were given. However, only one final state is expected because the circuit is for a flash lamp.

Since the set of endogenous actions $\mathbf{Act}_{En} = \{Activate, Release\}$, we can write $Activate \cup Release$ as $\bigcup_{a \in \mathbf{Act}_{En}} a$. In fact, $\bigcup_{a \in \mathbf{Act}_{En}} a$ can be the general form of tail for any exogenous actions with running once of endogenous actions if the action description suitably specifies the system.

Example 5.2 (*Running autonomously*) Consider the circuit of a Christmas light shown in Figure 4 (due to [Shanahan 1999]):

This system can be specified by the following action description:

$$\Sigma = \left\{ \begin{array}{ll} [sw_1 \wedge sw_2]light & sw_2 \rightarrow [Activate]\neg sw_2 \\ [\neg sw_1 \vee \neg sw_2]\neg light & \neg sw_2 \rightarrow [Release]sw_2 \\ \neg sw_i \rightarrow [Toggle_i]sw_i & sw_1 \wedge sw_2 \rightarrow \langle Activate \rangle \top \\ sw_i \rightarrow [Toggle_i]\neg sw_i & \neg sw_1 \vee \neg sw_2 \rightarrow \langle Release \rangle \top \\ \langle Toggle_i \rangle \top & i = 1, 2 \end{array} \right\}$$

Then the tail of any exogenous action is

$$\textbf{while } \top \textbf{ do } \mathit{Activate} \cup \mathit{Release}$$

i.e., **while** $\top$ **do** $\bigcup_{a \in \mathbf{Act}_{En}} a$. So, if the initial state is $S_0 = \{\neg sw_1, sw_2, \neg light\}$, then there is no final state of $Toggle_1$. The intermediate states are

$S_1 = \{sw_1, sw_2, light\}$,
$S_2 = \{sw_1, \neg sw_2, \neg light\}$,
$S_3 = \{sw_1, sw_2, light\}$,
...

The state evolution correctly reflects the flicker of a Christmas light.

This circuit was considered in [Shanahan 1999] as an example with vicious cycles since it yields an inconsistent scenario based on the inference mechanism in [Shanahan 1999].

The reader is invited to write an action description and tails for the exogenous actions of the circuit in [Denecker *et al.* 1998].

From the examples we find that the tail of an exogenous action in an automatic system can be independent on considered exogenous actions but is determined by the structure of the system: *the ramification involved in the system.* We even have a unified form of tails for any automatic system with a certain mode of running if the action description for the system is properly designed. A typical form of tail for the systems in the way of autonomously running is

$$\textbf{while } \neg \mathit{Interrupt\_Condition} \textbf{ do } \bigcup_{a \in \mathbf{Act}_{En}} a,$$

meaning that the system carries on running any feasible action till the interrupt condition holds.

We remark that programming a system is not an approach designed only for the benchmark examples but a general approach to express and specify ramifications involved in any dynamic system. In fact, effect propagations in a dynamic system can be abstracted into two categories: *zero-delay* and *nonzero-delay* (also see [Denecker et al. 1998]). The first one can be expressed by causal laws and modeled with logic-based causal theory, such as $EPDL^-$, McCain and Turner's causal theory. The second one can be represented and specified with operation-based action theory, such as Thielscher's approach and programming systems. With the extended propositional dynamic logic $EPDL$, we can unify the treatment of these two categories.

6 Conclusion and Related Work

This paper proposed a general solution to the ramification problem in the presence of nondeterministic effects and nondeterministic choice of actions. We first showed how the causation-based approach can be captured in the extended propositional dynamic logic. Then by taking advantage of the expressive power in compound actions and programs of dynamic logic, ramifications involved in an automatic system were programed and specified.

Although our approach shares many characteristics with other causation based approaches such as [McCain and Turner 1995] [McCain and Turner 1997] [Thielscher 1997] [Shanahan 1999] and etc., some subtle features differentiate ours from them.

We have shown that our definition of possible next states is different from both of McCain and Turner's causal theories ([McCain and Turner 1995] [McCain and Turner 1997]) by Proposition 4.5 and 4.6. In fact, we endowed the *EPDL* with relatively low power for causal propagation because causality is a very sensitive kind of relationship. However, since our approach is built on the axiomatized *EPDL*, it is easy to extend by adding new axioms to *EPDL* that capture particular philosophical motivations or application requirements.

The difference between Thielscher's approach and ours can be seen from the benchmark example in Section 5. In [Thielscher 1997], there are two possible next states of S_0 under action $Toggle_1$ with the action description Σ. Such multiplicity of states was explained in [Thielscher 1998] as a non-deterministic behavior of the circuit. From our point of view, however, for this circuit the states of this system should be determined by the initial state and the actions we perform. The aforementioned multiplicity is caused by obfuscating zero-delay abstraction ([Denecker et al. 1998]). In fact, different state changes reveal different types of circuits which require different descriptions of action.[7] Our approach is different from Shanahan's because our explanation of the circuit in Example 5.2 is quite different from the one in [Shanahan 1999] where the circuit was treated as an abnormal dynamic system.

[7] Image that we add more switches, relays and lights, this circuit could be as complex as a computer. Therefore there could be millions of possible states.

References

Denecker, M., D. T. Dupré, and K. van Belleghem. 1998. An inductive definition approach to ramifications. *Electronic Transactions on AI* 2:25–67.

Giacomo, G. De, and M. Lenzerini. 1995. PDL-based framework for reasoning about actions. In *Topics in Artificial Intelligence*, ed. M. Gori and G. Soda, 103–114. LNAI 992.

Giunchiglia, Enrico, G. Neelakantan Kartha, and Vladimir Lifschitz. 1997. Representing Action: Indeterminacy and Ramifications. *Artificial Intelligence* 95(2):409–438.

Goldblatt, R. 1987. *Logics of Time and Computation*. Stanford Univ. Press.

Goranko, V., and S. Passy. 1992. Using the Universal Modality: Gain and Questions. *Journal of Logic and Computation* 2 (1):5–30.

Groeneveld, W. 1995. *Logical Investigations into Dynamic Semantics*. Doctoral Dissertation, ILLC, University of Amsterdam.

Harel, D. 1979. *First-order dynamic logic*. Springer-Verlag.

Kakas, A., and R. Miller. 1997. Reasoning about actions, narrative and ramification. *Electronic Transactions on Artificial Intelligence* 1:39–72.

Levesque, Hector J., Raymond Reiter, Ives Lespérance, Fangzhen Lin, and Richard B. Scherl. 1997. GOLOG: A Logic Programming Language for Dynamic Domains. *Journal of Logic Programming* 31(1–3):59–83.

Lin, F. 1995. Embracing causality in specifying the indirect effects of actions. In *IJCAI-95*, 1985–1991.

McCain, N., and H. Turner. 1995. A causal theory of ramifications and qualifications. In *IJCAI-95*, 1978–1984.

McCain, N., and H. Turner. 1997. Causal theories of action and change. In *AAAI-97*, 460–465.

Prendinger, H., and G. Schurz. 1996. Reasoning about action and change: A dynamic logic approach. *J. of Logic, Language, and Information* 5:209–245.

Reiter, R. 1991. The frame problem in the situation calculus: a simple solution (sometimes) and a completeness result for goal regression. In *Artificial Intelligence and Mathematical Theory of Computation*, ed. V. Lifschitz, 359–380. Academic Press.

Sandewall, E. 1996. Assessments of ramification methods that use static domain constraints. In *KR-96*, 99–110.

Shanahan, M. 1999. The ramification problem in the event calculus. In *IJCAI'99*, 140–146.

Thielscher, M. 1997. Ramification and causality,. *Artificial Intelligence* 89:317–364.

Thielscher, M. 1998. Reasoning about actions: steady versus stabilizing state constraints. *Artificial Intelligence* 104:339–355.

Zhang, D., and N. Foo. 2000. EPDL: a logic for causal reasoning. In *University of New South Wales, Technical Report, UNSW-CSE-TR-0006*, Available at: `http://www.cse.unsw.edu.au/school/research/tr.html#0006`.

Norman Foo
School of Computer Science and Engineering
The University of New South Wales
NSW 2052, Australia
E-mail: norman@cse.unsw.edu.au
http://www.cse.unsw.edu.au/~norman

Dongmo Zhang
School of Computer Science and Engineering
The University of New South Wales
NSW 2052, Australia
E-mail: dongmo@cse.unsw.edu.au
http://www.cse.unsw.edu.au/~dongmo

11

From Bisimulation Quantifiers to Classifying Toposes

SILVIO GHILARDI AND MAREK ZAWADOWSKI

ABSTRACT. According to (Ghilardi and Zawadowski 1995a), it is known that Pitts' quantifiers (Pitts 1992) are semantically characterizable through direct and dual direct image in the topos of sheaves over the site given by the opposite category of finite Heyting algebras (endowed with the canonical topology). We investigate in this paper the geometric infinitary theory classified by that topos, within an appropriate general context from Universal Algebra (covering in this way also relevant cases arising from modal logic).

1 Motivations

A.M. Pitts in (Pitts 1992) proved a surprising theorem saying that the second order intuitionistic propositional calculus IPC^2 can be interpreted into ordinary intuitionistic propositional calculus IPC. More precisely, he proved that for each propositional variable p and for each formula α of IPC, there exist formulas $\exists^p \alpha$ and $\forall^p \alpha$ of IPC (effectively computable from α) containing only variables not equal to p which occur in α, and such that for any formula β not involving p, we have

$$\vdash_{IPC} \exists^p \alpha \to \beta \qquad \text{iff} \qquad \vdash_{IPC} \alpha \to \beta$$

and

$$\vdash_{IPC} \beta \to \forall^p \alpha \qquad \text{iff} \qquad \vdash_{IPC} \beta \to \alpha.$$

In categorical terms, the above theorem says that the opposite of the category of finitely presented Heyting algebras HA_{fp} is a *Heyting category* and in model-theoretic terms the above theorem says that the theory of Heyting algebras *admits a model completion*

Advances in Modal Logic, Volume 3
F. Wolter, H. Wansing, M. de Rijke, and M. Zakharyaschev, eds.

(Ghilardi and Zawadowski 1997). Thus Pitts' theorem is a quite stimulating result as it states a purely proof theoretic fact having many different interpretations.

Pitts quantifiers $\exists^p\alpha$ and $\forall^p\alpha$ are quite hard to compute directly; nevertheless they have a nice semantic characterization, which we are going to recall. We see here a Kripke model (over a finite set of propositional variables $\vec{p}$) as a monotonic map $u : (P, \leq) \longrightarrow (\mathcal{P}(\vec{p}), \subseteq)$ having a finite rooted poset as a domain. We also note by u_p the restriction of u to the rooted poset $\{q \,|\, p \leq q\}$. Truth of formulas can be introduced in the standard way; notation $u \models \alpha$ means that α is true in the root of u (such notation applies only in case u is a Kripke model over a finite set of propositional variables including all variables occurring in α). An open map between posets is a customary open map in topological sense, once posets are seen as topological spaces with upward closed subsets as open subsets. Given two such Kripke models $u : (P, \leq) \longrightarrow (\mathcal{P}(\vec{p}), \subseteq)$ and $u' : (Q, \leq) \longrightarrow (\mathcal{P}(\vec{p}, q), \subseteq)$, we say that u' is an *extension of* u iff there is an open surjective map $f : (Q, \leq) \longrightarrow (P, \leq)$, such that the square

$$\begin{array}{ccc} (Q, \leq) & \xrightarrow{f} & (P, \leq) \\ \Big\downarrow{\scriptstyle u'} & & \Big\downarrow{\scriptstyle u} \\ \mathcal{P}(\vec{p}, q) & \xrightarrow[r_{\vec{p}}]{} & \mathcal{P}(\vec{p}) \end{array}$$

commutes, where $r_{\vec{p}}$, is the map of 'intersecting with $\vec{p}$'.

Now, we can state the clauses for satisfaction determining Pitts' quantifiers; it can be shown that $\exists^q\alpha(\vec{p}, q)$ and $\forall^q\alpha(\vec{p}, q)$ are the formulas uniquely characterized (up to provable equivalence) by the following conditions:

- $u \models \exists^q\alpha(\vec{p}, q)$ iff
  there is an extension u' of u such that $u' \models \alpha(\vec{p}, q)$;
- $u \models \forall^q\alpha(\vec{p}, q)$ iff
  for any $p \in dom(u)$ and for any extension u' of u_p, we have $u' \models \alpha(\vec{p}, q)$.

The above clauses can be equivalently stated in slightly different forms; for instance, one may replace our 'extensions' by bisimulations relations as in (Visser 1996) (this is why Pitts quantifiers are sometimes called *bisimulation quantifiers*). We prefer the above formulation, because it has a clear *sheaf-theoretic* meaning: let us denote by $|\alpha(\vec{p})|$ the set of all $\vec{p}$-models of the formula $\alpha(\vec{p})$. It turns out that this is a sheaf (actually, a subsheaf of the sheaf of all $\vec{p}$-models) on the site $(\mathbf{T^f}, \mathbf{J^f})$ given by finite posets and open maps (with the canonical Grothendieck topology). Now the above clauses say that $|\exists^q \alpha(\vec{p}, q)|$ and $|\forall^q \alpha(\vec{p}, q)|$ are exactly ***the sheaf image and the sheaf dual image*** of $|\alpha(\vec{p}, q)|$ along the natural transformation given by 'composition with' $r_{\vec{p}}$. Actually, Pitts theorem is proved in (Ghilardi and Zawadowski 1995a) (in its categorical formulation) through the above mentioned semantic characterization of Pitts' quantifiers, namely by showing the existence of a conservative embedding (which preserves the relevant structure and also something more) of HA_{fp}^{op} into the sheaf topos $Sh(\mathbf{T^f}, \mathbf{J^f})$.

What we are going to do in the present paper is a kind of reverse investigation: we start *from geometry* (namely from the topos $Sh(\mathbf{T^f}, \mathbf{J^f})$ itself) and we give it a *logical* meaning, by characterizing the coherent infinitary theory implicit in it. This is always possible with Grothendieck toposes, because they always are ***classifying toposes*** of such theories. We recall here the related definitions. If $\mathcal{E}$ and $\mathcal{H}$ are toposes[1] then a *geometric morphism* $f : \mathcal{E} \longrightarrow \mathcal{H}$ is a pair of adjoint functors

$$\mathcal{E} \underset{f_*}{\overset{f^*}{\rightleftarrows}} \mathcal{H}$$

$f^* \dashv f_*$ such that f^* preserves finite limits. We call f^* the *algebraic part* and f_* the *geometric part* of the geometric morphism f. f^* preserves *geometric (or coherent) logic* i.e. the formulas, called *geometric formulas*, which are built up from atomic formulas of a first order language with equality by means of the existential quantifier and the following connectives: $\top$, $\bot$, $\wedge$, $\bigvee$ ($\bigvee$ is infinitary disjunction of formulas having altogether finitely many variables). A *geometric theory* is a theory T having axioms of form $\forall_{\vec{x}}(\gamma_1(\vec{x}) \Rightarrow \gamma_2(\vec{x}))$, where both $\gamma_1(\vec{x})$ and $\gamma_2(\vec{x})$ are geometric formulas. As toposes are endowed with all the relevant structure, it makes sense to speak of internal models of a geometric theory in a topos $\mathcal{E}$.

The important fact is that for any topos $\mathcal{E}$ there is a geometric theory T (the theory classified by $\mathcal{E}$) such that for any other topos $\mathcal{H}$, the category of geometric morphisms $\mathcal{H} \longrightarrow \mathcal{E}$ is naturally equivalent to the category of models of T in $\mathcal{H}$. This means that there is a *generic model* $\mathcal{G}_T$ of T in $\mathcal{E}$ such that for any model M of T in another topos $\mathcal{H}$ there is a unique (up to isomorphism) geometric morphism $f : \mathcal{H} \longrightarrow \mathcal{E}$ such that $f^*(\mathcal{G}_T) \cong M$.

[1]In this paper, by topos, we always mean a Grothendieck topos, i.e. a topos of sheaves over some site.

The converse is also true: for any geometric theory T there is a topos $\mathcal{E}(T)$ classifying T. If a theory T is equational, then $\mathcal{E}(T)$ is particularly simple, namely it is the topos of presheaves on the dual of the category of finitely presented T-algebras $Alg(T)^{op}_{fp}$, i.e. it is $Set^{Alg(T)_{fp}}$. This is because models of T in a topos $\mathcal{E}$ correspond to finite limits preserving (lex, for short) functors from $Alg(T)^{op}_{fp}$ into $\mathcal{E}$; moreover it is a general fact that any lex functor $Alg(T)^{op}_{fp} \longrightarrow \mathcal{E}$ extends uniquely to (the algebraic part of) a geometric morphism $\mathcal{E} \longrightarrow Set^{Alg(T)_{fp}}$.

If a geometric theory T' is obtained from an equational theory T by some additional geometric axioms in the same language, then $\mathcal{E}(T')$ is a topos of sheaves $Sh(Alg(T)^{op}_{fp}, J)$ for some Grothendieck topology J "generated by the additional axioms". For example, for an axiom

$$\forall_{\vec{x}}\,(\varphi(\vec{x}) \Rightarrow \bigvee_{i\in I} \exists_{\vec{y}}\, \psi_i(\vec{x},\vec{y}))$$

where $\varphi(\vec{x})$ and $\psi_i(\vec{x},\vec{y})$ for $i \in I$ are conjunctions of equations, we add a cover of $\mathcal{F}(\vec{x})/\varphi(\vec{x})$[2]

$$\{\mathcal{F}(\vec{x})/\varphi(\vec{x}) \longrightarrow \mathcal{F}(\vec{x},\vec{y})/\psi_i(\vec{x},\vec{y}) \,\&\, \varphi(\vec{x})\}_{i\in I}$$

and the topology J is the least topology on $Alg(T)^{op}_{fp}$ containing all such covers.

So putting it the other way around, a topos $Sh(Alg(T)^{op}_{fp}, J)$ classifies some special models of T, namely those for which the corresponding lex functors from $Alg(T)^{op}_{fp}$ preserve J-covers. Thus the additional set of axioms of the geometric theory T' containing T classified by $Sh(Alg(T)^{op}_{fp}, J)$ is any set of geometric axioms which "generates the topology J".

Turning to our main aim (identifying the geometric theory classified by the topos $Sh(\mathbf{T}^{\mathbf{f}}, \mathbf{J}^{\mathbf{f}})$ which gives the appropriate semantics to Pitts' quantifiers), we cannot directly apply the above observations because $\mathbf{T}^{\mathbf{f}}$ is not the opposite of a category of finitely presented algebras (notice that it is not even left exact), so we have to use an indirect strategy, passing through Grothendieck's comparison lemma (Artin et al. 1972). We nevertheless reach our goal in Section 4 by giving the appropriate axiomatization; we show in Section 5 that ordinary set-theoretic models of the theory we obtain are just *prime, locally finite and algebraically closed Heyting algebras*. A full classification of such models in the countable case is given in Section 6, where it is shown that only two such models exist up to isomorphism, differing from each other by the

[2]This is the free algebra over $\vec{x}$ divided by the congruence generated by the pairs of elements corresponding to the conjuncts of φ (see Section 2).

existence of a penultimate element, a condition which is elementary but not expressible in geometric logic.

We shall put ourselves in an appropriate context from Universal Algebra, covering not only the case of Heyting algebras, but also other equational classes related to propositional logics (where Pitts' theorem may not hold, see (Ghilardi and Zawadowski 1995b)). For space reasons, we must assume from the reader some familarity with topos theoretic techniques (to be found e.g. in (Makkai and Reyes 1977) or in (MacLane and Moerdijk 1992)).

2 Preliminaries

We fix for the whole paper an *equational* first order theory T. We use the letters $E, F, \ldots$ (or, indifferently, also the greek letters $\varphi, \psi, \ldots$) for formulas which are (possibly empty) conjunctions of equations (briefly, *e-formulas*). $Alg(T)$ is the category of algebras which are models of T (briefly, T-algebras) and $Alg(T)_{fin}$ is the full subcategory determined by finite algebras. T-algebras will be indicated by $A, B, \ldots$ and morphisms among T-algebras by $f, g, \ldots$. We shall make a systematic confusion between an algebra and its carrier set. Elements of an algebra are indicated with letters $a_1, a_2, \ldots, b_1, b_2, \ldots$; the letters $\vec{a}, \vec{b}, \ldots$ are used for strings of elements. When we expand the language of T with parameters arising from a given algebra, we do not distinguish between elements and their names in the expanded language. A pair given by a set of new constants and a set of e-formulas in the related expanded language is called a *presentation*. Given such a presentation $\langle X, \Delta \rangle$, there are an algebra (unique up to isomorphism) denoted by $\mathcal{F}(X)/\Delta$ and an interpretation for the new constants $i : X \longrightarrow \mathcal{F}(X)/\Delta$ such that $\mathcal{F}(X)/\Delta \models \Delta$ and such that for every other pair $\langle A, i' \rangle$ with the same property, there is a unique morphism $f : \mathcal{F}(X)/\Delta \longrightarrow A$ such that $f \circ i = i'$. We recall how $\mathcal{F}(X)/\Delta$ is built; we take all closed terms $t(\vec{a})$ with parameters from X and we introduce the equivalence relation:

$$t_1 \sim t_2 \quad \text{iff} \quad \vdash_T E_1 \,\&\cdots\&\, E_n \Rightarrow t_1 = t_2$$

for some $E_1, \ldots, E_n \in \Delta$. Equivalence classes can be turned into an algebra $\mathcal{F}(X)/\Delta$ which is a model of T by interpreting all function symbols on representative elements. The interpretation $i : X \longrightarrow \mathcal{F}(X)/\Delta$ is obtained by associating every $a \in X$ with its own equivalence class.

We say that $\langle X, \Delta \rangle$ is a presentation for (any algebra isomorphic to) $\mathcal{F}(X)/\Delta$. Each algebra admits a presentation, the canonical one is given by the carrier set and the positive diagram. If Δ is empty, $\mathcal{F}(X)$ is the free algebra over X; if both X and Δ are finite (in this case Δ can be taken to be a singleton because e-formulas are closed under finite conjunctions) the presentation is said to be finite. An algebra is *finitely presented* iff it admits a finite presentation: we can represent a finitely presented algebra as $\mathcal{F}(\vec{x})/E(\vec{x})$, meaning that we see its elements as terms in the variables $\vec{x}$ quotiented by provable equality under assumption $E(\vec{x})$. The notion of finitely presented algebra can be introduced also in a purely categorical way, cf. (Gabriel and Ulmer 1971), (Makkai and Pitts 1987): A is finitely presented iff the set-valued representable functor represented by A preserves filtered colimits. We denote by $Alg(T)_{fp}$ the full subcategory of $Alg(T)$ given by the finitely presented algebras. We use also the following notation: if $A \in Alg(T)$ and $E(\vec{a})$ is the e-formula with parameters from A given by $t_1(\vec{a}) = t'_1(\vec{a}) \,\&\cdots\&\, t_n(\vec{a}) = t'_n(\vec{a})$, $A/E(\vec{a})$ is the quotient of A by the congruence generated by the pairs $\langle t_i(\vec{a}), t'_i(\vec{a})\rangle$ $(i = 1, \ldots, n)$. Notice that if A admits the presentation $\mathcal{F}(X)/\Delta$ and if $\vec{a} \subseteq X$, then $A/E(\vec{a})$ admits the presentation $\mathcal{F}(X)/(\Delta \cup \{E(\vec{a})\})$.

In the whole paper we shall assume that our equational theory T is a FLEA (finitary, logical, equational, amalgamable) i.e. that

(i) T has *equationally definable principal congruences* ((EDPC) for short), which means (Köhler and Pigozzi 1980) that there exists an e-formula $I(x_1, x_2, x_3, x_4)$ such that for every algebra A and for every 4-tuple of elements a_1, a_2, a_3, a_4 from it we have that

$$A/a_1 = a_2 \models a_3 = a_4 \;\text{ iff }\; A \models I(a_1, a_2, a_3, a_4).$$

(ii) T has *equationally definable principal meets* ((EDPM) for short), which means (Blok and Pigozzi 1986) [3] that there exist e-formulas $J(x_1, x_2, x_3, x_4)$ and J_0 (J_0 is variable-free) such that: for every algebra A and for every 6-tuple of elements $a_1, a_2, a_3, a_4, b_1, b_2$ from it we have that

$$(A/a_1 = a_2 \models b_1 = b_2 \;\text{ and }\; A/a_3 = a_4 \models b_1 = b_2)$$
$$\text{iff} \qquad A/J(a_1, a_2, a_3, a_4) \models b_1 = b_2;$$

and moreover that

$$A/J_0 \models b_1 = b_2.$$

(iii) amalgamation property (AP) holds for $Alg(T)_{fin}$;

(iv) T has finite model property (fmp);

(v) the signature of T contains only finitely many function symbols.

The main examples of FLEA's are equational theories arising from propositional logics. For example Boolean algebras, Heyting algebras,

[3]The definition given in (Blok and Pigozzi 1986) is slightly different because presence of the formula J_0 (expressing the trivial total congruence) is not required.

interior algebras and diagonalizable algebras are FLEAs, and should be considered as our main reference examples.

Note that $Alg(T)$ need not satisfy the amalgamation property, even if we assumed this property for $Alg(T)_{fin}$ and even if we shall be able to prove it in next section for $Alg(T)_{fp}$ (see (Maksimova 1982) for a counterexample).

We shall use hypotheses (i)-(ii) in the following equivalent form (equivalence follows basically from results in (Köhler and Pigozzi 1980), (Blok and Pigozzi 1986), see also (Ghilardi and Zawadowski 200?) for details). Let $\mathbf{T}$ be the opposite category of $Alg(T)_{fp}$: now T has (EDPC) and (EDPM) iff *regular subobjects in* $\mathbf{T}$ *have a Heyting algebra structure which is pullback stable* (recall that a subobject - seen as an equivalence class of monomorphisms - is regular iff it is the equalizer of a pair of arrows).

Given that (EDPC) holds, we can use hypothesis (iii) in the following form: *any fp algebra embeds into a product of finite algebras.*

Finally, hypothesis (iv) makes clear that *finite algebras are fp.*

3 Introducing Sheaves

Recall that $\mathbf{T}$ denotes the dual of the category of finitely presented T-algebras (for our FLEA T) and let $\mathbf{T^f}$ denote the dual of the category of finite T-algebras. In this section we show that the jointly epic families of morphisms on both categories, denoted by $\mathbf{J}^\infty$ and $\mathbf{J^f}$, respectively, form Grothendieck topologies. Moreover they provide two different presentations of the same topos. This will be useful in the next section, as it reduces the problem of characterizing the geometric theory classified by $Sh(\mathbf{T^f}, \mathbf{J^f})$ to the problem of characterizing the theory classified by a site based on a customary category of finitely presented algebras.

First we state a technical lemma.

Lemma 3.1 (i) *A family* $\{g_i : C_i \longrightarrow C\}_{i \in I}$ *in a category with finite limits is jointly epic iff for every regular subobject* $S \hookrightarrow C$,

$$(\forall\, i \in I \quad \top \leq g_i^*(S)) \quad \Rightarrow \quad \top \leq S$$

(here $\top$ *denotes the biggest regular subobject, i.e. identity).*

(ii) *Suppose we are given in* $\mathbf{T}$ *a jointly epic family* $\{f_i : A_i \longrightarrow A\}_i$ *and a regular mono* $B \hookrightarrow A$. *Then the pulled back family* $\{\bar{f}_i : A_i \times_A B \longrightarrow B\}_i$ *is jointly epic as well.*

(iii) *Suppose we are given in* $\mathbf{T}$ *a finite jointly epic family* $\{A_i \hookrightarrow A\}_i$ *entirely composed of regular monos and an arrow* $f : B \longrightarrow A$. *Then the pulled back family is jointly epic as well.*

(iv) *Consider a pushout square among finitely presented* T*-algebras*

$$\begin{array}{ccc} A & \xrightarrow{m} & B \\ {\scriptstyle g}\downarrow & & \downarrow{\scriptstyle g'} \\ A' & \xrightarrow[m']{} & B' \end{array}$$

If A, A' are finite and m monic, so is m'.

(v) *If $m : A \longrightarrow B$, $g : A \longrightarrow A'$ are morphisms among finite algebras and if m is mono, then there are a finite algebra B', a monomorphism $m' : A' \longrightarrow B'$ and a morphism $g' : B \longrightarrow B'$ such that $m' \circ g = g' \circ m$.*

Proof. (i) is easy.

Ad (ii). We use (i) and the fact that **T** has stable implications among regular subobjects.

By taking composition with $B \hookrightarrow A$, [4] a regular subobject of $S \hookrightarrow B$ can be seen as a regular subobject of A; as such it satisfies the inequality $S \leq B$. Suppose that $\top \leq \bar{f}_i^*(S)$; if we see S as a regular subobject of A, this implies $f_i^*(B) \leq f_i^*(S)$. By existence and stability of implications, we get $\top \leq f_i^*(B \rightarrow S)$. As the f_i's are jointly epic, we get that B is equal to S (as a regular subobject of A), as required.

Ad (iii). By (i) and by the existence of joins among regular subobjects, the family $\{A_i \hookrightarrow A\}_i$ is jointly epic iff $\top \leq \bigvee_i A_i$. Thus (iii) immediately follows from the stability of joins.

Ad (iv). Since any morphism between finite algebras has regular epi/mono factorization, by (ii), we can suppose that g is mono, as well. Then, we can use fmp in order to produce $h : B \longrightarrow C$ such that C is finite and $h \circ m$ still monic. The claim now follows from amalgamation property for finite algebras and the universal property of pushouts.

Ad (v). Since T has fmp, B' in (iv) can be embedded into a product of finite algebras. As A' and B are finite the product can taken finite as well. ⊣

Proposition 3.2 *The families of jointly epic coverings $\mathbf{J}^\infty$ and $\mathbf{J}^\mathbf{f}$ on categories **T** and $\mathbf{T}^\mathbf{f}$, respectively, form Grothendieck topologies. Moreover, the toposes $Sh(\mathbf{T}, \mathbf{J}^\infty)$ and $Sh(\mathbf{T}^\mathbf{f}, \mathbf{J}^\mathbf{f})$ are equivalent.*

[4] Notice that regular monic compose in **T**, because **T** is the opposite of the category of finitely presented T-algebras (recall that regular epis among fp algebras are exactly surjective morphisms).

Proof. Using Lemma 3.1 we prove that $\mathbf{J}^\infty$ is a Grothendieck topology. Let

$$\{A \xrightarrow{a_i} A_i\}_{i\in I}$$

be a jointly mono family in $\mathbf{T}^{op}$, $A \xrightarrow{f} B$ a morphism in $\mathbf{T}^{op}$. We need to show that the family

$$\{B \xrightarrow{b_i} B_i\}_{i\in I} \tag{1}$$

obtained by taking pushout along f, is jointly mono in $\mathbf{T}^{op}$.

First, we show that we can assume that B is a finite algebra. In fact, since T has fmp, there is a jointly monic family $\{b'_j : B \longrightarrow B'_j\}_{j\in J}$, where the B'_j are finite. By stability for finite algebras, we obtain (for every $j \in J$) jointly monic families

$$\{B'_j \xrightarrow{c_{ij}} C_{ij}\}_{i\in I},$$

hence the composition $\{c_{ij} \circ b'_j : B \longrightarrow C_{ij}\}_{i\in I, j\in J}$ is jointly monic too and so is the family (1), which is first component by the universal property of pushouts.

Now, we assume that B is finite. Then, we can form a regular epi-mono factorization in $\mathbf{T}^{op}$

$$A \longrightarrow Q \longrightarrow B$$

Note that Q is finite as well. For $i \in I$ we can construct pushouts in $\mathbf{T}^{op}$

$$\begin{array}{ccc} A & \longrightarrow & A_i \\ \downarrow & & \downarrow \\ Q & \longrightarrow & Q_i \\ \downarrow & & \downarrow \\ B & \longrightarrow & B_i \end{array}$$

The family

$$\{Q \longrightarrow Q_i\}_{i\in I} \tag{2}$$

is jointly mono by Lemma 3.1(ii) above.

Now, we show that (1) is jointly mono. Note that there is a finite set $I_0 \subseteq I$ such that

$$\{Q \longrightarrow Q_i\}_{i\in I_0} \tag{3}$$

is jointly mono. We can form pushouts for $i \in I_0$

$$\begin{array}{ccccc} Q & \longrightarrow & R_i & \longrightarrow & Q_i \\ \downarrow & & \downarrow & & \downarrow \\ B & \longrightarrow & S_i & \longrightarrow & B_i \end{array}$$

with the top row being regular epi-mono factorization of $Q \longrightarrow Q_i$. Now the family $\{B \longrightarrow S_i\}$ is jointly mono by Lemma 3.1(iii) and for every i the morphism $S_i \longrightarrow B_i$ is mono by Lemma 3.1(iv). So the family (1) is jointly mono as well.

This shows that $\mathbf{J}^\infty$ is a Grothendieck topology. As we just proved that jointly monic families are stable under pushouts, it is not difficult to see, using fmp of T, that $\mathbf{J}^\mathbf{f}$ is a Grothendieck topology, as well.

Finally, since the embedding $\mathbf{T}^\mathbf{f} \longrightarrow \mathbf{T}$ is full and faithful, a family of morphisms $\{A \xrightarrow{a_i} A_i\}_{i\in I}$ in $\mathbf{T}^\mathbf{f}$ is a cover in $\mathbf{J}^\mathbf{f}$ iff it is a cover in $\mathbf{J}^\infty$, and, by fmp, every object in $\mathbf{T}^\infty$ has a cover consisting of object from $\mathbf{T}^\mathbf{f}$, the comparison lemma, cf. (Artin et al. 1972) p. 288, applies. Therefore the toposes $Sh(\mathbf{T}, \mathbf{J}^\infty)$ and $Sh(\mathbf{T}^\mathbf{f}, \mathbf{J}^\mathbf{f})$ are equivalent. ⊣

We remark that if the strong amalgamation property holds[5] for finite algebras, then the topology $\mathbf{J}^\mathbf{f}$ is canonical; even in this case, however, (witness the case of Heyting algebras) $\mathbf{J}^\infty$ may not be canonical.

4 $Sh(\mathbf{T}^\mathbf{f}, \mathbf{J}^\mathbf{f})$ as a Classifying Topos

By Proposition 3.2, for any FLEA T, $Sh(\mathbf{T}^\mathbf{f}, \mathbf{J}^\mathbf{f})$ is a Grothendieck topos. We shall describe below a geometric theory $T^\dagger$ classified by this topos. In the next section we shall discuss in more familiar way the models of $Sh(\mathbf{T}^\mathbf{f}, \mathbf{J}^\mathbf{f})$ in *Set*, i.e. the ordinary models of $T^\dagger$.

An e-formula $\varphi(\vec{x})$ is *finite* iff the T-algebra $\mathcal{F}(\vec{x})/\varphi(\vec{x})$ is finite.

Let Φ_n, for $n \in \omega$, be the family of e-formulas

$$\Phi_n = \{\varphi(\vec{x}) | \vec{x} = x_1, \ldots, x_n \text{ and } \varphi(\vec{x}) \text{ is finite }\}$$

The theory $T^\dagger$ has the following axioms:

(i) axioms of T

(ii) $\bigvee \Phi_n$, for every $n \in \omega$

(iii) $\psi(\vec{x}) \Rightarrow \exists \vec{y} \varphi(\vec{x}, \vec{y})$,
for every monomorphism $\mathcal{F}(\vec{x})/\psi(\vec{x}) \longrightarrow \mathcal{F}(\vec{x}, \vec{y})/\varphi(\vec{x}, \vec{y})$ between finite T-algebras that sends $\vec{x}$ to $\vec{x}$

(iv) $J_0 \Rightarrow x_1 = x_2$

[5]By strong amalgamation we mean the possibility of completing a pair of monomorphisms having the same domain into a commutative square of monomorphisms, which *moreover* is a pullback.

(v) $J(x_1, x_2, x_3, x_4) \Rightarrow (x_1 = x_2 \vee x_3 = x_4)$

where $J(x_1, x_2, x_3, x_4)$ and J_0 are the e-formulas introduced in Section 2.

Let E_1, E_2 be the e-formulas $\bigwedge_{i=1}^{n} (t_i = t'_i)$ and $\bigwedge_{j=1}^{m} (u_j = u'_j)$, respectively. By $\bar{J}(E_1, E_2)$ we denote the following e-formula

$$\bigwedge_{i=1}^{n} \bigwedge_{j=1}^{m} J(t_i, t'_i, u_j, u'_j)$$

This formula expresses joins of regular subobjects in $\mathbf{T}$,[6] hence we obtain

Lemma 4.1 *For any e-formulas E_1, E_2, E we have*

$$\vdash_T \bar{J}(E_1, E_2) \Rightarrow E \quad \textit{iff} \quad (\vdash_T E_1 \Rightarrow E \textit{ and } \vdash_T E_2 \Rightarrow E).$$

We have

Theorem 4.2 *$Sh(\mathbf{T^f}, \mathbf{J^f})$ classifies the theory $T^\dagger$.*

Proof. By Proposition 3.2, $Sh(\mathbf{T^f}, \mathbf{J^f})$ is equivalent to $Sh(\mathbf{T}, \mathbf{J}^\infty)$. So we need to show that for any topos $\mathcal{E}$ the left exact functors from $(\mathbf{T}, \mathbf{J}^\infty)$ to $\mathcal{E}$ that are $\mathbf{J}^\infty$-continuous (i.e. send $\mathbf{J}^\infty$-covers to epimorphic families in $\mathcal{E}$), correspond to the models of the theory specified in the Theorem. Clearly, left exact functors from $\mathbf{T}$ to $\mathcal{E}$ correspond to T-models in $\mathcal{E}$. We shall show that a left exact functor $H : \mathbf{T} \longrightarrow \mathcal{E}$ is $\mathbf{J}^\infty$-continuous iff it preserves the following four types of covers corresponding to the axioms of the theory $T^\dagger$:

(4) $$\{\mathcal{F}(\vec{x}) \longrightarrow \mathcal{F}(\vec{x})/\varphi(\vec{x}) \mid \varphi(\vec{x}) \in \Phi_n\}$$

where $n \in \omega$ and $\vec{x} = x_1, \dots, x_n$;

(5) $$\{m : \mathcal{F}(\vec{x})/\varphi(\vec{x}) \longrightarrow \mathcal{F}(\vec{x}, \vec{y})/\psi(\vec{x}, \vec{y})\}$$

where φ and ψ are finite e-formulas and m is a mono;

(6) empty cover on $\mathcal{F}(\emptyset)/J_0$

(7) $$\{\mathcal{F}(\vec{x}, \vec{y})/J(\vec{x}, \vec{y}) \longrightarrow \mathcal{F}(\vec{x}, \vec{y})/x_i = y_i\}_{i=\{1,2\}}$$

where $\vec{x} = x_1, x_2$ and $\vec{y} = y_1, y_2$.

[6] This is a purely lattice-theoretic fact. First, keep in mind that regular subobjects in $\mathbf{T}$ are nothing but finitely generated congruences in $Alg_{fp}(T)$ (with reversed inclusion partial order, because we passed to the opposite category). Now we know that, thanks to our assumptions on T, regular subobjects in $\mathbf{T}$ are Heyting algebras; moreover each regular subobject is a finite meet of regular subobjects corresponding to principal congruences. Binary joins of pricipal congruences are expressed by J, hence binary joins of arbitrary finitely generated congruences must be expressed as shown above, by distributivity law.

Clearly all the above covers are $\mathbf{J}^\infty$-covers. Let us fix a left exact functor $H : \mathbf{T} \longrightarrow \mathcal{E}$ that preserves the covers (4), (5), (6), (7). By an *H-cover* we shall mean a cover in $\mathbf{T}$ which is sent by H to a jointly epimorphic family in $\mathcal{E}$. Thus, we need to show that any $\mathbf{J}^\infty$-cover is an H-cover.

Let

$$\{\mathcal{F}(\vec{x})/\psi(\vec{x}) \longrightarrow \mathcal{F}(\vec{x},\vec{y_i})/\psi_i(\vec{x},\vec{y_i})\}_{i\in I} \tag{8}$$

be an arbitrary $\mathbf{J}^\infty$-cover, where $\vec{x} = x_1,\ldots,x_n$ for some $n \in \omega$.

Since (4) is an H-cover, it is enough to show that for $\varphi \in \Phi_n$, the cover

$$\{\mathcal{F}(\vec{x})/\psi(\vec{x}),\varphi(\vec{x}) \longrightarrow \mathcal{F}(\vec{x},\vec{y_i})/\psi_i(\vec{x},\vec{y_i}),\varphi(\vec{x})\}_{i\in I} \tag{9}$$

is an H-cover. Since $\varphi(\vec{x})$ is finite, the cover (9) contains a finite subfamily which is a $\mathbf{J}^\infty$-cover. Thus, since T has fmp, it is enough to show that the covers of form (8) with ψ, ψ_i and I all finite are H-covers. As each morphism between finite T-algebras has an epi/regular mono factorization in $\mathbf{T}$ and (5) in an H-cover, to show that (8) is an H-cover is enough to show that the $\mathbf{J}^\infty$-cover (entierely formed by onto morphisms)

$$\{\mathcal{F}(\vec{x})/\psi(\vec{x}) \longrightarrow \mathcal{F}(\vec{x})/\psi_i(\vec{x})\}_{i=1,\ldots,n} \tag{10}$$

with ψ, ψ_i, I all finite, is a H-cover (notice also that with respect to (8), the additional variables $\vec{y_i}$ are dropped).

We shall show by induction on $n \in \omega$ that the $\mathbf{J}^\infty$-covers (10) are H-covers.

If $n = 0$ then (10) is the empty cover on $\mathcal{F}(\vec{x})/\psi(\vec{x})$. As every $\mathbf{J}^\infty$-cover is jointly mono $\mathcal{F}(\vec{x})/\psi(\vec{x})$ must be the terminal object[7]. Hence $\mathcal{F}(\vec{x})/\psi(\vec{x})$ is isomorphic to $\mathcal{F}(\emptyset)/J_0$. As (6) is an H-cover so is (10) in this case.

If $n = 1$ then (10) contains just one isomorphism and hence it is preserved by any functor.

If $n > 1$, then to show that (10) is an H-cover, it is enough to show that the following two covers

$$\{\mathcal{F}(\vec{x},\vec{y})/\bar{J}(\psi_1,\psi_2) \longrightarrow \mathcal{F}(\vec{x},\vec{y})/\psi_i\}_{i=\{1,2\}} \tag{11}$$

and

$$\{\mathcal{F}(\vec{x})/\psi(\vec{x}) \longrightarrow \mathcal{F}(\vec{x})/\psi'_i(\vec{x})\}_{i=2,\ldots,n} \tag{12}$$

[7] In fact, empty cover is jointly mono on any subobject of the terminal object. But due to existence of an e-formula without free variables J_0, the language of T must contain at least one constant. As a consequence T-algebras in any topos must be inhabited.

where

$$\psi_i'(\vec{x}) = \begin{cases} \bar{J}(\psi_1, \psi_2) & \text{if } n = 2 \\ \psi_i(\vec{x}) & \text{if } n > 2 \end{cases}$$

are H-covers. The cover (12) is jointly surjective as a consequence of Lemma 4.1, thus, as it contains less than n elements, it is an H-cover by inductive hypothesis.

The fact that (11) is an H-cover is the content of the following Claim.

Claim. For any e-formulas $E(\vec{x})$ and $F(\vec{x})$ the cover

(13) $$\{\mathcal{F}(\vec{x})/\bar{J}(E, F) \longrightarrow \mathcal{F}(\vec{x})/E,\ \ \mathcal{F}(\vec{x})/\bar{J}(E, F) \longrightarrow \mathcal{F}(\vec{x})/F\}$$

is an H-cover.

Proof of the Claim. First, we make two observations:

- since H-covers are stable under pushouts and (7) is an H-cover, so is the cover given by the two arrows

  (14) $$\{\mathcal{F}(\vec{x})/J(e, f), E \longrightarrow \mathcal{F}(\vec{x})/e, E,$$
  $$\mathcal{F}(\vec{x})/J(e, f), E \longrightarrow \mathcal{F}(\vec{x})/f, E\}$$

  for any equations e, f and for any e-formula E, all with free variables among $\vec{x}$ (if e is $t = t'$ and f is $u = u'$, the notation $J(e, f)$ means $J(t, t', u, u')$);
- secondly, if E and F are e-formulas with free variables among $\vec{x}$, then obviously

  (15) $$\vdash_T E \Rightarrow F \quad \text{implies} \quad \mathcal{F}(\vec{x})/E, F = \mathcal{F}(\vec{x})/E$$

Now we shall prove the claim by induction on number of equations in e-formulas $E = \bigwedge_{i=1}^{m} e_i$ and $F = \bigwedge_{j=1}^{n} f_j$, where e_i and f_j are equations whose variables are among $\vec{x}$.

If $m = 1$ and $n = 1$ then the claim immediately follows from the first observation.

If $m = 1$ and $n > 1$ then we have two covers

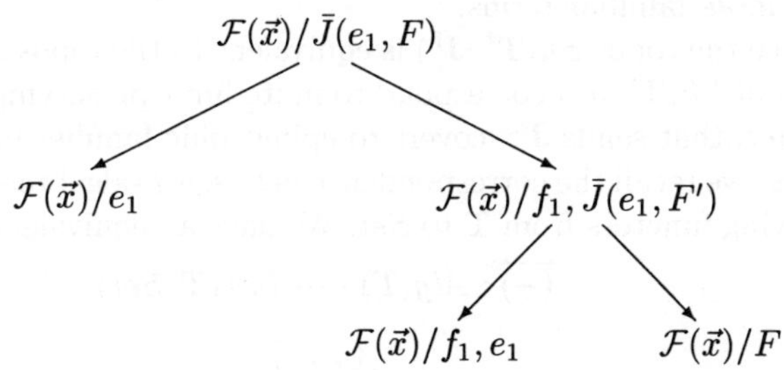

where F' is the e-formula $\bigwedge_{j=2}^{n} f_j$. The upper cover is an H-cover by the above observations and the lower cover is an H-cover by inductive hypothesis, since F' contain less than n equations. Since the composition of H-covers is an H-cover and the morphism $\mathcal{F}(\vec{x})/\bar{J}(e_1, F) \longrightarrow \mathcal{F}(\vec{x})/e_1, f_1$ factorises through $\mathcal{F}(\vec{x})/\bar{J}(e_1, F) \longrightarrow \mathcal{F}(\vec{x})/e_1$ the cover (13) is an H-cover in this case.

The case $m > 1$ and $n = 1$ is similar.

Now assume that $m > 1$ and $n > 1$. Then we have three covers

$$\begin{array}{ccccccc} & & & \mathcal{F}(\vec{x})/\bar{J}(E,F) & & & \\ & & \swarrow & & \searrow & & \\ & \mathcal{F}(\vec{x})/e_1, \bar{J}(E',F) & & & & \mathcal{F}(\vec{x})/f_1, \bar{J}(E,F') & \\ \swarrow & & \searrow & & \swarrow & & \searrow \\ \mathcal{F}(\vec{x})/E & & \mathcal{F}(\vec{x})/e_1, F & & \mathcal{F}(\vec{x})/f_1, E & & \mathcal{F}(\vec{x})/F \end{array}$$

where F' is as above and E' is the e-formula $\bigwedge_{i=2}^{m} e_i$. The upper cover is an H-cover by the two observations above and the lower covers are H-covers by these observations and inductive hypothesis. Now, if we take the cover composed from these three covers and eliminate from it, as before, these two morphisms that can be factorised by other morphisms we get again an H-cover which is exactly the cover (13). This ends the proof of both the Claim and the Theorem. $\dashv$

5 Points

In view of Theorem 4.2, we know that the models of $Sh(\mathbf{T}^{\mathbf{f}}, \mathbf{J}^{\mathbf{f}})$ in any Grothendieck topos $\mathcal{E}$, i.e. geometric morphisms from $\mathcal{E}$ to $Sh(\mathbf{T}^{\mathbf{f}}, \mathbf{J}^{\mathbf{f}})$, are exactly models of $T^{\dagger}$ in $\mathcal{E}$. So applying it to the case $\mathcal{E} = Set$, we obtain that *points* (namely models in *Set*) of $Sh(\mathbf{T}^{\mathbf{f}}, \mathbf{J}^{\mathbf{f}})$ are (correspond to) ordinary models of $T^{\dagger}$. In this section we shall make this correspondence explicit and as a by-product we shall describe the models of $T^{\dagger}$ in *Set* in more familiar terms.

Since the topos $Sh(\mathbf{T}^{\mathbf{f}}, \mathbf{J}^{\mathbf{f}})$ is equivalent to the topos $Sh(\mathbf{T}, \mathbf{J}^{\infty})$, the models of $Sh(\mathbf{T}^{\mathbf{f}}, \mathbf{J}^{\mathbf{f}})$ correspond to finite limit preserving functors from $\mathbf{T}$ to *Set*, that sends $\mathbf{J}^{\infty}$-covers to epimorphic families in *Set*.

First we recall the correspondence between T-algebras and finite limit preserving functors from $\mathbf{T}$ to *Set*. We have an equivalence of categories

$$\overline{(-)} : Alg(T) \longrightarrow Lex(\mathbf{T}, Set)$$

$$A \mapsto \overline{A}$$

where $Lex(\mathbf{T}, Set)$ denotes the category of finite limits preserving functors from $\mathbf{T}$ to Set and natural transformations between them, and the functor

$$\overline{A} : \mathbf{T} \longrightarrow Set$$

is a hom-functor, i.e. it sends a morphism

$$\mathcal{F}(\vec{x})/E(\vec{x}) \longrightarrow \mathcal{F}(\vec{x},\vec{y})/F(\vec{x},\vec{y})$$

in $\mathbf{T}$ to a function

$$(16) \qquad Hom(\mathcal{F}(\vec{x},\vec{y})/F(\vec{x},\vec{y}), A) \longrightarrow Hom(\mathcal{F}(\vec{x})/E(\vec{x}), A).$$

Since, a homomorphism from $\mathcal{F}(\vec{x})/E(\vec{x})$ to A correspond to a tuple of elements $\vec{a}$ from A such that $A \models E(\vec{a})$ the function (16) can be identified with the function

$$\{\vec{a},\vec{b} \in A : A \models F(\vec{a},\vec{b})\} \longrightarrow \{\vec{a} \in A : A \models E(\vec{a})\}$$

such that

$$\vec{a},\vec{b} \mapsto \vec{a}.$$

The functor $\overline{A}$ preserves a cover

$$\{\mathcal{F}(\vec{x})/E(\vec{x}) \longrightarrow \mathcal{F}(\vec{x},\vec{y^i})/F_i(\vec{x},\vec{y^i})\}_{i\in I}$$

iff the family of functions

$$\{\{\vec{a},\vec{b^i} \in A : A \models F_i(\vec{a},\vec{b^i})\} \longrightarrow \{\vec{a} \in A : A \models E(\vec{a})\}\}_{i\in I}$$

is jointly surjective, i.e. iff for all $\vec{a} \in A$

$$A \models E(\vec{a}) \;\Rightarrow\; \exists_i \exists_{\vec{b^i}\in A}\, A \models F_i(\vec{a},\vec{b^i})$$

$\overline{A}$ preserves all $\mathbf{J}^\infty$-covers iff it preserves any set of covers that generates the topology $\mathbf{J}^\infty$. In the proof of theorem 4.2 we have specified a set of covers that corresponds to axioms of the theory $T^\dagger$. Thus to describe the models of $T^\dagger$ we need to identify those T-algebras A for which the functors $\overline{A}$ preserves the covers (4), (5), (6), (7).

If $\overline{A}$ preserves the cover (4) then for any $n \in \omega$, and any n-tuple $\vec{a}$ in A there is a finite formula $\varphi(\vec{x})$ with n variables such that $A \models \varphi(\vec{a})$. This means that every finite tuple of elements of A generates a finite subalgebra of A. A T-algebra with this property is called *locally finite.*

A model M of an equational theory T is *algebraically closed*, if any finite set of equations with parameters from M having a solution in an extension of M has also a solution in M itself. In (Ghilardi and Zawadowski 1997), we have shown that this condition is elementarily expressible if $\mathbf{T}$ is r-regular (i.e. iff taking epis and regular monos we get a stable factorization system in $\mathbf{T}$). However, for locally finite algebras, as we shall see below, being algebraically closed is equivalent to the following elementary condition with no further assumptions on the theory T. We say that M has the *finite lifting property* iff for every monomorphism $h : B \longrightarrow C$ of finite T-algebras, and for any homomorphism $f : B \longrightarrow M$ there is a homomorphisms $g : C \longrightarrow M$, such that the diagram

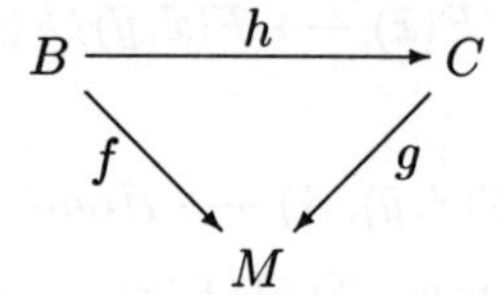

commutes. Thus A has finite lifting property iff $\overline{A}$ preserves the $\mathbf{J}^{\infty}$-covers (5).

We have

Proposition 5.1 *If the T-algebra A is locally finite, then A is algebraically closed iff it has the finite lifting property.*

Proof. Suppose that A has the finite lifting property and consider a monomorphism $m : A \longrightarrow B$ and an e-formula $\varphi(\vec{x}, \vec{y})$ such that $B \models \varphi(m(\vec{a}), \vec{b})$ for $\vec{a} \in A$ and $\vec{b} \in B$. As A is locally finite, the subalgebra generated by $\vec{a}$ is of the form $\mathcal{F}(\vec{x})/\psi(\vec{x})$, where ψ is a finite formula. As our aim consists in looking for a solution to $\varphi(\vec{a}, \vec{y})$ in A, we can freely assume that $\vdash_T \varphi(\vec{x}, \vec{y}) \Rightarrow \psi(\vec{x})$ (otherwise we can take as new φ just $\varphi \& \psi$). We thus have a commutative square

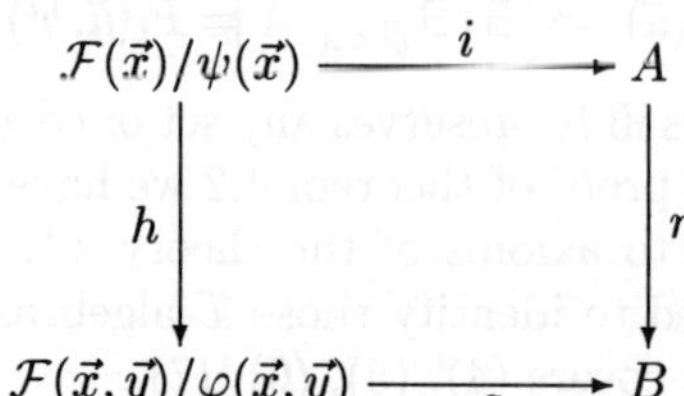

where i, h, g are the morphisms identified by $i(\vec{x}) = \vec{a}$, $h(\vec{x}) = \vec{x}$ and $g(\vec{x}) = m(\vec{a}), g(\vec{y}) = \vec{b}$, respectively. Notice that h is mono as the first component of a monomorphism. By fmp, we can find a finite algebra C and a morphism $f : \mathcal{F}(\vec{x}, \vec{y})/\varphi(\vec{x}, \vec{y}) \longrightarrow C$ such that $f \circ h$ is a monomorphism and then we can apply the definition of finite lifting property in order to find a solution in A to $\varphi(\vec{a}, \vec{y})$.

Suppose conversely that A is algebraically closed, that $m : B \longrightarrow C$ is a monomorphism among finite algebras and that $f : B \longrightarrow A$ is any homomorphism. Take the pushout

$$\begin{array}{ccc} B & \xrightarrow{f} & A \\ {\scriptstyle m}\downarrow & & \downarrow \\ C & \longrightarrow & P \end{array}$$

introduce suitable presentations for B, C and apply Lemma 5.2 below, in order to find the lifted homomorphism. $\dashv$

Lemma 5.2 *The pushout in $Alg(T)$ of a mono $m : B \longrightarrow C$ along any morphism $f : B \longrightarrow A$ is mono again, in case A is locally finite and B, C are finite.*

Proof. First notice that the Lemma holds in case A is finite as a consequence of Lemma 3.1(v). If A is locally finite, it is a filtered colimit of finite algebras. Hence the claim follows from the above case and from the fact that a filtered colimit of monos is mono in any category of algebras. $\dashv$

The claim in the proof of theorem 4.2 shows that if $\overline{A}$ preserves the cover (7) it preserves the more general covers (13), as well, i.e. for any pair of e-formulas $E_1(\vec{x})$ and $E_2(\vec{x})$ the pair of inclusions

$$\{\{\vec{a} \in A : A \models E_i(\vec{a})\} \longrightarrow \{\vec{a} \in A : A \models \bar{J}(E_1(\vec{a}), E_2(\vec{a}))\}\}_{i \in \{1,2\}}$$

is jointly surjective. The condition

$$A \models \bar{J}(E_1(\vec{a}), E_2(\vec{a}))$$

means that the congruence on A generated by $\bar{J}(E_1(\vec{a}), E_2(\vec{a}))$ is the identity and similarly the condition $A \models E_i(\vec{a})$ means that the congruence on A generated by $E_i(\vec{a})$ is the identity. On the other hand, the congruence generated by $\bar{J}(E_1(\vec{a}), E_2(\vec{a}))$ is the intersection of the congruences generated by $E_1(\vec{a})$ and $E_2(\vec{a})$. Thus preservation of (13) means that whenever the intersection of two finitely generated congruences is the identity congruence then so is one of the congruences.

$\overline{A}$ preserves (6) iff the set $\overline{A}(\mathcal{F}(\emptyset)/J_0)$ is empty, i.e. there are no homomorphisms from $\mathcal{F}(\emptyset)/J_0$ to A. As J_0 generates maximum congruence on any T-algebra (the intersection of the empty set of congruences), this means that the maximum congruence on A is not the identity congruence, i.e. the identity congruence on A is not the intersection of the empty set of congruences on A. In other words, A cannot be degenerated (it must have more than one element).

We say, that a T-algebra A is *prime* iff whenever the intersection of finitely many finitely generated congruences is the identity congruence, then one of such congruences is the identity congruence. So the above considerations shows that A is prime iff $\overline{A}$ preserves covers (6) and (7).

In this way, from Theorem 4.2 and the above, we get the following characterization of ordinary models of $T^{\dagger}$.

Theorem 5.3 *The models of* $Sh(\mathbf{T^f},\mathbf{J^f})$ *in Set, i.e. the category* $GTop(Set, Sh(\mathbf{T^f},\mathbf{J^f}))$ *of geometric morphisms from Set to* $Sh(\mathbf{T^f},\mathbf{J^f})$ *and natural transformations, is equivalent to the category of locally finite, algebraically closed, prime* T*-algebras.*

Before we give some examples illustrating this corollary, we prove a characterization of countable models of $T^{\dagger}$ which provides a method of constructing such models.

Proposition 5.4 *An at most countable* T*-algebra is locally finite, prime and algebraically closed iff it is the colimit of a chain*

$$A_0 \longrightarrow A_1 \longrightarrow A_2 \longrightarrow \ldots \longrightarrow A_n \longrightarrow \ldots \tag{17}$$

of finite subdirectly irreducible T*-algebras and monomorphisms with the following property:*

- *for any finite subdirectly irreducible* T*-algebra* B *and any monomorphism* $h : A_n \longrightarrow B$ *there are* $m \geq n$ *and a morphism* $g : B \longrightarrow A_m$ *such that* $g \circ h : A_n \longrightarrow A_m$ *is the embedding as in the chain above.*

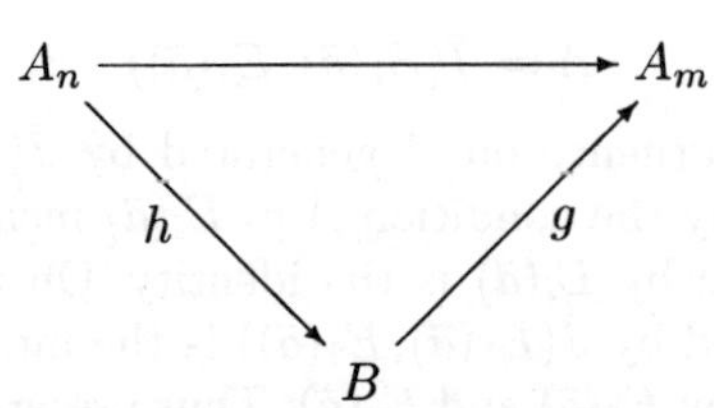

Proof. . Local finiteness means to be filtered colimits of monos among finite algebras; by countability we can restrict to chain colimits and by primality such finite algebras should be prime too (notice that to be prime is the same as to be subdirectly irreducible as far as finite algebras are concerned). We need only to show that the finite lifting property is the same as the property mentioned in the claim. One side is trivial: simply take the inclusion of A_n into the colimit as the morphism to be lifted. For the converse, take finite algebras C_1, C_2, a monomorphism $h : C_1 \longrightarrow C_2$ and a morphism into the colimit, which we can suppose to factor through a certain $f : C_1 \longrightarrow A_n$. By Lemma 3.1 (v), we can produce a commutative square

$$\begin{array}{ccc} C_1 & \xrightarrow{h} & C_2 \\ {\scriptstyle f}\downarrow & & \downarrow \\ A_n & \longrightarrow & C \end{array}$$

with C finite and the bottom arrow monic. Consider now a maximal congruence $\approx$ on C such that the composition

$$A_n \longrightarrow C \longrightarrow C/\approx$$

is still monic. Now $C/\approx$ is subdirectly irreducible,[8] so that the property mentioned in the claim of the Proposition applies, thus producing the lifting we are interested in. $\dashv$

Example 5.5 Consider **S5**-algebras, which are Boolean algebras endowed with a finite join preserving, idempotent, increasing operator $\Diamond$ satisfying the condition $\Diamond a \leq \Box\Diamond a$ (where $\Box$ is defined as $\neg\Diamond\neg$). Prime algebras are those satisfying the condition

$$1 = \Box a_1 \vee \cdots \vee \Box a_n \quad \Rightarrow \quad \exists i \; (1 = a_i),$$

for $n \in \omega$. Hence they are non-degenerated Boolean algebras in which $\Diamond a$ is taken to be 0 if $a = 0$ and 1 if $a \neq 0$. However, morphisms must be injective Boolean morphisms, because $\Diamond$ is to be preserved. Thus, by applying Proposition 5.4, [9] we easily realize that prime, algebraically closed and locally finite **S5**-algebras are nothing but atomless Boolean algebras. The opposite category of finitely presented **S5** algebras is the category having as objects finite sets endowed with an equivalence relation and as arrows the functions mapping each equivalence class onto an equivalence class. So sheaves over this category endowed with canonical topology classify atomless Boolean algebras (as is well known).[10]

Example 5.6 Consider Dummett algebras, which are Heyting algebras satisfying the condition $1 = (a \to b) \vee (b \to a)$. Prime Dummett algebras are simply chains with a minimum and a maximum element. Morphisms $f : C_1 \longrightarrow C_2$ among such chains are order-preserving maps which also preserve implication, i.e. they must satisfy the requirement

$$f(a) = f(b) \quad \Rightarrow \quad (a = b \text{ or } f(a) = 1).$$

[8] This is due to the fact that A_n is such: any two proper congruences of $C/\approx$ restrict to proper congruences over A_n, so their intersection, containing the minimum proper congruence of A_n, cannot be the identity congruence.

[9] Proposition 5.4 holds for the uncountable case too, by using arbitrary colimits instead of only chain colimits.

[10] We thank A. Joyal for drawing our attention to this example.

This means that such maps are injective, apart from the fact that they can collapse an upper segment to the top element. By applying Proposition 5.4, we realize that there are only two countable, prime, algebraically closed and locally finite Dummett algebras, namely the rational unit interval and the rational unit interval endowed with an extra element larger than 1. The two models differ for the existence of a penultimate element (which is an elementary but not a coherent condition). In the next section we shall see that for Heyting algebras the situation is surprisingly the same.

6 The Case of Heyting Algebras

We shall be interested in countable, prime, algebraically closed and locally finite Heyting algebras. Any such algebra is the colimit of a chain diagram of finite algebras and monomorphisms. We recall that the category of finite Heyting algebras is dual to the category of finite posets and open maps. A poset is said to be *non-degenerated* iff it has more than one element.

Proposition 6.1 *Countable, prime, locally finite and algebraically closed Heyting algebras are exactly colimits of duals of chains*

(C) $$\cdots P_{i+1} \xrightarrow{f_i} P_i \cdots$$

of non-degenerated finite rooted posets and open surjective morphisms satisfying the following condition:

(∗) *for every finite rooted poset Q, for every i, for every open surjective morphism $q : Q \longrightarrow P_i$, there are $j \geq i$ and an open morphism $r : P_j \longrightarrow Q$ such that $q \circ r = f_{ij}$ (where $f_{ij} = f_{j-1} \circ \cdots \circ f_i$).*

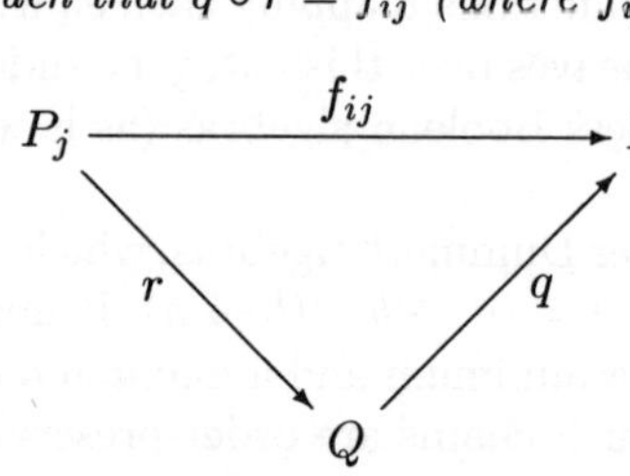

Proof. Immediate by Proposition 5.4 (the P_i are taken to be non-degenerated in order to avoid the colimit to be the two-element Heyting algebra). ⊣

We shall first make some comments about the above Proposition.

Remarks

1. From now on all posets are assumed to be finite.

2. r needs not be surjective, however (by the definition of openness, which implies that an open morphism between rooted posets is surjective just in case the root is in the image) it must be surjective in case q is what we call an *expansion morphism* (i.e. a surjective open morphism among rooted posets in which the root of the domain is the only element mapped onto the root of the codomain).
3. In case property $(*)$ is satisfied with the additional requirement that r is surjective, we say that the chain (C) is *strong*.
4. In view of the fact that we are interested in the limit of the dual diagram of (C), we consider any infinite subchain of (C) isomorphic to (C) itself. Moreover, we consider (C) and (C') isomorphic in case there exists for every i, morphisms $h_i : P'_i \longrightarrow P_i$ and $k_i : P_{i+1} \longrightarrow P'_i$ such that $h_i \circ k_i = f_i$ and $k_i \circ h_{i+1} = f'_i$ (this is the dual of the customary 'double chain argument', cf. (Chang and Keisler 1990)).

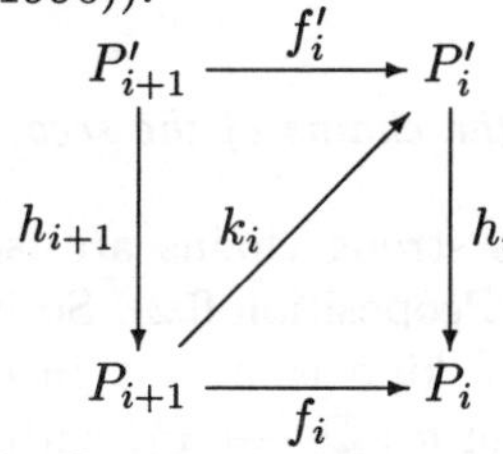

5. We say that (C) is of *the first kind* whenever all but finitely many of the f_i's are expansion morphisms, we say that it is of *the second kind* otherwise. Notice that chains of the second kind, up to isomorphisms as in remark 4, can be assumed not to contain expansion morphisms at all.

We shall prove that all chains of the same kind are isomorphic.

Proposition 6.2 *All the chains of the first kind are isomorphic.*

Proof. We can freely suppose that in (C) and (C'), both of the first kind, all the f_i are expansion morphisms (we can do that by removing an initial segment). We can also suppose that P'_0 is the two-element poset 2; as P_1 has more than one element, there is an expansion morphism $q : P_1 \longrightarrow P'_0$. Build $h_1 : P'_1 \longrightarrow P_1$ such that $q \circ h_1 = f'_0$ by using $(*)$ (we remove during the construction some edges of the chain, so we can freely suppose that P'_1 is the domain of h_1). By remark 2, h_1 is surjective; moreover it is an expansion morphism (because it is the first component of the expansion morphism f'_0). Build now $k_1 : P_2 \longrightarrow P'_1$ in the same way and go on. Finally, apply remark 4. $\dashv$

Before going on, we need a couple of easy Lemmas. Let P^+ be the poset obtained from the rooted poset P by adding a new root and let $c_P : P^+ \longrightarrow P$ be the morphism (called contraction morphism) which collapses the old and the new root.

Lemma 6.3 *Any open surjective morphism $f : Q \longrightarrow P$ among rooted posets which is not an expansion morphism factors through an expansion morphism $\bar{f}$ followed by the contraction morphism c_P.*

Proof. Define $\bar{f} : Q \longrightarrow P^+$ by $\bar{f}(p) = f(p)$ if p is not the root, $\bar{f}(p)$ equal to the root of P^+ otherwise. ⊣

Lemma 6.4 *The pullback (taken in the category of posets) of an expansion morphism along any open surjective morphism among rooted posets is an expansion morphism as well.*

Proposition 6.5 *All the chains of the second kind are isomorphic.*

Proof. Notice that two strong chains are isomorphic (by a simplified version of the proof of Peoposition 6.2). So it is sufficient to show that any chain of the second kind is strong. In order to prove it, consider any surjective morphism $q : Q \longrightarrow P_i$, with rooted domain. If q is an expansion morphism, see remark 2. Otherwise, let us apply Lemma 6.3 to both q and f_i (we can suppose that f_i is not an expansion morphism, because the chain is of the second kind, see remark 5). Thus we get morphisms $\bar{q}$ and $\bar{f}_i$ such that $q = c_{P_i} \circ q$ and $f_i = c_{P_i} \circ \bar{f}_i$. Secondly, apply Lemma 6.4 to the pullback of $\bar{q}$ and $\bar{f}_i$: we obtain that $p_2 : Q \times_{P_i^+} P_{i+1} \longrightarrow P_{i+1}$ is an expansion morphism, hence by remark 2, there is for some j a surjective morphism $h : P_j \longrightarrow Q \times_{P_i^+} P_{i+1}$ such that $p_2 \circ h = f_{i+1,j}$.

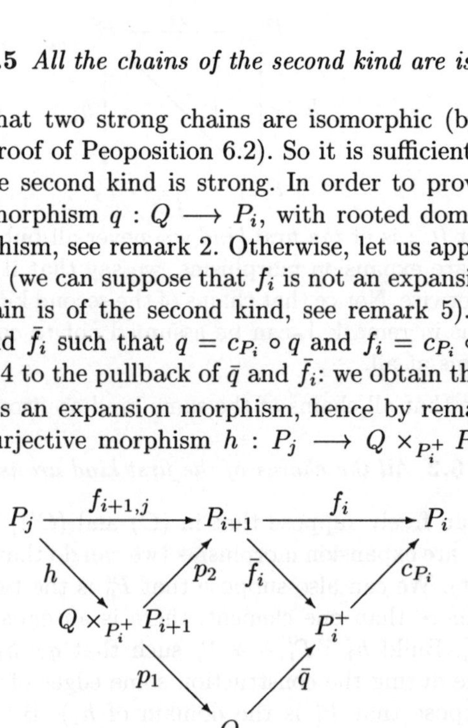

Now taking $p_1 \circ h$ we get a surjective morphism with the desired property. ⊣

Thus there are at most two Heyting algebras satisfying our requirements. Notice that chains of the first kind and of the second kind cannot be isomorphic because they produce non isomorphic Heyting algebras in the limit: such Heyting algebras differ because only one of them has a penultimate element (i.e. it is subdirectly irreducible).

Theorem 6.6 *There are exactly two countable, prime, locally finite and algebraically closed Heyting algebras (up to isomorphism).*

Proof. By Propositions 6.2 and 6.5, it is sufficient to prove the existence of a chain of the first kind and of a chain of the second kind. Here we use the following fact: given finitely many open surjective maps $\{P_i \longrightarrow P\}_i$, their pullback as posets $\{Q \longrightarrow P_i\}_i$ consists on open surjective maps as well, moreover Q is rooted if all P_i are rooted. For a given poset Q, let $\{q_i^Q : R_i^Q \longrightarrow Q\}_i$ be an enumeration of open surjective maps with target Q and for a given rooted poset P, let $\{r_i^P : R_i^P \longrightarrow P\}_i$ be an enumeration of open surjective maps with target P and having a rooted poset as domain.

We first produce a strong chain as follows: we begin with the two element poset as P_1 and at stage i we take as P_i the iterated pullback of the diagrams (for $k < i$)

$$\begin{array}{ccc} & & P_{i-1} \\ & & \Big\downarrow {\scriptstyle f_{i-k,i-1}} \\ R_k^{P_{i-k}} & \xrightarrow[r_k^{P_{i-k}}]{} & P_{i-k} \end{array}$$

(the map $P_i \longrightarrow P_{i-1}$ is the composition of the $(i-1)$-th projection with $r_1^{P_{i-1}}$).

Condition $(*)$ is clearly satisfied, with r surjective (the map $q : Q \longrightarrow P_i$ will appear as $r_k^{P_i}$ for some k and will be considered at stage $i+k$, so that the required r will be i-th projection at that stage).

In order to produce a first kind chain, let us first produce a chain

$$(C) \qquad \cdots\ P_{i+1} \xrightarrow{f_i} P_i\ \cdots$$

as above but by using the enumeration $\{q_i^Q\}_i$ of non necessarily rooted posets. Such chain will be made of non necessarily rooted posets and will satisfy condition $(*)$ with r surjective and Q non necessarily rooted. Consider now the endo-functor $(-)^+$ of the category of posets and order-preserving functions: this functor adds a new root to each object and expands morphisms by mapping the new root into the new root (notice that it gives open maps if applied to surjective open maps). Apply the functor $(-)^+$ to the chain (C). This modified chain is of the first kind: in fact, given a surjective open map $q : Q \longrightarrow P_i^+$, we can find r as required by $(*)$ by arguing as follows. First consider a maximal $p \in Q$ mapped onto the root of P_i^+; let $\bar{Q} \hookrightarrow Q$ be the open inclusion of the upward closed subset $\bar{Q} = \{p' \in Q \,|\, p' > p\}$ into Q and let $\bar{q}$ be q restricted to $\bar{Q}$ in the domain and to P_i in the codomain. By the construction of (C), there are j and $\bar{r} : P_j \longrightarrow \bar{Q}$ open and surjective such that $\bar{q} \circ \bar{r} = f_{ij}$. The required r is obtained by composing $\bar{r}^+$ with the inclusion of $\bar{Q}^+$ into Q. ⊣

7 Modal Algebras

The results contained in Section 5 hold for varieties of algebras corresponding to normal modal systems over $K4$ with finite model property and amalgamation property for finite algebras. However the classification analysis of last Section might become more problematic. We discuss here the case of varieties of *interior algebras* (also called closure algebras, or topological Boolean algebras, or $S4$-algebras): these are Boolean algebras endowed with a finite meet preserving, idempotent and decreasing operator $\Box$. The category of finite interior algebras is dual to the category of finite topological spaces (i.e. of finite *preordered* sets) and open maps. Such preordered sets $(P, \leq)$ are said to be *rooted* in case they contain an element (which is not anymore unique) ρ such that $\rho \leq p$ holds for all $p \in P$. The *cluster* of $p \in P$ is the subset $cl(p) = \{p' \in P \,|\, p \leq p' \;\&\; p' \leq p\}$.

Amalgamation property for finite algebras is very rare among varieties of interior algebras: although, strictly speaking, in presence of finite model property, this is a weaker condition than amalgamability *tout court*,[11] all the limitative results of (Maksimova 1979) apply, showing that only finitely many such varieties exist (up to 50!).

Given a variety L of interior algebras with fmp and amalgamation for finite algebras, an L-preordered set is just a finite preordered set whose dual interior algebra belongs to L. From Proposition 5.4 we immmediately get that countable, prime, locally finite and algebraically closed interior L-algebras are exactly colimits of duals of chains

$$(C) \qquad \cdots P_{i+1} \xrightarrow{f_i} P_i \cdots$$

of finite rooted L-preordered sets and open surjective morphisms satisfy-

[11] If we can amalgamate finite algebras A, B (over a certain finite algebra C) into an algebra D, then we can manifacture D to be finite as follows. First, we can take D to be just pushout, which turns out to be finitely presented, hence embeddable into a product $\Pi_i D_i$ of finite algebras D_i (by fmp); then, we can drop all indices but finitely many and still keep amalgamation, because A and B are finite.

ing condition $(*)$ from Proposition 6.1 (with obviously 'poset' everywhere replaced by 'L-preordered set').

We can now repeat word by word Remarks 1-5 from the previous Section; we only have to slightly modify the definition of an expansion morphism as follows: say that $f : Q_1 \longrightarrow Q_2$ is an *expansion* morphism iff it is an open surjective morphism among rooted preordered sets such that in case $f(q)$ belongs to the root-cluster of Q_2, then q belongs to the root-cluster of Q_1 too.

In some cases, classification of countable, prime, locally finite and algebraically closed L-algebras is quite easy: for instance, nothing changes from the analysis of previous Section in case we take as L the variety Grz whose finite algebras are just the dual of finite posets. If we identify L by limiting the cardinality of non-maximal clusters to 1 (this is one of the cases considered in (Maksimova 1979)), a third countable, algebraically closed, prime and locally finite L-algebra appears, namely the colimit of the dual of a chain like

$$\cdots\ C_{i+1} \xrightarrow{f_i} C_i\ \cdots$$

where C_i is a one-cluster preordered set (the chain should be carefully defined in such a way that each $p \in C_i$ has at least k preimages along f_{ij_k} for sufficiently large j_k). Notice that this third L-algebra differ from the other two because it is simple.[12]

However, if we take L to be just the full variety of interior algebras, complications arise. The construction P^+ we used in Lemma 6.3 does not have a clear analogous, because to have a convenient factorization of a surjective open (not expansion) morphism $f : Q \longrightarrow P$ among finite rooted preordered sets, we need to take into consideration the size of the root cluster of Q (which has no relation with the size of the root cluster of P, it can be smaller or greater). We shall prove that there are infinitely many non isomorphic countable, algebraically closed, prime and locally finite interior algebras.

The *spectrum* of an interior algebra B is the preordered set of its ultrafilters (preorder is: $U \leq V$ iff $\forall x \in B$, $\Box x \in U \Rightarrow x \in V$). As taking ultrafilters of a Boolean algebra is a representable functor, it changes colimits into limits; so the spectrum of the colimit of the dual of a chain (C) like the above one turns out to be the inverse limit of the P_i's with pointwise preorder: such limit is formed by the successions

[12] To prove the details of the above mentioned facts, little modifications to the arguments of the previous Section are needed (we leave them to the reader). For instance, pullbacks in the category of finite posets cannot be simply replaced by pullbacks in the category of finite preordered sets (they should be manifactured a little).

$\langle p_i \in P_i \rangle$ such that $f_i(p_{i+1}) = p_i$ for all i. It is clear then that *if two chains have non-isomorphic limit, the colimits of their duals are not isomorphic as interior algebras*, because they have different spectra (and taking spectrum is a functor).

An *n-marked preordered set* is a finite rooted preordered set with specified distinct elements $a_1, \ldots, a_n$ all belonging to the cluster of the root. A morphism of such n-marked preordered sets must preserve the marked elements. A strictly n-marked preordered set is an n-marked preordered set such that the cluster of the root consists only of the marked elements. A *n-marked chain* is a chain (C) as above entierely formed by strictly n-marked preordered sets and n-marked morphisms. Notice that an n-marked and an m-marked chain do not have isomorphic limit if $n \neq m$ (their limits are not isomorphic because the cluster of the root in one case has n elements and in the other case has m elements).

Theorem 7.1 *There are infinitely many non isomorphic locally finite, countable, prime and algebraically closed interior algebras.*

Proof. It is sufficient to show that there is an n-marked chain

$$(C) \qquad \cdots\ P_{i+1} \xrightarrow{f_i} P_i\ \cdots$$

of finite rooted preordered sets and open surjective morphisms satisfying the usual condition:

(∗) for every finite rooted preordered set Q, for every i, for every *expansion* morphism $q : Q \longrightarrow P_i$, there are $j \geq i$ and an open morphism $r : P_j \longrightarrow Q$ such that $q \circ r = f_{ij}$ (where $f_{ij} = f_{j-1} \circ \cdots \circ f_i$).

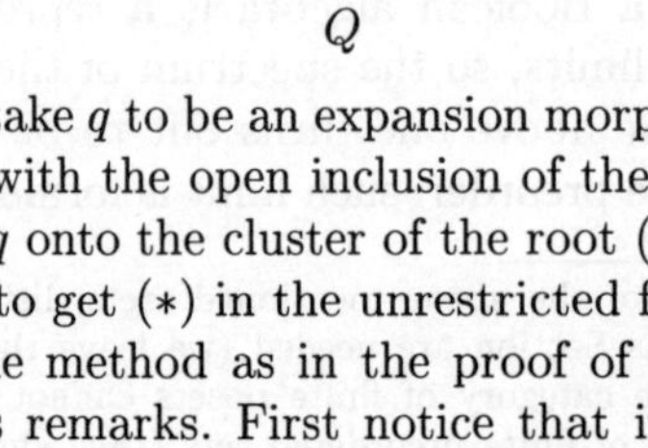

Notice that we can take q to be an expansion morphism: if it is not such, we can compose it with the open inclusion of the cone over a maximal cluster mapped by q onto the cluster of the root (and then apply (∗) in the restricted form to get (∗) in the unrestricted form).

We use the same method as in the proof of Theorem 6.6, but we need some previous remarks. First notice that if $f : Q \longrightarrow P$ is an expansion morphism and if P is n-marked, we can mark Q and f as well. Secondly, we need the following construction: if P is n-marked, let P^{+n} be the n-marked preordered set obtained by putting over P a new root-cluster with n-elements. We clearly have an open marked surjection $P^{+n} \longrightarrow P$ (notice that the root cluster of P may have size quite larger than n but this does not matter). Thirdly, notice that the pullback (in the category of preordered sets and order-preserving maps) of n-marked preordered sets and n-marked open surjective maps is n-marked as well (in the obvious way) and entierely formed by n-marked open surjective maps.

For a given n-marked preordered set Q, let $\{q_i^Q : R_i^Q \longrightarrow Q\}_i$ be an enumeration of expansion n-marked maps with target Q.

We produce an n-chain as follows: we begin with the one-cluster n-elements preordered set as P_1 and at stage i we take as P_i the n-marked preordered set P^{+n} where P is iterated pullback of the diagrams (for $k < i$)

$$\begin{array}{ccc} & & P_{i-1} \\ & & \Big\downarrow f_{i-k,i-1} \\ R_k^{P_{i-k}} & \xrightarrow[r_k^{P_{i-k}}]{} & P_{i-k} \end{array}$$

(the map $P_i \longrightarrow P_{i-1}$ is the composition of $P^{+n} \longrightarrow P$ with the $(i-1)$-th projection and with $r_1^{P_{i-1}}$).

Condition $(*)$ is clearly satisfied, because the expansion morphism $q : Q \longrightarrow P_i$ appearing in it can be n-marked, so it is $r_k^{P_i}$ for some k: it will be considered at stage $i+k$, so that the required r will be i-th projection at that stage (composed with a suitable open surjection of the kind $P^{+n} \longrightarrow P$). $\dashv$

References

Artin, M., A. Grothendieck, and J.L. Verdier. 1972. *Séminaire de Géometrie Algébrique 4*. Springer LNM, No. 269.

Blok, W.J., and D. Pigozzi. 1986. A finite basis theorem for quasivarieties. *Algebra Universalis* 22:1–13.

Chang, C.C., and H.J. Keisler. 1990. *Model theory*. North Holland.

Gabriel, P., and F. Ulmer. 1971. *Lokal prasentierbare Kategorien*. Springer LNM, No. 221.

Ghilardi, S., and M. Zawadowski. 1995a. A sheaf representation and duality for finitely presented Heyting algebras. *Journal of of Symbolic Logic* 60:911–939.

Ghilardi, S., and M. Zawadowski. 1995b. Undefinability of propositional quantifiers in the modal system *S*4. *Studia Logica* 55:259–271.

Ghilardi, S., and M. Zawadowski. 1997. Model completions and r-Heyting categories. *Annals of Pure and Applied Logic* 88:27–46.

Ghilardi, S., and M. Zawadowski. 200? ***Model completions and propositional logic.*** Book in preparation.

Köhler, P., and D. Pigozzi. 1980. Varieties with equationally definable principal congruences. *Algebra Universalis* 11:213–219.

MacLane, S., and I. Moerdijk. 1992. *Sheaves in geometry and logic.* Springer.

Makkai, M., and A.M. Pitts. 1987. Some results on locally finitely presentable categories. *Transactions of AMS* 299:473–495.

Makkai, M., and G.E. Reyes. 1977. *First order categorical logic.* Springer LNM, Vol. 611.

Maksimova, L.L. 1979. Interpolation Theorems in Modal Logics and Amalgamable Varieties of Topological Boolean Algebras. *Algebra i Logika* 18(5):556–586.

Maksimova, L.L. 1982. Absence of the interpolation property in the consistent normal extensions of the Dummett logic. *Algebra i Logika* 21(6):690–694.

Pitts, A.M. 1992. On an interpretation of second order quantification in first order intuitionistic propositional logic. *Journal of Symbolic Logic* 57:33–52.

Visser, A. 1996. Bisimulations, Model Descriptions and Propositional Quantifiers. Technical Report 161. Logic Group Preprint Series.

Silvio Ghilardi
Department of Computer Science
Università degli Studi di Milano
via Comelico 39
20135 Milano
Italy
E-mail: `ghilardi@dsi.unimi.it`

Marek Zawadowski
Department of Mathematics
Universytet Warszawski
ul. S. Banacha 2
00-913 Warszawa
Poland
E-mail: `zawado@mimuw.edu.pl`

12

Sahlqvist Formulas Unleashed in Polyadic Modal Languages

VALENTIN GORANKO AND DIMITER VAKARELOV

ABSTRACT. We propose a generalization of Sahlqvist formulas to polyadic modal languages by representing such languages in a combinatorial PDL style and thus, in particular, developing what we believe to be the right syntactic approach to Sahlqvist formulas at all. The class of polyadic Sahlqvist formulas PSF defined here expands essentially the so far known one. We prove first-order definability and canonicity for the class PSF.

Introduction

The quest for general frame-completeness results in modal logic has driven research ever since the first examples of incomplete modal logics were discovered by Thomason and Fine in the mid 1970's. One of the most general result of the sort was Sahlqvist theorem Sahlqvist 1975 where he proved two notable results for a large, syntactically defined class of modal formulas, called now Sahlqvist formulas: first, the *correspondence result:* they all define first-order conditions on Kripke frames and those conditions can be effectively "computed" from the modal formulas; and second, the *completeness result:* all these formulas are canonical, i.e. valid in their respective canonical frames, hence axiomatize completely the classes of frames determined by them. Some landmarks in the study of Sahlqvist formulas (SF) include, besides Sahlqvist's paper itself:

- Lemmon's conjecture in Lemmon and Scott 1977 on first-order definability and canonicity of a substantial subclass of Sahlqvist formulae, proven in Goldblatt 1976a, Goldblatt 1976b and independently confirmed and extended by Sahlqvist's result.

Advances in Modal Logic, Volume 3
F. Wolter, H. Wansing, M. de Rijke, and M. Zakharyaschev, eds.

- the systematic development in Benthem 1983 of Sahlqvist's algorithm for computing the first-order equivalent of a Sahlqvist formula, based on the substitution method;
- the modern approach to Sahlqvist formulas proposed in Sambin and Vaccaro 1989 , based on topological properties of descriptive frames allowing for unified treatment of first-order definability and canonicity.
- Kracht 1993 where the class of first-order formulas corresponding to Sahlqvist formulas was studied and a calculus computing the modal equivalent of such formulas is proposed, and further developed in Kracht 1999. (See also Blackburn et al. 2000 for a good exposition of Kracht's calculus.)
- Jónsson 1994 where an algebraic proof of canonicity of Sahlqvist formulas is given, building on ideas from Jónsson and Tarski 1952a. Actually, the latter can be regarded as a precursor and the earliest essential reference on the topic.
- Ghilardi and Meloni 1997, using new algebracic-topological ideas, prove an analogue of Sahlqvist's canonicity theorem for intermediate and intuitionistic modal logics.

Nice expositions of the ideas and technicalities around Sahlqvist's theorem can be found in Sambin and Vaccaro 1989, Gabbay et al. 1994, Blackburn et al. 2000, Chagrov and Zakharyaschev 1997.

Most of the results on Sahlqvist formulas are confined in the scope of *monadic* modal (or polymodal) logic. As de Rijke 1992 has shown, attempts for easy generalization for polyadic modal languages can fail badly. On the other hand, everyone who has worked on the topic must have felt that the phenomenon extends beyond the realm of the monadic languages, and actually, de Rijke and Venema in de Rijke 1993, Venema 1993, and de Rijke and Venema 1995, as well as Kracht and Wolter 1997 have proposed generalizations to the polyadic case. Blackburn et al. 2000 offers an insightful account of the first-order definability and canonicity results for the class of polyadic Sahlqvist formulas considered there.

All known to us approaches, however, follow the tradition of treating Sahlqvist formulas as *implications*[1], where the consequent is a positive formula, and the antecedent satisfies some syntactic restrictions, so essentially the only avenue for extension of the class of Sahlqvist formulas has been to relax appropriately the restrictions on the antecedent. This may be the right way to go in the *monadic* case. The polyadic languages,

[1] A notable exception is Thm 9.10 from Benthem 1983 which defines a class of formulas syntactically larger, yet easily reducible to Sahlqvist formulas as defined here.

however, offer a new perspective to the phenomenon, which can only be seen if the power of the polyadicity is fully unleashed, and this is what we are trying to demonstrate in this paper.

To start with, a well known observation which so far considered as trivial is that the disjunction and conjunction are actually dual to each other *dyadic modal operators* : $A \vee B = [=](A, B)$ and $A \wedge B = \langle=\rangle(A, B)$ with very simple Kripke semantics:

$$M, x \models [=](A, B) \text{ iff } \forall y_1, y_2(R^{=}xy_1y_2 \to (M, y_1 \models A \vee M, y_2 \models B)),$$

where $R^{=}xyz$ holds iff $x = y = z$.

Using this observation one can eliminate all Boolean connectives, except for the negation, from a polyadic modal language, and replace them by respective modalities. This may seem just a gimmick, but in fact it offers a new way of dealing with (polyadic) modal formulas, and in particular with Sahlqvist formulas, as *generalized modalities.*[2] Thus, one can work with a (almost) *purely modal polyadic language* defined in the style of PDL with test, where there is nothing else besides atoms and their negations, and polyadic diamonds and boxes. Here we use this style to develop a new treatment of Sahlqvist formulas, which, apart from extending their class, sheds a better light on their essence. In particular, this treatment allows us to give a more general definition of the class of Sahlqvist formulas in the polyadic case, which smoothly and naturally subsumes and even expands the monadic subclass, and also, to compactify the first-order definability and canonicity proofs.

The idea in a nutshell is that in purely modal polyadic languages, Sahlqvist formulas are represented by polyadic boxes the arguments of which are either positive formulas or formulas behaving as unary boxes over parametrically defined composite modal terms, which allow for first-order definable minimal valuations satisfying them. For simple polyadic Sahlqvist formulas these valuations are computed at once, while the general case requires an inductive procedure. Furthermore, these minimal valuations are closed sets in an appropriate topology, which yields the canonicity.

The structure of the paper is as follows. The preliminary section 1 briefly introduces the largest class SF of Sahlqvist formulas in polyadic modal languages, known to us from the literature (see de Rijke and Venema 1995). In section 2 we introduce the *purely modal polyadic languages*,

[2]We have meanwhile found that the idea of composing boxes in polyadic languages has also been used in Givant and Venema 1999 to describe a class of equations preserved under completions of Boolean algebras with operators. This class is essentailly an algebraic translation of the class of polyadic Sahlqvist formulas defined in de Rijke and Venema 1995.

define the class SPSF of *simple polyadic Sahlqvist formulas* then the full class of *polyadic Sahlqvist formulas* which in turn substantially extends SPSF. We show that the class of polyadic Sahlqvist formulas introduced here extends the so far known polyadic class both syntactically and semantically, and extends the monadic fragment at least syntactically. In section 3 we prove first-order definability of the polyadic Sahlqvist formulas. Section 4 is devoted to polyadic descriptive frames and contains the technical results needed to prove the canonicity of polyadic Sahlqvist formulas in section 5.

1 Sahlqvist formulas in classical polyadic languages.

We assume basic familiarity with the syntax and semantics of the standard polyadic modal languages, a state-of-the-art reference for which is e.g. Blackburn et al. 2000, from where we quote some of the definitions below.

Consider an arbitrarily fixed polyadic modal language. **Positive and negative formulas** of the language are defined as usual.

Definition 1.1 Boxed atom is a formula $\mathbf{L}_1...\mathbf{L}_n p$ where $\mathbf{L}_1, ..., \mathbf{L}_n$ is a (possibly empty) string of unary boxes and p is a propositional variable.

Sahlqvist antecedent: a formula constructed from propositional constants, boxed atoms and negative formulas by applying $\vee$, $\wedge$, and diamonds of arbitrary arities.

Definite Sahlqvist antecedent is a Sahlqvist antecedent obtained without applying $\vee$ (i.e. constructed from propositional constants, boxed atoms and negative formulas by applying only $\wedge$ and diamonds of arbitrary arities).

(Definite) Sahlqvist implication: $A \to B$ where A is a (definite) Sahlqvist antecedent and B is a positive formula.

The following definition is combined from Blackburn et al. 2000 and de Rijke and Venema 1995.

Definition 1.2 (Definite) Sahlqvist formula ((D)SF): a formula constructed from (definite) Sahlqvist implications by freely applying unary boxes and conjunctions, and applying polyadic boxes and disjunctions to formulas sharing no common variables.

Basic Sahlqvist formula is a definite Sahlqvist formula obtained without applying conjunctions.

This class of polyadic Sahlqvist formulas will be denoted by dRV. The following easy observations will be used in the next section.

Proposition 1.3

1. *Every Sahlqvist implication is equivalent to a conjunction of definite Sahlqvist implications.*
2. *Every Sahlqvist formula from dRV is equivalent to a conjunction of basic Sahlqvist formulas.*

Remark 1.4 Kracht and Wolter 1997 too define a similar class of polyadic Sahlqvist formulas, using also definable operators like $\clubsuit(A, B) = \Box(A, B) \wedge \Box(\neg A, B) \wedge \Box(A, \neg B)$ which is actually equivalent to $\Box(A, \bot) \wedge \Box(\bot, B)$, so that class does not extend dRV.

2 Sahlqvist formulas in polyadic modal languages.

2.1 Purely modal polyadic languages.

Definition 2.1 A purely modal polyadic language $\mathcal{L}_\tau$ contains propositional variables, negation $\neg$, and a **modal similarity type** τ consisting of a set of **basic modal terms** (modalities) with pre-assigned finite arities, including a 0-ary modality ι_0 a unary one ι_1 and a binary one ι_2.

The intuition behind the 3 distinguished modalities above is simple: ι_0 will be interpreted as the constant $\top$ and its dual as $\bot$; ι_1 will be the self-dual identity; ι_2 will be $\vee$ and its dual — $\wedge$. Treating these connectives as modalities, besides allowing for elegance and uniformity, will provide suitable technical framework for working with Sahlqvist formulas.

Before we give the definition of a formula, let us stipulate that **constant formula** will mean a formula containing no variables.

Definition 2.2 By simultaneous mutual induction we define the set of **modal terms** $MT(\tau)$ and their **arity function** ρ, and the set of **(purely) modal formulas** $MF(\tau)$ as follows:

(MT ii) Every basic modal term is a modal term of the predefined arity.

(MT ii) Every constant formula is a 0-ary modal term.

(MT iii) If $n > 0, \alpha, \beta_1, ..., \beta_n \in MT(\tau)$ and $\rho(\alpha) = n$, then $\alpha(\beta_1, ..., \beta_n) \in MT(\tau)$ and $\rho(\alpha(\beta_1, ..., \beta_n)) = \rho(\beta_1) + ... + \rho(\beta_n)$.

Modal terms of arity 0 will be called **modal constants**.

(MF i) Every propositional variable is a modal formula.

(MF ii) Every modal constant is a modal formula.

(MF iii) If A is a formula then $\neg A$ is a formula;

(MF iv) If $A_1, ..., A_n$ are formulas, α is a modal term and $\rho(\alpha) = n > 0$, then $[\alpha](A_1, ..., A_n)$ is a modal formula.

Note that constant formulas and 0-ary terms are regarded as both modal terms and formulas. This ambiguity of the syntax should not cause confusion if properly handled, and we have put up with it for the sake of technical simplicity and convenience.

For technical purposes we extend the series of ι's with n-ary modalities ι_n: inductively as follows: $\iota_{n+1} = \iota_2(\iota_1, \iota_n)$ for $n > 1$. Furthermore, again for technical convenience, we can assume that the language contains *transposers*: operators θ_{ij} which swap the i-th and j-th argument of a modal term. We will not treat these transposers formally, but assuming them in the language will allow us not to be concerned with the specific ordering of the arguments in a modal formula.

Some notation on formulas:

$\langle\alpha\rangle(A_1, ..., A_n) = \neg[\alpha](\neg A_1, ..., \neg A_n)$;

$\top = \iota_0, \bot = \neg\iota_0$;

$A \vee B = [\iota_2](A, B), A \wedge B = \langle\iota_2\rangle(A, B)$, and respectively $A_1 \vee ... \vee A_n = [\iota_n](A_1, ..., A_n), A_1 \wedge ... \wedge A_n = \langle\iota_n\rangle(A_1, ..., A_n)$;

$A \to B = \neg A \vee B$.

Positive and negative occurrences of variables and **positive and negative formulas** are defined as usual.

One effect of the mutual definition of modal terms and formulas is that it allows construction of *parametrized* modal terms. For instance, if α is a unary term and β is a binary one, then $\gamma = \beta(\neg[\beta]([\alpha](\bot), \top), \iota_1)$ is a unary modal term. Indeed, the formula $[\gamma]p$ can be essentially identified with $[\beta](\neg[\beta]([\alpha](\bot), \top), p)$, but the advantage of our approach is that this formula can now be treated just like a normal unary box-modality, which will be of important technical convenience.

The **semantics** of purely modal languages is a straightforward combination of the standard Kripke semantics for polyadic modal languages and PDL-type of polymodal languages, taking into account the fact that conjunctions and disjunctions are now treated as modalities. In particular, a **τ-frame** is a structure $\langle W, \{R_\alpha\}_{\alpha \in MT(\tau)}\rangle$ where $R_\alpha \subseteq W^{\rho(\alpha)+1}$ is defined recursively by:

- $R_{\iota_0} = W,\ R_{\iota_1} = \{(x, x) | x \in W\},\ R_{\iota_2} = \{(x, x, x) | x \in W\}$.
- $R_{\alpha(\beta_1,...,\beta_n)} = \{(x, x_{11}, ..., x_{1b_1}, ..., x_{n1}, ..., x_{nb_n}) \subseteq W^{b_1+...+b_n+1} | \exists y_1...y_n(R_\alpha x y_1...y_n \wedge \bigwedge_{i=1}^n R_{\beta_i} y_i x_{i1}...x_{ib_i})\}$, where $\rho(\beta_i) = b_i, i = 1, ...n$.

Note that $R_{\iota_n} = \{(x, \cdots, x) \in W^{n+1} | x \in W\}$.

Now, the *truth definition* of a formula at a state of a Kripke model extends the classical modal case with the clause (where $\overline{y} = y_1 ... y_n$):

- $M, x \models [\alpha](A_1, ..., A_n)$ iff $\forall \overline{y}(R_\alpha x y_1 ... y_n \to \bigvee_{i=1}^{n} M, y_i \models A_i)$.
  In particular, $M, x \models \alpha$ iff $R_\alpha x$ for any modal constant α.

It follows immediately from the definitions that
$[\alpha(\beta_1, ..., \beta_n)](A_{11}, ..., A_{1n_1}, ..., A_{m1}, ..., A_{mn_m})$ is equivalent to
$[\alpha]([\beta_1](A_{11}, ..., A_{1n_1}), ..., [\beta_n](A_{m1}, ..., A_{mn_m}))$.

Accordingly, the **standard translation** ST generalizes the one for monadic languages with the clauses:

- $ST(\sigma) = R_\sigma(x)$ for every modal constant σ;
- $ST([\alpha](A_1, ..., A_n) = \forall \overline{y}(R_\alpha x y_1 ... y_n \to \bigvee_{i=1}^{n} ST(A_i)(y_i/x))$

Again, note that all propositional logical connectives, as defined here, have their standard interpretation. Furthermore, the purely modal polyadic languages are equally expressive as the traditional ones.

2.2 Simple Polyadic Sahlqvist formulas.

Definition 2.3 Essentially positive formula (EPF) of a variable p is a formula $A = [\alpha](p)$ or $A = [\alpha](p, A_1, ..., A_n)$ where $A_1, ..., A_n$ are negative formulas not containing p. The variable p in such a formula is called **the essential variable of** A, while all other variables are **inessential** in the formula.

Definition 2.4 A set of essentially positive formulas is: **independent** if no essential variable in a formula from the set occurs as an inessential variable in any formula from the set; **separated** if all EPFs have *different* essential variables; **strongly independent** if it is independent and separated.

Of course, the essential variable in an EPF need not be the first argument, but to simplify the notation we shall put it usually in first position.

Definition 2.5 Simple Polyadic Sahlqvist formula (SPSF) is any modal constant σ, or $A = [\alpha](A_1, ..., A_n)$ where α is an n-ary modal term and each formula A_i is either positive, or a negation of an essentially positive formula, and the subset of essentially positive formulas in $\{A_1, ..., A_n\}$ is independent.

Of course, we can close the class of SPSFs under conjunctions, but for technical reasons we will keep it as is.

2.3 Pre-processing

Sahlqvist formulas are traditionally (here as well) defined *syntactically*, while the idea behind them is *semantic*. Thus, inessential syntactic manipulations, such as tautological transformations, can take a formula in or out of the class of these formulas without chainging its semantic properties. Here we briefly discuss how a purely modal polyadic formula can be *pre-processed* in a search for its representation as a SPSF, and how to present SPSFs in a canonical form. By $A \sim B$ we shall indicate that the formulas A and B are equivalent with respect to frame definability.

1. Tautological transformations can present the formula as a polyadic box (if necessary, by prefixing with $[\iota_1]$).

2. All *monotone* variables, that have only positive/negative occurrences in the formula, can be replaced resp. by $\bot/\top$. Some easy logical equivalences can further simplify the formula.

3. Boxes can be pulled outwards and composed. Eventually the formula can be written as $[\alpha](A_1, ..., A_n)$, where $A_1, ..., A_n$ are negated boxes or variables.

4. At this stage one can identify if the formula is a SPSF (or to a general PSF, to be introduced later) and if not, one can try to convert it into one, by an appropriate substitution, typically reversing the polarities of a variable p by replacing it with $\neg p$. For example, $[\iota_2](\neg[\alpha](p,q),[\beta](p,q)) \sim [\iota_2(\iota_1,\beta)](\neg[\alpha](p,q),p,q) \sim [\iota_2(\iota_1,\beta)](\neg[\alpha](\neg p,\neg q),\neg p,\neg q) \sim [\iota_2(\iota_1,\beta)](\langle\alpha\rangle(p,q),\neg p,\neg q)$.

5. Now, assuming the formula is a SPSF, every EPF B can be regarded as a *unary* box over its essential variable, where all negative arguments of B are regarded as parameters. Computing the relation corresponding to this composite box is straightforward from the definition.

6. Finally, an SPSF can be transformed to one, equivalent with respect to frame validity, with a *strongly independent* set of EPFs. This can be done by means of successive *splitting* a common essential variable of two EPFs into two different copies, illustrated by the following example:
   $[\alpha](\neg[\beta_1]p,\neg[\beta_2]p,P(p,\overline{q})) \sim [\alpha](\neg[\beta_1]p_1,\neg[\beta_2]p_2,P(p_1 \vee p_2,\overline{q}))$,
   where p_1, p_2 are variables not occurring in A.

Actually, all these transformations can be easily performed *deductively* in the minimal polyadic modal logic.

Clearly, a series of such transformations can convert any SPSF into one with strongly independent set of EPFs, so every SPSF can be transformed into an $\sim$-equivalent one in a *canonical* form:

$$[\alpha](\neg[\beta_1]p_1, \ldots \neg[\beta_n]p_n, C_1, \ldots C_k)$$

where $\beta_1, \ldots \beta_n$ are unary modal terms, $p_1, \ldots, p_n$ are different propositional variables, and $C_1, \ldots C_k$ are positive formulas. In particular, each of n and k above can be 0, and the canonical form may become a constant formula.

Hereafter, whenever suitable, we can assume that any SPSF has been pre-processed this way.

2.4 SPSF subsumes dRV

Lemma 2.6 *Every definite Sahlqvist antecedent A is equivalent to a negation of a SPSF in which all EPFs are boxed atoms.*

Proof. Induction on A:

- The cases of A atomic, boxed atom or a negative formula are trivial, (note that every atomic formula is equivalent to a negation of a positive formula);
- $A = A_1 \wedge A_2$,where $A_1 \equiv \neg B_1$ and $A_2 \equiv \neg B_2$ for some SPSFs B_1 and B_2. Then $A \equiv \neg[\iota_2](B_1, B_2)$. Note that all EPFs in $[\iota_2](B_1, B_2)$ are boxed atoms, hence still an independent set.
- $A = \langle\alpha\rangle(A_1, ..., A_n)$. Then $A \equiv \neg[\alpha](\neg A_1, ..., \neg A_n)$ where each $\neg A_1$ is equivalent to a SPSF in which all EPFs are boxed atoms, hence so is $[\alpha](\neg A_1, ..., \neg A_n)$.

⊣

Proposition 2.7 *Every formula from dRV is equivalent to a conjunction of SPSFs.*

Proof. Immediately from Lemmata 1.3 and 2.6. Note that if $A \to C$ is a SF and $A \equiv \neg B$ for some SPSF B then $A \to C \equiv [\iota_2](B, C)$ is a SPSF, and also that applying disjunctions and polyadic boxes to SFs not sharing variables preserves the independence of the essential variables. ⊣

Actually, SPSF *properly* extends dRV. A simple example is

$$[\mathbf{2}](\neg[\mathbf{2}](\bot, p), \langle\mathbf{2}\rangle(p, \top)),$$

where $\mathbf{2}$ is a binary modality. It defines the frame condition $\forall xyz(R_2xyz \to \exists uvw(R_2yuv \wedge R_2zvw))$.

Moreover, SPSF extend, at least syntactically the class of monadic SF as they allow the antecedent of a SF can take not only boxed atoms but any formulas of the type: $\Box_1(A_1 \rightarrow \Box_2(A_2 \rightarrow ...\Box_n(A_n \rightarrow p)...)$ where $A_1, ..., A_n$ are positive formulas not containing p. Indeed, this is equivalent to $\Box_1(\neg A_1 \vee \Box_2(\neg A_2 \vee ...\Box_n(\neg A_n \vee p)...)$ which can be written as an EPF $[\alpha](\neg A_1, \neg A_2, ..., \neg A_n, p)$.

2.5 Polyadic Sahlqvist formulas.

Let $A = [\alpha](\neg B_1, ..., \neg B_n, C_1, ..., C_k)$ where $B_1, ..., B_n$ are EPFs with essential variables resp. $p_1, ..., p_n$ which after pre-processing can be assumed different, and $C_1, ..., C_k$ are positive. In general, this need not have the virtues of a Sahlqvist formula. For instance, de Rijke 1992 has shown that $[\iota_2](\neg[\mathbf{2}]([\mathbf{2}](p,p),p), \langle\mathbf{2}\rangle(\langle\mathbf{2}\rangle(p,p),p))$ is not FO definable. A simpler example is $[\iota_2](\neg[\mathbf{2}](p,p), \langle\mathbf{2}\rangle(p,p))$which defines the non-elementary frame condition "*For every x the binary relation R_x on the remaining two variables y and z has an unoriented cycle of odd length.*"

Definition 2.8 Dependency digraph of the set of essential variables $\{p_1, ..., p_n\}$ of A is a digraph $G = \langle \mathbf{V}_A, \mathbf{E}_A \rangle$ where $\mathbf{V}_A = \{p_1, ..., p_n\}$, and $p_i \mathbf{E}_A p_j$ iff p_i occurs as an inessential variable in a formula from $B_1, ..., B_n$ with an essential variable p_j. A digraph is called **acyclic** if it does not contain oriented cycles.

Definition 2.9 A Polyadic Sahlqvist formula (PSF) is any modal constant σ, or $A = [\alpha](A_1, ..., A_n)$ where α is an n-ary modal term and each formula A_i is either positive, or a negation of an essentially positive formula, and the dependency digraph of the set of essential variables in A is acyclic.

A variable in a PSF is **essential** if it has at least one occurrence as an essential variable in the formula, otherwise it is inessential.

The particular case when there are no arcs in the dependency digraph corresponds to the class of SPSFs.

The class Polyadic Sahlqvist formulas PSF can be considered closed under conjunctions.

Remark 2.10 PSFs can be assumed pre-processed, just like PSFSs, to a *canonical form* in which there are no inessential variables and every variable is essential in only one EPF. However, unlike SPSFs, not all EPFs can be made unary boxes.

Note that this pre-processing does not alter the (a)cyclicity of the dependency graph of the formula.

Example 2.11 The formula $[\mathbf{3}](\neg[\mathbf{1}]p, \neg[\mathbf{2}](\neg p, q), \langle\mathbf{1}\rangle[\mathbf{1}]q)$ is a PSF but not a simple one.

PSF generalizes *syntactically* the classical monadic Sahlqvist formulas, but we do not know yet if it essentially extends their *semantic conditions.*

Example 2.12 Consider:

$$D = p \wedge \Box(\Diamond p \to \Box q) \to \Diamond\Box\Box q.$$

This is not a classical Sahlqvist formula and by now we do not know if it is semantically equivalent to one. In particular, its local FO correspondent:

$$FO(D) = \exists y(Rxy \wedge \forall z(R^2yz \to \exists u(Rxu \wedge Rux \wedge Ruz)))$$

is not a Kracht formula (see Kracht 1999 and Blackburn et al. 2000, Sect. 3.7). On the other hand, D written in a purely modal polyadic language becomes a PSF:

$$D = [\iota_3](\neg p, \neg[\alpha(\iota_2(\alpha,\alpha))](\neg p, q), \langle\alpha\rangle[\alpha]q),$$

where α is the modal term corresponding to the $\Box$.

However, this formula has a Sahlqvist equivalent in the classical *tense language:*

$$D_t = p \to FGGP(Fp \wedge Pp).$$

3 First-order definability of polyadic Sahlqvist formulas

Theorem 3.1 *Every polyadic Sahlqvist formula is first-order definable.*

Proof. We adapt the Sahlqvist – van Benthem's algorithm. First we will demonstrate it on a simple PSF and then will extend it to arbitrary PSFs.

Let $A = [\alpha](\neg B_1, ..., \neg B_n, C_1, ..., C_k)$ be a pre-processed PSF, where $\{B_1, ..., B_n\}$ is a separated set of EPFs and $C_1, ..., C_k$ are positive formulas. Let $\overline{\mathbf{q}} = q_1, ..., q_n$ be the variables occurring in A, $\overline{\mathbf{Q}} = Q_1, ..., Q_n$ be the respective unary predicate variables, and $\overline{\mathbf{y}} = y_1, ..., y_{n+k}$ be a string of fresh different individual variables. By $\overline{\mathrm{ST}(A)}$ we denote the second-order closure of $\mathrm{ST}(A)$, which corresponds to frame validity. Then $\overline{\mathrm{ST}(A)} =$
$\forall x\forall\overline{\mathbf{Q}}\forall\overline{\mathbf{y}}\left(R_\alpha xy_1...y_{n+k} \to (\bigvee_{i=1}^n \neg\mathrm{ST}(B_i)(y_i/x) \vee \bigvee_{i=1}^k \mathrm{ST}(C_i)(y_{n+i})\right)$
$\equiv \forall x\forall\overline{\mathbf{Q}}\forall\overline{\mathbf{y}}\,(R_\alpha xy_1...y_{n+k} \wedge \bigwedge_{i=1}^n \mathrm{ST}(B_i)(y_i/x) \to POS(y_{n+1}, ..., y_{n+k}))$
for some positive formula POS.

First, let us consider the particular case when A is a SPSF, so all EPFs are unary boxes: $B_j = [\beta_j](q_j)$ for some modal term β_j and essential variable q_j. Then

$$\mathrm{ST}(B_j)(y_j/x) \equiv \forall z_j \left(R_{\beta_j} y_j z_j \to Q_j(z_j)\right).$$

We now define the **minimal valuation** V_m which makes these true:

$$V_m(q_j) = R_{\beta_j}(y_j).$$

It is now easy to see that for any frame $\mathbf{F}$ and $x \in \mathbf{F}$:

$$\mathbf{F}, x \models \forall \overline{\mathbf{Q}}\mathrm{ST}(A) \text{ iff } \mathbf{F}, x \models \mathrm{ST}(A)(\mathbf{V}_m(\overline{\mathbf{q}})/\overline{\mathbf{Q}}),$$

hence

$$\mathbf{F}, x \models A \text{ iff } \mathbf{F}, \mathbf{V}_m, x \models A.$$

Indeed, $\overline{\mathrm{ST}(A)} = \forall x \forall \overline{\mathbf{y}} \forall \overline{\mathbf{Q}}(ANT(\overline{\mathbf{Q}}) \to POS(\overline{\mathbf{Q}}))$ where $\overline{\mathbf{Q}}$ is the string of predicates corresponding to all essential variables and $ANT(\overline{\mathbf{Q}}) = R_\alpha x y_1 ... y_{n+k} \wedge \bigwedge_{i=1}^n \mathrm{ST}(B_i)(y_i/x)$.

Note that, once the parameters $x, y_1, ..., y_{n+k}$ are fixed consistently with ANT, the valuation V_m is *the minimal one* (in set-theoretic sense) which makes each B_i, and hence $ANT(\overline{\mathbf{Q}})$, true. Therefore if $\mathbf{F}, \mathbf{V}_m, x \models A$ then $POS(V_m(\overline{\mathbf{q}}))$ is true. Now, take any valuation V. If it falsifies any B_i, then ANT is rendered false, so the whole formula is true. Otherwise, $V_m(q) \subseteq V(q)$ for every essential variable q. Then, by monotonicity of positive formulas, $POS(V_m(\overline{\mathbf{q}})) \to POS(V(\overline{\mathbf{q}}))$ is valid, hence $POS(V(\overline{\mathbf{q}}))$ is true, so the formula turns out true again.

Thus, A defines the following first-order condition on frames:

$$FO(A) = \forall x \forall \overline{\mathbf{y}} \left(R_\alpha x y_1 ... y_{n+k} \to POS(\mathbf{V}_m(\overline{\mathbf{q}})/\overline{\mathbf{Q}})\right).$$

Now, the proof for the general case of a PSF A essentially repeats the one above. The key concern again is to define the right minimal valuation. Let G_A be the dependency digraph of A. First, note that since G_A does not contain cycles, it defines a *strict partial ordering* $\prec$ between the vertices: $q_i \prec q_j$ iff there is an arc path leading from q_i to q_j. Following that partial ordering, a minimal valuation can be defined on the set of essential variables inductively as follows.

Suppose all $\prec$-predecessors (if any) of an essential variable q have already been assigned values. Let the string of these predecessors be $\overline{\mathbf{q}}_q$ and let the string of second-order variables corresponding to them be $\overline{\mathbf{Q}}_q$.

Take any EPF B_j with an essential variable q_j. Then, as before:
$\mathrm{ST}(B_j)(y_j/x) \equiv$
$\forall z_j \forall \overline{\mathbf{u}_j} \left(R_{\beta_j} y_j z_j u_{j1} ... u_{jn_j} \wedge \bigwedge_{i=1}^{n_j} \mathrm{ST}(P_{ji})(u_{ji}/x) \to Q_j(z_j)\right).$

Note that all variables occurring in any $\mathrm{ST}(P_{ji})(u_{ji}/x)$ above correspond to predecessors of q_j, so they are amongst $\overline{\mathbf{Q}}_{q_j}$ and hence they have already been assigned their minimal values.

Then we put

$$V_m(q_j) = \left\{ z \,\middle|\, \exists \overline{\mathbf{u}_j}(R_{\beta_j} y_j z u_{j1} ... u_{jn_j} \wedge \bigwedge_{i=1}^{n_j} \mathrm{ST}(P_{ji})(V_m(\overline{\mathbf{q}}_j)/\overline{\mathbf{Q}}_{q_j})(u_{ji}))) \right\}.$$

In particular, if $\overline{\mathbf{q}}_{q_j}$ is empty, i.e. q_j is $\prec$-minimal, then $V_m(q_j)$ is defined as before.

Now, an inductive argument on $\prec$ proves that V_m has indeed the properties of the minimal valuation needed to prove first-order definability of A as in the case of a SPSF. ⊣

Example 3.2 Let us compute the PSF formula from example 2.11: $B = [\mathbf{3}](\neg[\mathbf{1}]p, \neg[\mathbf{2}](\neg p, q), \langle\mathbf{1}\rangle[\mathbf{1}]q)$. Since $p \prec q$, we first compute $V_m(p) = R_1(y_1)$. Then $V_m(q) = \{z | \exists s(R_2 y_2 s z \wedge R_1 y_1 s)$. Thus, $FO(B) = \forall x y_1 y_2 y_3 (R_3 x y_1 y_2 y_3 \to \exists v(R_1 y_3 v \wedge \forall w(R_1 v w \to \exists s(R_2 y_2 s w \wedge R_1 y_1 s))))$.

Remark 3.3 Note that in the latter example above, once $V_m(p)$ is determined, then $[\mathbf{2}](\neg p, q)$ can be regarded as a unary box $[\alpha](q)$ where $\alpha = \mathbf{2}(\neg V_m(p), \iota_1)$ is a unary *parametrized modal term*, the relation of which can be accordingly computed: $R_\alpha xy$ iff $\exists s(R_2 x s y \wedge V_m(p)(s))$. This trick will be essential in the proof of canonicity of PSFs.

4 Polyadic descriptive frames

In this section we obtain results about descriptive frames for polyadic modal languages which will be used further. Most of these will be generalizations of well-known properties of monadic descriptive frames, but we will establish some important relations between them and will present them in a way suitable for purely modal languages.

Let us fix an arbitrary purely modal language $\mathcal{L}(\tau)$.

Definition 4.1 Given a τ-frame $F = \langle W, \{R_\alpha\}_{\alpha \in MT(\tau)} \rangle$, every n-ary modal term $\beta \in MT(\tau)$ defines an n-ary **operator** $\langle\beta\rangle$ on $\mathcal{P}(W)$ as follows:

$$\langle\beta\rangle(X_1, ..., X_n) = \{x \in W | R_\beta x x_1 ... x_n \text{ for some } x_1 \in X_1, ..., x_n \in X_n\}.$$

In particular, $\langle\beta\rangle = R_\beta$ for every 0-ary term β.

Note that every operator $\langle\beta\rangle$ is monotone on each of its arguments.

Definition 4.2 A **general frame** for $\mathcal{L}(\tau)$ (**general** τ **-frame**) is a structure $\langle W, \{R_\alpha\}_{\alpha \in MT(\tau)}, \mathbb{W}\rangle$ extending a τ-frame with a Boolean algebra of **admissible subsets** of $\mathcal{P}(W)$, closed under the operators corresponding to the basic modal terms, and therefore under all operators $\langle \beta \rangle$.

Every general τ-frame $\mathfrak{F} = \langle W, \{R_\alpha\}_{\alpha \in MT(\tau)}, \mathbb{W}\rangle$ determines a **topological space** $T(\mathfrak{F})$ with a base of clopen sets $\mathbb{W}$. $T(\mathfrak{F})$ is a Hausdorff space with some additional properties which will be obtained further. For detailed study of this topology, its properties and usage in modal logic, see Sambin and Vaccaro 1988 and for topological treatment of Sahlqvist formulas see Sambin and Vaccaro 1989.

Hereafter, a **closed** set in a general τ-frame will mean a subset of the domain closed with respect of the above mentioned topology, i.e. an intersection of a family of admissible sets.

We are going to extend the set of modal terms and respective class of operators to allow parametrization with closed sets.

Definition 4.3 Let $\mathfrak{F} = \langle W, \{R_\alpha\}_{\alpha \in MT(\tau)}, \mathbb{W}\rangle$ be a general τ-frame. We define the set $PMT(\tau, \mathfrak{F})$ of **parametrized modal terms associated with** $\mathfrak{F}$ and their respective operators on $\mathcal{P}(W)$ inductively as follows:

- $MT(\tau) \subseteq PMT(\tau, \mathfrak{F})$;
- For every $(n+1)$-ary term $\beta \in PMT(\tau, \mathfrak{F})$ and a closed set Z in $T(\mathfrak{F})$, $\beta(Z)$ is an n-ary term in $PMT(\tau, \mathfrak{F})$ such that $\langle \beta(Z)\rangle(X_1, ..., X_n) = \langle \beta \rangle(X_1, ..., X_n, Z)$.

It is easy to see that $R_{\beta(Z)}x, x_1, ..., x_n$ iff there exists $x_{n+1} \in Z$ such that $R_\beta x x_1 ... x_n x_{n+1}$.

Definition 4.4 Let $\mathfrak{F} = \langle W, \{R_\alpha\}_{\alpha \in MT(\tau)}, \mathbb{W}\rangle$ be a general τ-frame and $\beta \in PMT(\tau, \mathfrak{F})$. The relation R_β is **tight in** $\mathfrak{F}$ if the following condition holds: for any $x, x_1, ..., x_n \in W$,

$R_\beta x, x_1, ..., x_n$ iff

$\forall X_1, ..., X_n \in \mathbb{W}(x_1 \in X_1, ..., x_n \in X_n \Rightarrow x \in \langle \beta \rangle(X_1, ..., X_n))$.

Note that this condition is equivalent to: for every $x \in W$,

$$x \in \bigcap \{\langle \beta \rangle(X_1, ..., X_n) | X_1, ..., X_n \in \mathbb{W} \ \& \ x_1 \in X_1, ..., x_n \in X_n\}.$$

In particular, every R_β for a 0-ary term β is tight.

Definition 4.5 A family of sets $\mathcal{F}$ has the **finite intersection property (FIP)** if the intersection of every finite subfamily of $\mathcal{F}$ is non-empty.

Definition 4.6 A general τ-frame $\langle W, \{R_\alpha\}_{\alpha \in MT(\tau)}, \mathbb{W}\rangle$ is:

- **differentiated** if for every $x, y \in W$, if $x \neq y$ then there is $X \in \mathbb{W}$ such that $x \in X$ and $y \notin X$;
- **tight** if for every basic modal term β the relation R_β is tight in $\mathfrak{F}$;
- **compact** if every family of admissible sets in $\mathfrak{F}$ with FIP has a non-empty intersection.
- **refined** if it is differentiated and tight.
- **descriptive** if it is refined and compact.

Note that:

- the property of being differentiated is expressed by the tightness of R_{ι_1}, and so it becomes redundant. We keep it in the definition mainly to respect the tradition.
- compactness of a general τ-frame $\mathfrak{F}$ as defined above is equivalent to the standard topological notion of compactness of $T(\mathfrak{F})$, i.e. every family of *closed* sets with the FIP has a non-empty intersection.
- by (a weaker version of) Tychonov's theorem, if $\mathfrak{F}$ is compact then for every $n \in N$, the product space $(T(\mathfrak{F}))^n$ is compact, too.

Hereafter, closedness of Cartesian products of sets will mean closedness in the respective product topology.

In the rest of this section we show that every descriptive τ-frame has the properties that will be necessary to prove the canonicity of any PSF.

Lemma 4.7 *In any differentiated τ-frame $\mathfrak{F} = \langle W, \{R_\alpha\}_{\alpha \in MT(\tau)}, \mathbb{W}\rangle$, for any n-ary term $\beta \in PMT(\tau, \mathfrak{F})$, R_β is tight iff for every $x \in W$ the set $R_\beta(x) = \{(x_1, ..., x_n) | R_\beta x x_1 ... x_n\}$ is closed.*

Proof. For 0-ary modal terms β each of these conditions is trivially true, so we can assume that $\rho(\beta) > 0$.

First, note that
$\forall X_1, ..., X_n \in \mathbb{W}(x_1 \in X_1, ..., x_n \in X_n \Rightarrow x \in \langle\beta\rangle(X_1, ..., X_n))$ iff
$\forall X_1, ..., X_n \in \mathbb{W}(x \in [\beta](-X_1, ..., -X_n) \Rightarrow (x_1, ..., x_n) \in -(X_1 \times ... \times X_n))$.

Therefore, R_β is tight iff for every $x \in W$,
$R_\beta(x) = \bigcap\{-(X_1 \times ... \times X_n) | X_1, ..., X_n \in \mathbb{W} \,\&\, x \in [\beta](-X_1, ..., -X_n)\}$.
⊣

Definition 4.8 A family $\mathcal{F}$ of subsets of a set X is called **downwards directed** if $\mathcal{F}$ contains a subset of the intersection of any two (and hence, any finitely many) members of $\mathcal{F}$.

Lemma 4.9 *In any differentiated and compact general τ-frame $\mathfrak{F} = \langle W, \{R_\alpha\}_{\alpha \in MT(\tau)}, \mathbb{W}\rangle$, the following are equivalent for any n-ary term $\beta \in PMT(\tau, \mathfrak{F})$:*

(i) R_β is tight.

(ii) (Esakia's lemma) For any downwards directed family $\{X_{1i} \times ... \times X_{ni}\}_{i \in I}$ of closed subsets of W^n,

$$\bigcap_{i \in I}\{\langle\beta\rangle(X_{1i}, ..., X_{ni})\} = \langle\beta\rangle(\bigcap_{i \in I} X_{1i}, ..., \bigcap_{i \in I} X_{ni}).$$

(iii) For every $x \in W$ the set $R_\beta(x)$ is closed.

Proof. Again, the non-trivial case is $\rho(\beta) > 0$.

(i) $\Rightarrow$ (iii): Lemma 4.7.

(iii) $\Rightarrow$ (ii): The inclusion $\supseteq$ follows from the monotonicity of $\langle\beta\rangle$.

For the converse inclusion, let $x \in \bigcap_{i \in I}\{\langle\beta\rangle(X_{1i}, ..., X_{ni})\}$. Then, due to the downwards directedness, the family of closed sets

$$R_\beta(x) \cup \{X_{1i} \times ... \times X_{ni}\}_{i \in I}$$

has the FIP, so it has a non-empty intersection, i.e. there is a tuple $(x_1, ..., x_n)$ such that $R_\beta x x_1 ... x_n$ and $(x_1, ..., x_n) \in \bigcap_{i \in I}\{(X_{1i} \times ... \times X_{ni}\} = \bigcap_{i \in I} X_{1i} \times ... \times \bigcap_{i \in I} X_{ni}$.

Therefore, $x \in \langle\beta\rangle(\bigcap_{i \in I} X_{1i}, ..., \bigcap_{i \in I} X_{ni})$.

(ii) $\Rightarrow$ (i):

The implication from left to right in the tightness condition for R_β holds by definition. For the converse, it suffices to note that by (ii):

$\bigcap\{\langle\beta\rangle(X_1, ..., X_n) | X_1, ..., X_n \in \mathbb{W} \;\&\; x_1 \in X_1, ..., x_n \in X_n\} = \langle\beta\rangle(\{x_1\}, ..., \{x_n\})$. ⊣

Lemma 4.10 *In every descriptive τ-frame $\mathfrak{F}$, each of the conditions of Lemma 4.9 holds for every term $\beta \in PMT(\tau, \mathfrak{F})$.*

Proof. We shall prove that Esakia's lemma holds for every β, by induction on β.

First, we will complete the induction on $MT(\tau)$.

For the basic terms tightness holds by definition, and hence the claim holds by Lemma 4.9.

The inductive step for $\beta = \alpha(\alpha_0, ..., \alpha_m)$ is quite straightforward. Finally, suppose the claim holds for some $(n+1)$-ary term $\beta \in PMT(\tau, \mathfrak{F})$ and let Z be a closed set in $T(\mathfrak{F})$. Then

$\langle\beta(Z)\rangle(\bigcap_{i \in I} X_{1i}, \cdots, \bigcap_{i \in I} X_{ni})$

$= \langle\beta\rangle(\bigcap_{i \in I} X_{1i}, \cdots, \bigcap_{i \in I} X_{ni}, \bigcap_{i \in I} Z)$

$= \bigcap_{i \in I}\langle\beta\rangle(X_{1i}, \cdots, X_{ni}, Z)$

$= \bigcap_{i \in I}\langle\beta(Z)\rangle(X_{1i}, \cdots, X_{ni})$. ⊣

Lemma 4.11 *In any descriptive general τ-frame $\mathfrak{F}$, for every positive formula $A(p_1, ..., p_n)$ the corresponding operator in $\mathfrak{F}$ $\lambda X_1...X_n.A(X_1, ..., X_n)$ satisfies Esakia's lemma: for any downwards directed family of closed sets $\{X_{1i} \times ... \times X_{ni}\}_{i \in I}$,*

$$\bigcap_{i\in I}\{A(X_{1i}, ..., X_{ni})\} = A(\bigcap_{i\in I} X_{1i}, ..., \bigcap_{i\in I} X_{ni}).$$

In particular, for every positive formula $A(p_1, ..., p_n)$ the operator $\lambda X_1...X_n.A(X_1, ..., X_n)$ is closed, i.e. $A(X_1, ..., X_n)$ is closed whenever $X_1, ..., X_n$ are closed.

Proof. First, note that every positive formula A is equivalent to a formula built from propositional variables and modal constants applying only polyadic boxes and diamonds.

We shall prove the statement by induction on A, assuming it is constructed as above.

For propositional variables and modal constants the claim is trivial. For $\langle\beta\rangle(X_1, ..., X_n)$ the inductive step is the Esakia's lemma.

Finally, for $[\beta](X_1, ..., X_n)$ the inductive step follows from the identity

$$[\beta](\bigcap_{i\in I} X_{1i}, ..., \bigcap_{i\in I} X_{ni}) = \bigcap_{i_1\in I, ..., i_n\in I} [\beta](X_{1i_1}, ..., X_{ni_n})$$

which easily follows from the definition $[\beta](X_1, ..., X_n)$, combined with the equality

$$\bigcap_{i_1\in I, ..., i_n\in I} [\beta](X_{1i_1}, ..., X_{ni_n}) = \bigcap_{i\in I}[\beta](X_{1i}, ..., X_{ni})$$

which follows from the downward directedness. ⊣

5 Canonicity of polyadic Sahlqvist formulas

Theorem 5.1 *Every polyadic Sahlqvist formula is canonical.*

Proof. We show that every PSF is d-persistent following the scheme of the proof of canonicity of Sahlqvist formulas presented in Blackburn et al. 2000, to which the reader is referred for those technical details which would not differ in the general case presented here.

Let $A = [\alpha](\neg B_1, ..., \neg B_n, C_1, ..., C_k)$ be a pre-processed PSF with EPFs $B_1, ..., B_n$ and different essential variables resp. $q_1, ..., q_n$. Let the dependency digraph of A determine a partial order $\prec$ on these variables.

Let $\mathfrak{F} = \langle \mathbf{F}, \mathbb{W}\rangle$ be a descriptive general frame such that $\mathfrak{F} \models A$.

As we showed in the proof of theorem 3.1, if $\mathbf{F}, V_m \models A$ for the minimal valuation V_m, defined as before, then $\mathbf{F}, V \models A$ for *any* valuation V, so it suffices to prove that $\mathbf{F}, V_m \models A$. The problem is that *the minimal valuation need not be admissible* in $\mathfrak{F}$. However, it will suffice to show the following:

(C1) V_m is closed i.e. an intersection of admissible valuations.

(C2) For every closed valuation U in $\mathfrak{F}$ and a positive formula P, $U(P) = \bigcap_{U \preceq V} V(P)$ where the intersection ranges over all admissible valuations V which extend U.

For (C1), we can restrict our consideration to the variables occurring in A, i.e. the essential variables $q_1, ..., q_n$. We shall prove by $\prec$-induction that every valuation

$$V_m(q_j) = \left\{ z \middle| \exists \overline{\mathbf{u}_j}(R_{\beta_j} y_j z u_{j1} ... u_{jn_j} \wedge \bigwedge_{i=1}^{n_j} \mathrm{ST}(P_{ji})(V_m(\overline{\mathbf{q}}_{q_j})/\overline{\mathbf{Q}}_{q_j})(u_{ji})) \right\}$$

is of the type $R_\beta(y_j)$ for some $\beta \in PMT(\tau, \mathfrak{F})$, and hence, by Lemma 4.10, is closed in $\mathfrak{F}$.

For the $\prec$-minimal variables the claim is immediate, because their respective EPFs are unary boxes.

Now suppose the claim holds for all predecessors $\overline{\mathbf{q}}_p$ of the variable $p = q_j$, i.e. for every $q_i \in \overline{\mathbf{q}}_p$, $V_m(q_i) = R_{\beta_{q_i}}$ for some $\beta_{q_i} \in PMT(\tau, \mathfrak{F})$, and hence is closed.

Let $n_j = n$ and denote $C_i = P_{ji}(V_m(\overline{\mathbf{q}}_p))$ for $i = 1, ..., n$. Note that each C_i is closed by the inductive hypothesis and Lemma 4.11, since P_{ji} is positive.

Consider the unary term $\gamma = \beta(C_n)...(C_2)(C_1) \in PMT(\tau, \mathfrak{F})$, i.e. such that $[\gamma](A) = [\beta](A, C_1, ..., C_n)$. Then for any $z \in W$, $R_\gamma yz$ holds iff there exist $u_1, ..., u_n$ such that $R_\beta yzu_1...u_n$ and $u_i \in C_i$ for $i = 1, ..., n$. Therefore $V_m(p_j) = R_\gamma(y_j)$.

(C2) follows from Lemma 4.11.

Now, to complete the proof, let us see why $\mathbf{F}, V_m \models A$. As in the proof of theorem 3.1, let $\mathrm{ST}(A) = \forall x \forall \overline{\mathbf{y}} \forall \overline{\mathbf{Q}}(ANT(\overline{\mathbf{Q}}) \to POS(\overline{\mathbf{Q}}))$. Fix the parameters $x, \overline{y}$ consistently with ANT (otherwise the formula turns vacuously true) and take any admissible valuation U defined inductively on $\prec$ and extending V_m. It will render ANT true, hence POS true, because $\mathfrak{F} \models A$. Then, by (C2), POS will be true for V_m. $\dashv$

Remark 5.2 The conditions (C1) and (C2) in the proof above hold trivially for Kripke frames (i.e. full general frames), which allows for simultaneous treatment of first-order definability and canonicity of PSFs, in the spirit of Sambin and Vaccaro 1989 and Kracht 1993.

6 Concluding remarks

This work is certainly not the last word on Sahlqvist formulas. On the contrary, we believe that the approach developed here will allow for further generalizations and insights on them. An ultimate goal in this direction would be an explicit *semantic* characterization of the class of *all* formulas for which the proof of Sahlqvist's theorem work, through a deeper analysis of their semantic and topological nature.

An important question is to what extent polyadic Sahlqvist formulas can successfully co-operate in axiomatizations with non-orthodox rules of inference, a line of research initiated in Venema 1993. In a separate paper we study polyadic extensions of the class of Sahlqvist formulas in extended languages and additional rules.

Last but not least, proof systems for modal logic are still to catch up with the semantic richness and harmony of Sahlqvist formulas and make them equally useful in a deductive framework. In particular, "displaying" Sahlqvist formulas is a challenge to proof theorists.

Acknowledgements

We thank Johan van Benthem, Marcus Kracht, Silvio Ghilardi, and Yde Venema for useful discussions and comments. The work of both authors was partly supported by a research grant GUN 2034353 of the National Research Foundation of South Africa. The work of Valentin Goranko was partly done during his sabbatical leave, while visiting the University of Victoria, BC, Canada and supported in part by NSERC. He would like to thank Bruce Kapron who invited him for that visit.

References

Benthem, J. van. 1983. *Modal Logic and Classical Logic*. Naples: Bibliopolis.

Blackburn, P., M. de Rijke, and Y. Venema. 2000. *Modal Logic*. Cambridge Tracts in Theoretical Computer Science.

Chagrov, A., and M. Zakharyaschev. 1997. *Modal Logic*. Oxford Logic Guides, Vol. 35. Oxford: Oxford University Press.

de Rijke, M. 1992. How not to generalize Sahlqvist's Theorem. technical note.

de Rijke, M. 1993. *Extending Modal Logic*. Doctoral dissertation, ILLC, University of Amsterdam.

de Rijke, M., and Y. Venema. 1995. Sahlqvist's Theorem for Boolean Algebras with Operators with an Application to Cylindric Algebras. *Studia Logica* 54:61–78.

Gabbay, D.M., I. Hodkinson, and M. Reynolds. 1994. *Temporal Logic: Mathematical Foundations and Computational Aspects*. Oxford University Press.

Ghilardi, S., and G. Meloni. 1997. Constructive Canonicity in Non-classical Logics. *Annals of Pure and Applied Logic* 86:1–32.

Givant, S., and Y. Venema. 1999. The Preservation of Sahlqvist's equations of Boolean Algebras with operators. *Algebra Universalis* 41:47–84.

Goldblatt, R.I. 1976a. Metamathematics of Modal Logic I. *Reports on Mathematical Logic* 6:41–78.

Goldblatt, R.I. 1976b. Metamathematics of Modal Logic II. *Reports on Mathematical Logic* 7:21–52.

Jónsson, B. 1994. On the canonicity of Sahlqvist identities. *Studia Logica* 53:473–491.

Jónsson, B., and A. Tarski. 1952a. Boolean Algebras with Operators, Part I. *American Journal of Mathematics* 73:891–939.

Jónsson, J., and A. Tarski. 1952b. Boolean Algebras with Operators, Part II. *American Journal of Mathematics* 74:127–162.

Kanger, S. (ed.). 1975. *Proceedings of the Third Scandinavian Logic Symposium. Uppsala 1973.* Amsterdam. North-Holland.

Kracht, M. 1993. How Completeness and Correspondence Theory Got Married. In *Diamonds and Defaults*, ed. M. de Rijke, 175–214. Kluwer, Dordrecht.

Kracht, M. 1999. *Tools and Techniques in Modal Logic.* Elsevier.

Kracht, M., and F. Wolter. 1997. Simulation and Transfer Results in Modal Logic - A Survey. *Studia Logica* 59:149–177.

Lemmon, E.J., and D.S. Scott. 1977. *The "Lemmon Notes": An Introduction to Modal Logic.* Oxford: Blackwell.

Sahlqvist, H.1975. Completeness and Correspondence in the First and Second Order Semantics for Modal Logic. In Kanger Kanger 1975, 110–143.

Sambin, G., and V. Vaccaro. 1988. Topology and Duality in Modal Logic. *Annals of Pure and Applied Logic* 37:249–296.

Sambin, G., and V. Vaccaro. 1989. A New Proof of Sahlqvist's Theorem on Modal Definability and Completeness. *Journal of Symbolic Logic* 54:992–999.

Venema, Y. 1993. Derivation rules as anti-axioms in modal logic. *Journal of Symbolic Logic* 58:1003–1034.

Valentin Goranko
Department of Mathematics, Rand Afrikaans University
PO Box 524, Auckland Park 2006, Johannesburg, South Africa
E-mail: `vfg@na.rau.ac.za`
URL: `http://general.rau.ac.za/maths/goranko/`

Dimiter Vakarelov
Department of Mathematical Logic with Laboratory for Applied Logic
Faculty of Mathematics and Computer Science, Sofia University
blvd James Bouchier 5, 1126 Sofia, Bulgaria
E-mail: `dvak@fmi.uni-sofia.bg`

13

Normal Products of Modal Logics

YASUSI HASIMOTO

ABSTRACT. In this paper, we will propose a new concept of products of modal logics, which will resemble the products familiar from measure theory and topology. Our products of modal logics can be defined either by means of *normal products of general frames* or by means of *normal products of modal algebras.* It enables us to develop a duality theory between these two. This brings about a desired effect that the definition of the normal product of modal logics L_1 and L_2 is not affected by the choice of classes of general frames (or, modal algebras) which determine L_1 and L_2. We also show some transfer results, including transfer of the finite model property.

1 Introduction

In recent years, products of modal logics have been studied intensively, for example, in Gabbay and Shehtman 1998, Marx and Venema 1997, Reynolds 1997 and Wolter 2000. These products, first introduced in Shehtman 1978, have been defined through products of Kripke frames. Shehtman's product of modal logics L_1 and L_2 is the modal logic determined by the class of Kripke frames $\mathcal{F} \times \mathcal{G}$ such that $\mathcal{F}$ and $\mathcal{G}$ validate L_1 and L_2, respectively.

In this paper, we will propose a new concept of products of modal logics, which we will call a *normal product.* Normal products will resemble the products familiar from measure theory and topology, and will be defined as a generalization of *products of algebras of sets* (see § 2). Our products of modal logics can be defined either by means of *normal products of general frames* or by means of *normal products of modal algebras.* It enables us to develop a duality theory between these two, as shown in § 3. This brings about a desired effect that the definition of the

Advances in Modal Logic, Volume 3
F. Wolter, H. Wansing, M. de Rijke, and M. Zakharyaschev, eds.

normal product of modal logics L_1 and L_2 is not affected by the choice of classes of general frames (or, modal algebras) which determine L_1 and L_2. Note, that this is not the case for usual products of modal logics, as pointed out in Reynolds 1997. We also show some transfer results, including transfer of the finite model property, in § 4.

One conceptual difference between normal products and usual products is, that in the usual case, when L_1 and L_2 are m-modal and n-modal logics, then their product $L_1 \times L_2$ is a $(m+n)$-modal logic. On the other hand, the normal product can be defined only when both L_1 and L_2 are m-modal logics and their normal product is also a m-modal logic. This difference, however, is not essential. In fact, when a m-modal logic L_1 and a n-modal logic L_2 are given, we first consider two $(m+n)$-modal logics $L_1{}^\circ$ and ${}^\circ L_2$ and then take their normal product. Here, $L_1{}^\circ$ and ${}^\circ L_2$ are essentially the same as L_1 and L_2, respectively, but in addition they have n and m *dummy* modal operators, respectively (see § 5 for the detailed definition). We will denote it as $L_1 \otimes L_2$ and call it the *shifted product* of L_1 and L_2. Relations between shifted and usual products will be discussed in the last section.

As the definition of our normal products is quite general, by a slight modification of definitions, we can also introduce normal products of two superintuitionistic logics, two intuitionistic modal logics, and even of *infinitely many* of them, and obtain results similar to these from the present paper.

In this paper, we are going to assume a certain familiarity with Gabbay and Shehtman 1998, and basically follow the terminology in Chagrov and Zakharyaschev 1997.

2 Normal products of general frames

In this section, we will define normal products of general frames, following the standard method used in measure theory and topology (see e.g. Neveu 1965). For simplicity, we will consider only general frames for mono-modal logics, but it is easily seen that every definition and result can be naturally extended to m-modal case for any given $m > 1$.

A pair (W, R) is a *Kripke frame* if W is a nonempty set and R is a binary relation on W. A set $\mathcal{P}$ of subsets of W is called a *modal algebra* on a Kripke frame (W, R) if (i) $\emptyset \in \mathcal{P}$, (ii) $X, Y \in \mathcal{P} \Rightarrow X \cap Y \in \mathcal{P}$, (iii) $X \in \mathcal{P} \Rightarrow -X \in \mathcal{P}$, (iv) $X \in \mathcal{P} \Rightarrow \Diamond X \in \mathcal{P}$, where $\Diamond X = \{x \in W \mid \exists y \in X \text{ such that } xRy\}$. A modal algebra on (W, R) contains W and is closed under $\cup$, $\Box$, where $\Box X = \{x \in W \mid \forall y \in W(xRy \Rightarrow y \in X)\}$.

Clearly, $\mathcal{P}(W)$ is an example of a modal algebra on (W, R). A set $\mathcal{S}$ of subsets of W is called a *modal semi-algebra* on a Kripke frame (W, R) if (i) $\emptyset \in \mathcal{S}$, (ii) $X, Y \in \mathcal{S} \Rightarrow X \cap Y \in \mathcal{S}$, (iii) $X \in \mathcal{S} \Rightarrow -X$ is a union of finitely many members of $\mathcal{S}$, (iv) $X \in \mathcal{S} \Rightarrow \Diamond X \in \mathcal{S}$.

Suppose that $\mathcal{P}$ and $\mathcal{Q}$ are modal algebras on (W, R) and (V, S), respectively. We consider the product $(W \times V, R \times S)$ of Kripke frames (W, R) and (V, S), where $W \times V$ is the direct product of W and V and $R \times S$ is the product relation of R and S, i.e., a binary relation on $W \times V$ defined by

$$(w_1, v_1) R \times S (w_2, v_2) \text{ if and only if } w_1 R w_2 \text{ and } v_1 S v_2,$$

for $w_1, w_2 \in W$ and $v_1, v_2 \in V$.

Now, consider the set $\{X \times Y \mid X \in \mathcal{P} \text{ and } Y \in \mathcal{Q}\}$, which is a set of subsets of $W \times V$. This set is not always a modal algebra on $(W \times V, R \times S)$, but it is always a modal semi-algebra on it. For a given Kripke frame (W, R) and a set $\mathcal{S}$ of subsets of W, the smallest modal algebra on (W, R) containing $\mathcal{S}$ is called the *modal algebra generated by* $\mathcal{S}$. Then, the following lemma holds.

Lemma 2.1 *Suppose that $\mathcal{S}$ is a modal semi-algebra on a Kripke frame (W, R). Then the modal algebra generated by $\mathcal{S}$ is $\{\bigcup_{i=1}^{n} X_i \mid X_i \in \mathcal{S}$ and some $n < \omega\}$.*

Proof. Since $\{\bigcup_{i=1}^{n} X_i \mid X_i \in \mathcal{S} \text{ and some } n < \omega\}$ contains $\mathcal{S}$, we will show that $\{\bigcup_{i=1}^{n} X_i \mid X_i \in \mathcal{S} \text{ and some } n < \omega\}$ satisfies conditions (i)–(iv) of modal algebras. Since $\mathcal{S}$ contains $\emptyset$, (i) holds. Since $\bigcup_{i=1}^{n} X_i \cap \bigcup_{j=1}^{m} Y_j = \bigcup_{i=1}^{n}\bigcup_{j=1}^{m} X_i \cap Y_j$, (ii) holds. Since $-\bigcup_{i=1}^{n} X_i = \bigcap_{i=1}^{n} -X_i$ and $-X$ is a union of finitely many members of $\mathcal{S}$, (ii) implies (iii). Since $\Diamond \bigcup_{i=1}^{n} X_i = \bigcup_{i=1}^{n} \Diamond X_i$, (iv) holds. ⊣

For modal algebras $\mathcal{P}$ on (W, R) and $\mathcal{Q}$ on (V, S), define $\mathcal{P} \divideontimes \mathcal{Q}$ to be the modal algebra on $(W \times V, R \times S)$ generated by $\{X \times Y \mid X \in \mathcal{P} \text{ and } Y \in \mathcal{Q}\}$.

Corollary 2.2

$$\mathcal{P} \circledast \mathcal{Q} = \{\bigcup_{i\in I}(X_i \times Y_i) \mid X_i \in \mathcal{P} \textit{ and } Y_i \in \mathcal{Q} \textit{ for a finite } I\}.$$

Now, we come to the definition of normal product of general frames. Recall, that a triple $(W, R, \mathcal{P})$ is a *general frame* if (W, R) is a Kripke frame and $\mathcal{P}$ is a modal algebra on (W, R). In the following, $\mathcal{F}_\lambda$ and $\mathcal{G}_\mu$ (with or without indices) denote general frames of the form $(W_\lambda, R_\lambda, \mathcal{P}_\lambda)$ and $(V_\mu, S_\mu, \mathcal{Q}_\mu)$, respectively.

For given general frames $\mathcal{F} = (W, R, \mathcal{P})$ and $\mathcal{G} = (V, S, \mathcal{Q})$, the *normal product* $\mathcal{F} \circledast \mathcal{G}$ of $\mathcal{F}$ and $\mathcal{G}$ is a general frame $(W \times V, R \times S, \mathcal{P} \circledast \mathcal{Q})$. Note that the normal product thus obtained is also a general frame for mono-modal logics.

In the following, we will characterize the operations in $\mathcal{F} \circledast \mathcal{G}$. Suppose that both $\bigcup_{i\in I}(X_i \times Y_i)$ and $\bigcup_{j\in J}(T_j \times U_j)$ are elements of $\mathcal{P} \circledast \mathcal{Q}$.

(1) $$\emptyset = \emptyset \times \emptyset (= \emptyset \times V = W \times \emptyset).$$

(2) $$\bigcup_{i\in I}(X_i \times Y_i) \cap \bigcup_{j\in J}(T_j \times U_j) = \bigcup_{(i,j)\in I\times J}(X_i \cap T_j) \times (Y_i \cap U_j).$$

(3) $$\bigcup_{i\in I}(X_i \times Y_i) \cup \bigcup_{j\in J}(T_j \times U_j) = \bigcup_{k\in I\sqcup J} A_k \times B_k,$$

where $I \sqcup J = \{(i, 0) \mid i \in I\} \cup \{(j, 1) \mid j \in J\}, A_{(i,0)} = X_i, A_{(j,1)} = T_j, B_{(i,0)} = Y_j, B_{(j,1)} = U_j$.

(4) $$-\bigcup_{i\in I}(X_i \times Y_i) = \bigcup_{J\in\mathcal{P}(I)}(-\bigcup_{i\in J} X_i) \times (-\bigcup_{i\in -J} Y_i).$$

(5) $$\Diamond\bigcup_{i\in I}(X_i \times Y_i) = \bigcup_{i\in I}(\Diamond X_i \times \Diamond Y_i).$$

(6) $$\Box\bigcup_{i\in I}(X_i \times Y_i) = \bigcup_{C\subseteq\mathcal{P}(I)}(\Box\bigcap_{J\in C}\bigcup_{i\in J} X_i) \times (\Box\bigcup_{J\in C}\bigcap_{i\in J} Y_i).$$

Moreover, the following holds in $\mathcal{P} \circledast \mathcal{Q}$.

$$(7) \quad \bigcup_{i \in I}(X_i \times Y_i) = \bigcup_{j \in J}(T_j \times U_j)$$

$$\iff \quad \forall i \in I, \exists \mathcal{C} \subseteq \mathcal{P}(J) \text{ such that } X_i \subseteq \bigcap_{K \in \mathcal{C}} \bigcup_{j \in K} T_j, Y_i \subseteq \bigcup_{K \in \mathcal{C}} \bigcap_{j \in K} U_j,$$

$$\text{and} \quad \forall j \in J, \exists \mathcal{C} \subseteq \mathcal{P}(I) \text{ such that } T_j \subseteq \bigcap_{K \in \mathcal{C}} \bigcup_{i \in K} X_i, U_j \subseteq \bigcup_{K \in \mathcal{C}} \bigcap_{i \in K} Y_i.$$

The general frames of the form $\mathcal{F} = (W, R, \mathcal{P}(W))$ will be called *Kripke frames* and denoted as before by $\mathcal{F} = (W, R)$. Note that for Kripke frames $\mathcal{F}$ and $\mathcal{G}$, $\mathcal{F} \circledast \mathcal{G}$ is a general frame $(W \times V, R \times S, \mathcal{P}(W) \circledast \mathcal{P}(V))$ while $\mathcal{F} \times \mathcal{G}$ is a Kripke frame $(W \times V, R \times S)$.

Theorem 2.3 *Let $\mathcal{F}$ and $\mathcal{G}$ be Kripke frames. Then, $\mathcal{F} \circledast \mathcal{G}$ is a Kripke frame if and only if either $\mathcal{F}$ or $\mathcal{G}$ is finite. In other words, $\mathcal{P}(W) \circledast \mathcal{P}(V) = \mathcal{P}(W \times V)$ if and only if either W or V is finite.*

Proof. Suppose that W is finite. For $Z \in \mathcal{P}(W \times V)$, we can show $Z = \bigcup_{w \in W} (\{w\} \times Z_w)$, where $Z_w = \{v \in V \mid (w, v) \in Z\}$. Conversely, suppose that both W and V are infinite. Then we can take distinct sequences $\{w_i\}_{i<\omega} \subseteq W$ and $\{v_i\}_{i<\omega} \subseteq V$. Hence we can show that $\{(w_i, v_i) \mid i < \omega\} \notin \mathcal{P}(W) \circledast \mathcal{P}(V)$. ⊣

Recall that for given general frames $\mathcal{F} = (W, R, \mathcal{P})$ and $\mathcal{G} = (V, S, \mathcal{Q})$, a map f from W onto V is a *reduction* of $\mathcal{F}$ to $\mathcal{G}$ if the following conditions are satisfied for all $w, w' \in W$ and all $Y \in \mathcal{Q}$: (i) $wRw' \Rightarrow f(w)Sf(w')$, (ii) $f(w)Sf(w') \Rightarrow \exists u \in W$ such that wRu and $f(u) = f(w')$, (iii) $f^{-1}(Y) \in \mathcal{P}$.

It is well-known that each projection from a product of topological spaces to any of its component spaces is continuous. As in the realm of modal logic reductions are natural counterparts of continuous maps in topology, we can ask whether each projection from a normal product of general frames to any of its component frames is a reduction. The following theorem gives a necessary and sufficient condition. Here, $S(v)$ and $R(w)$ denote sets $\{v' \mid vSv'\}$ and $\{w' \mid wRw'\}$, respectively.

Theorem 2.4 *A projection $\pi : W \times V \to W$ is a reduction of $\mathcal{F} \circledast \mathcal{G}$ to $\mathcal{F}$ iff either $S(v) \neq \emptyset$ for each $v \in V$ or $R(w) = \emptyset$ for each $w \in W$.*

Proof. Since it is easily seen that a projection π always satisfies the conditions (i),(iii) of reduction, we will consider condition (ii).
A projection π satisfies the condition (ii),

iff for all $w, w' \in W$ and $v, v' \in V$ $wRw' \Rightarrow \exists u \in V : (w, v)R \times S(w', u)$,
iff for all $w, w' \in W$ and $v, v' \in V$ either $w \not R w'$ or $\exists u \in V : vSu$,
iff either $S(v) \neq \emptyset$ for each $v \in V$ or $R(w) = \emptyset$ for each $w \in W$. ⊣

The next theorem gathers together several essential properties of normal products.

Theorem 2.5

1. *If f_i is a reduction of $\mathcal{F}_i$ to $\mathcal{G}_i$ for $i = 1, 2$, then the* product map *$f_1 \divideontimes f_2$ is a reduction of $\mathcal{F}_1 \divideontimes \mathcal{F}_2$ to $\mathcal{G}_1 \divideontimes \mathcal{G}_2$, where $f_1 \divideontimes f_2$ is defined by $(f_1 \divideontimes f_2)(w_1, w_2) = (f_1(w_1), f_2(w_2))$ for each $w_1 \in W_1$ and $w_2 \in W_2$.*
2. *If $\mathcal{G}_i$ is a generated subframe of $\mathcal{F}_i$ for $i = 1, 2$, then $\mathcal{G}_1 \divideontimes \mathcal{G}_2$ is also a generated subframe of $\mathcal{F}_1 \divideontimes \mathcal{F}_2$.*
3. *The identity map id : $\sum_{\lambda \in \Lambda} \sum_{\mu \in M} (W_\lambda \times V_\mu) \to (\sum_{\lambda \in \Lambda} W_\lambda) \times (\sum_{\mu \in M} V_\mu)$ is a reduction of $\sum_{\lambda \in \Lambda} \sum_{\mu \in M} (\mathcal{F}_\lambda \divideontimes \mathcal{G}_\mu)$ to $(\sum_{\lambda \in \Lambda} \mathcal{F}_\lambda) \divideontimes (\sum_{\mu \in M} \mathcal{G}_\mu)$, where $\sum$ denotes disjoint unions.*

Proof. Here, we will only show the part of algebras of sets. The rest is routine.
For 1, $(f_1 \divideontimes f_2)^{-1}(\bigcup_{i \in I}(X_i \times Y_i)) = \bigcup_{i \in I}(f_1^{-1}(X_i) \times f_2^{-1}(Y_i)) \in \mathcal{P}_1 \divideontimes \mathcal{P}_2$.
For 2, $(\bigcup_{i \in I}(X_i \times Y_i)) \cap (V_1 \times V_2) = \bigcup_{i \in I}((X_i \cap V_1) \times (Y_i \cap V_2)) \in \mathcal{Q}_1 \divideontimes \mathcal{Q}_2$.
For 3, $\bigcup_{i \in I}(\sum_{\lambda \in \Lambda} X^i_\lambda) \times (\sum_{\mu \in M} Y^i_\mu) = \sum_{\lambda \in \Lambda} \sum_{\mu \in M} \bigcup_{i \in I}(X^i_\lambda \times Y^i_\mu) \in \sum_{\lambda \in \Lambda} \sum_{\mu \in M}(\mathcal{P}_\lambda \divideontimes \mathcal{Q}_\mu)$.
⊣

3 Normal products of modal algebras and duality theory

In the previous section, we have defined normal products of general frames employing products $\mathcal{P} \divideontimes \mathcal{Q}$ of modal algebras $\mathcal{P}$ and $\mathcal{Q}$ on Kripke frames. In this section, we will define *normal products of arbitrary modal algebras*. A structure $\boldsymbol{A} = (A, \wedge, \vee, \neg, 0, 1, \Box)$ is a *modal algebra* if $(A, \wedge, \vee, \neg, 0, 1)$ is a Boolean algebra and $\Box$ is unary operator on A satisfying $\Box 1 = 1$ and $\Box(x \wedge y) = \Box x \wedge \Box y$ for $x, y \in A$. Any modal algebra $\mathcal{P}$ on a Kripke frame (W, R) can be regarded as a modal algebra in the present sense, if we identify $\mathcal{P}$ with $(\mathcal{P}, \cap, \cup, -, \emptyset, W, \Box)$, where $\Box$ is the operator defined in § 2. Below, we will take this identification for granted.

For a given modal algebra $\boldsymbol{A}$, let $\boldsymbol{A}_+$ stand, as usual, for the *dual* of $\boldsymbol{A}$. Also, for a given general frame $\mathcal{F}$, let $\mathcal{F}^+$ stand for the *dual* of $\mathcal{F}$.

For modal algebras $\boldsymbol{A}$ and $\boldsymbol{B}$, we define their *normal product* $\boldsymbol{A} \divideontimes \boldsymbol{B}$ as $(\boldsymbol{A}_+ \divideontimes \boldsymbol{B}_+)^+$. We can show that for modal algebras $\mathcal{P}$ and $\mathcal{Q}$ on some Kripke frames, normal product $\mathcal{P} \divideontimes \mathcal{Q}$ in the present sense is isomorphic to $\mathcal{P} \divideontimes \mathcal{Q}$ in the sense of the previous section. Though our definition of the normal product $\boldsymbol{A} \divideontimes \boldsymbol{B}$ is stated by means of the normal product of general frames $\boldsymbol{A}_+$ and $\boldsymbol{B}_+$, we can also give a direct definition by representing equations (1)–(6) from the previous section in algebraic terms, and introducing an equivalence relation $\sim$ following (7), over the set of all finite sets of $A \times B$. We will not, however, enter into further details.

Corollary 3.1 *For general frames $\mathcal{F}$ and $\mathcal{G}$,*

$$(\mathcal{F} \divideontimes \mathcal{G})^+ \cong \mathcal{F}^+ \divideontimes \mathcal{G}^+. \tag{8}$$

Recall that a general frame $\mathcal{F}$ is said to be *descriptive* if $\mathcal{F} \cong (\mathcal{F}^+)_+$. Note that a general frame $\mathcal{F}$ is descriptive iff

- differentiated $\forall x, y \in W(x \neq y \Rightarrow \exists X \in \mathcal{P} : x \in X, y \notin X)$,
- tight $\forall x, y \in W(x \not R y \Rightarrow \exists X \in \mathcal{P} : x \in \Box X, y \notin X)$,
- compact $\forall \mathcal{A} \subseteq \mathcal{P}(\mathcal{A}$ has finite intersection property $\Rightarrow \bigcap \mathcal{A} \neq \emptyset)$,

where $\mathcal{A} \subseteq \mathcal{P}(W)$ has *finite intersection property* if $\bigcap \mathcal{A}_0 \neq \emptyset$ for finite $\mathcal{A}_0 \subseteq \mathcal{A}$.

Theorem 3.2 *If both general frames $\mathcal{F}, \mathcal{G}$ are descriptive, then their normal product $\mathcal{F} \divideontimes \mathcal{G}$ is also descriptive.*

Proof. It is easily shown that it is differentiated and tight. We can show compactness applying the method of Tikhonov's theorem. For the case of finitely many products of modal frames we can prove without Zorn's Lemma. First, note that we can express an element of $\mathcal{P}$ as the form of $\bigcap_{i \in I}(X_i \times V \cup W \times Y_i)$. Therefore it is enough to show that for any $\mathcal{A} = \{X_\lambda \times V \cup W \times Y_\lambda \mid X_\lambda \in \mathcal{P}, Y_\lambda \in \mathcal{Q}, \lambda \in \Lambda\}$, ($\mathcal{A}$ has finite intersection property $\Rightarrow \bigcap \mathcal{A} \neq \emptyset$). Let us define that $\mathcal{A}_w = \{X_\lambda \times V \cup W \times Y_\lambda \in \mathcal{A} \mid X_\lambda \times V \cup W \times Y_\lambda \not\supseteq \{w\} \times V\}$ and $\mathcal{B}_w = \{Y_\lambda \mid X_\lambda \times V \cup W \times Y_\lambda \in \mathcal{A}_w\}$ for each $w \in W$. Suppose that $\mathcal{B}_w$ doesn't have finite intersection property for any $w \in W$. Hence for each $w \in W$ there exist $X_1^w, \dots, X_{n_w}^w, Y_1^w, \dots, Y_{n_w}^w$ such that $Y_1^w \cap \dots \cap Y_{n_w}^w = \emptyset$ and $X_{i_w}^w \times V \cup W \times Y_{i_w}^w \in \mathcal{A}_w$. Therefore, $\bigcap_{i_w=1}^{n_w} X_{i_w}^w \times V \cup W \times Y_{i_w}^w \supseteq (\bigcup_{i_w=1}^{n_w} X_{i_w}^w) \times V$. Since $\{(\bigcup_{i_w=1}^{n_w} X_{i_w}^w) \times V \mid$

$w \in W\}$ also has finite intersection property by finite intersection property of $\mathcal{A}$, $\bigcap_{w\in W}(\bigcup_{i_w=1}^{n_w} X_{i_w}^w) \times V \neq \emptyset$ by compactness of $\mathcal{F}$. But since $(\bigcup_{i_w=1}^{n_w} X_{i_w}^w) \times V \supseteq -(\{w\} \times V)$, we have $\bigcap_{w\in W}(\bigcup_{i_w=1}^{n_w} X_{i_w}^w) \times V = \emptyset$. It is contradiction. Therefore, $\mathcal{B}_w$ has finite intersection property for some w. Since $\bigcap \mathcal{A} \supseteq \{w\} \times (\bigcap \mathcal{A}_w) \neq \emptyset$, $\mathcal{F} \circledast \mathcal{G}$ is also compact by compactness of $\mathcal{G}$. ⊣

Corollary 3.3 *For modal algebras* $\boldsymbol{A}$ *and* $\boldsymbol{B}$,

$$(\boldsymbol{A} \circledast \boldsymbol{B})_+ \cong \boldsymbol{A}_+ \circledast \boldsymbol{B}_+. \tag{9}$$

We can show the duality of Theorem 2.5 as follows.

Theorem 3.4

1. *If* $\boldsymbol{A}_2$ *is a homomorphic image of* $\boldsymbol{A}_1$ *and* $\boldsymbol{B}_2$ *is a homomorphic image of* $\boldsymbol{B}_1$, *then* $\boldsymbol{A}_2 \circledast \boldsymbol{B}_2$ *is also a homomorphic image of* $\boldsymbol{A}_1 \circledast \boldsymbol{B}_1$.
2. *If* $\boldsymbol{A}_1$ *is a subalgebra of* $\boldsymbol{A}_2$ *and* $\boldsymbol{B}_1$ *is a subalgebra of* $\boldsymbol{B}_2$, *then* $\boldsymbol{A}_1 \circledast \boldsymbol{B}_1$ *is isomorphic to a subalgebra of* $\boldsymbol{A}_2 \circledast \boldsymbol{B}_2$.
3. $\prod_{\lambda\in\Lambda} \boldsymbol{A}_\lambda \circledast \prod_{\mu\in M} \boldsymbol{B}_\mu$ *is isomorphic to a subalgebra of* $\prod_{(\lambda,\mu)\in\Lambda\times M} \boldsymbol{A}_\lambda \circledast \boldsymbol{B}_\mu$, *where* $\prod$ *denotes direct products.*

Let $\mathcal{K}$ be a class of modal algebras. As usual, $\boldsymbol{V}(\mathcal{K})$, $\boldsymbol{H}(\mathcal{K})$, $\boldsymbol{I}(\mathcal{K})$, $\boldsymbol{S}(\mathcal{K})$ and $\boldsymbol{P}(\mathcal{K})$ denote, respectively: the variety generated by $\mathcal{K}$, the class of homomorphic images of algebras in $\mathcal{K}$, the class of isomorphic copies of algebras in $\mathcal{K}$, the class of subalgebras of algebras in $\mathcal{K}$ and the class of direct products of nonempty families of algebras in $\mathcal{K}$, respectively. For classes $\mathcal{K}_1$ and $\mathcal{K}_2$ of modal algebras, $\mathcal{K}_1 \circledast \mathcal{K}_2$ is the class $\{\boldsymbol{A}_1 \circledast \boldsymbol{A}_2 \mid \boldsymbol{A}_1 \in \mathcal{K}_1, \boldsymbol{A}_2 \in \mathcal{K}_2\}$. By Theorem 3.4, we have the following.

Theorem 3.5 *For classes* $\mathcal{K}_1, \mathcal{K}_2$ *of modal algebras,*

1. $\boldsymbol{H}(\mathcal{K}_1) \circledast \boldsymbol{H}(\mathcal{K}_2) \subseteq \boldsymbol{H}(\mathcal{K}_1 \circledast \mathcal{K}_2)$,
2. $\boldsymbol{S}(\mathcal{K}_1) \circledast \boldsymbol{S}(\mathcal{K}_2) \subseteq \boldsymbol{IS}(\mathcal{K}_1 \circledast \mathcal{K}_2)$,
3. $\boldsymbol{P}(\mathcal{K}_1) \circledast \boldsymbol{P}(\mathcal{K}_2) \subseteq \boldsymbol{ISP}(\mathcal{K}_1 \circledast \mathcal{K}_2)$.

Hence,

4. $\boldsymbol{V}\big(\boldsymbol{V}(\mathcal{K}_1) \circledast \boldsymbol{V}(\mathcal{K}_2)\big) = \boldsymbol{V}(\mathcal{K}_1 \circledast \mathcal{K}_2)$.

4 Normal products of modal logics

For a given modal logic L, $\boldsymbol{F}(L)$ is the class of all general frames which validate L, and $\boldsymbol{V}(L)$ is the class of all modal algebras which validate L. For a class $\mathcal{K}$ of modal algebras (a class $\mathcal{C}$ of general frames), $\boldsymbol{L}(\mathcal{K})$ ($\boldsymbol{L}(\mathcal{C})$) denotes the set of formulas valid in every modal algebra in $\mathcal{K}$ (in every general frame in $\mathcal{C}$). For classes $\mathcal{C}_1$ and $\mathcal{C}_2$ of general frames, $\mathcal{C}_1 \circledast \mathcal{C}_2$ is the class $\{\mathcal{F}_1 \circledast \mathcal{F}_2 \mid \mathcal{F}_1 \in \mathcal{C}_1 \text{ and } \mathcal{F}_2 \in \mathcal{C}_2\}$.

Now, for modal logics L_1, L_2, define the *normal product* $L_1 \circledast L_2$ to be $\boldsymbol{L}(\boldsymbol{F}(L_1) \circledast \boldsymbol{F}(L_2))$. By (8) and (9), $L_1 \circledast L_2 = \boldsymbol{L}(\boldsymbol{V}(L_1) \circledast \boldsymbol{V}(L_2))$ holds. Thus, we may also take the latter as the definition of $L_1 \circledast L_2$. By Theorem 3.5(4), we have the following.

Theorem 4.1 *Let $\mathcal{K}_1$ and $\mathcal{K}_2$ be classes of modal algebras, and let $\mathcal{C}_1$ and $\mathcal{C}_2$ be classes of general frames. Then*

1. $\boldsymbol{L}(\mathcal{K}_1) \circledast \boldsymbol{L}(\mathcal{K}_2) = \boldsymbol{L}(\mathcal{K}_1 \circledast \mathcal{K}_2)$,
2. $\boldsymbol{L}(\mathcal{C}_1) \circledast \boldsymbol{L}(\mathcal{C}_2) = \boldsymbol{L}(\mathcal{C}_1 \circledast \mathcal{C}_2)$.

This theorem says that the logic $\boldsymbol{L}(\mathcal{C}_1 \circledast \mathcal{C}_2)$ is uniquely determined by the logics $\boldsymbol{L}(\mathcal{C}_1)$ and $\boldsymbol{L}(\mathcal{C}_2)$. More precisely, if $\boldsymbol{L}(\mathcal{C}_1) = \boldsymbol{L}(\mathcal{C}_1')$ and $\boldsymbol{L}(\mathcal{C}_2) = \boldsymbol{L}(\mathcal{C}_2')$, then $\boldsymbol{L}(\mathcal{C}_1 \circledast \mathcal{C}_2) = \boldsymbol{L}(\mathcal{C}_1' \circledast \mathcal{C}_2')$. Since the normal product of general frames is associative, the normal product of modal logics is also associative, by the above theorem.

Following Gabbay and Shehtman 1998, we say that a logic L has the *product f.m.p.* if L is a logic of some class of finite products. It is obvious that if L has the product f.m.p., then L has the f.m.p.. By Theorem 2.3 and Theorem 4.1, we have the following.

Theorem 4.2

1. *If both L_1 and L_2 have the f.m.p., then $L_1 \circledast L_2$ has the product f.m.p.*
2. *If L_1 has the f.m.p. and L_2 is Kripke complete, then $L_1 \circledast L_2$ is also Kripke complete.*

We say that L is *product-persistent* if, for every Kripke frames $\mathcal{F}$ and $\mathcal{G}$, $\mathcal{F} \circledast \mathcal{G} \models L$ implies $\mathcal{F} \times \mathcal{G} \models L$.

As usual we define a Sahlqvist formula in the following way:

A formula φ is called a *Sahlqvist formula* if φ is a formula which is equivalent to a conjunction of formulas of the form $\Box^m(\psi \to \chi)$, where $m \geq 0$, χ is constructed by $\wedge, \vee, \Box, \Diamond, \top, \bot$ from propositional variables and ψ is obtained from propositional variables and their negations applying $\wedge, \vee, \Box, \Diamond, \top, \bot$ in such a way that no positive occurrence of a variable is in a subformula of the form either $\psi_1 \vee \psi_2$ or $\Diamond\psi_1$ with the scope of any $\Box$.

Theorem 4.3 *Suppose that L is a product-persistent logic and Γ is any set of Sahlqvist formulas. Then the logic $L \oplus \Gamma$ is also product-persistent.*

Proof. The point of proof is that a $(R\times S)$*-expression* $(R\times S)^{k_1}(w_1, v_1) \cup \cdots \cup (R\times S)^{k_n}(w_n, v_n) = \bigcup_{i=1}^{n} R^{k_i}(w_i) \times S^{k_i}(v_i)$ belongs to $\mathcal{P} \divideontimes \mathcal{Q}$. (Refer Sambin and Vaccaro 1989.) $\dashv$

Recall that, given a formula $\varphi(p_1, \ldots, p_n)$ (whose variables are listed among $p_1, \ldots, p_n$), a general frame $\mathcal{F} = (W, R, \mathcal{P})$ and sets $X_1, \ldots, X_n$ in $\mathcal{P}$, we denote by $\varphi(X_1, \ldots, X_n)$ the set of points in $\mathcal{F}$ at which φ is true under the valuation v defined by $v(p_i) = X_i$, for $i = 1, \ldots, n$, i.e., $\varphi(X_1, \ldots, X_n) = v(\varphi)$.

Theorem 4.4 *Suppose that L_1 is a modal logic and L_2 is an extension of the logic* **D**. *Then $L_1 \supseteq L_1 \divideontimes L_2$.*
Moreover, if both L_1 and L_2 are extensions of the logic **D**,

$$L_1 \cap L_2 \supseteq L_1 \divideontimes L_2.$$

Proof. By induction on construction of formulas, we can show that for any formula φ,
$\varphi(X_1 \times V, \ldots, X_n \times V) = \varphi(X_1, \ldots, X_n) \times V$. Therefore this implies that $L_1 \divideontimes L_2$ is included in L_1. $\dashv$

A formula φ is *preserved* under normal products if, whenever φ belongs to $L_1 \cap L_2$, then φ belongs also to $L_1 \divideontimes L_2$. The next theorem shows that restricted Sahlqvist formulas are preserved.

Theorem 4.5 *Let φ be a formula which is equivalent to a conjunction of formulas of the form $\Box^m(\psi \to \chi)$, where $m \geq 0$, χ is constructed by $\wedge, \Box, \Diamond, \top, \bot$ from propositional variables and ψ is obtained from propositional variables applying $\wedge, \vee, \Box, \Diamond, \top, \bot$ in such a way that no subformula of the form either $\psi_1 \vee \psi_2$ or $\Diamond\psi_1$ occurs in the scope of any $\Box$ and that ψ_1 and ψ_2 have no common propositional variable in a subformula of the form $\psi_1 \wedge \psi_2$. Then φ is preserved under normal products.*

Proof. By induction on construction of formulas, we can show
$\chi(\bigcup_{i_1\in I_1} (X^1_{i_1} \times Y^1_{i_1}), \ldots, \bigcup_{i_n\in I_n} (X^n_{i_n} \times Y^n_{i_n})) \supseteq \bigcup_{i_1\in I_1,\ldots,i_n\in I_n} \chi(X^1_{i_1}, \ldots, X^n_{i_n}) \times \chi(Y^1_{i_1}, \ldots, Y^n_{i_n})$ and

$\psi(\bigcup_{i_1\in I_1}(X^1_{i_1}\times Y^1_{i_1}),\dots,\bigcup_{i_n\in I_n}(X^n_{i_n}\times Y^n_{i_n}))\subseteq\bigcup_{i_1\in I_1,\dots,i_n\in I_n}\psi(X^1_{i_1},\dots,X^n_{i_n})\times$ $\psi(Y^1_{i_1},\dots,Y^n_{i_n})$. Hence, since we can show $\varphi(\bigcup_{i_1\in I_1}(X^1_{i_1}\times Y^1_{i_1}),\dots,\bigcup_{i_n\in I_n}(X^n_{i_n}\times Y^n_{i_n}))\supseteq\bigcap_{i_1\in I_1,\dots,i_n\in I_n}\varphi(X^1_{i_1},\dots,X^n_{i_n})\times$ $\varphi(Y^1_{i_1},\dots,Y^n_{i_n})$, we have our theorem. ⊣

5 Shifted products

As we have noticed in the Introduction, the normal product of two m-modal logics is a m-modal logic. On the other hand, products introduced in Gabbay and Shehtman 1998 have a property that if L_1 and L_2 are m-modal and n-modal logics, the product $L_1 \times L_2$ is a $(m+n)$-modal logic. To compare this type of products with our normal products, we will introduce yet another kind of products, called *shifted products*. Though we will discuss only mono-modal cases for brevity, but we can extend the results to (n, m)-modal cases easily.

Suppose that $\boldsymbol{A} = (A, \wedge, \vee, \neg, 0, 1, \Box_1)$ and $\boldsymbol{B} = (B, \wedge, \vee, \neg, 0, 1, \Box_2)$ are mono-modal algebras. Define bi-modal algebras $\boldsymbol{A}^\circ$ and ${}^\circ\boldsymbol{B}$ as $\boldsymbol{A}^\circ = (A, \wedge, \vee, \neg, 0, 1, \Box_1, \Box_2)$ with $\Box_2 a = a$ for all $a \in A$ and ${}^\circ\boldsymbol{B} = (B, \wedge, \vee, \neg, 0, 1, \Box_1, \Box_2)$ with $\Box_1 b = b$ for all $b \in B$. For general frames $\mathcal{F}$ and $\mathcal{G}$, define general frames $\mathcal{F}^\circ$ and ${}^\circ\mathcal{G}$, with two binary relations, as $\mathcal{F}^\circ = (W, R, \Delta, \mathcal{P})$ and ${}^\circ\mathcal{G} = (V, \Delta, S, \mathcal{Q})$, where Δ denotes the diagonal relation. For a class $\mathcal{K}$ of mono-modal algebras and a class $\mathcal{C}$ of general frames, $\mathcal{K}^\circ$, ${}^\circ\mathcal{K}$, $\mathcal{C}^\circ$ and ${}^\circ\mathcal{C}$ are defined in a natural way. Now, define the *shifted product* $\boldsymbol{A} \otimes \boldsymbol{B}$ of $\boldsymbol{A}$ and $\boldsymbol{B}$ (the shifted product $\mathcal{F} \otimes \mathcal{G}$ of $\mathcal{F}$ and $\mathcal{G}$) by $\boldsymbol{A} \otimes \boldsymbol{B} = \boldsymbol{A}^\circ \divideontimes {}^\circ\boldsymbol{B}$ ($\mathcal{F} \otimes \mathcal{G} = \mathcal{F}^\circ \divideontimes {}^\circ\mathcal{G}$). We can define *shifted product* of classes of modal algebras and those of classes of general frames in an obvious way. Then, we define the shifted product $L_1 \otimes L_2$ of modal logics L_1 and L_2 by $L_1 \otimes L_2 = \boldsymbol{L}(\boldsymbol{F}(L_1) \otimes \boldsymbol{F}(L_2))$.

Theorem 5.1 *Let $\mathcal{K}_1$ and $\mathcal{K}_2$ be classes of modal algebras, and $\mathcal{C}_1$ and $\mathcal{C}_2$ be classes of general frames. Then*

1. $\boldsymbol{L}(\mathcal{K}_1) \otimes \boldsymbol{L}(\mathcal{K}_2) = \boldsymbol{L}(\mathcal{K}_1 \otimes \mathcal{K}_2)$,
2. $\boldsymbol{L}(\mathcal{C}_1) \otimes \boldsymbol{L}(\mathcal{C}_2) = \boldsymbol{L}(\mathcal{C}_1 \otimes \mathcal{C}_2)$.

Proof. Since $\boldsymbol{V}(\mathcal{K}^\circ) = \boldsymbol{V}(\mathcal{K})^\circ$ and $\boldsymbol{V}({}^\circ\mathcal{K}) = {}^\circ\boldsymbol{V}(\mathcal{K})$ hold, Theorem 4.1 implies this theorem. ⊣

Suppose that L_1 and L_2 are modal logics. Define bi-modal logics $L_1{}^\circ$ and ${}^\circ L_2$ as $L_1{}^\circ = L_1 * \mathbf{Triv}$ and ${}^\circ L_2 = \mathbf{Triv} * L_2$, where $\mathbf{Triv} = \mathbf{K} \oplus \Box p \leftrightarrow p$ and $*$ denotes the *fusion* of logics.

Theorem 5.2 *For modal logics L_1 and L_2,*

$$L_1 \otimes L_2 = L_1{}^\circ \circledast {}^\circ L_2.$$

Proof. Since $\boldsymbol{V}(L^\circ) = \boldsymbol{V}(L)^\circ$ and $\boldsymbol{V}({}^\circ L) = {}^\circ\boldsymbol{V}(L)$ hold, by definition we have the result. $\dashv$

Define a translation $\mathbf{T}$ from the set of formulas in the language of $L_1 \circledast L_2$ (with one modal operator $\Box$) to the set of formulas in the language of $L_1 \otimes L_2$ (with two modal operators $\Box_1$ and $\Box_2$) inductively as follows. (i) $\mathbf{T}(p) = p$ for all propositional variables, (ii) $\mathbf{T}(\varphi \wedge \psi) = \mathbf{T}(\varphi) \wedge \mathbf{T}(\psi)$, (iii) $\mathbf{T}(\varphi \vee \psi) = \mathbf{T}(\varphi) \vee \mathbf{T}(\psi)$, (iv) $\mathbf{T}(\neg\varphi) = \neg\mathbf{T}(\varphi)$, (v) $\mathbf{T}(\bot) = \bot$, (vi) $\mathbf{T}(\top) = \top$, (vii) $\mathbf{T}(\Box\varphi) = \Box_1\Box_2\mathbf{T}(\varphi)$. Then, we have the following.

Theorem 5.3 *A formula φ is a theorem of $L_1 \circledast L_2$ iff $\mathbf{T}(\varphi)$ is a theorem of $L_1 \otimes L_2$.*

Proof. Since $\Diamond_1\Diamond_2\bigcup_{i\in I}(X_i \times Y_i) = \Diamond_1\bigcup_{i\in I}(X_i \times \Diamond_2 Y_i) = \bigcup_{i\in I}(\Diamond_1 X_i \times \Diamond_2 Y_i)$ $= \Diamond\bigcup_{i\in I}(X_i \times Y_i)$ holds, we can show this result by induction on construction of formulas. $\dashv$

After Gabbay and Shehtman 1998, we will define the bi-modal logic $[L_1, L_2]$ to be $(L_1 * L_2) \oplus \Box_1\Box_2 p \leftrightarrow \Box_2\Box_1 p \oplus \Diamond_1\Box_2 p \to \Box_2\Diamond_1 p$. The following theorem lists some properties of $\otimes$ and enables us to compare them with the corresponding properties of $\times$.

Theorem 5.4 *Let L_1, L_2 be modal logics. Then*

1. *If L_1 and L_2 are consistent logics, then $L_1 \otimes L_2$ is a conservative extension of L_1 and L_2,*
2. *$L_1 \otimes L_2 \supseteq [L_1, L_2]$,*
3. *If L_1 and L_2 have the f.m.p., then $L_1 \otimes L_2$ has the product f.m.p.,*
4. *If L_1 has the f.m.p. and L_2 is Kripke complete, then $L_1 \otimes L_2$ is also Kripke complete.*

Proof. For 1, by induction on construction of formulas, we can show that for any formula φ in the language of L_1, $\varphi(\sum_{i_1\in I_1}(X^1_{i_1} \times Y^1_{i_1}), \ldots, \sum_{i_n \in I_n}(X^n_{i_n} \times Y^n_{i_n})) = \sum_{i_1\in I_1,\ldots,i_n\in I_n}(\varphi(X^1_{i_1}, \ldots, X^n_{i_n}) \times \bigcap_{j=1}^{n} Y^j_{i_j})$, where for any $j = 1, \ldots, n$, $V = \sum_{i_j \in I_j} Y^j_{i_j}$ holds. Therefore this implies that $L_1 \otimes L_2$ is a conservative extension of L_1.

For 2, since both L_1° and ${}^\circ L_2$ contain $\{\Box_1\Box_2 p \leftrightarrow \Box_2\Box_1 p, \Diamond_1\Box_2 p \to \Box_2\Diamond_1 p\}$, by Theorem 4.5 $L_1 \otimes L_2 \supseteq [L_1, L_2]$ holds.
3 and 4 are shown similarly to Theorem 4.2. $\dashv$

6 Conclusion

Here we will compare properties of our $\otimes$ and the usual product $\times$ (cf. Table 1 in Gabbay and Shehtman 1998). For a modal logic L, $\boldsymbol{F}^K(L)$ is the class of all Kripke frames which validate L. For modal logics L_1 and L_2, the usual product $L_1 \times L_2$ is defined by $\boldsymbol{L}(\boldsymbol{F}^K(L_1)^\circ \times {}^\circ\boldsymbol{F}^K(L_2))$.

By *$\otimes$-transfer* (*$\times$-transfer*) of a property $\mathcal{P}$, we mean that $\mathcal{P}$ carries over from modal logics L_1 and L_2 to the product of $L_1 \otimes L_2$ ($L_1 \times L_2$).

Clearly, $\times$-transfer of Kripke completeness holds, whereas $\otimes$-transfer of Kripke completeness remains an open question. We have, however, shown a partial result in Theorem 5.4(4). On the other hand, while $\times$-transfer of the f.m.p. does not hold, its $\otimes$-transfer holds as shown in Theorem 5.4(3). It is easy to see that both $\otimes$-transfer and $\times$-transfer of tabularity hold.

We say that the modal logics L_1 and L_2 are *$\otimes$-product-matching* (*$\times$-product-matching*) if $L_1 \otimes L_2 = [L_1, L_2]$ ($L_1 \times L_2 = [L_1, L_2]$). A *PTC-logic* is a modal logic axiomatized by a set of *PTC-formulas*, i.e., either *pseudo-transitive* formulas $\Diamond^m\Box p \to \Box^n p$ for $m, n \geq 0$ or closed formulas. It is shown in Gabbay and Shehtman 1998 that every pair of PTC-logic is $\times$-product-matching. For a given modal logic L, $\boldsymbol{F}_f(L)$ is the class of all finite general frames which validate L, and $\boldsymbol{F}_f^K(L)$ is the class of all finite Kripke frames which validate L. For modal logics L_1 and L_2, $L_1 \otimes_f L_2$ ($L_1 \times_f L_2$) denotes $\boldsymbol{L}(\boldsymbol{F}_f(L_1) \otimes \boldsymbol{F}_f(L_2))$ ($\boldsymbol{L}(\boldsymbol{F}_f^K(L_1) \times \boldsymbol{F}_f^K(L_2))$). It is easy to see that $L_1 \otimes_f L_2 \supseteq L_1 \otimes L_2 \supseteq [L_1, L_2]$ and $L_1 \times_f L_2 \supseteq L_1 \times L_2 \supseteq [L_1, L_2]$ hold. The next theorem follows from Theorem 2.3.

Theorem 6.1 *Let L_1 and L_2 be modal logics. Then*

$$L_1 \otimes_f L_2 = L_1 \times_f L_2.$$

As a corollary, we obtain:

Corollary 6.2 *Let L_1 and L_2 have the f.m.p. which are $\times$-product-matching (e.g. $\mathbf{K}, \mathbf{D}, \mathbf{S4}, \mathbf{S5}$ and so on). Then L_1 and L_2 are $\otimes$-product-matching if and only if $L_1 \times L_2$ has the product f.m.p. if and only if $L_1 \otimes L_2 = L_1 \times L_2$.*

Since both $\mathbf{K} \times \mathbf{K}$ and $\mathbf{S5} \times \mathbf{S5}$ have the product f.m.p. (cf. Gabbay and Shehtman 2000) but $\mathbf{K4} \times \mathbf{K4}$ doesn't, $\mathbf{K} \otimes \mathbf{K} = \mathbf{K} \times \mathbf{K}$,

$\mathbf{S5} \otimes \mathbf{S5} = \mathbf{S5} \times \mathbf{S5}$ and $\mathbf{K4} \otimes \mathbf{K4} \neq \mathbf{K4} \times \mathbf{K4}$.

Theorem 6.3 *Let L_1 and L_2 be Kripke complete. Then $L_1 \otimes L_2 \supseteq L_1 \times L_2$.*

Proof. Since L_1 and L_2 be Kripke complete, $L_1 \otimes L_2 = \boldsymbol{L}(\boldsymbol{F}^K(L_1) \otimes \boldsymbol{F}^K(L_2))$. Hence since $\boldsymbol{L}(\boldsymbol{F}^K(L_1) \otimes \boldsymbol{F}^K(L_2)) \supseteq \boldsymbol{L}(\boldsymbol{F}^K(L_1)^\circ \times {}^\circ\boldsymbol{F}^K(L_2))$ holds, $L_1 \otimes L_2 \supseteq L_1 \times L_2$ holds. ⊣

Theorem 6.4 *Let φ be a Sahlqvist formula. Then if φ belongs to $L_1 \otimes L_2$, φ belongs also to $L_1 \times L_2$.*

Proof. Since φ belongs to $\boldsymbol{L}(\boldsymbol{F}(L_1)^\circ \divideontimes {}^\circ\boldsymbol{F}(L_2))$, φ belongs also to $\boldsymbol{L}(\boldsymbol{F}^K(L_1)^\circ \divideontimes {}^\circ\boldsymbol{F}^K(L_2))$. Therefore by Theorem 4.3, φ belongs also to $\boldsymbol{L}(\boldsymbol{F}^K(L_1)^\circ \times {}^\circ\boldsymbol{F}^K(L_2))$. ⊣

Numerous interesting problems on normal products remain unexplored yet. For instance, as we have mentioned in the Introduction, we can define normal products of various logics, like superintuitionistic logics and intuitionistic modal logics, and obtain similar results about them. We intend to discuss these elsewhere.

Acknowledgements

I would like to thank my adviser Hiroakira Ono and Tomasz Kowalski for helpful comments and suggestions.

References

Chagrov, A., and M. Zakharyaschev. 1997. *Modal Logic.* Oxford Logic Guide, Vol. 35. Oxford University Press.

Gabbay, D., and V. Shehtman. 1998. Products of modal logics, Part 1. *L. J. of IGPL* 6(1):73–146.

Gabbay, D., and V. Shehtman. 2000. Products of modal logics. Part 2. *L. J. of IGPL* 8(2):165–210.

Marx, M., and Y. Venema. 1997. *Multi-dimensional modal logics.* Applied Logic Series, Vol. 4. Kluwer Academic Publishers.

Neveu, J. 1965. *Mathematical foundation of the calculus of probability.* San Francisco: Holden-Day.

Reynolds, M. 1997. A decidable temporal logic of parallelism. *Notre Dame Journal of Formal Logic* 38(3):419–436.

Sambin, G., and V. Vaccaro. 1989. A new proof of Sahlqvist's theorem on modal definability and completeness. *The Journal of Symbolic Logic* 54(3):992–999.

Shehtman, V. 1978. Two-dimensional modal logics. *Math. Zametki* 23(5):759–772.

Wolter, F. 2000. The product of converse PDL and polymodal K. *Journal of Logic and Computation* 10:223–251.

Yasusi Hasimoto
Graduate School of Information Science,
Japan Advanced Institute of Science and Technology
Tatsunokuchi, Ishikawa, 923-1292, Japan
E-mail: yasusi-h@jaist.ac.jp
URL: http://www.jaist.ac.jp/~yasusi-h/

14

A Tableau Algorithm for the Clique Guarded Fragment

COLIN HIRSCH AND STEPHAN TOBIES

ABSTRACT. We describe a "modal style" tableau algorithm that decides satisfiability for the clique guarded fragment. As a corollary of constructions used to prove the correctness of the algorithm, we obtain a new proof for the generalised tree model property of the clique guarded fragment.

1 Introduction

The Guarded Fragment of first-order logic, introduced by Andréka, van Benthem, and Németi (1998), has been a successful attempt to transfer many good properties of modal, temporal, and description logics to a larger fragment of predicate logic. Among these are decidability, the finite model property, invariance under an appropriate variant of bisimulation, and other nice model theoretic properties (Andréka et al. 1998, Grädel 1999b).

The Guarded Fragment (GF) is obtained from full first-order logic through relativisation of quantifiers by so-called guard formulas. Every appearance of a quantifier in GF must be of the form

$$\exists \mathbf{y}(\alpha(\mathbf{x},\mathbf{y}) \land \phi(\mathbf{x},\mathbf{y})) \text{ or } \forall \mathbf{y}(\alpha(\mathbf{x},\mathbf{y}) \to \phi(\mathbf{x},\mathbf{y})),$$

where α is an atomic formula, the *guard*, that contains all free variables of ϕ. This generalises quantification in modal and temporal logics, where quantification is restricted to those elements reachable via some accessibility relation.

By allowing for more general formulas as guards while preserving the idea of quantification only over elements that are *close together* in the model, one obtains generalisations of GF which are still well-behaved in

Advances in Modal Logic, Volume 3
F. Wolter, H. Wansing, M. de Rijke, and M. Zakharyaschev, eds.

the above sense. Most importantly, one can obtain the *loosely guarded fragment* (LGF) (van Benthem 1997) and the *clique guarded fragment* (CGF) (Grädel 1999a), for which decidability, invariance under clique guarded bisimulation, and some other properties have been shown in (Grädel 1999a). Suffice to say that both LGF and CGF properly extend GF, and, with respect to sentences, CGF properly extends LGF, as shown in (Grädel 1999b).

GF, LGF, and CGF are decidable and known to be 2-ExpTime complete, which is shown by Grädel (1999a, 1999b) using game and automata-based approaches. While these approaches yield (worst-case) optimal complexity results for many logics, they appear to be unsuitable as a starting point for an efficient implementation—their worst-case complexity is actually their any-case complexity. By contrast many decidability results for modal or description logics are based on tableau algorithms (see Ladner 1977, Halpern and Moses 1992, Donini et al. 1997, or Horrocks et al. 1999 for examples) and some of the fastest implementations of modal satisfiability procedures are based on tableau calculi (Horrocks et al. 2000). Unlike automata algorithms, the average-case behaviour in practice is so good that finding *really* hard problems to test these implementations has become a problem in itself.

In this paper, we generalise the principles from tableau algorithms for modal logics in order to develop a tableau algorithm for CGF. To the best of our knowledge, this is the first algorithm for CGF that can be used as the basis for an efficient implementation[1]

Recall the conjecture by Vardi that the tree model property is the main reason for the decidability of many modal style logics (Vardi 1997). As pointed out in (Grädel 1999b), the generalised tree model property explains the similarly robust decidability of guarded logics, and can be seen as a strong indication that guarded logics are a generalisation of modal logics that retain the essence of modal logics. This becomes even more evident when regarding the respective fixed-point extensions (Grädel 1999a). The generalised tree model property of CGF is also essential for our tableau algorithm. Indeed, as a corollary of the constructions used to show the soundness of our algorithm, we obtain an alternative proof for the fact that CGF has the generalised tree model property.

[1]There are resolution based decision procedures for GF and LGF (Ganzinger and de Nivelle 1999) that are readily implemented using the saturation theorem prover SPASS (Weidenbach 1997). While CGF-SAT can be easily reduced to LGF-SAT, we believe that dedicated CGF algorithms are more efficient, c.f. before Definition 3.3.

2 Preliminaries

For the definitions of GF and LGF we refer the reader to (Grädel 1999a). The *clique guarded fragment* CGF of first-order logic can be obtained in two equivalent ways, by either semantically or syntactically restricting the range of the first-order quantifiers. In the following we will use bold letters to refer to tuples of elements of the universe $(\mathbf{a}, \mathbf{b}, \dots)$ resp. tuples of variables $(\mathbf{x}, \mathbf{y}, \dots)$.

Definition 2.1 (Semantic CGF) Let τ be a relational vocabulary. For a τ-structure $\mathfrak{A}$ with universe A, the *Gaifman graph* of $\mathfrak{A}$ is defined as the undirected graph $G(\mathfrak{A}) = (A, E^{\mathfrak{A}})$ with

$$E^{\mathfrak{A}} = \{(a, a') \ : \ a \neq a', \text{there exists } R \in \tau \text{ and } \mathbf{a} \in R^{\mathfrak{A}} \text{ which contains both } a \text{ and } a'\}.$$

Under *clique guarded semantics* we understand the modification of standard first order semantics, where, instead of ranging over all elements of the universe, a quantifier is restricted to elements that form a clique in the Gaifman graph, including the binding for the free variables of the matrix formula. More precisely, let $\mathfrak{A}$ be a τ-structure and ρ an environment mapping variables to elements of A. We define the model relation inductively over the structure of formulas as the usual FO semantics with the exception

$$\mathfrak{A}, \rho \models \forall y.\phi(\mathbf{x}, y) \text{ iff for all } a \in A \text{ such that } \rho(\mathbf{x}) \cup \{a\} \text{ forms a clique in } G(\mathfrak{A}) \text{ it is the case that } \mathfrak{A}, \rho[x \mapsto a] \models \phi \ ,$$

and a similar definition for the existential case. With CGF we denote first order logic restricted to clique guarded semantics.

Definition 2.2 (Syntactic CGF) Let τ be a relational vocabulary. A formula α is a *clique-formula* for a set $\mathbf{x} \subseteq \text{free}(\alpha)$ if α is a (possibly empty if $\mathbf{x}$ contains only one variable) conjunction of atoms (excluding equality statements) such that each two distinct elements from $\mathbf{x}$ coexist in at least one atom, each atom contains at least an element from $\mathbf{x}$, and each element from $\text{free}(\alpha) \setminus \mathbf{x}$ occurs exactly once in α. In the following, we will identify a clique-formula α with the set of its conjuncts.

The *syntactic* CGF is inductively defined as follows.

1. Every relational atomic formula $Rx_{i_1} \dots x_{i_m}$ or $x_i = x_j$ belongs to CGF.
2. CGF is closed under boolean operations.

3. If $\mathbf{x}, \mathbf{y}, \mathbf{z}$ are tuples of variables, $\alpha(\mathbf{x}, \mathbf{y}, \mathbf{z})$ is a *clique-formula* for $\mathbf{x} \cup \mathbf{y}$ and $\phi(\mathbf{x}, \mathbf{y})$ is a formula in CGF such that $\text{free}(\phi) \subseteq \mathbf{x} \cup \mathbf{y}$,

$$\begin{array}{ll} \text{then} & \exists \mathbf{yz}.(\alpha(\mathbf{x}, \mathbf{y}, \mathbf{z}) \wedge \phi(\mathbf{x}, \mathbf{y})) \\ \text{and} & \forall \mathbf{yz}.(\alpha(\mathbf{x}, \mathbf{y}, \mathbf{z}) \rightarrow \phi(\mathbf{x}, \mathbf{y})) \end{array}$$

belong to CGF.

We will use $(\exists \mathbf{yz}.\alpha(\mathbf{x}, \mathbf{y}, \mathbf{z}))\phi(\mathbf{x}, \mathbf{y})$ and $(\forall \mathbf{yz}.\alpha(\mathbf{x}, \mathbf{y}, \mathbf{z}))\phi(\mathbf{x}, \mathbf{y})$ as alternative notations for $\exists \mathbf{yz}.(\alpha(\mathbf{x}, \mathbf{y}, \mathbf{z}) \wedge \phi(\mathbf{x}, \mathbf{y}))$ and $\forall \mathbf{yz}.(\alpha(\mathbf{x}, \mathbf{y}, \mathbf{z}) \rightarrow \phi(\mathbf{x}, \mathbf{y}))$, respectively. A formula of the form $\forall \mathbf{yz}.(\alpha(\mathbf{x}, \mathbf{y}, \mathbf{z}) \rightarrow \phi(\mathbf{x}, \mathbf{y}))$ is called *universally quantified.*

The following Lemma can be shown by elementary formula manipulations that exploit that every $z \in \mathbf{z}$ occurs exactly once in α.

Lemma 2.3 *Let $\alpha(\mathbf{x}, \mathbf{y}, \mathbf{z})$ be a clique-formula for $\mathbf{x}, \mathbf{y}$. Then*

$$\begin{aligned} &\forall \mathbf{yz}.(\alpha(\mathbf{x}, \mathbf{y}, \mathbf{z}) \rightarrow \phi(\mathbf{x}, \mathbf{y})) \\ \equiv\ &\forall \mathbf{y}.(\exists \mathbf{z}.\alpha(\mathbf{x}, \mathbf{y}, \mathbf{z}) \rightarrow \phi(\mathbf{x}, \mathbf{y})). \end{aligned}$$

The use of the name CGF for both the semantic and the syntactic clique guarded fragment is justified by the following Lemma.

Lemma 2.4 *Over any finite relational vocabulary the syntactic and semantic versions of the* CGF *are equally expressive.*

Proof sketch: By some elementary equivalence transformations, every syntactically clique guarded formula can be brought into a form where switching from standard semantics to clique guarded semantics does not change its meaning. Conversely, for any finite signature there is a finite disjunction $clique(\mathbf{x}, y, \mathbf{z})$ of clique-formulas for $\mathbf{x}, y$ such that $\mathbf{a}, b$ form a clique in $G(\mathfrak{A})$ iff $\mathfrak{A} \models \exists \mathbf{z}.clique(\mathbf{a}, b, \mathbf{z})$. By guarding every quantifier with such a formula and applying some elementary formula transformations and Lemma 2.3, we get, for every FO formula ψ, a syntactically clique guarded formula that is equivalent to ψ under clique guarded semantics. If we fix a finite relational vocabulary, this transformation is polynomial in the number of variables of the formula, or, more precisely, the maximal number of free variables of all subformulas. $\dashv$

In the following we will only consider the syntactic variant of the clique guarded fragment.

Definition 2.5 (NNF, Closure, Width) In the following, all formulas are assumed to be in negation normal form (NNF), where negation occurs only in front of atomic formulas. Every formula in CGF can be transformed into NNF in linear time by pushing negation inwards using DeMorgan's law and the duality of the quantifiers.

For a sentence $\psi \in$ CGF in NNF, let $\mathrm{cl}(\psi)$ be the smallest set that contains ψ and is closed under sub-formulas. Let C be a set of constants. With $\mathrm{cl}(\psi, C)$ we denote the set

$$\mathrm{cl}(\psi, C) = \{\phi(\mathbf{a}) \ : \ \mathbf{a} \subseteq C, \phi(\mathbf{x}) \in \mathrm{cl}(\psi)\}.$$

The *width* of a formula $\psi \in$ CGF is defined by

$$\mathrm{width}(\psi) := \max\{|\mathrm{free}(\phi)| \ : \ \phi \in \mathrm{cl}(\psi)\}.$$

3 A Tableau Algorithm for CGF

For various modal and description logics, decidability can be shown by means of tableau algorithms, where satisfiability of a formula ψ is decided by a syntactically guided search for a model for ψ. Examples for these kind of algorithms can be found, e.g., in (Ladner 1977, Halpern and Moses 1992, Horrocks et al. 1999). Models are usually represented by a graph in which the nodes correspond to worlds and the edges correspond to the accessibility relations in the model. Each node is labeled with a set of formulas that this node must satisfy, and new edges and nodes are created as required by existential modalities. Since many modal and description logics have the tree model property, the graphs generated by these algorithms are trees, which allows for simpler algorithms and easier implementation and optimisation of these algorithms. Indeed, some of the fastest implementations of modal or description logics satisfiability algorithms use tableau calculi (Horrocks et al. 2000).

For many modal or description logics, e.g. K or $\mathcal{ALC}$, termination of these algorithms is due to the fact that the modal depth of the formulas appearing at a node strictly decreases with every step from the root of the tree. For other logics, e.g., $\mathsf{K4}$, K with the universal modality, or the expressive DL $\mathcal{SHIQ}$, this is no longer true and termination has to be enforced by other means. One possibility for this is *blocking*, i.e., stopping the creation of new successor nodes below a node v if there already is an ancestor node w that is labeled with similar formulas as v. Intuitively, in this case the model can fold back from the predecessor of v to w, creating a cycle. Unraveling of these cycles recovers an (infinite) tree model. Since the algorithms guarantee that the formulas occurring in the label of the nodes stem from a finite set (usually the sub-formulas of the input formula), every growing path will eventually contain a blocked node, preventing further growth of this path and (together with a bound on the degree of the tree) ensuring termination of the algorithm. All these notions will also be encountered in our tableau algorithm for CGF, an indication for the "modal nature" of CGF, as it is amenable to the same techniques used successfully for modal logics.

Our investigation of a tableau algorithm for CGF starts with the observation that CGF also has some kind of tree model property.

Definition 3.1 Let τ be a relational vocabulary. A τ-structure $\mathfrak{A}$ has *tree width* k if $k \in \mathbb{N}$ is minimal with the following property.

There exists a directed tree $T = (V, E)$ and a function $f : V \to 2^A$ such that

- for every $v \in V$, $|f(v)| \leq k + 1$,
- for every $R \in \tau$ and $\mathbf{a} \in R^{\mathfrak{A}}$, there exists $v \in V$ with $\mathbf{a} \subseteq f(v)$, and
- for every $a \in A$, the set $V_a = \{v \in V \ : \ a \in f(v)\}$ induces a subtree of T.

Every node v of T induces a substructure $\mathfrak{F}(v) \subseteq \mathfrak{A}$ of cardinality at most $k+1$. The tuple $\langle T, (\mathfrak{F}(v))_{v \in T}\rangle$ is called a *tree decomposition* of $\mathfrak{A}$.

A logic $\mathcal{L}$ has the *generalised tree model property* if there exists a computable function t, assigning to every sentence $\psi \in \mathcal{L}$ a natural number $t(\psi)$ such that, if ψ is satisfiable, then ψ has a model of tree width at most $t(\psi)$.

Fact 3.2 (Tree Model Prop. for CGF**)** Every satisfiable sentence $\psi \in$ CGF of width k has a countable model of tree width at most $k - 1$.

This is a simple corollary of (Grädel 1999a), Theorem 4, where the same result is given for μCGF, that is CGF extended by a least fixed point operator.

Fact 3.2 is the starting point for our definition of a *completion tree* for a formula $\psi \in$ CGF. A node v of such a tree no longer stands for a single element of the model (as in the modal case), but rather for a substructure $\mathfrak{F}(v)$ of a tree decomposition of the model. To this purpose, we label every node v with a set $\mathbf{C}(v)$ of constants (the elements of the substructure) and a subset of $cl(\psi, \mathbf{C}(v))$, reflecting the formulas that must hold true for these elements.

To deal with *auxiliary elements*—elements helping to form a clique in $G(\mathfrak{A})$ that are not part of this clique themselves—we will use the auxiliary constant symbol $*$ as a placeholder for unspecified elements in atoms. The intention is to keep the number of constants at each node as small as possible. The $*$ will be used for the extra elements occuring in clique formulas that are not part of the clique itself.

The following definitions are useful when dealing with these generalised atoms.

Definition 3.3 Let K denote an infinite set of constants and $* \notin K$. For any set of constants $C \subseteq K$ we set $C^* = C \cup \{*\}$. We use $t_1, t_2, \ldots$ to range over elements of K^*. The relation $\geq^*$ is defined by

$$Rt_1 \ldots t_n \geq^* Rt'_1 \ldots t'_n \text{ iff for all } i \in \{1 \ldots n\}$$
$$\text{either } t_i = * \text{ or } t_i = t'_i.$$

For an atom β and a set of formulas Φ we define $\beta \in^* \Phi$ iff there is a $\beta' \in \Phi$ with $\beta \geq^* \beta'$.

For a set of constants $C \subseteq K$ and an atom $\beta = Rt_1 \ldots t_n$, we define

$$\beta \circledast C = Rt'_1 \ldots t'_n \quad \text{where } t'_i = \begin{cases} t_i & \text{if } t_i \in C \\ * & \text{otherwise} \end{cases}$$

We use the notation $\mathbf{a}^*$ to indicate that the tuple $\mathbf{a}^*$ may contain $*$'s. Obviously, $\geq^*$ is transitive and reflexive, and $\beta \circledast C \geq^* \beta$ for all atoms β and sets of constants C.

While these are all syntactic notions, they have a semantic counterpart that clarifies the intuition of $*$ standing for an unspecified element. Let $\mathbf{a}'$ denote the tuple obtained from a tuple $\mathbf{a}^*$ by replacing every occurrence of $*$ in $\mathbf{a}^*$ with a distinct fresh variable, and let $\mathbf{z}$ be precisely the variables used in this replacement. For an atom β, we define

$$\mathfrak{A} \models \beta(\mathbf{a}^*) \text{ iff } \mathfrak{A} \models \exists \mathbf{z}.\beta(\mathbf{a}').$$

It is easy to see that

$$\beta(\mathbf{a}) \geq^* \beta(\mathbf{b}) \quad \text{implies} \quad \beta(\mathbf{b}) \models \beta(\mathbf{a})$$
$$\beta(\mathbf{a}) \in^* \Phi \quad \text{implies} \quad \Phi \models \beta(\mathbf{a})$$

because, if $\mathbf{a} \geq^* \mathbf{b}$, then $\mathbf{b}$ is obtained from $\mathbf{a}$ by replacing some $*$ with constants, which provide witnesses for the existential quantifier.

We further write $\Phi|_C$ to denote the subset of Φ containing all formulas that only use constants in C.

Definition 3.4 (Compl. Tree, Tableau) Let $\psi \in$ CGF be a closed formula in NNF. A *completion tree* $\mathbf{T} = (\mathbf{V}, \mathbf{E}, \mathbf{C}, \Delta, \mathbf{N})$ for ψ is a vertex labeled tree $(\mathbf{V}, \mathbf{E})$ with the labeling function $\mathbf{C}$ labeling each node $v \in \mathbf{V}$ with a subset of K, Δ labeling each node $v \in \mathbf{V}$ with a subset of $cl(\psi, \mathbf{C}(v)^*)$ where all formulas $\beta(\mathbf{x}, *, \ldots, *) \in \Delta(v)$ using $*$ are atoms (excluding equality statements), and the function $\mathbf{N} : \mathbf{V} \to \mathbb{N}$ mapping each node to a distinct natural number, with the additional property that, if v is an ancestor of w, then $\mathbf{N}(v) < \mathbf{N}(w)$.

A constant $c \in K$ is called *shared* between two nodes $v_1, v_2 \in \mathbf{V}$, if $c \in \mathbf{C}(v_1) \cap \mathbf{C}(v_2)$, and $c \in \mathbf{C}(w)$ for all nodes w on the (unique, undirected, possibly empty) shortest path connecting v_1 to v_2.

A node $v \in \mathbf{V}$ is called *directly blocked*[2] by a node $w \in \mathbf{V}$, if w is not blocked, $\mathbf{N}(w) < \mathbf{N}(v)$, and there is an injective mapping π from $\mathbf{C}(v)$ into $\mathbf{C}(w)$ such that, for all constants $c \in \mathbf{C}(v)$ that are shared between v and w, $\pi(c) = c$, and $\pi(\Delta(v)) = \Delta(w)|_{\pi(\mathbf{C}(v)^*)}$. Here and throughout this paper we use the convention $\pi(*) = *$ for every function π that verifies a blocking.

A node is called *blocked* if it is directly blocked or if its predecessor is blocked.

A completion tree $\mathbf{T}$ *contains a clash* if there is a node $v \in \mathbf{V}$ such that

- for a constant $c \in \mathbf{C}(v)$, $c \neq c \in \Delta(v)$, or
- there is an atomic formula β and a tuple of constants $\mathbf{a} \subseteq \mathbf{C}(v)$ such that $\{\beta(\mathbf{a}), \neg\beta(\mathbf{a})\} \subseteq \Delta(v)$.

Otherwise, $\mathbf{T}$ is called *clash-free*. A completion tree $\mathbf{T}$ is called *complete* if none of the *completion rules* given in Figure 1 can be applied to $\mathbf{T}$. A complete and clash-free completion tree for ψ is called a *tableau* for ψ.

To test ψ for satisfiability, the tableau algorithm creates an initial tree with only a single node v_0, $\Delta(v_0) = \{\psi\}$ and $\mathbf{C}(v_0) = \{a_0\}$ for an arbitrary constant a_0. The rules from Figure 1 are applied until either a clash occurs, producing output "ψ `unsatisfiable`", or the tree is complete, in which case "ψ `satisfiable`" is output.

The set $\mathbf{C}(v_0)$ is initialised with a non-empty set of constants to make sure that empty structures are excluded.

While our notion of tableaux has many similarities to the tableaux appearing in (Grädel and Walukiewicz 1999), there are two important differences that make the version used here more suitable as basis for a tableau algorithm.

We will see that every completion tree generated by the tableau algorithm is finite. Conversely, tableaux in (Grädel and Walukiewicz 1999), in general, can be infinite.

Also, in (Grädel and Walukiewicz 1999) every node is labeled with a complete $(\psi, \mathbf{C}(v))$-type, i.e., every formula $\phi \in \mathrm{cl}(\psi, \mathbf{C}(v))$ is explicitly asserted true of false at v. Conversely, a completion tree contains only assertions about relevant formulas. This implies a lower degree of non-determinism in the algorithm, which is important for an efficient implementation.

[2]The definition of blocking is recursive. This does not cause any problems because the status of a node v only depends on its label and the status of nodes w with $\mathbf{N}(w) < \mathbf{N}(v)$. The recursion terminates at the root node, where the $\mathbf{N}$-value is minimal.

R$\wedge$: if $\phi \wedge \theta \in \Delta(v)$ and $\{\phi, \theta\} \not\subseteq \Delta(v)$
then $\Delta(v) := \Delta(v) \cup \{\phi, \theta\}$

R$\vee$: if $\phi \vee \theta \in \Delta(v)$ and $\{\phi, \theta\} \cap \Delta(v) = \emptyset$
then $\Delta(v) := \Delta(v) \cup \{\chi\}$ for $\chi \in \{\phi, \theta\}$ (chosen non-deterministically)

R$=$: if $a = b \in \Delta(v), a \neq b$
then for all w that share a with v,
$\mathbf{C}(w) := (\mathbf{C}(w) \setminus \{a\}) \cup \{b\}$ and $\Delta(w) := \Delta(w)[a \mapsto b]$

R$\forall$: if $(\forall \mathbf{yz}.\alpha(\mathbf{a}, \mathbf{y}, \mathbf{z}))\phi(\mathbf{a}, \mathbf{y}) \in \Delta(v)$, there exists a $\mathbf{b} \subseteq \mathbf{C}(v)$
such that for all atoms $\beta(\mathbf{x}, \mathbf{y}, \mathbf{z}) \in \alpha, \beta(\mathbf{a}, \mathbf{b}, * \cdots *) \in^* \Delta(v)$,
and $\phi(\mathbf{a}, \mathbf{b}) \notin \Delta(v)$
then $\Delta(v) := \Delta(v) \cup \{\phi(\mathbf{a}, \mathbf{b})\}$

R$\exists$: if $(\exists \mathbf{yz}.\alpha(\mathbf{a}, \mathbf{y}, \mathbf{z}))\phi(\mathbf{a}, \mathbf{y}) \in \Delta(v)$ and for every $\mathbf{b}, \mathbf{c} \subseteq \mathbf{C}(v)$,
$\{\alpha(\mathbf{a}, \mathbf{b}, \mathbf{c}), \phi(\mathbf{a}, \mathbf{b})\} \not\subseteq \Delta(v)$ and there is no child w of v
with $\{\alpha(\mathbf{a}, \mathbf{b}, \mathbf{c}), \phi(\mathbf{a}, \mathbf{b})\} \subseteq \Delta(w)$ for some $\mathbf{b}, \mathbf{c} \subseteq \mathbf{C}(w)$
and v is not blocked
then let $\mathbf{b}, \mathbf{c}$ be sequences of distinct and fresh constants that
match the lengths of $\mathbf{y}, \mathbf{z}$,
create a child w of v with $\mathbf{C}(w) := \mathbf{a} \cup \mathbf{b} \cup \mathbf{c}$ and
$\Delta(w) := \{\alpha(\mathbf{a}, \mathbf{b}, \mathbf{c}), \phi(\mathbf{a}, \mathbf{b})\}$, and
let $\mathbf{N}(w) = 1 + \max\{\mathbf{N}(v) : v \in \mathbf{V} \setminus \{w\}\}$

R$\updownarrow$: if $\beta(\mathbf{a}^*) \in \Delta(v), \beta$ atomic, not an equality, w is a neighbour of v
with $\mathbf{a}^* \cap \mathbf{C}(w) \neq \emptyset$, and $\beta(\mathbf{a}^*) \Cap \mathbf{C}(w) \notin \Delta(w)$
then $\Delta(w) := \Delta(w) \cup \{\beta(\mathbf{a}^*) \Cap \mathbf{C}(w)\}$

R$\updownarrow\forall$: if $\phi(\mathbf{a}) \in \Delta(v), \phi(\mathbf{a})$ is univ. quantified, and w is a neighbour of v
with $\mathbf{a} \subseteq \mathbf{C}(w)$ and $\phi(\mathbf{a}) \notin \Delta(w)$
then $\Delta(w) := \Delta(w) \cup \{\phi(\mathbf{a})\}$

FIGURE 1 The Completion Rules for CGF

Theorem 3.5 *The tableau algorithm is a (non-deterministic) decision procedure for* CGF*-satisfiability.*

Proof: This is an immediate consequence of the following facts established in the subsequent sections.

1. Every sequence of rule applications terminates after a finite number of steps. (*Termination*, Lemma 3.7)
2. If ψ is satisfiable, then the rules can be applied to generate a tableau for ψ. (*Completeness*, Lemma 3.8)
3. If the algorithm constructs a tableau for ψ, then ψ is satisfiable (*Soundness*, Lemma 3.11). $\dashv$

3.1 Termination

The following technical lemma is a consequence of the completion rules and the blocking condition.

Lemma 3.6 *Let* $\psi \in \mathrm{CGF}$ *be a sentence in NNF with* $|\psi| = n$, $\mathrm{width}(\psi) = m$, *and* $\mathbf{T}$ *a completion tree generated for* ψ *by application of the rules in Figure 1. For every node* $v \in \mathbf{T}$,

1. $|\mathbf{C}(v)| \leq m$
2. $|\Delta(v)| \leq n \times (m+1)^m$
3. *Any* $\ell > 2^{n\times(m+1)^m}$ *distinct nodes in* $\mathbf{T}$ *contain a blocked node.*

Proof: Nodes are only generated when intialising the tree (with a singleq constant) and by the R∃-rule and no constants are added to a $\mathbf{C}(v)$ once v has been generated (but some may be removed by application of the R=-rule).

When triggered by the formula $(\exists \mathbf{yz}.\alpha(\mathbf{a},\mathbf{y},\mathbf{z}))\phi(\mathbf{a},\mathbf{y})$, the R∃-rule initializes $\mathbf{C}(w)$ such that it contains $\mathbf{a}$ and another constant for every variable in $\mathbf{x}$ and $\mathbf{y}$. Hence,

$$|\mathbf{C}(w)| \leq |\mathbf{a} \cup \mathbf{y} \cup \mathbf{z}| \leq |\mathrm{free}(\alpha)| \leq \mathrm{width}(\psi).$$

The set $\Delta(v)$ is a subset of $cl(\psi, \mathbf{C}(v)^*)$, for which $|cl(\psi, \mathbf{C}(v))| \leq n \times (m+1)^m$ holds because there are at most n formulas in $cl(\psi)$, each of which has at most m free variables. There are at most $(|\mathbf{C}(v)|+1)^m$ distinct sequences of length m with constants from $\mathbf{C}(v)^*$.

Let $v_1, \ldots, v_\ell$ be $\ell > 2^{n\times(m+1)^m}$ distinct nodes. For every v_i, we will construct an injective mapping $\pi_i : \mathbf{C}(v_i) \to \{1, \ldots m\}$ such that, if a constant a is shared between two nodes v_i, v_j, then $\pi_i(a) = \pi_j(a)$.

Let $u_1, \cdots, u_k$ denote the nodes of a subtree of $\mathbf{T}$ that contains every node v_i and that is rooted at u_1. By induction over the distance to u_1, we define an injective mapping $\nu_i : \mathbf{C}(u_i) \to \{1, \ldots, m\}$ for every $i \in \{1, \ldots, k\}$ as follows. For ν_1 we pick an arbitrary injective function from $\mathbf{C}(u_1)$ to $\{1, \ldots, m\}$. For a node u_i let u_j be the predecessor of u_i in $\mathbf{T}$ and ν_j the corresponding function, which has already been defined because u_j has a smaller distance to u_1 than u_i. For ν_i we choose an arbitrary injective function such that $\nu_i(a) = \nu_j(a)$ for all $a \in \mathbf{C}(u_i) \cap \mathbf{C}(u_j)$.

All mappings ν_i are injective. For any constant a the set $\mathbf{V}_a := \{v \in \mathbf{V} \mid a \in \mathbf{C}(v)\}$ induces a subtree of $\mathbf{T}$. If $u_i, u_j \in \mathbf{V}_a$ are neighbours, the definition above ensures $\nu_i(a) = \nu_j(a)$. By induction over the length of the shortest connecting path we obtain the same for arbitrary $u_i, u_j \in \mathbf{V}_a$.

For every node v_i there is a j_i such that $v_i = u_{j_i}$ and we set $\pi_i = \nu_{j_i}$. There are at most $2^{n\times(m+1)^m}$ distinct subsets of $cl(\psi, \{1, \ldots, m, *\})$. Hence, there must be two nodes v_i, v_j such that $\pi_i(\Delta(v_i)) = \pi_j(\Delta(v_j))$ and, w.l.o.g., $\mathbf{N}(v_i) < \mathbf{N}(v_j)$. This implies that v_j is blocked by v_i via $\pi := \pi_i^{-1} \circ \pi_j$. Note that for π to be well-defined, π_i must be injective. By construction, π preserves shared constants. Since $\pi_i(\Delta(v_i)) = \pi_j(\Delta(v_j))$, $\pi(\Delta(v_j)) = \Delta(v_i)|_{\pi(\mathbf{C}(v_j))}$ holds. $\dashv$

Lemma 3.7 (Termination) *Let* $\psi \in$ CGF *be a sentence in NNF. Any sequence of rule applications of the tableau algorithm starting from the initial tree terminates.*

Proof: For any completion tree $\mathbf{T}$ generated by the tableau algorithm, we define $\|\cdot\| : \mathbf{V} \mapsto \mathbb{N}^3$ by

$$\|v\| := (|\mathbf{C}(v)|, \;\; n \times (m+1)^m - |\Delta(v)|,$$
$$|\{\phi \in \Delta(v) : \phi \text{ triggers the R}\exists\text{-rule for } v\}|).$$

The lexicographic order $\prec$ on $\mathbb{N}^3$ is well-founded, i.e. it has no infinite decreasing chains. Any rule application decreases $\|v\|$ w.r.t. $\prec$ for at least one node v, and never increases $\|v\|$ w.r.t. $\prec$ for an existing node v. However it may create new successors, one at a time. Since $\prec$ is well-founded, there can only be a finite number of applications of rules to every node in $\mathbf{T}$ and hence a finite number of successors and an infinite sequence of rule applications would generate a tree of infinite depth.

Yet, as a corollary of Lemma 3.6, we have that the depth of $\mathbf{T}$ is bounded by $2^{n\times(m+1)^m}$. For assume that there is a path of length $> 2^{n\times(m+1)^m}$ in $\mathbf{T}$ with deepest node v. By the time v has been created (by an application of the R$\exists$-rule to its predecessor u), the path from the root of $\mathbf{T}$ to u contains at least $2^{n\times(m+1)^m}$ nodes, and hence a blocked node. This implies that u is blocked too, and the R$\exists$-rule cannot be applied to create v. $\dashv$

3.2 Completeness

Lemma 3.8 *Let* $\psi \in$ CGF *be a closed formula in* NNF. *If* ψ *is satisfiable, then there is a sequence of rule applications starting from the initial tree that yields a tableau.*

Proof: Since ψ is satisfiable, there is a model $\mathfrak{A}$ of ψ. We will use $\mathfrak{A}$ to guide the application of the non-deterministic R$\vee$-rule. For this we incremently define a function $g : \bigcup\{\mathbf{C}(v) \mid v \in \mathbf{V}\} \to A$ such that for all $v \in \mathbf{V} : \mathfrak{A} \models g(\Delta(v))$. We refer to this property by (§).

The set $\Delta(v)$ can contain atomic formulas $\alpha(\mathbf{a}^*)$, where $*$ occurs at some positions of $\mathbf{a}^*$. The constant $*$ is not mapped to an element of A by g. We deal with this as described just after Definition 3.3 by setting

$$\mathfrak{A} \models g(\alpha(\mathbf{a}^*)) \;\text{ iff }\; \mathfrak{A} \models \exists\mathbf{z}.g(\alpha(\mathbf{a}')).$$

Claim 3.9 *If, for a completion tree* $\mathbf{T}$, *there exists a function* g, *such that* (§) *holds and a rule is applicable to* $\mathbf{T}$, *then it can be applied in a way that maintains* (§).

- For the R∧- and the R∨-rule this is obvious.
- If the R=-rule is applicable to $v \in \mathbf{V}$ with $a = b \in \Delta(v)$, then, since $\mathfrak{A} \models g(a) = g(b)$, $g(a) = g(b)$ must hold. Hence, for every node w that shares a with v, $g(\Delta(w)) = g(\Delta(w)[a \mapsto b])$ and the rule can be applied without violating (§).
- If the R∀-rule is applicable to $v \in \mathbf{V}$ with $(\forall \mathbf{yz}.\alpha(\mathbf{a},\mathbf{y},\mathbf{z}))\phi(\mathbf{a},\mathbf{y}) \in \Delta(v)$ and $\mathbf{b} \subseteq \mathbf{C}(v)$ with $\beta(\mathbf{a},\mathbf{b},* \cdots *) \in^* \Delta(v)$ for all atoms $\beta(\mathbf{x},\mathbf{y},\mathbf{z}) \in \alpha$, then, from the definition of $\in^*$, there is a tuple $\mathbf{c}^* \subseteq \mathbf{C}(v)^*$, such that $\beta(\mathbf{a},\mathbf{b},* \cdots *) \geq^* \beta(\mathbf{a},\mathbf{b},\mathbf{c}^*)$ and $\beta(\mathbf{a},\mathbf{b},\mathbf{c}^*) \in \Delta(v)$. From (§) we get that $\mathfrak{A} \models \exists \mathbf{z}.\beta(g(\mathbf{a}),g(\mathbf{b}),\mathbf{z})$ and since every $z \in \mathbf{z}$ appears exactly once in α, also $\mathfrak{A} \models \exists \mathbf{z}.\alpha(g(\mathbf{a}),g(\mathbf{b}),\mathbf{z})$. Hence, we have
$$\{\mathfrak{A} \models \{\forall \mathbf{yz}.\alpha(g(\mathbf{a}),\mathbf{y},\mathbf{z})) \to \phi(g(\mathbf{a}),\mathbf{y}),\\ \exists \mathbf{z}.\alpha(g(\mathbf{a}),g(\mathbf{b}),\mathbf{z})\}$$
which, by Lemma 2.3, implies $\mathfrak{A} \models \phi(g(\mathbf{a}),g(\mathbf{b}))$ and hence $\phi(\mathbf{a},\mathbf{b})$ can be added to $\Delta(v)$ without violating (§).
- If the R∃-rule is applicable to $v \in \mathbf{V}$ with $(\exists \mathbf{yz}.\alpha(\mathbf{a},\mathbf{y},\mathbf{z}))\phi(\mathbf{a},\mathbf{y})$, then this implies
$$\mathfrak{A} \models g((\exists \mathbf{yz}.\alpha(\mathbf{a},\mathbf{y},\mathbf{z}))\phi(\mathbf{a},\mathbf{y})).$$
Hence, there are sequences $\mathbf{b}',\mathbf{c}' \subseteq A$ of elements such that $\mathfrak{A} \models \{\alpha(g(\mathbf{a}),\mathbf{b}',\mathbf{c}'),\phi(g(\mathbf{a}),\mathbf{b}')\}$. If we define g such that $g(\mathbf{b}) = \mathbf{b}'$ and $g(\mathbf{c}) = \mathbf{c}'$, then obviously $\mathfrak{A} \models \{g(\alpha(\mathbf{a},\mathbf{b},\mathbf{c}),g(\phi(\mathbf{a},\mathbf{b}))\}$. Note, that this might involve setting $g(b_1) = g(b_2)$ for some $b_1, b_2 \in \mathbf{b}$. With this construction the resulting extended completion-tree $\mathbf{T}$ and extended function g again satisfy (§).
- If the R↕-rule is applicable to $v \in \mathbf{V}$ with $\beta(\mathbf{a}^*) \in \Delta(v)$ and a neighbour w with $\mathbf{a}^* \cap \mathbf{C}(w) \neq \emptyset$, then it adds $\beta(\mathbf{a}^*) \circledast \mathbf{C}(w)$ to $\Delta(w)$. From (§) we get that $\mathfrak{A} \models \beta(g(\mathbf{a}^*))$, and since $\beta(\mathbf{a}^*) \circledast \mathbf{C}(w) \geq^* \beta(\mathbf{a}^*)$, this implies $\mathfrak{A} \models \beta(g(\mathbf{a}^*)) \circledast \mathbf{C}(w)$. Hence, adding $\beta(\mathbf{a}^*) \circledast \mathbf{C}(w)$ to $\Delta(w)$ does not violate (§).
- If the R↕∀-rule is applicable to a node $v \in \mathbf{V}$ with a universally quantified formula $\phi(\mathbf{a}) \in \Delta(v)$ and a neighbour w which shares $\mathbf{a}$ with v, (§) yields $\mathfrak{A} \models \phi(g(\mathbf{a}))$. Hence, adding $\phi(\mathbf{a})$ to $\Delta(w)$ does not violate (§).

Claim 3.10 *A completion-tree* $\mathbf{T}$ *for which a function* g *exists such that* (§) *holds is clash free.*

Assume that $\mathbf{T}$ contains a clash, namely, there is a node $v \in \mathbf{V}$ such that either $a \neq a \in \mathbf{V}(v)$—implying $\mathfrak{A} \models g(a) \neq g(a)$—, or that there is a sequence $\mathbf{a} \subseteq \mathbf{C}(v)$, and an atomic formula β such that $\{\beta(\mathbf{a}), \neg\beta(\mathbf{a})\} \subseteq \Delta(v)$. From (§), $\mathfrak{A} \models \{\beta(g(\mathbf{a})), \neg\beta(g(\mathbf{a}))\}$ would follow, also a contradiction.

These claims yield Lemma 3.8 as follows. Let $\mathbf{T}$ be a tableau for ψ. Since $\mathfrak{A} \models \psi$, (§) is satisfied for the initial tree together with the function g mapping a_0 to an arbitrary element of the universe of $\mathfrak{A}$. By Lemma 3.7, any sequence of applications is finite, and from Claim 3.9 we get that there is a sequence of rule-applications that maintains (§). By Claim 3.10, this sequence results in a tableau. This completes the proof of Lemma 3.8. $\dashv$

Lemma 3.8 involves two different kinds of non-determinism, namely, the choice which rule to apply to which constraint (as several rules might be applicable simultaneously), and which disjunct to choose in an application of the R$\vee$-rule. While the latter choice is *don't-know* non-deterministic, i.e., for a satisfiable formula only certain choices will lead to the discovery of a tableau, the former choice is *don't-care* non-deterministic. This means that arbitrary choices of which rule to apply next will lead to the discovery of a tableau for a satisfiable formula. For an implementation of the tableau algorithm this has the following consequences. Exhaustive search is necessary to deal with all possible expansions of the R$\vee$-rule, but arbitrary strategies of choosing which rule to apply next, and where to apply it, will lead to a correct implementation, although the efficiency of the implementation will strongly depend on a sophisticated strategy.

3.3 Correctness

In order to prove the correctness of the tableau algorithm we have to show that the existence of a tableau for ψ implies satisfiability of ψ. To this purpose, we will construct a model from a tableau. From the construction employed in the proof we obtain an alternative proof of Fact 3.2.

Lemma 3.11 *Let $\psi \in$ CGF$[\tau]$ with $k =$ width(ψ) and let $\mathbf{T}$ be a tableau for ψ generated by the tableau algorithm. Then ψ is satisfiable and has a model of tree width at most $k-1$.*

Proof: Let $\mathbf{T} = (\mathbf{V}, \mathbf{E}, \mathbf{C}, \Delta, \mathbf{N})$ a tableau for ψ. For every direct blocking situation we fix a mapping π verifying this blocking. Using an unraveling construction, we will construct a model $\mathfrak{A}$ for ψ of width at most $k-1$ from $\mathbf{T}$. First, we "unravel" blocking situations in $\mathbf{T}$ by successively replacing every blocked node with a copy of the subtree of $\mathbf{T}$ rooted at the blocking node. Formally, this is achieved by the following path construction. We define

$$\mathbf{V}_u = \{v \in \mathbf{V} \ : \ v \text{ is not blocked or directly blocked }\}.$$

Since from now on we only deal with nodes from $\mathbf{V}_u$, every blocking is direct and we will no longer explicitly mention this fact.

The set $\mathsf{Paths}(\mathbf{T})$ is inductively defined by[3]

- $[\frac{v_0}{v_0}] \in \mathsf{Paths}(\mathbf{T})$ for the root v_0 of $\mathbf{T}$,
- if $[\frac{v_1}{v_1'} \dots \frac{v_n}{v_n'}] \in \mathsf{Paths}(\mathbf{T})$, w is a successor of v_n and w is not blocked, then $[\frac{v_1}{v_1'} \dots \frac{v_n}{v_n'} \frac{w}{w}] \in \mathsf{Paths}(\mathbf{T})$,
- if $[\frac{v_1}{v_1'} \dots \frac{v_n}{v_n'}] \in \mathsf{Paths}(\mathbf{T})$, w is a successor of v_n blocked by the node $u \in \mathbf{V}$, then $[\frac{v_1}{v_1'} \dots \frac{v_n}{v_n'} \frac{u}{w}] \in \mathsf{Paths}(\mathbf{T})$.

The set $\mathsf{Paths}(\mathbf{T})$ forms a tree, with p' being a successor of p if p' is obtained from p by concatenating one element $\frac{u}{w}$ at the end. We define the auxiliary functions $\mathsf{Tail}, \mathsf{Tail}'$ by setting $\mathsf{Tail}(p) = v_n$ and $\mathsf{Tail}'(p) = v_n'$ for every path $p = [\frac{v_1}{v_1'} \dots \frac{v_n}{v_n'}]$.

Intuitively, for every node v of $\mathbf{T}$, the paths $p \in \mathsf{Paths}(\mathbf{T})$ with $v = \mathsf{Tail}(p)$ stand for distinct copies of v created by the unraveling. The universe of $\mathfrak{A}$ consists of (classes of) constants labeling nodes in $\mathbf{T}$ paired with the paths at whose Tail they appear to distinguish constants occuring at different copies of a node of $\mathbf{T}$. Formally, we define

$$\mathbf{C}(\mathbf{T}) = \{(a, p) \ : \ p \in \mathsf{Paths}(\mathbf{T}) \wedge a \in \mathbf{C}(\mathsf{Tail}(p))\}.$$

Constants appearing at consecutive nodes of $\mathbf{T}$ stand for the same element and the same holds for constants related by a mapping π verifying a block. Hence, to obtain the universe of $\mathfrak{A}$, we factorise $\mathbf{C}(\mathbf{T})$ as follows. Let $\sim$ be the smallest symmetric relation on $\mathbf{C}(\mathbf{T})$ satisfying

- $(a,p) \sim (a,q)$ if q is a successor of p in $\mathsf{Paths}(\mathbf{T})$, $\mathsf{Tail}'(q)$ is an unblocked successor of $\mathsf{Tail}(p)$, and $a \in \mathbf{C}(\mathsf{Tail}(p)) \cap \mathbf{C}(\mathsf{Tail}'(q))$,
- $(a,p) \sim (b,q)$ if q is a successor of p in $\mathsf{Paths}(\mathbf{T})$, $\mathsf{Tail}'(q)$ is a blocked successor of $\mathsf{Tail}(p)$, $a \in \mathbf{C}(\mathsf{Tail}(p)) \cap \mathbf{C}(\mathsf{Tail}'(q))$, and $\pi(a) = b$ for the function π that verifies that $\mathsf{Tail}'(q)$ is blocked by $\mathsf{Tail}(q)$.

With $\approx$ we denote the reflexive, transitive closure of $\sim$. Since $(a,p) \sim (b,q)$ iff p, q are neighbours in $\mathsf{Paths}(\mathbf{T})$, for every (a,p), the set

$$\mathsf{Paths}([a,p]_\approx) := \{q \mid \exists b.(b,q) \in [a,p]_\approx\}$$

[3]This complicated form of unraveling, where we record both blocked and blocking node is necessary because there might be a situation where two successors v_1, v_2 of a node are directly blocked by the same node w.

is a subtree of $\mathsf{Paths}(\mathbf{T})$.

The classes of $\mathbf{C}(\mathbf{T})/\approx$ will be the elements of the universe of $\mathfrak{A}$. First we need to prove some technicalities for this construction.

Claim 3.12 *Let $p \in \mathsf{Paths}(\mathbf{T})$ and $a, b \in \mathbf{C}(\mathsf{Tail}(p))$. Then $(a,p) \approx (b,p)$ iff $a = b$.*

Assume the claim does not hold and let $a \neq b$ with $(a,p) \approx (b,p)$. By definition of $\sim$, $(a,p) \not\sim (b,p)$ must hold. Hence, there must be a path $(c_1,p_1) \sim \cdots \sim (c_k,p_k)$ such that $a = c_1$, $b = c_k$, and $p = p_1 = p_k$. W.l.o.g., assume we have picked a, b, p such that this path has minimal length k. Such a minimal path must be of length $k = 3$, for if we assume a path of length $k > 3$, there must be $2 \leq i < j \leq k-1$ such that $p_i = p_j$, because the relation $\sim$ is defined along paths in the tree $\mathsf{Paths}(\mathbf{T})$. If $c_i = c_j$ then we can shorten the path between position i and j and obtain a shorter path. If $c_i \neq c_j$, then the path $(c_i,p_i) \sim \cdots \sim (c_j,p_j)$ is also a shorter path with the same properties. Hence, a minimal path must be of the form $(a,p) \sim (c,q) \sim (b,p)$. If $\mathsf{Tail}'(q)$ is not blocked, by the definition of $\sim$, $a = c = b$ must hold. Hence, since $a \neq b$, $\mathsf{Tail}'(q)$ must be blocked by $\mathsf{Tail}(q)$. From the definition of $\sim$ we have $a, b \in \mathbf{C}(\mathsf{Tail}'(q))$ and $\pi(a) = c = \pi(b)$ for the function π verifying that $\mathsf{Tail}'(q)$ is blocked by $\mathsf{Tail}(q)$. Since π must be injective, this is a contradiction.

Since the set $\mathsf{Paths}(\mathbf{T})$ is a tree, and as a consequence of Claim 3.12, we get the following.

Claim 3.13 *Let $p, q \in \mathsf{Paths}(\mathbf{T})$ with $p = [\frac{v_1}{v_1'} \cdots \frac{v_n}{v_n'}]$, $q = [\frac{v_1}{v_1'} \cdots \frac{v_n}{v_n'}\frac{w}{w'}]$. If, for $a \in \mathbf{C}(v_n), b \in \mathbf{C}(w)$, $(a,p) \approx (b,q)$ then $(a,p) \sim (b,q)$.*

If $(a,p) \approx (b,q)$ then there must be a path $(c_1,p_1) \sim \cdots \sim (c_k,p_k)$ such that $a = c_1$, $b = c_k$, $p = p_1$, and $q = p_k$. Since $\sim$ is only defined along paths in the tree $\mathsf{Paths}(\mathbf{T})$, there must be a step from p to q (or, dually, from q to p) in this path, more precisely, there must be an $i \in \{1, \ldots k-1\}$ such that $p_i = p$ and $p_{i+1} = q$ holds. Hence, we have the situation

$$(a,p) \approx (c_i,p) \sim (c_{i+1},q) \approx (b,q).$$

Claim 3.12 implies $a = c_i$ and $b = c_{i+1}$ and hence $(a,p) \sim (b,q)$.

Using Claim 3.13, we can show that the blocking condition and the R$\updownarrow$- and R$\updownarrow\forall$-rule work as desired.

Claim 3.14 *Let $p, q \in \mathsf{Paths}(\mathbf{T})$, $\mathbf{a} \subseteq \mathbf{C}(\mathsf{Tail}(p)), \mathbf{b} \subseteq \mathbf{C}(\mathsf{Tail}(q))$ and $(\mathbf{a},p) \approx (\mathbf{b},q)$.*

- *For every atom* β, $\beta(\mathbf{a}, * \cdots *) \in^* \Delta(\mathsf{Tail}(p))$ *iff* $\beta(\mathbf{b}, * \cdots *) \in^* \Delta(\mathsf{Tail}(q))$.
- *For every universally quantified* ϕ, $\phi(\mathbf{a}) \in \Delta(\mathsf{Tail}(p))$ *iff* $\phi(\mathbf{b}) \in \Delta(\mathsf{Tail}(q))$.

Since both propositions are symmetric, we only need to prove one direction. If $(\mathbf{a}, p) \approx (\mathbf{b}, q)$ with $\mathbf{a} = a_1 a_2 \ldots a_m$ and $\mathbf{b} = b_1 b_2 \ldots b_m$, then

$$\{p, q\} \subseteq \bigcap_{i=1}^{m} \mathsf{Paths}([a_i, p]_{\approx})$$

and this intersection contains a path $p_1, \ldots, p_k$ in $\mathsf{Paths}(\mathbf{T})$ such that there are tuples of constants $\mathbf{c}_1, \ldots, \mathbf{c}_k$ with $(\mathbf{c}_1, p_1) \approx \cdots \approx (\mathbf{c}_k, p_k)$, $p = p_1$, $q = p_k$, $\mathbf{a} = \mathbf{c}_1$, and $\mathbf{b} = \mathbf{c}_k$. From Claim 3.13, we get that for any two neighbours p_i, p_{i+1} in $\mathsf{Paths}(\mathbf{T})$, $(\mathbf{c}_i, p_i) \approx (\mathbf{c}_{i+1}, p_{i+1})$ implies $(\mathbf{c}_i, p_i) \sim (\mathbf{c}_{i+1}, p_{i+1})$.

By two similar inductions on i with $1 \leq i \leq k$ we show that if $\beta(\mathbf{a}, * \cdots *) \in^* \Delta(\mathsf{Tail}(p))$ then $\beta(\mathbf{c}_i, * \cdots *) \in^* \Delta(\mathsf{Tail}(p_i))$ and if $\phi(\mathbf{a}) \in \Delta(\mathsf{Tail}(p))$ then $\phi(\mathbf{c}_i) \in \Delta(\mathsf{Tail}(\mathbf{c}_i))$.

For $i = 1$ in both cases nothing has to be shown. Now assume that the we have shown these properties up to i. W.l.o.g., assume p_{i+1} is a successor of p_i in the tree $\mathsf{Paths}(\mathbf{T})$. The other case is handled dually. There are two possibilities:

- $\mathsf{Tail}'(p_{i+1})$ is not blocked. Then $\mathsf{Tail}(p_{i+1}) = \mathsf{Tail}'(p_{i+1})$ and by the definition of $\sim$, $\mathsf{Tail}(p_{i+1})$ is a successor of $\mathsf{Tail}(p_i)$ in $\mathbf{T}$ and $\mathbf{c}_i = \mathbf{c}_{i+1}$ holds.
  If $\beta(\mathbf{a}, * \cdots *) \in^* \Delta(\mathsf{Tail}(p))$ then $\beta(\mathbf{c}_i, * \cdots *) \in^* \Delta(\mathsf{Tail}(p_i))$ holds by induction and due to the R$\updownarrow$-rule this implies $\beta(\mathbf{c}_{i+1}, * \cdots *) \in^* \Delta(\mathsf{Tail}(p_{i+1}))$.
  If $\phi(\mathbf{a}) \in \Delta(\mathsf{Tail}(p))$ then by induction $\phi(\mathbf{c}_i) \in \Delta(\mathsf{Tail}(\mathbf{c}_i))$ and due to the R$\updownarrow\forall$-rule this implies $\phi(\mathbf{c}_{i+1}) \in \Delta(\mathsf{Tail}(p_{i+1}))$.
- $\mathsf{Tail}'(p_{i+1})$ is blocked by $\mathsf{Tail}(p_{i+1})$ (with function π) and $\mathsf{Tail}'(p_{i+1})$ is a successor of $\mathsf{Tail}(p_i)$ in $\mathbf{T}$. Then, by definition of $\sim$ we have $\mathbf{c}_{i+1} = \pi(\mathbf{c}_i)$ and $\mathbf{c}_i \subseteq \mathbf{C}(\mathsf{Tail}(p_i)) \cap \mathbf{C}(\mathsf{Tail}'(p_{i+1}))$.
  If $\beta(\mathbf{a}, * \cdots *) \in^* \Delta(\mathsf{Tail}(p))$ then $\beta(\mathbf{c}_i, * \cdots *) \in^* \Delta(\mathsf{Tail}(p_i))$ holds by induction and due to the R$\updownarrow$-rule this implies $\beta(\mathbf{c}_i, * \cdots *) \in^* \Delta(\mathsf{Tail}'(p_i))$. $\mathsf{Tail}(p_{i+1})$ blocks $\mathsf{Tail}'(p_{i+1})$, which implies
  $$\pi(\beta(\mathbf{c}_i, * \cdots *)) = \beta(\mathbf{c}_{i+1}, * \cdots *) \in^* \Delta(\mathsf{Tail}(p_{i+1})).$$
  If $\phi(\mathbf{a}) \in \Delta(\mathsf{Tail}(p))$ then by induction $\phi(\mathbf{c}_i) \in \Delta(\mathsf{Tail}(\mathbf{c}_i))$ and due to the R$\updownarrow\forall$-rule this implies $\phi(c_i) \in \mathsf{Tail}'(p_{i+1})$. Since $\mathsf{Tail}(p_{i+1})$ blocks $\mathsf{Tail}'(p_{i+1})$, $\pi(\phi(\mathbf{c}_i)) = \phi(\mathbf{c}_{i+1}) \in \Delta(\mathsf{Tail}(p_{i+1}))$ holds.

We now define the structure $\mathfrak{A}$ over the universe $A = \mathbf{C}(\mathbf{T})/\approx$. For a relation $R \in \tau$ of arity m, $R^{\mathfrak{A}}$ is defined to be the set of tuples $([a_1, p_1]_\approx, \ldots, [a_m, p_m]_\approx)$ for which there exists a path $p \in \mathsf{Paths}(\mathbf{T})$ and constants $c_1, \ldots c_m$ such that $(c_i, p) \approx (a_i, p_i)$ and $Rc_1 \ldots c_m \in \Delta(\mathsf{Tail}(p))$.

It remains to show that this construction yields $\mathfrak{A} \models \psi$. This is a consequence of the following claim.

Claim 3.15 *For every path* $p \in \mathsf{Paths}(\mathbf{T})$ *and* $\mathbf{a} \subseteq \mathbf{C}(\mathsf{Tail}(p))$, *if* $\phi(\mathbf{a}) \in \Delta(\mathsf{Tail}(p))$, *then* $\mathfrak{A} \models \phi([\mathbf{a}, p]_\approx)$.

We show this claim by induction on the structure of ϕ. If $\phi(\mathbf{a}) = Ra_1 \ldots a_m \in \Delta(\mathsf{Tail}(p))$, then the claim holds immediately by construction of $\mathfrak{A}$.

Assume $\phi(\mathbf{a}) = \neg R\mathbf{a} \in \Delta(\mathsf{Tail}(p))$, but $[\mathbf{a}, p]_\approx \in R^{\mathfrak{A}}$. Then, by the definition of $\mathfrak{A}$, there must be a path p' and constants $\mathbf{c}$ such that $(\mathbf{a}, p) \approx (\mathbf{c}, p')$ and $R\mathbf{c} \in \Delta(\mathsf{Tail}(p'))$. From Claim 3.14 we have that $(\mathbf{a}, p) \approx (\mathbf{c}, p')$ implies $R\mathbf{a} \in \Delta(\mathsf{Tail}(p))$ and hence $\mathbf{T}$ contains the clash $\{R\mathbf{a}, \neg R\mathbf{a}\} \subseteq \Delta(\mathsf{Tail}(p))$, a contradiction to the fact that $\mathbf{T}$ is clash-free. Thus, $[\mathbf{a}, p]_\approx \notin R^{\mathfrak{A}}$.

Assume $\phi(\mathbf{a}) = a \neq b \in \Delta(\mathsf{Tail}(p))$ but $[a, p]_\approx = [b, p]_\approx$. From Claim 3.12 we get that this implies $a = b$ and hence $\mathbf{T}$ contains the clash $a \neq a \in \Delta(\mathsf{Tail}(p))$. Again, this is a contradiction to the fact that $\mathbf{T}$ is clash-free and $[a, p]_\approx \neq [b, p]_\approx$ must hold.

For positive Boolean combinations the claim is immediate due to the R$\wedge$- and R$\vee$-rule.

Let $\phi(\mathbf{a}) = (\forall \mathbf{yz}.\alpha(\mathbf{a}, \mathbf{y}, \mathbf{z}))\chi(\mathbf{a}, \mathbf{y}) \in \Delta(\mathsf{Tail}(p))$ and $\mathbf{b}, \mathbf{p}, \mathbf{c}, \mathbf{q}$ arbitrarily chosen with

(1) $\qquad \mathfrak{A} \models \alpha([\mathbf{a}, p]_\approx, [\mathbf{b}, \mathbf{p}]_\approx, [\mathbf{c}, \mathbf{q}]_\approx)$.

We need to show that also $\mathfrak{A} \models \chi([\mathbf{a}, p]_\approx, [\mathbf{b}, \mathbf{p}]_\approx)$ holds. In order to bring completeness of $\mathbf{T}$ and the R$\forall$-rule into play, we show that information about the fact that (1) holds is present at a *single* node in $\mathbf{T}$ where it triggers the R$\forall$-rule. We rely on the fact that universal quantifiers must be guarded.

Every $y_i \in \mathbf{y}$ coexists with every other variable $y_j \in \mathbf{y}$ in at least one atom $\beta^{(i,j)} \in \alpha(\mathbf{a}, \mathbf{y}, \mathbf{z})$ and with every element $a_\ell \in \mathbf{a}$ in at least one atom $\gamma^{(i,\ell)} \in \alpha(\mathbf{a}, \mathbf{y}, \mathbf{z})$. For any two distinct variables y_i, y_j, $\mathfrak{A} \models \beta^{(i,j)}([\mathbf{a}, p]_\approx, [\mathbf{b}, \mathbf{p}]_\approx, [\mathbf{c}, \mathbf{q}]_\approx)$ holds and this can only be the case if there is a path $q^{(i,j)}$ and constants $d^{(i,j)}, e^{(i,j)}$ such that $(b_i, p_i) \approx (c^{(i,j)}, q^{(i,j)})$ and $(b_j, p_j) \approx (d^{(i,j)}, q^{(i,j)})$.

Similarly, for every element $[b_i, p_i]_\approx \in [\mathbf{b}, \mathbf{p}]_\approx$ and every element (a_ℓ, p) there exists a path $r^{(i,\ell)}$ and constants $f^{(i,\ell)}, g^{(i,\ell)}$ such that

$(b_i, p_i) \approx (f^{(i,\ell)}, r^{(i,\ell)})$ and $(a_\ell, p) \approx (g^{(i,\ell)}, r^{(i,\ell)})$. For every i and ℓ, $\mathsf{Paths}([b_i, p_i]_\approx)$ and $\mathsf{Paths}([a_\ell, p]_\approx)$ are subtrees of $\mathsf{Paths}(\mathbf{T})$.

The tree $\mathsf{Paths}([b_i, p_i]_\approx)$ overlaps with the tree $\mathsf{Paths}([b_j, p_j]_\approx)$ at $q^{(i,j)}$ and with the tree $\mathsf{Paths}([a_\ell, p]_\approx)$ at $r^{(i,\ell)}$. From this it follows (Golumbic 1980) that there exists a common path

$$s \in \bigcap_i \mathsf{Paths}([b_i, p_i]_\approx) \cap \bigcap_\ell \mathsf{Paths}([a_\ell, p]_\approx).$$

Thus, there are tuples $\mathbf{a}'$, $\mathbf{b}'$ such that

$$(\mathbf{a}', s) \approx (\mathbf{a}, p) \text{ and } (\mathbf{b}', s) \approx (\mathbf{b}, \mathbf{p}). \tag{2}$$

We now show that the preconditions of the R$\forall$-rule are satisfied at $\mathsf{Tail}(s)$ for the formula $(\forall \mathbf{yz}.\alpha(\mathbf{a}', \mathbf{y}, \mathbf{z}))\chi(\mathbf{a}', \mathbf{y})$ and the tuple $\mathbf{b}'$. First, due to Claim 3.14, $(\forall \mathbf{yz}.\alpha(\mathbf{a}', \mathbf{y}, \mathbf{z}))\chi(\mathbf{a}', \mathbf{y}) \in \Delta(\mathsf{Tail}(s))$ holds because $(\mathbf{a}, p) \approx (\mathbf{a}', s)$ and $(\forall \mathbf{yz}.\alpha(\mathbf{a}, \mathbf{y}, \mathbf{z}))\chi(\mathbf{a}, \mathbf{y}) \in \Delta(\mathsf{Tail}(p))$.

For every $\beta(\mathbf{x}, \mathbf{y}, \mathbf{z}) \in \alpha(\mathbf{x}, \mathbf{y}, \mathbf{z})$, $\beta(\mathbf{a}', \mathbf{b}', * \cdots *) \in^* \Delta(\mathsf{Tail}(s))$ holds as follows: from (1,2) we get

$$\mathfrak{A} \models \beta([\mathbf{a}', s]_\approx, [\mathbf{b}', s]_\approx, [\mathbf{c}, \mathbf{q}]_\approx).$$

Since β is an atom, this implies the existence of a path t and tuples $\mathbf{a}'', \mathbf{b}'', \mathbf{c}'$ with

$$\begin{array}{c} (\mathbf{a}', s) \approx (\mathbf{a}'', t),\ (\mathbf{b}', s) \approx (\mathbf{b}'', t),\ (\mathbf{c}, \mathbf{q}) \approx (\mathbf{c}', t) \\ \text{and} \qquad \beta(\mathbf{a}'', \mathbf{b}'', \mathbf{c}') \in \Delta(\mathsf{Tail}(t)) \end{array} \tag{3}$$

It holds that $\beta(\mathbf{a}'', \mathbf{b}'', * \cdots *) \geq^* \beta(\mathbf{a}'', \mathbf{b}'', \mathbf{c}')$ and since $\beta(\mathbf{a}'', \mathbf{b}'', \mathbf{c}') \in \Delta(\mathsf{Tail}(t))$, $\beta(\mathbf{a}'', \mathbf{b}'', * \cdots *) \in^* \Delta(\mathsf{Tail}(t))$. Thus, by Claim 3.14 it holds that $\beta(\mathbf{a}', \mathbf{b}', * \cdots *) \in^* \Delta(\mathsf{Tail}(s))$.

Since this is true for every atom β, the preconditions of the R$\forall$-rule are satisfied and the completeness of $\mathbf{T}$ yields $\chi(\mathbf{a}', \mathbf{b}') \in \Delta(\mathsf{Tail}(s))$. By induction, $\mathfrak{A} \models \chi([\mathbf{a}', s]_\approx, [\mathbf{b}', s]_\approx)$ holds and together with (2) this implies $\mathfrak{A} \models \chi([\mathbf{a}, p]_\approx, [\mathbf{b}, \mathbf{p}]_\approx)$. Since $\mathbf{a}, \mathbf{p}, \mathbf{c}, \mathbf{q}$ have been chose arbitrarily, $\mathfrak{A} \models \phi([\mathbf{a}, p]_\approx)$ holds.

If $\phi(\mathbf{a}) = (\exists \mathbf{yz}.\alpha(\mathbf{a}, \mathbf{y}, \mathbf{z}))\chi(\mathbf{a}, \mathbf{y}) \in \Delta(\mathsf{Tail}(p))$, there are two possibilities:

- there are $\mathbf{b}, \mathbf{c} \subseteq \mathbf{C}(\mathsf{Tail}(p))$ such that $\{\alpha(\mathbf{a}, \mathbf{b}, \mathbf{c})\} \subseteq \Delta(\mathsf{Tail}(p))$ and $\chi(\mathbf{a}, \mathbf{b}) \in \Delta(\mathbf{a}, \mathbf{b})$. Then by induction we have

$$\mathfrak{A} \models \{\alpha([\mathbf{a}, p]_\approx, [\mathbf{b}, p]_\approx, [\mathbf{c}, p]_\approx), \chi([\mathbf{a}, p]_\approx, [\mathbf{b}, p]_\approx)\}$$

  and hence $\mathfrak{A} \models \phi([\mathbf{a}, p]_\approx)$.

- there are no such $\mathbf{b}, \mathbf{c} \subseteq \mathbf{C}(\mathsf{Tail}(p))$, then there is a successor w of $\mathsf{Tail}(p)$ and $\mathbf{b}, \mathbf{c} \subseteq \mathbf{C}(w)$ with $\{\alpha(\mathbf{a}, \mathbf{b}, \mathbf{c}), \chi(\mathbf{a}, \mathbf{b})\} \subseteq \Delta(w)$. The node w can be blocked or not.

If w is not blocked, then $p' = [p, \frac{w}{w}] \in \mathsf{Paths}(\mathbf{T})$ and by induction

$$\mathfrak{A} \models \{\alpha([\mathbf{a}, p']_{\approx}, [\mathbf{b}, p']_{\approx}, [\mathbf{c}, p']_{\approx}), \chi([\mathbf{a}, p']_{\approx}, [\mathbf{b}, p']_{\approx})\}.$$

From the definition of $\approx$ we have, $(\mathbf{a}, p') \approx (\mathbf{a}, p)$ and hence $\mathfrak{A} \models \phi([\mathbf{a}, p]_{\approx})$.

If w is blocked by a node u (with function π) then $p' = [p, \frac{u}{w}] \in \mathsf{Paths}(\mathbf{T})$. From the blocking condition, we have that u is unblocked and $\pi\{\alpha(\mathbf{a}, \mathbf{b}, \mathbf{c}), \chi(\mathbf{a}, \mathbf{b})\}) \subseteq \Delta(u)$. Hence, by induction

$$\mathfrak{A} \models \{ \begin{array}{l} \alpha([\pi(\mathbf{a}), p']_{\approx}, [\pi(\mathbf{b}), p']_{\approx}, [\pi(\mathbf{c}), p']_{\approx}), \\ \chi([\pi(\mathbf{a}), p']_{\approx}, [\pi(\mathbf{b}), p']_{\approx}) \end{array} \},$$

and by definition of $\approx$ we have that $(\mathbf{a}, p) \approx (\pi(\mathbf{a}), p')$ and hence, $\mathfrak{A} \models \phi([\mathbf{a}, p]_{\approx})$.

As a special instance of Claim 3.15 we get that $\mathfrak{A} \models \psi$. From Lemma 3.6, we get that, for every node $v \in \mathbf{V}$, $|\mathbf{C}(v)| \leq \text{width}(\psi)$ and hence the tree $\mathsf{Paths}(\mathbf{T})$ together with the function $f : \mathsf{Paths}(\mathbf{T}) \to \mathbf{C}(\mathbf{T})$ with $f(p) = \mathbf{C}(\mathsf{Tail}(p))$ provides a tree decomposition of $\mathfrak{A}$ of width $\leq \text{width}(\psi) - 1$. This completes the proof of Lemma 3.11. ⊣

Corollary 3.16 CGF, *and hence also* LGF *and* GF *have the generalised tree model property.*

Proof: Let $\psi \in \text{CGF}[\tau]$ be satisfiable. Then, from Lemma 3.8 we get that there is a tableau $\mathbf{T}$ for ψ. By Lemma 3.11, $\mathbf{T}$ induces a model for ψ of tree width at most $\text{width}(\psi) - 1$. Note that we have never relied on Fact 3.2 to obtain any of the results in this paper and hence have indeed given an alternative proof for the generalised tree model property of CGF. For LGF and GF, observe that the embedding of these logics into CGF may increase the width of the sentence but not by more than a recursive amount. ⊣

4 Conclusion

We have developed a tableau algorithm for CGF, which we hope can serve as basis for an efficient implementation of a decision procedure for CGF. This hope is justified by the fact that some of the most efficient implementations of modal or description logic reasoners are based on tableau calculi similar to the one for CGF presented in this paper. As a corollary from the constructions used to prove the correctness we obtain a new proof of the fact that GF/LGF/CGF have the generalised tree model property.

Acknowledgements

We would like to thank Andrei Voronkov and an anonymous reviewer for helpful suggestions. The second author is supported by the DFG, Project No. GR 1324/3–1.

References

Andréka, H., J. van Benthem, and I. Németi. 1998. Modal Languages and Bounded Fragments of Predicate Logic. *Journal of Philosophical Logic* 27:217–274.

Donini, F. M., M. Lenzerini, D. Nardi, and W. Nutt. 1997. The Complexity of Concept Languages. *Information and Computation* 134(1):1–58.

Ganzinger, H., and H. de Nivelle. 1999. A Superposition Decision Procedure for the Guarded Fragment with Equality. In *Proc. 14th IEEE Symp. on Logic in Computer Science*, 295–303.

Golumbic, M. 1980. *Algorithmic Graph Theory and Perfect Graphs.* New York, NY: Academic Press.

Grädel, E. 1999a. Decision procedures for guarded logics. In *Automated Deduction - CADE16. Proceedings of 16th International Conference on Automated Deduction, Trento, 1999.* Lecture Notes in Artificial Intelligence, No. 1632. Springer-Verlag.

Grädel, E. 1999b. On the restraining power of guards. *Journal of Symbolic Logic* 64(4):1719–1742.

Grädel, E., and I. Walukiewicz. 1999. Guarded Fixed Point Logic. In *Proc. 14th IEEE Symp. on Logic in Computer Science*, 45–54.

Halpern, J. Y., and Y. Moses. 1992. A Guide to Completeness and Complexity for Model Logics of Knowledge and Belief. *Artificial Intelligence* 54(3):319–379.

Horrocks, I., P. F. Patel-Schneider, and R. Sebastiani. 2000. An Analysis of Empirical Testing for Modal Decision Procedures. *Logic Journal of the IGPL* 8(3):293–323.

Horrocks, I., U. Sattler, and S. Tobies. 1999. Practical Reasoning for Expressive Description Logics. In *Proceedings of the 6th International Conference on Logic for Programming and Automated Reasoning (LPAR'99)*, ed. H. Ganzinger, D. McAllester, and A .Voronkov, 161–180. Lecture Notes in Artificial Intelligence, No. 1705. Springer-Verlag, September.

Ladner, R. 1977. The computational complexity of provability in systems of propositional modal logic. *SIAM Journal on Computing* 6:467–480.

van Benthem, J. 1997. Dynamic bits and pieces. ILLC Research Report LP-97-01. University of Amsterdam.

Vardi, M. 1997. Why is modal logic so robustly decidable? In *Descriptive Complexity and Finite Models*, ed. N. Immerman and P. Kolaitis. DIMACS Series in Discrete Mathematics and Theoretical Computer Science, Vol. 31, 149–184. AMS.

Weidenbach, C. 1997. SPASS—Version 0.49. *J. of Automated Reasoning* 18(2):247–252.

Colin Hirsch
Mathematische Grundlagen der Informatik, RWTH Aachen
Ahornstr. 55
52064 Aachen
Germany
E-mail: `hirsch@cs.rwth-aachen.de`
URL: `http://www-mgi.informatik.rwth-aachen.de/~hirsch/`

Stephan Tobies
LuFG Theoretical Computer Science, RWTH Aachen
Ahronstr. 55
52064 Aachen
Germany
E-mail: `tobies@cs.rwth-aachen.de`
URL: `http://www-lti.informatik.rwth-aachen.de/~tobies/`

15

Epistemic Logic of Shallow Depths and Game Theoretical Applications

MAMORU KANEKO AND NOBU-YUKI SUZUKI

ABSTRACT. This paper presents a multi-modal epistemic logic GL_{EF} of shallow depths. Logic GL_{EF} is a fragment of multi-modal KD^n, and enables us to consider interpersonal inferential epistemic complexities. The subscripts E and F are called epistemic structures, which restrict interpersonal epistemic depths involved in formulae and proofs. We give its Gentzen-type sequent formulation as well as its semantics. Also, we provide some meta-theorems in order to evaluate interpersonal inferential epistemic complexities of the provability of a given sequent. These meta-theorems are applied to some game theoretical problems.

1 Introduction

Decision making in a game needs *interpersonal* epistemic introspections as well as *intrapersonal* ones. Although the latter is certainly important, we focus on the former in this paper. In particular, we discuss epistemic complexities of interpersonal inferences and their applications to game theoretical problems.

Multi-modal epistemic logics have shown to be useful for considerations of game theoretical decision making. In literature, various multi-modal epistemic logics have been developed, including products and fusions (cf. Gabbay and Shehtman (1998) and Wolter (1998)) as well as common knowledge extensions (for recent textbooks, see Fagin et al. (1995) and Meyer and van der Hoek (1995)). These extensions already involve common knowledge to a certain degree. That is, arbitrarily finite depths and/or infinite depths of interpersonal introspections are involved. On the other hand, human interpersonal epistemic introspec-

Advances in Modal Logic, Volume 3
F. Wolter, H. Wansing, M. de Rijke, and M. Zakharyaschev, eds.

tions typically stop at very shallow levels. It is the purpose of our research project (Kaneko and Suzuki (1999a), (1999b)) to take this limitation of human reasoning seriously and to develop logical systems so that they reflect limitations on interpersonal epistemic introspections. In this paper, we are interested particularly in limitations of interpersonal inferential complexities.

To make our motivation more specific, we discuss a difficulty involved, by referring to epistemic logic KD^n. Logic KD^n is defined as Classical Logic + Axiom K: $\mathrm{B}_i(A \supset C) \supset (\mathrm{B}_i(A) \supset \mathrm{B}_i(C))$; Axiom D: $\neg\mathrm{B}_i(\neg A \wedge A)$; and Necessitation Rule: if A is provable, so is $\mathrm{B}_i(A)$, where $\mathrm{B}_1, ..., \mathrm{B}_n$ are the *belief operators* of players $1, ..., n$.[1] In KD^n, formulae having nested occurrences of $\mathrm{B}_1, ..., \mathrm{B}_n$ in any depths are allowed, and the Necessitation Rule may be applied arbitrarily many times in proofs. In this sense, KD^n involves already common knowledge, though a common knowledge extension of KD^n is needed to capture the entirety of common knowledge. Rather than considering extensions of KD^n, we are interested in fragments of KD^n. For meaningful considerations of such fragments, we have needed to develop proof-theoretic and model-theoretic apparatuses, which are undertaken in Kaneko and Suzuki (1999a, 1999b). By these developments, we are now able to consider suitable fragments of KD^n having finite depths of nested occurrences of belief operators.

In considering such a fragment of KD^n, we make a distinction between *descriptive epistemic structure* E and *inferential epistemic structure* F. To illustrate this distinction, consider two provable formulae $\mathrm{B}_1(p) \supset \mathrm{B}_1(p)$ and $\mathrm{B}_1(p \supset p)$ in KD^n. Both include B_1 without nesting. Nevertheless, the first is provable without the Necessitation Rule, but a proof of the second requires it. These examples suggest to distinguish between the structure of occurrences of $\mathrm{B}_1, ..., \mathrm{B}_n$ in formulae and the applications of the Necessitation Rule for interpersonal inferences.

The restrictions mentioned above are formulated by means of *epistemic structures* E and F, which are the subscripts of epistemic logic GL_{EF} of shallow depths. These E and F enable us to capture the difference pointed out above between $\mathrm{B}_1(p) \supset \mathrm{B}_1(p)$ and $\mathrm{B}_1(p \supset p)$ in the way that the same E suffices for the description of formulae but their derivations require different inferential F's.

[1] In the literature of epistemic logics, systems $\mathrm{S4}^n$ and $\mathrm{S5}^n$ are typically discussed (cf., Hintikka (1962), Fagin et al. (1995), Meyer and van der Hoek (1995)). In these systems, Axiom T: $\mathrm{B}_i(A) \supset A$ is assumed, i.e., beliefs are true from the viewpoint of the outside thinker. If we replace Axiom T by Axiom D, we obtain system $\mathrm{KD4}^n$. Without Axiom T, we could discuss *false* beliefs, which would be important in game theoretical applications. Although we adopt the KD-type logic to avoid some complications caused by Axiom 4, our results hold with some modifications for the KD4-type logic of shallow depths. The S4-type logic can be discussed inside the KD4-type logic.

Epistemic structures E and F define our logical system GL_{EF}. The former, E, restricts admissible formulae, and the latter, F, does admissible proofs. As suggested by the above example, it may be the case that some provable formula has large depths of nested occurrences of belief operators but needs only shallow depths for its proof. From the viewpoint of considering human reasoning, it is more important to give restrictions on the latter F.

Kaneko and Suzuki (1999a, 1999b) have developed logical systems GL_{EF} and their semantics without differentiating F from E. Nevertheless, their developments can be modified without much difficulty to allow F different from E. Therefore, we can borrow their basic results.

The purpose of limiting interpersonal introspections leads us to consider a *minimal* inferential epistemic structure F for a (provable) formula. For example, the outside investigator (observer) can derive $\mathrm{B}_1(p) \supset \mathrm{B}_1(p)$ without reading player 1's mind, while he needs to assume some logical ability for player 1 to derive $\mathrm{B}_1(p \supset p)$. The focus of this paper is to give some results to facilitate the considerations of minimal inferential epistemic structures. These results will be applied to some game theoretical problems. Extensive discussions on game theoretical applications are given in Kaneko and Suzuki (2001).

2 Set of Formulae $\mathcal{P}$ and Epistemic Structures

We adopt the following list of primitive symbols:

propositional variables: $p, q, r, ...$;

logical connective symbols: $\neg$ (not), $\supset$ (implies), $\bigwedge$ (and), $\bigvee$ (or) ;

unary belief operator symbols: B_1, $\mathrm{B}_2, ..., \mathrm{B}_n$;

parentheses: (,); *comma*: , and *braces* $\{,\}$.

The set of propositional variables is denoted by PV. Based on this list of symbols, we define *formulae* inductively in the standard manner, except that $\bigwedge$ and $\bigvee$ are applied to a finite nonempty set Φ of formulae.[2] We denote the set of all formulae by $\mathcal{P}$. The subscripts of $\mathrm{B}_1, ..., \mathrm{B}_n$ are the names of players (agents), and the set $\{1, ..., n\}$ is denoted by N. We say that a formula A is *nonepistemic* if no B_i occurs in A for $i \in N$.

The notions of the epistemic depths of formulae and epistemic structures play central roles in our discussion of an epistemic logic of shallow depths. First, let $N^{<\omega} := \{(i_1, ..., i_m) : i_1, ..., i_m \in N \text{ and } m \geq 0\}$, where

[2]Hence, we need commas and braces as primitive symbols. Also, we need a stipulation of identifying a finite set.

This deviates slightly from the standard formulation of formulae, but this makes certain game theoretical considerations simpler.

$N^{<\omega}$ contains the null sequence ϵ, i.e., the sequence of length 0. For $e = (i_1, ..., i_m) \in N^{<\omega}$, $B_{i_1}...B_{i_m}(A)$ is denoted by $B_e(A)$, and $B_\epsilon(A)$ is stipulated to be A. We define the concatenation: $e \circ e' = (i_1, ..., i_m, j_1, ..., j_k)$ for $e = (i_1, ..., i_m)$, $e' = (j_1, ..., j_k) \in N^{<\omega}$, and stipulate $e \circ \epsilon = \epsilon \circ e = e$. We write $(i) \circ e$ and $e \circ (i)$ as $i \circ e$ and $e \circ i$, respectively.

We define the *epistemic depth* $\delta(A)$ of a formula A inductively by:

D0: $\delta(p) = \{\epsilon\}$ for any $p \in PV$;

D1: $\delta(\neg C) = \delta(C)$;

D2: $\delta(C \supset D) = \delta(C) \cup \delta(D)$;

D3: $\delta(\bigwedge \Phi) = \delta(\bigvee \Phi) = \bigcup_{C \in \Phi} \delta(C)$;

D4: $\delta(B_i(C)) = \{i \circ e : e \in \delta(C)\}$.

Then $\delta(A) \subseteq N^{<\omega}$. For example, $\delta(p \supset B_2B_3(q)) = \{\epsilon, (2,3)\}$, where $p, q \in PV$. We define $\delta(\Gamma) = \bigcup_{C \in \Gamma} \delta(C)$ for a set Γ of formulae.

We say that a nonempty subset E of $N^{<\omega}$ is an *epistemic structure* if

(1) $$(i_1, ..., i_m) \in E \text{ implies } (i_1, ..., i_{m-1}) \in E.$$

When $m = 0$, $(i_1, ..., i_m)$ is the null symbol ϵ. Trivial examples of epistemic structure are $N^{<\omega}$ and $\{\epsilon\}$. Less trivial examples are $\{\epsilon, (1), (2)\}$ and $\{\epsilon, (2), (2, 1)\}$.

We define $\mathcal{P}_E = \{A \in \mathcal{P} : \delta(A) \subseteq E\}$. For example, any formula in $\mathcal{P}_{\{\epsilon,(1),(2)\}}$ may have B_1 and B_2 without their nested occurrences. Since the null symbol ϵ always belongs to E, all the nonepistemic formulae are included in $\mathcal{P}_E$. For $e \in E$, we define $\mathcal{P}_E(e) = \{C : e \circ \delta(C) \subseteq E\} = \{C : e \circ e' \in E \text{ for all } e' \in \delta(C)\}$. Also, we use the convention to write $B_i(\Phi)$ for $\{B_i(A) : A \in \Phi\}$.

3 Epistemic Logic GL_{EF} of Shallow Depths

In this paper, we give our logical system GL_{EF} in the Gentzen-style sequent form.

Let E and F be two epistemic structures with $F \subseteq E$, where E and F will be used as *descriptive* and *inferential* epistemic structures. We give restrictions in terms of E and F, respectively, on admissible formulae and on admissible proofs.

To facilitate considerations of interpersonal inferential interactions, we introduce a thought sequent. Let $e = (i_1, ..., i_m) \in E$, and Γ, Θ finite subsets of $\mathcal{P}_E$. Using auxiliary symbols [,], and $\to$, we introduce a new expression $B_e[\Gamma \to \Theta] := B_{i_1}...B_{i_m}[\Gamma \to \Theta]$, which we call a *thought sequent*. As in the standard sequent calculus, $\Gamma \to \Theta$ is intended to be

$\bigwedge\Gamma \supset \bigvee\Theta$, where $\bigwedge\emptyset$ and $\bigvee\emptyset$ are $\neg p \bigvee p$ and $\neg p \bigwedge p$ respectively. We say that a thought sequent $\mathrm{B}_e[\Gamma \to \Theta]$ is *admissible in* E if $e \circ \delta(\Gamma \cup \Theta) := \{e \circ e' : e' \in \delta(\Gamma \cup \Theta)\} \subseteq E$. Hence the formulae in Γ and Θ are in $\mathcal{P}_E(e)$. When $e = \epsilon$, we abbreviate $\mathrm{B}_\epsilon[\Gamma \to \Theta]$ as $\Gamma \to \Theta$. We abbreviate set-theoretical parentheses, for example, $\mathrm{B}_e[\{A\} \cup \Gamma \to \Theta \cup \{B\}]$ is written as $\mathrm{B}_e[A, \Gamma \to \Theta, B]$.

The outer $\mathrm{B}_e[\cdots] = \mathrm{B}_{i_1} \ldots \mathrm{B}_{i_m}[\cdots]$ expresses the viewpoint of the player to conduct logical deductions. For example, when $m = 2$, $\mathrm{B}_{i_1}\mathrm{B}_{i_2}[\cdots]$ expresses the idea that player i_2 in the scope of player i_1 conducts logical deductions. In particular, $\mathrm{B}_\epsilon[\cdots]$ corresponds to the viewpoint of the investigator (observer).

The logical deductions of player i_m in $e = (i_1, ..., i_m)$ are governed by one axiom schema and various inference rules.

Axiom (Initial Sequent): $\mathrm{B}_e[A \to A]$,

Structural Rules:

$$\frac{\mathrm{B}_e[\Gamma \to \Theta]}{\mathrm{B}_e[\Delta, \Gamma \to \Theta, \Lambda]} \text{(Th)} \qquad \frac{\mathrm{B}_e[\Gamma \to \Theta, A] \quad \mathrm{B}_e[A, \Delta \to \Lambda]}{\mathrm{B}_e[\Gamma, \Delta \to \Theta, \Lambda]} \text{(Cut)}$$

Operational Rules:

$$\frac{\mathrm{B}_e[\Gamma \to \Theta, A]}{\mathrm{B}_e[\neg A, \Gamma \to \Theta]} (\neg \to) \qquad \frac{\mathrm{B}_e[A, \Gamma \to \Theta]}{\mathrm{B}_e[\Gamma \to \Theta, \neg A]} (\to \neg)$$

$$\frac{\mathrm{B}_e[\Gamma \to \Theta, A] \quad \mathrm{B}_e[B, \Gamma \to \Theta]}{\mathrm{B}_e[A \supset B, \Gamma \to \Theta]} (\supset\to) \qquad \frac{\mathrm{B}_e[A, \Gamma \to \Theta, B]}{\mathrm{B}_e[\Gamma \to \Theta, A \supset B]} (\to\supset)$$

$$\frac{\mathrm{B}_e[A, \Gamma \to \Theta]}{\mathrm{B}_e[\bigwedge\Phi, \Gamma \to \Theta]} (\bigwedge \to) \qquad \frac{\{\mathrm{B}_e[\Gamma \to \Theta, A] : A \in \Phi\}}{\mathrm{B}_e[\Gamma \to \Theta, \bigwedge\Phi]} (\to \bigwedge)$$

$$\frac{\{\mathrm{B}_e[A, \Gamma \to \Theta] : A \in \Phi\}}{\mathrm{B}_e[\bigvee\Phi, \Gamma \to \Theta]} (\bigvee \to) \qquad \frac{\mathrm{B}_e[\Gamma \to \Theta, A]}{\mathrm{B}_e[\Gamma \to \Theta, \bigvee\Phi]} (\to \bigvee)$$

where $A \in \Phi$ in $(\bigwedge \to)$ and $(\to \bigvee)$.

Epistemic Rule (Distribution Rule):

$$\frac{\mathrm{B}_{e \circ i}[\Gamma \to \Theta]}{\mathrm{B}_e[\mathrm{B}_i(\Gamma) \to \mathrm{B}_i(\Theta)]} (\mathrm{B}_i \to \mathrm{B}_i), \text{ subject to } |\Theta| \leq 1,$$

where $|\Theta|$ is the cardinality of Θ.

The outer $B_e[\cdots]$'s of the upper and lower thought sequents in each of the structural and operational rules are identical, and these rules are the same as those in classic logic LK if the outer $B_e[\cdots]$ is ignored. That is, the innermost player i_m in $e = (i_1, ..., i_m)$ is capable of conducting logical deductions described by classical logic. The outer $B_e[\cdots]$ changes only at $(B_{i_m} \to B_{i_m})$, and the innermost i_m goes into the scope of $B_{(i_1,...,i_{m-1})}[\cdots]$. The length of $e = (i_1, ..., i_m)$ of the outer $B_e[\cdots]$ gets shorter at an occurrence of $(B_{i_m} \to B_{i_m})$ in a proof.

If we delete the restriction on Θ having at most one formula, then we would have the negative introspection, i.e., the system would be a KD5-type. In this paper, we consider only the KD-type GL_{EF}.

We say that $B_e[\Gamma \to \Theta]$ is *provable in* GL_{EF} if there is a proof P of $B_e[\Gamma \to \Theta]$ such that

(1): all thought sequents in P are admissible in E;

(2): e' belongs to F for any thought sequent $B_{e'}[\Delta \to \Lambda]$ in P.

We write $\vdash_{EF} B_e[\Gamma \to \Theta]$ if it is provable in GL_{EF}. Recall that when $e = \epsilon$, we abbreviate the outer $B_\epsilon[\cdots]$, that is, $\vdash_{EF} B_\epsilon[\Gamma \to \Theta]$ is written as $\vdash_{EF} \Gamma \to \Theta$. We say that A is *provable in* GL_{EF} if $\vdash_{EF} \to A$.

We say that an inferential epistemic structure F is *minimal for* $B_e[\Gamma \to \Theta]$ if $\vdash_{EF} B_e[\Gamma \to \Theta]$ but $\nvdash_{EF'} B_e[\Gamma \to \Theta]$ for any epistemic structure $F' \subsetneqq F$. A minimal F may not be unique.

The examples $B_1(p) \supset B_1(p)$ and $B_1(p \supset p)$ of Section 1 are proved as follows:

$$\frac{B_1(p) \to B_1(p)}{\to B_1(p) \supset B_1(p)}\ (\to\supset) \qquad \frac{\dfrac{B_1[p \to p]}{B_1[\ \to p \supset p]}\ (\to\supset)}{\to B_1(p \supset p)}\ (B_1 \to B_1).$$

The first is a proof in $GL_{\{\epsilon,(1)\}\{\epsilon\}}$, and the second is in $GL_{\{\epsilon,(1)\}\{\epsilon,(1)\}}$. In the first proof, the investigator derives $\to B_1(p) \supset B_1(p)$ without considering deductions by player 1, while in the second, the investigator once assumes that 1 conducts logical deductions. In the first example, it is easy to see that $F = \{\epsilon\}$ is minimal. However, we need a certain result to prove the minimality of $F = \{\epsilon, (1)\}$ for $\to B_1(p \supset p)$. It follows from this minimality, using the cut-elimination theorem given below, that $\to B_1(p \supset p) \bigvee B_2(p \supset p)$ has two minimal F's, i.e., $\{\epsilon, (1)\}$ and $\{\epsilon, (2)\}$.

Since these are too simple to see the structure of GL_{EF}, we give slightly more complex examples. The minimality of $F = \{\epsilon, (1)\}$ for $\to B_1(p \supset p)$ will be discussed as part of Example 3.1.(2).

Example 3.1 (1): The sequent $B_2B_1(p), B_2B_1(q) \to B_2(B_1(p) \bigwedge B_1(q))$ is admissible in $E = \{\epsilon, (2), (2,1)\}$. One possible choice of an inferential epistemic structure is $F = \{\epsilon, (2)\}$. A proof of this sequent is given:

$$\dfrac{\dfrac{\dfrac{B_2[B_1(p) \to B_1(p)]}{B_2[B_1(p), B_1(q) \to B_1(p)]}\text{(Th)} \quad \dfrac{B_2[B_1(q) \to B_1(q)]}{B_2[B_1(p), B_1(q) \to B_1(q)]}\text{(Th)}}{B_2[B_1(p), B_1(q) \to B_1(p) \wedge B_1(q)]}(\to \wedge)}{B_2B_1(p), B_2B_1(q) \to B_2(B_1(p) \wedge B_1(q))}(B_2 \to B_2)$$

This proof can be extended so that F coincides with E, for example, we add $(B_1 \to B_1)$ to the uppermost thought sequent $B_2[B_1(p) \to B_1(p)]$. Nevertheless, we are interested in a minimal inferential epistemic structure F, as mentioned in Section 1.

(2): The sequent $\to B_2B_1(p \supset p)$ needs also the descriptive epistemic structure $E = \{\epsilon, (2), (2,1)\}$. The following is a proof of $\to B_2B_1(p \supset p)$ in GL_{EF} with $F = E$.

$$\dfrac{\dfrac{\dfrac{B_2B_1[p \to p]}{B_2B_1[\ \to (p \supset p)]}(\to\supset)}{B_2[\ \to B_1(p \supset p)]}(B_1 \to B_1)}{\to B_2B_1(p \supset p)}(B_2 \to B_2)$$

In this case, the investigator assumes that player 2 reads the mind of player 1. The provability of this sequent is considerably different from that of $\to B_2B_1(p) \supset B_2B_1(p)$ in that the latter does not need to assume that either player reads the other's mind. This difference explains why we are interested in a minimal inferential epistemic structure. In the case of $\to B_2B_1(p \supset p)$, we will show below in this section that $F = \{\epsilon, (2), (2,1)\}$ is a minimal inferential epistemic structure.

Logic GL_{EF} enjoys cut-elimination, which is a basic proof-theoretic property of GL_{EF}.

Theorem 3.2 (Cut-Elimination) *If* $\vdash_{EF} \Gamma \to \Lambda$, *then there is a cut-free proof* P *of* $\Gamma \to \Lambda$ *in* GL_{EF}.

Allowing the outer $B_e[\cdots]$'s is the main difference from this cut-elimination theorem from the original one for Gentzen's LK. For the treatment of the outer $B_e[\cdots]$'s, see Kaneko and Nagashima (1997).

Sequent calculus KD^n is obtained from the above system by deleting all the outer $B_e[\cdots]$ from the axiom and inference rules, and by eliminating the restrictions on the admissibility of thought sequents in proofs. This sequent calculus, KD^n, is known to have the cut-elimination property. Using this cut-elimination, we can have the following theorem.

Theorem 3.3 (Relation to KD^n) *$\Gamma \to \Lambda$ is provable in KD^n if and only if $\vdash_{EE} \Gamma \to \Lambda$, where E is any epistemic structure including $\delta(\Gamma \cup \Lambda)$.*

Thus, the most relevant E is the epistemic structure generated by $\delta(\Gamma \cup \Lambda)$, and $F = E$ is large enough for $\Gamma \to \Lambda$. The above examples of proofs state that it may be possible to choose a smaller F than E.

By the cut-elimination theorem, we can prove that $F = \{\epsilon, (2)\}$ and $F = \{\epsilon, (2), (2,1)\}$ are the minimal inferential epistemic structures, respectively, in Example 3.1.(1) and (2). For this kind of purposes it would be useful to develop certain meta-theorems.

The following is a key theorem to evaluate an inferential epistemic structure F.

Theorem 3.4 (Epistemic Inferences) *Let Γ be a finite set of formulae with $\mathrm{B}_i(\Gamma) \cup \{\mathrm{B}_i(A)\} \subseteq \mathcal{P}_E$.*

(1): If $\vdash_{EF} \mathrm{B}_i(\Gamma) \to$ or $\vdash_{EF} \to \mathrm{B}_i(A)$, then $(i) \in F$.

(2): If $\vdash_{EF} \mathrm{B}_i(\Gamma) \to \mathrm{B}_i(A)$ and $A \notin \Gamma$, then $(i) \in F$.

(3): Let F be a minimal inferential epistemic structure for $\mathrm{B}_i(\Gamma) \to \mathrm{B}_i(A)$. Then $A \in \Gamma$ implies $F = \{\epsilon\}$.

(4): If $F = \{\epsilon\}$ and $\vdash_{EF} \mathrm{B}_i(\Gamma) \to \mathrm{B}_i(A)$, then $A \in \Gamma$.

Proof. (1): Suppose $\vdash_{EF} \to \mathrm{B}_i(A)$. Then there is a cut-free proof P of $\to \mathrm{B}_i(A)$ by Theorem 3.2. Then the lowermost inference of P is $(\mathrm{B}_i \to \mathrm{B}_i)$. The upper thought sequent of this inference is $\mathrm{B}_i[\ \to A]$. Hence $(i) \in F$. It is proved in the same way that if $\vdash_{EF} \mathrm{B}_i(\Gamma) \to$, then $(i) \in F$. (2): By (1), we can assume that $\nvdash_{EF} \mathrm{B}_i(\Gamma) \to$ and $\nvdash_{EF} \to \mathrm{B}_i(A)$. Since $\vdash_{EF} \mathrm{B}_i(\Gamma) \to \mathrm{B}_i(A)$, there is a cut-free proof P of $\mathrm{B}_i(\Gamma) \to \mathrm{B}_i(A)$ in GL_{EF} by Theorem 3.2. Then the lowermost inference must be either (Th) or $(\mathrm{B}_i \to \mathrm{B}_i)$. If this is $(\mathrm{B}_i \to \mathrm{B}_i)$, then $(i) \in F$. Consider the former case, i.e.,

$$\frac{\mathrm{B}_i(\Gamma') \to \mathrm{B}_i(A)}{\mathrm{B}_i(\Gamma) \to \mathrm{B}_i(A)}\ (\mathrm{Th}) \quad \text{or} \quad \frac{\mathrm{B}_i(\Gamma') \to}{\mathrm{B}_i(\Gamma) \to \mathrm{B}_i(A)}\ (\mathrm{Th}),$$

where Γ' is a nonempty subset of Γ. Then $A \notin \Gamma'$ by $A \notin \Gamma$. The latter case is excluded by the assumption $\nvdash_{EF} \mathrm{B}_i(\Gamma) \to$. Consider the former case. The second lowermost inference must be again either (Th) or $(\mathrm{B}_i \to \mathrm{B}_i)$. If this is $(\mathrm{B}_i \to \mathrm{B}_i)$, then $(i) \in F$. If it is (Th), then we consider the 3rd lowermost inference. We repeat this argument. Then we meet $(\mathrm{B}_i \to \mathrm{B}_i)$ or an initial sequent. If the uppermost sequent was the initial sequent, it must be $B_i(A) \to B_i(A)$, which is impossible by $A \notin \Gamma$. Hence we meet $(B_i \to B_i)$, which implies $(i) \in F$.

(3): Let $A \in \Gamma$. Then we have

$$\frac{B_i(A) \to B_i(A)}{B_i(\Gamma) \to B_i(A)} \text{(Th)}.$$

This means $\vdash_{E\{\epsilon\}} B_i(\Gamma) \to B_i(A)$. Since F is a minimal epistemic structure for $B_i(\Gamma) \to B_i(A)$, we have $F = \{\epsilon\}$.

(4): This follows (2). $\dashv$

Let us see the minimality of $F = \{\epsilon, (2), (2,1)\}$ for $\vdash_{EF} \to B_2B_1(p \supset p)$ in Example 3.1.(2), where $E = \{\epsilon, (2), (2,1)\}$. Applying Theorem 3.4.(1) to this sequent, we have $(2) \in F$. It remains to show $(2,1) \in F$. For this step, the following two lemmas are useful.

Lemma 3.5 *Consider a sequent* $B_i(\Gamma) \to B_i(\Theta)$ *with* $|\Theta| \le 1$. *If* $\vdash_{EF} B_i(\Gamma) \to B_i(\Theta)$ *and* $(i) \in F$, *then* $\vdash_{EF} B_i[\Gamma \to \Theta]$.

Proof. Suppose $\vdash_{EF} B_i(\Gamma) \to B_i(\Theta)$. If $\Theta = \{A\} \subseteq \Gamma$, then $\vdash_{EF} B_i[A \to A]$, which implies $\vdash_{EF} B_i[\Gamma \to \Theta]$. Now, let $\Theta \cap \Gamma = \emptyset$. Then there is a cut-free proof P of $B_i(\Gamma) \to B_i(\Theta)$ in GL_{EF}. Then the lowermost inference is either (Th) or $(B_i \to B_i)$. If it is $(B_i \to B_i)$, then the upper sequent of $(B_i \to B_i)$ is $B_i[\Gamma \to \Theta]$, which implies $\vdash_{EF} B_i[\Gamma \to \Theta]$. Consider the other case where the lowermost inference is (Th). Then the upper sequent is $B_i(\Gamma') \to B_i(\Theta')$ for some $\Gamma' \subseteq \Gamma$ and $\Theta' \subseteq \Theta$. We can apply the same argument to this sequent $B_i(\Gamma') \to B_i(\Theta')$. That is, the lowermost inference of a cut-free proof of $B_i(\Gamma') \to B_i(\Theta')$ is either (Th) or $(B_i \to B_i)$. Repeat the same argument, we would meet $(B_i \to B_i)$. In this case, the upper sequent of $(B_i \to B_i)$ is $B_i[\Gamma'' \to \Theta'']$ for some $\Gamma'' \subseteq \Gamma$ and $\Theta'' \subseteq \Theta$. Hence $\vdash_{EF} B_i[\Gamma'' \to \Theta'']$. By (Th), we have $\vdash_{EF} B_i[\Gamma \to \Theta]$. $\dashv$

For an epistemic structure E and $i \in N$, we write $E_{-i} = \{e : i \circ e \in E\}$. Then E_{-i} is also an epistemic structure.

Lemma 3.6 *If* $\vdash_{EF} B_i[\Gamma \to \Theta]$, *then* $\vdash_{E_{-i}F_{-i}} \Gamma \to \Theta$.

Proof. Suppose $\vdash_{EF} B_i[\Gamma \to \Theta]$. Then there is a proof P of $B_i[\Gamma \to \Theta]$. Then i_1 of any thought sequent $B_{i_1}...B_{i_m}[\Delta \to \Lambda]$ in P is i. We delete the outermost B_i from each $B_{i_1}...B_{i_m}[\Delta \to \Lambda]$ in P. Let P' be a tree obtained in this manner. Then P' is a proof of $\Gamma \to \Theta$ in $GL_{E_{-i}F_{-i}}$. $\dashv$

Let us return to the minimality of $F = \{\epsilon, (2), (2,1)\}$ for $\vdash_{EF} \to B_2B_1(p \supset p)$. By Lemma 3.5, we have $\vdash_{EF} B_2[\ \to B_1(p \supset p)]$. By

Lemma 3.6, we have $\vdash_{E_{-2}F_{-2}} \rightarrow B_1(p \supset p)$. Again, by Theorem 3.4.(1), we have $(1) \in F_{-2}$, which implies $(2,1) \in F$.

4 Semantics for $\mathbf{GL}_{EF}$

In this section, we consider the semantics for GL_{EF}, which is a modification of the one for the logic GL_{EE} of KD4-type given in Kaneko and Suzuki (1999a). We have the completeness result, which is also obtained by modifying the completeness for GL_{EE} taking the inferential epistemic structure F into account. The semantics helps often us evaluate the required inferential epistemic structure F. For the application purpose, we give two more meta-theorems given in Kaneko and Suzuki (1999b).

We modify the standard Kripke semantics to incorporate descriptive and inferential epistemic structures E and F. In the modification, E restricts the set of admissible formulae and F does a structure of a frame. An *F-frame* $\mathcal{F}$ is given as an $n+2$ tuple $(W; w_0; R_1, ..., R_n)$ satisfying:

F1: W is a nonempty set, and is partitioned into $\{W_e\}_{e \in F}$ with $W_\epsilon = \{w_0\}$;

F2: each R_i is a binary relation over W satisfying

F2a(F-structure): $w \in W_e$ and wR_iu imply $u \in W_{e \circ i}$;

F2b(F-seriality): $w \in W_e$ and $e \circ i \in F$ imply wR_iu for some $u \in W$.

The set W_e is the set of possible epistemic worlds corresponding to the thought sequents $B_e[\cdots]$. The following $(W; w_0; R_1, R_2)$ is an example for an F-frame with $F = \{\epsilon, (2), (2,1)\}$, where $W = \{w_0, w_1, w_2\}, W_\epsilon = \{w_0\}, W_{(2)} = \{w_1\}, W_{(2,1)} = \{w_2\}, R_1 = \{(w_1, w_2)\}$ and $R_2 = \{(w_0, w_1)\}$.

$$\begin{array}{c} w_2 : \mathcal{P}_E(2,1) \\ \uparrow_1 \\ w_1 : \mathcal{P}_E(2) \\ \uparrow_2 \\ w_0 : \mathcal{P}_E \end{array}$$

Diagram 4.1.

Let $\mathcal{F}$ be an F-frame, and E a descriptive epistemic structure with $F \subseteq E$. Recall $\mathcal{P}_E(e) = \{C : e \circ \delta(C) \subseteq E\}$ for $e \in F$. Then we associate the set of formulae $\mathcal{P}_E(e)$ with each possible world $w_e \in W_e$ and $e \in F$. The root node w_ϵ is associated with $\mathcal{P}_E(\epsilon) = \mathcal{P}_E$. This association is made, since we like to have the correspondence between the worlds in W_e in $\mathcal{F}$ and the sequents with the outer $B_e[\cdots]$. The formulae in $\mathcal{P}_E(e)$ are those in the scope of $B_e[\cdots]$.

When F is a proper subset of E, it may be the case that $w \in W_e$, wR_iu for no $u \in W_e$ but $B_i(A) \in \mathcal{P}_E(e)$. For example, when $E =$

$\{\epsilon, (1), (2), (2,1)\}$ in the example of Diagram 4.1, $\mathcal{P}_E$ has $\mathrm{B}_1(p)$ but there is no $u \in W$ with $w_0 R_1 u$. We cannot refer to any world for the valuation of $\mathrm{B}_1(p)$. In this case, we allow $\mathrm{B}_1(p)$ to take either truth value $\top$ or $\bot$, i.e., we treat it just like a propositional variable.

An *EF-assingment* σ in an F-frame $\mathcal{F}$ is a function from $\bigcup_{e \in F}(W_e \times (PV \cup \{\mathrm{B}_i(C) : \mathrm{B}_i(C) \in \mathcal{P}_E(e) \text{ and } e \circ i \notin F\}))$ to $\{\top, \bot\}$. An *EF-model* is a pair $(\mathcal{F}, \sigma)$ of an F-frame and an EF-assignment σ in $\mathcal{F}$. The *valuation relation* $(\mathcal{F}, \sigma, w) \vDash$ is defined inductively as follows: for any $w \in W_e$ and $e \in F$,

V0: for any $A \in PV \cup \{\mathrm{B}_i(C) \in \mathcal{P}_E(e) : e \circ i \notin F\}$, $(\mathcal{F}, \sigma, w) \vDash A$ iff $\sigma(w, A) = \top$;

V1: $(\mathcal{F}, \sigma, w) \vDash \neg A$ iff $(\mathcal{F}, \sigma, w) \nvDash A$;

V2: $(\mathcal{F}, \sigma, w) \vDash A \supset B$ iff $(\mathcal{F}, \sigma, w) \nvDash A$ or $(\mathcal{F}, \sigma, w) \vDash B$;

V3: $(\mathcal{F}, \sigma, w) \vDash \bigwedge \Phi$ iff $(\mathcal{F}, \sigma, w) \vDash A$ for all $A \in \Phi$;

V4: $(\mathcal{F}, \sigma, w) \vDash \bigvee \Phi$ iff $(\mathcal{F}, \sigma, w) \vDash A$ for some $A \in \Phi$;

V5: for any $\mathrm{B}_i(A) \in \mathcal{P}_E(e)$ with $e \circ i \in F$, $(\mathcal{F}, \sigma, w) \vDash \mathrm{B}_i(A)$ iff $(\mathcal{F}, \sigma, u) \vDash A$ for all u with $w R_i u$.

For finite subsets Γ, Θ of $\mathcal{P}_E(e)$ and $w \in W_e$, we write $(\mathcal{F}, \sigma, w) \vDash \Gamma \to \Theta$ if $(\mathcal{F}, \sigma, w) \vDash \bigwedge \Gamma \supset \bigvee \Theta$, where $\bigwedge \emptyset$ and $\bigvee \emptyset$ are $\neg p \vee p$ and $\neg p \wedge p$, respectively.

Example 4.1 Let $F = \{\epsilon, (2)\}$ and $E = \{\epsilon, (2), (2,1)\}$. One example of an F-frame is given by Diagram 4.2. It is written as $(W; w_0; R_1, R_2)$, where $W = W_\epsilon \cup W_{(2)} = \{w_0\} \cup \{w_1, w_2\}$, $R_1 = \emptyset$ and $R_2 = \{(w_0, w_1), (w_0, w_2)\}$.

$$w_1 : \mathcal{P}_E(2) \qquad\qquad w_2 : \mathcal{P}_E(2)$$
$$\nwarrow_2 \quad \nearrow_2$$
$$w_0 : \mathcal{P}_E$$

Diagram 4.2.

Let $PV := \{p, q\}$. Then $\sigma(w_0, \cdot)$ is defined over $\{p, q\}$, but $\sigma(w_1, \cdot)$, $\sigma(w_2, \cdot)$ are defined over $\{p, q\} \cup \{\mathrm{B}_1(A) : \mathrm{B}_1(A) \in \mathcal{P}_E(2)\}$. In world w_0, the valuations of $\mathrm{B}_2\mathrm{B}_1(p)$, $\mathrm{B}_2\mathrm{B}_1(q)$ and $\mathrm{B}_2(\mathrm{B}_1(p) \wedge \mathrm{B}_1(q))$ are defined by referring to w_1 and w_2. In world w_1 and w_2, however, $\mathrm{B}_1(p)$, $\mathrm{B}_1(q)$ are treated in the same way as propositional variables. Nevertheless, since in w_1 and w_2, $\mathrm{B}_1(p) \wedge \mathrm{B}_1(q)$ is true if both are true, we have $(\mathcal{F}, \sigma, w_0) \vDash \mathrm{B}_2\mathrm{B}_1(p), \mathrm{B}_2\mathrm{B}_1(q) \to \mathrm{B}_2(\mathrm{B}_1(p) \wedge \mathrm{B}_1(q))$ for any assignment σ.

We have the completeness theorem, which is obtained by modifying completeness for the logic GL_{EE} of KD4-type proved in Kaneko and Suzuki (1999a).

Theorem 4.2 (Completeness) *Let Γ, Θ be finite subsets of $\mathcal{P}_E$. Then* $\vdash_{EF} \Gamma \to \Theta$ *if and only if* $(\mathcal{F}, \sigma, w_0) \models \Gamma \to \Theta$ *for all EF-models* $(\mathcal{F}, \sigma)$.

To illustrate the usefulness of this completeness theorem, we return to the above example.

Example 4.3 In Example 3.1.(1), we showed that sequent $\mathrm{B}_2\mathrm{B}_1(p)$, $\mathrm{B}_2\mathrm{B}_1(q) \to \mathrm{B}_2(\mathrm{B}_1(p) \wedge \mathrm{B}_1(q))$ is provable in GL_{EF} with $E = \{\epsilon, (2), (2,1)\}$ and $F = \{\epsilon, (2)\}$. By the completeness theorem, we can show that this inferential F is minimal for this sequent. Suppose $F = \{\epsilon\}$. Then any F-frame has only one possible world w_0. In this case, $\sigma(w_0, \mathrm{B}_2\mathrm{B}_1(p))$, $\sigma(w_0, \mathrm{B}_2\mathrm{B}_1(q))$ and $\sigma(w_0, \mathrm{B}_2(\mathrm{B}_1(p) \wedge \mathrm{B}_1(q)))$ have independent truth values. Hence we find an EF-model $(\mathcal{F}, \sigma)$ so that $(\mathcal{F}, \sigma, w_0) \not\models \mathrm{B}_2\mathrm{B}_1(p)$, $\mathrm{B}_2\mathrm{B}_1(q) \to \mathrm{B}_2(\mathrm{B}_1(p) \wedge \mathrm{B}_1(q))$.

Kaneko and Suzuki (1999b) developed surgical operations of models for GL_{EE} and obtained various meta-theorems. Here, we refer to two theorems for the use in game theoretical applications in Section 5, which are obtained by modifying their results slightly for the present GL_{EF} of KD-type.

Let S be a subset of $N \cup \{0\}$, and A a formula in $\mathcal{P}$. We say that A is an *S-formula* if

(i): $e = (i_1, ..., i_m) \in \delta(A)$ and $m \geq 1$ imply $i_1 \in S$;

(ii): $\epsilon \in \delta(A)$ implies $0 \in S$.

For example, $\mathrm{B}_2\mathrm{B}_1(p \supset p)$ is a $\{2\}$-formula. Any $\{0\}$-formula is nonepistemic.

Theorem 4.4 (Epistemic Separation) *Let $S_1, ..., S_k$ be disjoint nonempty subsets of $N \cup \{0\}$. Let Γ_t be a finite set of S_t-formulae for $t = 1, ..., k$, and A_t an S_t-formula for $t = 1, ..., k$. Also, we let $E^t = \{(i_1, ..., i_m) \in E : i_1 \in S_t\} \cup \{\epsilon\}$ and $F^t = \{(i_1, ..., i_m) \in F : i_1 \in S_t\} \cup \{\epsilon\}$ for $t = 1, ..., k$.*

(1): $\vdash_{EF} \Gamma_1, ..., \Gamma_k \to A_1 \vee \cdots \vee A_k$ *if and only if* $\vdash_{E^tF^t} \Gamma_t \to A_t$ *for some $t = 1, ..., k$.*

(2): *Suppose that $\Gamma_1 \cup \cdots \cup \Gamma_k$ is consistent in GL_{EF}. Then* $\vdash_{EF} \Gamma_1, \ldots, \Gamma_k \to A_1 \wedge \cdots \wedge A_k$ *if and only if* $\vdash_{E^tF^t} \Gamma_t \to A_t$ *for all $t = 1, ..., k$.*

Theorem 4.5 (Epistemic Disjunction Property) *Let Γ and Φ be finite sets of formulae with $\Phi \neq \emptyset$. Then* $\vdash_{EF} \mathrm{B}_i(\Gamma) \to \bigvee \mathrm{B}_i(\Phi)$ *if and only if* $\vdash_{EF} \mathrm{B}_i(\Gamma) \to \mathrm{B}_i(A)$ *for some $A \in \Phi$.*

The restriction by F on a proof or a frame requires that we treat $\mathrm{B}_i(A)$ in the same way as propositional variable if $e \circ i \notin F$. The following remark may help us understand this requirement.

Remark 4.6 Suppose $\vdash_{EF} \Gamma \to \Lambda$. Then there is another sequent $\Gamma' \to \Lambda'$ such that (1) $\delta(\Gamma' \cup \Lambda') \subseteq F$, (2) $\vdash_{FF} \Gamma' \to \Lambda'$ and (3) $\Gamma \to \Lambda$ is obtained by the substitutions of all the occurrences of some propositional variables $p_1, ..., p_m$ in $\Gamma' \to \Lambda'$ with some formulae $\mathrm{B}_{i_1}(A_1), ..., \mathrm{B}_{i_m}(A_m)$.

5 Applications: Prediction-Decision Making in Games

In this section, we give applications of GL_{EF} and the meta-theorems given in the previous sections to game theoretical problems. More extensive considerations of such game theoretical problems are given in Kaneko and Suzuki (2001).

Consider a 2-person finite noncooperative game $g = (g_1, g_2)$. Each player $i = 1, 2$ has ℓ_i (pure) strategies ($\ell_i \geq 2$). The set of strategies for player i is denoted by $S_i := \{\mathbf{s}_{i1}, ..., \mathbf{s}_{i\ell_i}\}$. His *payoff function* is a real valued function g_i on $S := S_1 \times S_2$. A strategy $s_1 \in S_1$ is a *best response to* s_2 if $g_1(s_1, s_2) \geq g_1(t_1, s_2)$ for all $t_1 \in S_1$. We say that s_1 is a *dominant strategy* if s_1 is a best response to s_2 for any $s_2 \in S_2$. A dominant strategy for player 2 is defined in the parallel manner. See Myerson (1991) for these basic notions.

In the game g^1 of Table 5.1 (Prisoner's Dilemma), the second strategy $\mathbf{s}_{i2}$ for each player i is a dominant strategy. In the game g^2 of Table 5.2 which is obtained from g^1 by changing payoff 6 to 2 in the northeast corner, player 1 has the same dominant strategy $\mathbf{s}_{12}$ as in g^1, but 2 has no dominant ones.

	$\mathbf{s}_{21}$	$\mathbf{s}_{22}$
$\mathbf{s}_{11}$	$(5, 5)$	$(1, 6)$
$\mathbf{s}_{12}$	$(6, 1)$	$(3, 3)$

Table 5.1: $g^1 = (g^1_1, g^1_2)$

	$\mathbf{s}_{21}$	$\mathbf{s}_{22}$
$\mathbf{s}_{11}$	$(5, 5)$	$(1, \widetilde{2})$
$\mathbf{s}_{12}$	$(6, 1)$	$(3, 3)$

Table 5.2: $g^2 = (g^2_1, g^2_2)$

First, consider the following prediction-decision criteria.

DC1: Player i makes a decision by choosing a dominant strategy.

In game g^1, DC1 recommends each player i to choose $\mathbf{s}_{i2}$. However, in game g^2, DC1 is applied only to player 1 but not to player 2, since 2 has no dominant strategies. We consider another decision criterion for player 2.

DC21: Player $j \neq i$ first predicts that player i would choose a strategy following DC1, and then j would choose a best strategy to the predicted strategy.

Let us apply this criterion to game g^2 for $j = 2$: player 2 predicts that 1 would choose $\mathbf{s}_{12}$ following DC1, and then 2 would choose $\mathbf{s}_{22}$ as the best strategy to $\mathbf{s}_{12}$.

The above two criteria are related to the *procedure of iterated elimination of dominated strategies* (cf., Moulin (1982) and Myerson (1991)). This is more or less a standard example of a prediction-decision criterion in the game theory literature. The minimal depth of inferential epistemic structure appears to be related to the number of the rounds needed for the iterated eliminations of dominated strategies. Kaneko and Suzuki (2001) discuss a general definition of a prediction-decision criterion. A special case of the variant of the general definition is the method of iterated eliminations of dominated strategies. See Kaneko and Suzuki (2001).

We can consider a lot of other prediction-decision criteria which have never been discussed in the literature of game theory. Here, we show by examples that our logical approach enables us to take such prediction-decision criteria in the scope of our research. In particular, we discuss the subtlety of inferential epistemic structures required for decision making with such criteria.

We give the following somewhat extreme criterion, which suggests that there are a lot of variants of prediction-decision criteria.

DC0: Player i has the predetermined strategy $\mathbf{s}_{i1}$ as his decision, which we call the *pure default decision.*

This may sound too trivial if it is adopted for player i's decision making. However, it would not be so if this is adopted for player j's prediction on i's decision making. That is, the prediction on the other player's decision making is made without much deliberations on the other's subjective elements. We modify criterion DC21 into the following.

DC20: Player j first predicts that player i would choose strategy $\mathbf{s}_{i1}$ following DC0, and then j would choose a best strategy to the predicted strategy $\mathbf{s}_{i1}$.

This criterion is free from the symmetric assumption that a decision maker assumes payoff maximization for himself as well as for the other player in his prediction. In this sense, DC20 differs considerably from DC21. In game g^2, this criterion states that $j = 2$ chooses $\mathbf{s}_{21}$ as the best strategy to the default prediction $\mathbf{s}_{11}$.

Now, we look at how these criteria can be discussed in our logical approach. First, we replace the set of propositional variables by the set of atomic formulae defined by

strategy symbols: $\mathbf{s}_{11}, ..., \mathbf{s}_{1\ell_1}$ and $\mathbf{s}_{21}, ..., \mathbf{s}_{2\ell_2}$;

4-*ary preference symbols*: P_1 and P_2;

unary default decision symbols: d_1 and d_2.

That is, for $s, t \in S = S_1 \times S_2, s_i \in S_i$ and $i = 1, 2$, we call the expression $P_i(s : t)$ and $d_i(s_i)$ *atomic* formulae. Let AF be the set of all atomic formulae, and the set of formulae $\mathcal{P}$ is defined based on AF instead of PV. The expressions $P_i(s : t)$ and $d_i(s_i)$ are intended to mean, respectively, that player i weakly prefers strategy pair s to t and that s_i is a "default" decision of player i.

In the present language, the statement "s_1 is a best response to s_2 for player 1" is expressed by the formula $\bigwedge\{P_1(s_1, s_2 : t_1, s_2) : t_1 \in S_1\}$, which we denote by $\text{Best}_1(s_1 \mid s_2)$. The statement "$s_1$ is a dominant strategy for 1" is expressed by $\bigwedge\{\text{Best}_1(s_1 \mid s_2) : s_2 \in S_2\}$. This is equivalent to $\bigwedge\{P_1(s_1, s_2 : t_1, s_2) : t_1 \in S_1 \text{ and } s_2 \in S_2\}$, which we denote by $\text{Dom}_1(s_1)$. In the parallel manner, we define $\text{Best}_2(s_2 \mid s_1)$ and $\text{Dom}_2(s_2)$.

We express the payoff function g_i by the set of preferences (formulae):

(2) $$\hat{g}_i := \{P_i(s : t) : g_i(s) \geq g_i(t)\} \cup \{\neg P_i(s : t) : g_i(s) < g_i(t)\}.$$

Hence game $g = (g_1, g_2)$ is expressed as $\hat{g}_1 \cup \hat{g}_2$.

Now, we are in the state to formulate the above prediction-decision criteria. One possible formulation of a prediction-decision criterion for player i is given as a set of formulae indexed by $s_i \in S_i$:

(3) $$\mathcal{D}_i = \{D_i(s_i) : s_i \in S_i\}.$$

This means that if $D_i(s_i)$ holds, then s_i would be a decision. The capability of player i's decision making is expressed as the following sequent:

(4) $$\mathrm{B}_i(\Gamma_i) \rightarrow \bigvee\nolimits_{s_i} \mathrm{B}_i(D_i(s_i)),$$

where Γ_i is the finite set of player i's basic beliefs. That is, the existence of some decision s_i is derived from player i's beliefs Γ_i. This statement itself is made from the viewpoint of the investigator (observer). A general statement by the investigator is expressed as the provability of the following:

(5) $$\Gamma, \mathrm{B}_1(\Gamma_1), \mathrm{B}_2(\Gamma_2) \rightarrow \bigvee\nolimits_{s_1} \mathrm{B}_1(D_1(s_1)) \wedge \bigvee\nolimits_{s_2} \mathrm{B}_2(D_2(s_2)),$$

where Γ is a finite set of nonepistemic formulae, which expresses the objective situation such as $\hat{g}_1 \cup \hat{g}_2$.

The ultimate goal is to consider the provability of (5) from the viewpoint of the investigator. However, the following theorem states that for each player's decision making, it suffices to consider (4) separately and that we can ignore the objective part Γ.

Theorem 5.1 (Decomposition) *Suppose that* $\Gamma \cup \mathrm{B}_1(\Gamma_1) \cup \mathrm{B}_2(\Gamma_2)$ *is consistent in* GL_{EF}. *Then the following two statements are equivalent:*

(1): $\vdash_{EF} \Gamma, \mathrm{B}_1(\Gamma_1), \mathrm{B}_2(\Gamma_2) \to \bigvee_{s_1} \mathrm{B}_1(D_1(s_1)) \wedge \bigvee_{s_2} \mathrm{B}_2(D_2(s_2))$;

(2): *for* $i = 1, 2$, $\vdash_{E_iF_i} \mathrm{B}_i(\Gamma_i) \to \mathrm{B}_i(D_i(s_i))$ *for some* s_i,

where $E_i = \{(i_1, ..., i_m) \in E : i_1 = i\} \cup \{\epsilon\}$ *and* $F_i = \{(i_1, ..., i_m) \in F : i_1 = i\} \cup \{\epsilon\}$.

Proof. The derivation of (1) from (2) is straightforward. Suppose (1). We apply Theorem 4.4 to (1), and have $\vdash_{E_iF_i} \mathrm{B}_i(\Gamma_i) \to \bigvee_{s_i} \mathrm{B}_i(D_i(s_i))$ for $i = 1, 2$. Then, by Theorem 4.5, we have (2). ⊣

Now, we focus on the statement on each player's decision making.

DC1: This is formulated as $\mathcal{D}_i^1 = \{\mathrm{Dom}_i(s_i) : s_i \in S_i\}$. Applying this criterion to game g^1, we have the following derivation:

$$\cfrac{\cfrac{\left\{ \cfrac{\mathrm{B}_1[P_1(\mathbf{s}_{12}, s_2 : s_1, s_2) \to P_1(\mathbf{s}_{12}, s_2 : s_1, s_2)]}{\mathrm{B}_1[\hat{g}_1^1 \to P_1(\mathbf{s}_{12}, s_2 : s_1, s_2)]} \text{(Th)} \right\}_{s \in S}}{\mathrm{B}_1[\hat{g}_1^1 \to \mathrm{Dom}_1(\mathbf{s}_{12})]} (\to\bigwedge)}{\mathrm{B}_1(\hat{g}_1^1) \to \mathrm{B}_1(\mathrm{Dom}_1(\mathbf{s}_{12})).} (\mathrm{B}_1 \to \mathrm{B}_1)$$

That is, player 1 derives $\mathbf{s}_{12}$ as his decision from his basic beliefs $\hat{g}_1^1$. This is a proof in $\mathrm{GL}_{E_1F_1}$, where $E_1 = F_1 = \{\epsilon, (1)\}$. The parallel assertion holds for 2. In the game g^2 of Table 5.2, the above result holds only for player 1.

In fact, we can show that $F_1 = \{\epsilon, (1)\}$ is minimal for the above sequent. However, we will discuss the minimality of the suggested F_i only for DC20.

DC21: This criterion for player 2 is formulated $\mathcal{D}_2^{21} = \{D_2^{21}(s_2) : s_2 \in S_2\}$, where each $D_2^{21}(s_2)$ is given as

(6) $\bigvee_{s_1} \mathrm{B}_1(\mathrm{Dom}_1(s_1)) \wedge \bigwedge_{s_1} (\mathrm{B}_1(\mathrm{Dom}_1(s_1)) \supset \mathrm{Best}_2(s_2 \mid s_1))$.

In game g^2, we can prove

(7) $$\vdash_{E_2F_2} \mathrm{B}_2(\hat{g}_2^2), \mathrm{B}_2\mathrm{B}_1(\hat{g}_1^2) \to \mathrm{B}_2(D_2^{21}(\mathbf{s}_{22})),$$

where $E_2 = F_2 = \{\epsilon, (2), (2, 1)\}$. The set $\mathrm{B}_2(\hat{g}_2^2) \cup \mathrm{B}_2\mathrm{B}_1(\hat{g}_1^2)$ means that 2 has the beliefs on his own payoff function g_2^2 and believes that 1 has the beliefs on 1' payoff function g_1^2. The latter is needed to derive 2's prediction on 1's decision. Here, $F_2 = \{\epsilon, (2), (2,1)\}$ is minimal.

DC0: The default decision criterion $\mathcal{D}_i^0$ is given as

(8) $$\mathcal{D}_i^0 = \{d_i(s_i) : s_i \in S_i\}.$$

To derive a default decision, player i needs to assume that some is a default decision. Let $\Gamma_i^0 = \{d_i(\mathbf{s}_{i1})\} \cup \{\neg d_i(\mathbf{s}_{it}) : t = 2, ..., \ell_i\}$. Then

(9) $$\vdash_{E_iF_i} \mathrm{B}_i(\Gamma_i^0) \to \mathrm{B}_i(d_i(\mathbf{s}_{i1})),$$

(10) $$\vdash_{E_iF_i} \mathrm{B}_i(\Gamma_i^0) \to \mathrm{B}_i(\neg d_i(\mathbf{s}_{it})) \text{ for } t = 2, ..., \ell_i,$$

where $E_i = \{\epsilon, (i)\}$ and $F_i = \{\epsilon\}$. Here player i chooses $\mathbf{s}_{i1}$ *dogmatically* and is *conscious of* the other strategies not to be default decisions. Here, player i needs no beliefs on the game but does his dogmatic beliefs Γ_i^0. He conducts no logical reasonings to have his decision.

DC20: This decision criterion for player 2 is expressed as $\mathcal{D}_2^{20} = \{D_2^{20}(s_2) : s_2 \in S_2\}$, where each $D_2^{20}(s_2)$ is given as

(11) $$\bigvee_{s_1} \mathrm{B}_1(d_1(s_1)) \wedge \bigwedge_{s_1} (\mathrm{B}_1(d_1(s_1)) \supset \mathrm{Best}_2(s_2 \mid s_1))\,.$$

For this, 2 predicts that 1 would choose a default decision. In the game g^2 of Table 5.2, the following holds:

(12) $$\vdash_{E_2F_2} \mathrm{B}_2(\hat{g}_2^2), \mathrm{B}_2\mathrm{B}_1(\Gamma_1^0) \to \mathrm{B}_2(D_2^{20}(\mathbf{s}_{21})),$$

where $E_2 = F_2 = \{\epsilon, (2), (2,1)\}$.

For (12), $F_2 = \{\epsilon, (2), (2,1)\}$ is minimal. The reader may wonder why this minimal F_2 includes $(2,1)$, since 2 believes that 1 simply uses his dogmatic decision $\mathbf{s}_{11}$. To verify $D_2^{20}(\mathbf{s}_{21})$, however, 2 derives $\neg\mathrm{B}_1(d_1(\mathbf{s}_{12}))$ from $\mathrm{B}_1(\neg d_1(\mathbf{s}_{12}))$. Here, 2 needs to think about the relationship between $\neg\mathrm{B}_1(d_1(\mathbf{s}_{12}))$ and $\mathrm{B}_1(\neg d_1(\mathbf{s}_{12}))$.

Theorem 5.2 *$F_2 = \{\epsilon, (2), (2,1)\}$ is a unique minimal inferential epistemic structure for (12).*

Proof. It suffices to show $(2) \in F_2$ and $(2,1) \in F_2$. Suppose $\vdash_{E_2F_2}$ $\mathrm{B}_2(\hat{g}_2^2)$, $\mathrm{B}_2\mathrm{B}_1(\Gamma_1^0) \to \mathrm{B}_2(D_2^{20}(\mathbf{s}_{21}))$. By Theorem 3.4.(2), we have $(2) \in F_2$. By Lemma 3.5, we have $\vdash_{E_2F_2} \mathrm{B}_2[\hat{g}_2^2, \mathrm{B}_1(\Gamma_1^0) \to D_2^{20}(\mathbf{s}_{21})]$. Here, we denote $(E_2)_{-2}$ and $(F_2)_{-2}$ by E_{-2} and F_{-2}, respectively. Hence $\vdash_{E_{-2}F_{-2}}$ $\hat{g}_2^2, \mathrm{B}_1(\Gamma_1^0) \to D_2^{20}(\mathbf{s}_{21})$ by Lemma 3.6. Since $D_2^{20}(\mathbf{s}_{21})$ is given as (11), we have $\vdash_{E_{-2}F_{-2}} \hat{g}_2^2, \mathrm{B}_1(\Gamma_1^0) \to \neg\mathrm{B}_1(d_1(\mathbf{s}_{12})) \vee \mathrm{Best}_2(\mathbf{s}_{21} \mid \mathbf{s}_{12})$. By Theorem 4.4,

$$\vdash_{E_{-2}F_{-2}} \hat{g}_2^2 \to \mathrm{Best}_2(\mathbf{s}_{21} \mid \mathbf{s}_{12}) \text{ or } \vdash_{E_{-2}F_{-2}} \mathrm{B}_1(\Gamma_1^0) \to \neg\mathrm{B}_1(d_1(\mathbf{s}_{12})).$$

Due to Table 5.2, the first is not the case. From the second, we have $\vdash_{E_{-2}F_{-2}} \mathrm{B}_1(d_1(\mathbf{s}_{12})), \mathrm{B}_1(\Gamma_1^0) \to$. By Theorem 3.4.(1), we have $(1) \in F_{-2}$, which implies $(2,1) \in F_2$. $\dashv$

Simple Dogmatic Belief: Consider the following belief set Γ_i^{00} rather than $\Gamma_i^0 = \{d_i(\mathbf{s}_{i1})\} \cup \{\neg d_i(\mathbf{s}_{it}) : t = 2, ..., \ell_i\}$:

$$\Gamma_i^{00} = \{d_i(\mathbf{s}_{i1})\}. \tag{13}$$

In (13), player i is conscious only of his first strategy $\mathbf{s}_{i1}$ and ignores the other possibilities. In this case, the result for player i's own decision making is more or less the same as (9), i.e., $\vdash_{E_iF_i} \mathrm{B}_i(\Gamma_i^{00}) \to \mathrm{B}_i(d_i(\mathbf{s}_{i1}))$, where $E_i = \{\epsilon, (i)\}$ and $F_i = \{\epsilon\}$.

However, if $\mathrm{B}_1(\Gamma_1^{00})$ is used as a belief set for player 2, the result would be different from (12). First, we give our result in a sightly general manner. After the theorem, we discuss specific cases.

Theorem 5.3 *Let $g = (g_1, g_2)$ be a game where 2 has no dominant strategy, and Γ_2 a set of $\{1\}$-formulae. Then* $\vdash_{E_2F_2} \mathrm{B}_2(\hat{g}_2), \mathrm{B}_2(\Gamma_2) \to \mathrm{B}_2(D_2^{20}(s_2))$ *if and only if*

(1): $\vdash_{E_2F_2} \mathrm{B}_2[\Gamma_2 \to \mathrm{B}_1(d_1(s_1))]$ *for some* s_1;

(2): *for any* $s_1 \in S_1$, $\vdash_{E_2F_2} \mathrm{B}_2[\hat{g}_2 \to \mathrm{Best}_2(s_2 \mid s_1)]$ *or* $\vdash_{E_2F_2} \mathrm{B}_2[\Gamma_2 \to \neg\mathrm{B}_1(d_1(s_1))]$.

Proof. We prove the *only-if* part. Suppose that $\vdash_{EF} \mathrm{B}_2(\hat{g}_2), \mathrm{B}_2(\Gamma_2) \to \mathrm{B}_2(D_2^{20}(s_2))$. Since $D_2^{20}(s_2) \notin \hat{g}_2 \cup \Gamma_2$ by the assumption of the theorem, we have $(2) \in F_2$ by Theorem 3.4.(2). Then $\vdash_{E_2F_2} \mathrm{B}_2[\hat{g}_2, \Gamma_2 \to D_2^{20}(s_2)]$ by Lemma 3.5. Since $D_2^{20}(s_2)$ is given as (11), we have

$$\vdash_{E_2F_2} \mathrm{B}_2[\hat{g}_2, \Gamma_2 \to \textstyle\bigvee_{s_1} \mathrm{B}_1(d_1(s_1))] \tag{14}$$

$$\vdash_{E_2F_2} \mathrm{B}_2[\hat{g}_2, \Gamma_2 \to \mathrm{B}_1(d_1(s_1)) \supset \mathrm{Best}_2(s_2 \mid s_1)] \text{ for all } s_1 \in S_1. \tag{15}$$

First, consider (15). This is equivalent to that $\vdash_{E_2F_2} \mathrm{B}_2[\hat{g}_2, \Gamma_2 \to \neg\mathrm{B}_1(d_1(s_1)) \vee \mathrm{Best}_2(s_2 \mid s_1)]$ for all $s_1 \in S_1$. Let s_1 be an arbitrary one in S_1. By Lemma 3.6, we have $\vdash_{E_{-2}F_{-2}} \hat{g}_2, \Gamma_2 \to \neg\mathrm{B}_1(d_1(s_1)) \vee \mathrm{Best}_2(s_2 \mid s_1)$. By Theorem 4.4, we have

$$\vdash_{E_{-2}F_{-2}} \hat{g}_2 \to \mathrm{Best}_2(s_2 \mid s_1) \text{ or } \vdash_{E_{-2}F_{-2}} \Gamma_2 \to \neg\mathrm{B}_1(d_1(s_1)).$$

This implies the statement (2).

Second, consider (14). By Lemma 3.6, we have $\vdash_{E_{-2}F_{-2}} \hat{g}_2, \Gamma_2 \to \bigvee_{s_1} \mathrm{B}_1(d_1(s_1))$. Since Γ_2 is a set of $\{1\}$-formulae, we have $\vdash_{E_{-2}F_{-2}} \Gamma_2 \to \bigvee_{s_1} \mathrm{B}_1(d_1(s_1))$ or $\vdash_{E_{-2}F_{-2}} \hat{g}_2 \to$ by Theorem 4.4. Since $\hat{g}_2$ is a consistent set, the former is the case. Suppose $\nvdash_{E_{-2}F_{-2}} \Gamma_2 \to \mathrm{B}_1(d_1(s_1))$ for any s_1. By (2), $\vdash_{E_{-2}F_{-2}} \hat{g}_2 \to \mathrm{Best}_2(s_2 \mid s_1)$ for any $s_1 \in S_1$, which implies $\vdash_{E_{-2}F_{-2}} \hat{g}_2 \to \bigwedge_{s_1} \mathrm{Best}_2(s_2 \mid s_1)$. This means that 2 has a dominant strategy in g, which contradicts the assumption of the theorem. Thus $\vdash_{E_{-2}F_{-2}} \Gamma_2 \to \mathrm{B}_1(d_1(s_1))$ for some s_1. Hence $\vdash_{E_2F_2} \mathrm{B}_2[\Gamma_2 \to \mathrm{B}_1(d_1(s_1))]$ for some s_1. $\dashv$

Let us return to the simple dogmatic default belief set $\Gamma_1^{00} = \{d_1(\mathbf{s}_{11})\}$ in the game g^2 of Table 5.2. In contrast to (12), we have

$$\nvdash_{E_2F_2} \mathrm{B}_2(\hat{g}_2^2), \mathrm{B}_2\mathrm{B}_1(\Gamma_1^{00}) \to \mathrm{B}_2(D_2^{20}(\mathbf{s}_{21})) \text{ for } E_2 = F_2 = \{\epsilon, (2), (2,1)\}.$$

This unprovability is a consequence of Theorem 5.2: Suppose, on the contrary, that this sequent is provable. Then $\vdash_{E_2F_2} \mathrm{B}_2[\hat{g}_2^2 \to \mathrm{Best}_2(\mathbf{s}_{21} \mid \mathbf{s}_{12})]$ or $\vdash_{E_2F_2} \mathrm{B}_2[\mathrm{B}_1(\Gamma_1^{00}) \to \neg\mathrm{B}_1(d_1(\mathbf{s}_{12}))]$. However, neither is the case because of Table 5.2 and $\Gamma_1^{00} = \{d_1(\mathbf{s}_{11})\}$.

One possible way out from the above unprovability is to expand $\mathrm{B}_1(\Gamma_1^{00})$ into

(16) $$\Gamma_2^0 = \{\mathrm{B}_1(d_1(\mathbf{s}_{11})), \neg\mathrm{B}_1(d_1(\mathbf{s}_{12})), ..., \neg\mathrm{B}_1(d_1(\mathbf{s}_{1\ell_1}))\}.$$

Then we have the following:

(17) $$\vdash_{E_iF_i} \mathrm{B}_2(\hat{g}_2^2), \mathrm{B}_2(\Gamma_2^0) \to \mathrm{B}_2(D_2^{20}(\mathbf{s}_{21})),$$

where $E_2 = \{\epsilon, (2), (2,1)\}$ and $F_2 = \{\epsilon, (2)\}$. This F_2 is the minimal inferential epistemic structure. Here, player 2 takes into account that player 1 does not think about the second strategy $\mathbf{s}_{12}$ in the simple dogmatic default belief Γ_1^{00}.

6 Conclusion

We have obtained logical system GL_{EF} by restricting KD^n by means of descriptive and inferential epistemic structures E and F. We presented the Gentzen-style proof system and Kripke-type semantics for GL_{EF}. In this paper, we proved one theorem and two lemmas for the evaluation of a required inferential epistemic structure F for a given sequent. Using these results, we have evaluated the minimal inferential epistemic structures in some game theoretical examples. A more extensive application to game theory is presented in Kaneko and Suzuki (2001). We hope that our approach gives some light on multi-modal logic as well as game theory.

Acknowledgements

The authors thank F. Wolter and the referees for helpful comments on earlier versions of this paper. The authors are partially supported by Grant-in-Aids for Scientific Research No.106330003 and No.10440027,

Ministry of Education, Science and Culture. The second author is also supported in part by the Sumitomo Foundation.

References

Fagin, R., J. Y. Halpern, Y. Moses, and M. Y. Vardi. 1995. *Reasoning about Knowledge*. Cambridge: The MIT Press.

Gabbay, D., and V. Shehtman. 1998. Products of Modal Logics, Part I. *Journal of IGPL* 6:73–146.

Hintikka, J. 1962. *Knowledge and Belief*. Cornell University Press.

Kaneko, M., and T. Nagashima. 1997. Game Logic and its Applications II. *Studia Logica* 58:273–303.

Kaneko, M., and N.-Y. Suzuki. 1999a. Semantics of Epistemic Logics of Shallow Depths for Game Theory. IPPS. DP. 814. University of Tsukuba.

Kaneko, M., and N.-Y. Suzuki. 1999b. Epistemic Models of Shallow Depths and Decision Making in Games. IPPS. DP. 828. University of Tsukuba.

Kaneko, M., and N.-Y. Suzuki. 2001. Bounded Interpersonal Inferences and Decision Making. *Economic Theory*. To appear.

Meyer, J.-J. Ch., and W. van der Hoek. 1995. *Epistemic Logic for AI and Computer Science*. Cambridge University Press.

Moulin, H. 1982. *Game Theory for the Social Sciences*. New York: New York University Press.

Myerson, R. B. 1991. *Game Theory*. Cambridge: Harvard University Press.

Wolter, F. 1998. Fusions of Modal Logics Revisited. In *Advances in Modal Logic I*, ed. M. Kracht, M. de Rijke, H. Wansing, and M. Zakharyaschev. CSLI, Stanford.

Mamoru Kaneko
Institute of Policy and Planning Sciences
University of Tsukuba
Ibaraki 305-8573, JAPAN
E-mail: kaneko@shako.sk.tsukuba.ac.jp

Nobu-Yuki Suzuki
Department of Mathematics
Faculty of Science
Shizuoka University
Ohya, Shizuoka 422-8529, JAPAN
E-mail: smnsuzu@ipc.shizuoka.ac.jp

16

The Semantics of Modal Predicate Logic I. Counterpart-Frames

MARCUS KRACHT AND OLIVER KUTZ

ABSTRACT. We introduce a new semantics for modal predicate logic, with respect to which a rich class of first-order modal logics is complete, namely all normal first-order modal logics that are extensions of free quantified **K**. This logic is defined by combining positive free logic with equality $\mathbf{PFL}^{\doteq}$ and the propositional modal logic **K**. We then uniformly construct—for each modal predicate logic L—a canonical model whose theory is exactly L. This proves completeness with respect to so-called modal-structures. We add some remarks on canonicity and frame-completeness and finally show that if suitable modal algebras of 'admissible interpretations' are added to modal predicate frames, general frame-completeness is gained.

1 Introduction

In the past several years there has been a continuous interest in so-called generalized Kripke-semantics or Kripke-type-semantics. The main motivation for this revival of interest in the semantics for first-order modal logics basically comes from two different sources. First of all, it has been realized that standard Kripke-semantics is highly incomplete as witnessed e.g. by a theorem by Ghilardi (cf. Ghilardi 1991). This has led to several generalizations—such as functor semantics, Kripke-sheaves or metaframe-semantics—which were on the one hand mathematically rich enough to allow for general completeness results and on the other hand enabled one to understand the insufficiency of standard Kripke-semantics much better. [1]

[1] Cf. Skvortsov and Shehtman 1993 and Kutz 2001 for a brief overview.

Advances in Modal Logic, Volume 3
F. Wolter, H. Wansing, M. de Rijke, and M. Zakharyaschev, eds.

Secondly, there has been a growing awareness that standard semantics is also inadequate from a more philosophical point of view or that standard syntactic machinery is insufficient with respect to a proper treatment of e.g. singular terms or definite descriptions in modal contexts (cf. Fitting and Mendelsohn 1998). One should also mention here different attempts to modify or generalize the Counterpart-Theory by David Lewis.

The main problem obviously lies in the question of how to treat "modal individuals". Observe that in standard semantic approaches like constant or varying domain semantics, the treatment of variables and constants is quite distinct. While the former take *objects* as values that are supposed to be—once fixed—constant throughout all possible worlds (apart from changing all their properties) the latter can very well take *individual concepts* as values as is the case with non-rigid constants. This is hence the starting point for generalizations of the semantics. Note that there is also an alternative way of dealing with this asymmetry, i.e. to explicitly allow quantification over both individual objects and individual concepts as has been done in (Fitting 2001).

Now, there is still no consensus as to which generalized semantics is the 'right' one from a purely mathematical point of view. Furthermore, the high technicality of the proposed semantics has led to—as a matter of fact—almost complete neglect on behalf of philosophers, linguists etc., despite their far greater flexibility in handling e.g. problems of the individuation of individuals.

In this paper we first single out a rich class of first-order modal logics, which are normal extensions of our base logic **FK**. This base logic is obtained from a combination of positive free logic with identity, $\mathbf{PFL}^{\dot{=}}$, and the modal propositional logic **K**. We then present a semantics which is on the one hand closely related to the functor-semantics proposed by Ghilardi (cf. e.g. Ghilardi 1991) and on the other hand picks up some ideas from David Lewis' Counterpart-Theory (Lewis 1968) to deal e.g. with the failure of the principle of the necessity of identity. We follow a proposal by van Benthem and avoid all categorial terminology that is not strictly necessary (cf. van Benthem 1993), thus skirting around **QS4** as a base logic. Our approach to the semantics of modal predicate logics is hence quite elementary, and it is hoped that it will provide a useful tool for philosophers, linguists or people working in AI.

In the next paragraph we start off by saying a few words about free logic and by defining the class of modal predicate logics we are going to investigate. In §3 we define the semantics and state a soundness theorem for our base logic **FK**. In §4 we prove a general completeness theorem by constructing a canonical model $\mathfrak{M}_L$ for each modal predicate logic L. This proves completeness with respect to so-called modal structures, which are frames (in the sense defined here) together with an interpretation of the relation symbols. In §5 we sketch how to obtain frame-completeness results by showing the canonicity of certain axioms. In the last paragraph we show how to obtain general frame-completeness by enriching the canonical frame by a suitable modal algebra of 'admissible interpretations'.

2 Free Logic and Free Modal Logic

Our language consists in some signature for predicate logic, with symbols for relations only, and the set $Var := \{x_i : i \in \omega\}$ of variables. We omit constant symbols here because a proper treatment would introduce further complications that will be dealt with in the sequel to this paper. The set of formulae of this language is denoted by $\mathbb{L}$. $\mathbb{ML}$ is the language obtained by admitting any number of unary modal operators. For sake of exposition, we use just one modal operator, $\Box$. The base logic is not the usual quantified **K** (henceforth denoted by **QK**), but a weaker logic, denoted here by **FK**; its $\mathbb{L}$-fragment is denoted by $\mathbf{PFL}^{E!}$. For a more detailed treatment of the motivations behind the use of free logic the reader is referred to (Kutz 2000) and (Garson 1991). Here, a few brief remarks must suffice.

In the following, the existence symbol $E!$ abbreviates the formula $\exists y(y \doteq x)$, where y is a variable distinct from x. We first state the axioms of $\mathbf{PFL}^{E!}$. We then add the axioms for identity and then the axioms and rules of the propositional modal logic **K** to obtain **FK**.

Definition 2.1 (Axiomatization of $PFL^{E!}$) Let ϕ and ψ be modal formulae. Then all formulae of the following type are axioms:

TAUTOLOGIES: All instances of classical propositional tautologies are axioms.

VACUOUS QUANTIFICATION: If x is not free in ϕ, then the formula $\phi \to \forall x\phi$ is an axiom.

UNIVERSAL DISTRIBUTIVITY: $\forall x(\phi \to \psi) \to (\forall x\phi \to \forall x\psi)$.

E! 1: $\forall x E!(x)$.

E! 2: If y is free for x in $\phi(x, \bar{z})$ and $y \notin \bar{z}$, then $\forall x\phi(x, \bar{z}) \to (E!(y) \to \phi(y, \bar{z}))$ is an axiom.

Next, the following axioms for the identity symbol are added.

Definition 2.2 (Axioms for Equality) Let x and y be any variables and ϕ a formula such that x is free in ϕ and y is free for x in ϕ. $\phi(y//x)$

is shorthand for formulae which result from ϕ by replacing some (not necessarily all) occurrences of x by y. Then all formulae of the following type are axioms:

SELF-IDENTITY: $(x \doteq x)$

LEIBNIZ' LAW: $(x \doteq y) \to (\phi \to \phi(y//x))$

As usual, we have the rules of inference *Modus Ponens* (MP) and *Universal Generalization* ($\forall$). The axioms for $\mathbf{PFL}^{E!}$ together with the axioms for equality and the inference rules constitute the system $\mathbf{PFL}^{\doteq}$ of positive free logic with identity.

Before turning to the modal logic **FK**, a few short remarks on the use of free logic might be in order. First of all, since free logic is just a weakening of classical predicate logic, those who wish to deal with classical quantificational logic can simply install an extra axiom saying that all terms exist. This readily entails full classical universal instantiation. One can, on the other hand, use a *primitive* existence symbol $E!$, two (classical) quantifiers $\bigwedge$ and $\bigvee$ and define the (free) quantifiers $\forall$ and $\exists$ by restricting to the extension of the predicate $E!$ (the *domain of existence*).

Now, there is an important relationship between *existence* and *identity*, a fact which presumably goes back to Jaakko Hintikka.[2] If one formulates an axiom system with an existence symbol but without identity, the existence symbol is not *eliminable*, i.e. there is no formula $\phi(x)$, which does not contain the existence symbol, such that the formula $\phi(x) \leftrightarrow E!(x)$ is *provable*. This was shown by Bencivenga, Lambert and Meyer (cf. Bencivenga et al. 1982). But if the language is enriched by the identity symbol, the formula $E!(x)$ is provably equivalent to $\exists y(y \doteq x)$, whence $E!$ can be understood to be defined by this formula. In the literature, this fact has been called 'Hintikka's Theorem'.

Concerning the intrinsic reasons for using free logic, we just mention three basic points. Firstly, in standard quantified **K**, there are a number of theorems that do not seem to be valid under all possible interpretations. Consider for example the formula $\Box\exists x(x \doteq x)$, which states in the standard interpretation that it is *necessary that there are things.* Now it is perfectly imaginable that this claim be challenged by some philosopher. Also, the Converse Barcan Formula is easily proved in **QK**, while it is quite obviously disputable. Secondly, if one works for example in a varying domain setting, a formula like $\Diamond\phi(t)$ might contain a term t that denotes an object **a** in world w that does not exist in another ac-

[2]Compare his "Existential Presuppositions and Their Elimination" (Hintikka 1969).

cessible world v. If one then wants to assign a truth value to the formula $\Diamond\phi(t)$ at world w, one has the choice between giving up the ***principle of bivalence*** or adopting free logic by assigning truth values also to formulae that contain terms denoting non-existing things. Thirdly, adopting classical predicate logic also blocks a proper treatment of non-denoting terms and definite descriptions which is particularly important in the modal setting.

We now turn to the definition of our modal base logic. The logic **FK** is obtained by adding the axiom-schema *Box-Distribution* and the rule *Necessitation* (MN). The only subtle point concerns the axioms for identity.

Definition 2.3 (Modal Axioms for Equality) Let x appear free in the formula ϕ and let y be a variable that does not appear free within the scope of a modal operator in ϕ and that is free for x in ϕ. The notation $\phi(y//x)$ is shorthand for a formula that results from ϕ by replacing some occurrences of x where x appears free but *not* within the scope of a modal operator by y. Within the scope of a modal operator, either *all* or *none* of the occurrences of the variable x have to be replaced by y. Then all formulae of the following type are axioms:

SELF-IDENTITY: $(x \doteq x)$

MODAL LEIBNIZ' LAW: $(x \doteq y) \rightarrow (\phi \rightarrow \phi(y//x))$

By means of the above definition of the identity axioms, we restrict the Quinean principle of the 'substitutability of identicals' insofar as it is thus formulated so as to avoid all implications with respect to possible 'transworld identifications' of objects. That is to say, it avoids making assumptions about the behaviour of objects when one moves from one possible world to another. This suffices to establish 'classical circumstances' in the theory of a single world. Formulae on the other hand that entail 'transworld identifications' are assumed to be part of a given modal theory or otherwise of a logic stronger than the base logic.

By way of example, let us consider the following formulae, which are *not* correct instances of the Modal Leibniz' Law:

$$(x \doteq y) \rightarrow (\Box(x \doteq x) \rightarrow \Box(x \doteq y))$$

(Not all occurrences of the variable x are replaced by y, x is free within the scope of "$\Box$".)

$$(x \doteq y) \rightarrow (\Diamond(x \not\doteq y) \rightarrow \Diamond(y \not\doteq y))$$

(The variable y is free within the scope of a modal operator.)

$$(x \doteq y) \rightarrow (\exists y \Diamond(x \not\doteq y) \rightarrow \exists y \Diamond(y \not\doteq y))$$

(The variable y is not free for x.)

We illustrate this situation in the next section by giving countermodels for the formulae above.

However, the following formulae are correct instances:

$$(x \doteq y) \to (\Box(x \doteq x) \to \Box(y \doteq y))$$

(All occurrences of x are replaced by y.)

$$(x \doteq y) \to (\Diamond(x \not\doteq z) \to \Diamond(y \not\doteq z))$$

(The variable y is not free within the scope of a modal operator.)

$$(x \doteq y) \to ((x \doteq z) \to (y \doteq z))$$

(y is free for x and is not free within the scope of a modal operator.)

$$(x \doteq y) \to ((x \doteq x) \to (y \doteq x))$$

(y is free for x and is not free within the scope of a modal operator. Only one occurrence of x has been replaced by y.)

As the last examples have shown, we can easily establish that the equality symbol "$\doteq$" satisfies the axioms of an equivalence relation in any given world. The set of formulae derivable in the system **FK** is then denoted by $\mathbb{FK}$. The modal axioms for equality could also be thought of as a *minimal theory of identity* that accompanies every quantified modal logic. We now distinguish first and second order substitutions:

Definition 2.4 (First-Order Substitutions) Let $\phi(x, \bar{z})$ be a modal formula, in which x appears free and y is free for x. Then $\phi(y/x, \bar{z})$ is called a *first-order substitution instance*, if $\phi(y/x, \bar{z})$ is the result of replacing every free occurrence of x in ϕ by y.

Definition 2.5 (Second-order Substitutions) Let ϕ be a formula in which the n-place relation symbol P appears and let ψ be some modal formula. Then $(\psi/P)\phi$ is called a *second-order substitution instance*, if $(\psi/P)\phi$ is the result of replacing every occurrence of $P^n(\bar{y})$ in ϕ by $\psi(\bar{y}/\bar{x})$, possibly renaming some bound variables.

Note that assuming unrestricted second order substitution for a given logic L automatically extends the underlying modal theory of identity. For instance, given that $(x \doteq y) \to (P(x, x) \to P(x, y))$ is an admissible instance of Leibniz' Law, second order substitution yields $(x \doteq y) \to (\Box(x \doteq x) \to \Box(x \doteq y))$ and hence $(x \doteq y) \to \Box(x \doteq y)$. This is one of the motivations for distinguishing first- and second-order-closed logics.

Definition 2.6 (Modal Predicate Logics) A set of formulae L with $\mathbb{FK} \subseteq L \subseteq \mathbb{ML}$ is called a *first-order closed modal predicate logic*, if it is closed under the rules (MN), ($\forall$) and (MP) and L is also closed under first-order substitutions. If L is additionally closed under second-order substitutions, it is called a *second-order closed modal predicate logic.* If we speak of a modal predicate logic L simpliciter, L is assumed to be at least first-order closed.

A natural solution to the above problem of generating possibly unintended theorems involving equality is therefore to deal with second-order closed logics without identity and to add a modal theory of identity, or, alternatively, to incorporate the theory of identity into the logic while restricting substitution in an appropriate way.

Note that a second-order closed logic is a second-order logic in the sense that predicate symbols are treated as second-order variables without having explicit second-order quantification. Hence, predicate variables are treated as being implicitly universally quantified.

3 Semantics

In the following we very briefly introduce the relevant semantical concepts. First, we define first-order structures to interpret the free logical part of the system **FK**. We basically use the so-called inner-domain/outer-domain approach to free logic.

Definition 3.1 (First-Order Structures) A triple $\mathcal{S} = \langle U_\mathcal{S}, D_\mathcal{S}, I_\mathcal{S} \rangle$ is called a *first-order structure* if $U_\mathcal{S}$ is a *non-empty* set (the universe of the structure), $D_\mathcal{S}$ a (possibly empty) subset of $U_\mathcal{S}$ (the domain of existence) and $I_\mathcal{S}$ an *interpretation*, which assigns to each n-place relation symbol a subset of the n-dimensional cartesian product of $U_\mathcal{S}$. The class of all structures is then called $\mathcal{K}_{PFL}$.

The next step towards a definition of modal frames is to define an appropriate class of 'counterpart-relations'. The next definition is intended to replace the usual notion of the monotonicity of individual domains.

Definition 3.2 (Counterpart-Existence-Property) Let $\mathcal{S}$ and $\mathcal{T}$ be two structures and C a binary relation between $U_\mathcal{S}$ and $U_\mathcal{T}$. Then C has the *Counterpart-Existence-Property* with respect to $\mathcal{S}$ (CE-Property for short), if for each element $\mathbf{a} \in U_\mathcal{S}$ there is at least one element $\mathbf{b} \in U_\mathcal{T}$, such that $\langle \mathbf{a}, \mathbf{b} \rangle \in C$. C is then said to be a *CE-relation.* $C_{\mathcal{S},\mathcal{T}}$ stands for the set of all CE-relations between $U_\mathcal{S}$ and $U_\mathcal{T}$.

This last definition does not only take care of the *bivalence* of the semantics, it is also necessary to establish *normality*, as a look at the soundness proof below will reveal.

Definition 3.3 (Modal Structures and Counterpart-Frames) A pair $\mathfrak{f} = \langle \mathcal{W}, C \rangle$ is a *modal structure*, if $\mathcal{W} \neq \emptyset$ is a *non-empty* set of first-order structures and $C \subseteq C_{\mathcal{W}}$ is a subset of the set $C_{\mathcal{W}}$ of all CE-relations between (universes of) structures from $\mathcal{W}$. Members $\mathcal{S}$ of $\mathcal{W}$ are also called *(possible) worlds.* $\mathfrak{Co}_{\mathcal{S},\mathcal{T}}$ denotes the set of all CE-relations between $\mathcal{S}$ and $\mathcal{T}$ in the modal structure $\mathfrak{f}$. The class of all modal structures is called $\mathfrak{frK}$. If the interpretation function of the modal structure $\mathfrak{f}$ is omitted, we denote the result by $\mathfrak{F} = \langle U, C \rangle$ (where U is now just a family of pairs of the form $\langle U_{\mathcal{S}}, D_{\mathcal{S}} \rangle$ with $\mathcal{S} \in \mathcal{W}$) and call $\mathfrak{F}$ a *counterpart-frame* or, for brevity, a *modal frame.* The class of all counterpart-frames will be denoted by $\mathfrak{FrK}$.

We say that $\mathcal{S}$ *sees* $\mathcal{T}$ in $\mathfrak{f}$ if there is a CE-relation from $\mathcal{S}$ to $\mathcal{T}$ in $\mathfrak{f}$. A *valuation* is a function v which assigns to each variable x and possible world $\mathcal{S}$ an element from the universe $U_{\mathcal{S}}$ of $\mathcal{S}$. We write $v_{\mathcal{S}}$ for the valuation v at $\mathcal{S}$. A *modal model* $\mathfrak{M} = \langle \mathfrak{f}, v \rangle$ is a modal structure $\mathfrak{f}$ together with a valuation v. Finally, an *existential x-variant* of a valuation v in the world $\mathcal{S}$ is a valuation function $\widetilde{v}$ which is like v except that $\widetilde{v}$ assigns some element from $D_{\mathcal{S}}$ to the variable x at $\mathcal{S}$.

Definition 3.4 (Truth in a Modal Model) Let $\phi(y_1, \ldots, y_n)$ and $\psi(z_1, \ldots, z_m)$ be modal formulae with the free variables $y_1, \ldots, y_n$ and $z_1, \ldots, z_m$, respectively. Let $\mathfrak{f}$ be a modal structure, $\mathcal{S}$ a possible world and let v be a valuation. We define:

(a) $\langle \mathfrak{f}, v, \mathcal{S} \rangle \vDash x_i \doteq x_j :\Longleftrightarrow v_{\mathcal{S}}(x_i) = v_{\mathcal{S}}(x_j)$ in $U_{\mathcal{S}}$.

(b) $\langle \mathfrak{f}, v, \mathcal{S} \rangle \vDash R(y_1, \ldots, y_n) :\Longleftrightarrow \langle v_{\mathcal{S}}(y_1), \ldots, v_{\mathcal{S}}(y_n) \rangle \in I_{\mathcal{S}}(R)$.

(c) $\langle \mathfrak{f}, v, \mathcal{S} \rangle \vDash \neg\phi :\Longleftrightarrow \langle \mathfrak{f}, v, \mathcal{S} \rangle \nvDash \phi$.

(d) $\langle \mathfrak{f}, v, \mathcal{S} \rangle \vDash \phi \wedge \psi :\Longleftrightarrow \langle \mathfrak{f}, v, \mathcal{S} \rangle \vDash \phi$ and $\langle \mathfrak{f}, v, \mathcal{S} \rangle \vDash \psi$.

(e) $\langle \mathfrak{f}, v, \mathcal{S} \rangle \vDash \Diamond\phi(y_1, \ldots, y_n) :\Longleftrightarrow$ there is $\mathcal{S} \xrightarrow{C} \mathcal{T}$ and a $\langle y_1, \ldots, y_n \rangle$-variant $\widetilde{v}$, such that $\langle v_{\mathcal{S}}(y_i), \widetilde{v}_{\mathcal{T}}(y_i) \rangle \in C$ for $i = 1, \ldots, n$ and $\langle \mathfrak{f}, \widetilde{v}, \mathcal{T} \rangle \vDash \phi(y_1, \ldots, y_n)$.

(f) $\langle \mathfrak{f}, v, \mathcal{S} \rangle \vDash \exists x \phi(x) :\Longleftrightarrow$ there is an existential x-variant $\widetilde{v}$, such that $\langle \mathfrak{f}, \widetilde{v}, \mathcal{S} \rangle \vDash \phi(x)$.

Now the usual soundness theorems like e.g. a Coincidence Lemma can be proved, of which we shall make free use below. In particular, we can prove the following Soundness Theorem for **FK**. The only subtle points concern the normality of **FK** and the modal equality axioms.

Theorem 3.5 (Soundness) *Let $\mathfrak{F}$ be an arbitrary counterpart-frame. Then it holds that $\mathfrak{F} \vDash$* **FK**.

Proof. We have to prove that every formula that is derivable in **FK** is true in any modal model $\mathfrak{M}$. We only consider Leibniz' Law and Box-Distribution. The validity of the other axiom-schemes is easily verified.

MODAL LEIBNIZ' LAW Inductively we show the following: *Suppose that ϕ is an arbitrary modal formula and $\widetilde{\phi} := (x \doteq y) \to (\phi \to \phi(y//x))$ is an* admissible *instance of Leibniz' Law. Then $\widetilde{\phi}$ is valid.* From this, Leibniz' Law follows.

(i) This claim is trivial for atomic formulae and easy to verify for formulae of the form $\neg\psi$ or $(\psi \wedge \chi)$.

(ii) Let $\phi(x, y, \bar{z}) = \exists w \psi(w, x, y, \bar{z})$ and $\widetilde{\phi}(x, y, \bar{z})$ be an admissible instance of Leibniz' Law. Then

$$(x \doteq y) \to (\psi(w, x, y, \bar{z}) \to \psi(w, y//x, y, \bar{z})) \qquad (\star)$$

is also admissible, hence valid by induction hypotheses. Now suppose that $\langle v, \mathcal{S}\rangle \vDash (x \doteq y) \wedge \exists w \psi(w, x, y, \bar{z})$ in some model $\mathfrak{M}$. Then $v_{\mathcal{S}}(x) = v_{\mathcal{S}}(y)$ and $\langle \widetilde{v}, \mathcal{S}\rangle \vDash \psi(w, x, y, \bar{z})$ for some w-variant $\widetilde{v}$ of v. Because the variable w is different from x and y, $\widetilde{v}_{\mathcal{S}}(x) = \widetilde{v}_{\mathcal{S}}(y)$, whence $(\star)$ implies $\langle \widetilde{v}, \mathcal{S}\rangle \vDash \psi(w, y//x, y, \bar{z})$. It follows that $\langle v, \mathcal{S}\rangle \vDash \exists w \psi(w, y//x, y, \bar{z})$.

(iii) Let $\phi(x, \bar{z}) = \Diamond\psi(x, \bar{z})$ and $\widetilde{\phi}(x, \bar{z})$ be admissible. Then, by definition of 'admissible instance', $y \notin \bar{z}$ and x is within the scope of a modal operator. Therefore $\widetilde{\phi}(x, \bar{z})$ is of the form $(x \doteq y) \to (\Diamond\psi(x, \bar{z}) \to \Diamond\psi(y, \bar{z}))$, where every occurrence of x in $\psi(y, \bar{z}))$ is replaced by y. Now let $\mathfrak{M}$ be a model and $\mathcal{S}$ a world such that $\langle \mathcal{S}, v\rangle \vDash (x \doteq y) \wedge \Diamond\psi(x, \bar{z})$. Then $v_{\mathcal{S}}(x) = v_{\mathcal{S}}(y)$ and there is a world $\mathcal{T}$ and a $(x, \bar{z})$-variant $\widetilde{v}$, such that $\mathcal{S} \xrightarrow{C} \mathcal{T}$, $\langle v_{\mathcal{S}}(x), \widetilde{v}_{\mathcal{T}}(x)\rangle \in C$ and $\langle v_{\mathcal{S}}(z_i), \widetilde{v}_{\mathcal{T}}(z_i)\rangle \in C$ for all $z_i \in \bar{z}$ and $\langle \widetilde{v}, \mathcal{T}\rangle \vDash \psi(x, \bar{z})$. Now let $\widehat{v}$ be a $(y, \bar{z})$-variant of v such that $\widehat{v}_{\mathcal{T}}(y) = \widetilde{v}_{\mathcal{T}}(x)$ and $\widehat{v}_{\mathcal{T}}(z_i) = \widetilde{v}_{\mathcal{T}}(z_i)$. It may be the case that $\widetilde{v}_{\mathcal{T}}(x) \neq \widetilde{v}_{\mathcal{T}}(z_i)$ for all i. Hence the above definition of $\widehat{v}$ is only admissible, since—by assumption—y is different from all z_i. So the valuations $\widehat{v}$ and $\widetilde{v}$ coincide at $\mathcal{T}$ on the free variables of ψ and y is free for x in $\psi(x, \bar{z})$. By the Coincidence Lemma it follows that $\langle \widehat{v}, \mathcal{T}\rangle \vDash \psi(y, \bar{z})$ and because $v_{\mathcal{S}}(x) = v_{\mathcal{S}}(y)$ and $\widehat{v}$ is a $(y, \bar{z})$-variant of $v_{\mathcal{T}}$ with $\langle v_{\mathcal{S}}(u), \widehat{v}_{\mathcal{T}}(u) \in C$ for all $u \in y \cup \bar{z}$, we have that $\langle v, \mathcal{S}\rangle \vDash \Diamond\psi(y, \bar{z})$.

BOX-DISTRIBUTION Let $\phi(\bar{x})$ and $\psi(\bar{y})$ be modal formulae with free variables $\bar{x}$ and $\bar{y}$ respectively and suppose that $\langle v, \mathcal{S}\rangle \vDash \Box(\phi(\bar{x}) \to \psi(\bar{y})) \wedge \Box\phi(\bar{x})$. Suppose wlog that there is a world $\mathcal{T}$ in a modal model $\mathfrak{M}$ and a CE-relation C, such that $\mathcal{S} \xrightarrow{C} \mathcal{T}$. Now let $\widetilde{v}$ be any $\bar{y}$-variant of v such that $\langle v_{\mathcal{S}}(y_i), \widetilde{v}_{\mathcal{T}}(y_i)\rangle \in C$ for all $y_i \in \bar{y}$. Because C is a CE-

relation, there is—for every variable $x_j \in \bar{x} - \bar{y}$—an element $\mathbf{a}_j \in U_{\mathcal{T}}$, such that $\langle \upsilon_{\mathcal{S}}(x_j), \mathbf{a}_j \rangle \in C$. Now define a $\bar{x} \cup \bar{y}$-variant $\widehat{\upsilon}$ of υ by:

$$\widehat{\upsilon}_{\mathcal{T}}(x) = \begin{cases} \mathbf{a}_i & : \text{ if } x = x_i \in \bar{x} - \bar{y} \\ \widetilde{\upsilon}_{\mathcal{T}}(x) & : \text{ otherwise} \end{cases}$$

By assumption $\langle \widehat{\upsilon}, \mathcal{T} \rangle \vDash (\phi(\bar{x}) \to \psi(\bar{y})) \wedge \phi(\bar{x})$, hence $\langle \widehat{\upsilon}, \mathcal{T} \rangle \vDash \psi(\bar{y})$. Now the valuations $\widehat{\upsilon}$ and $\widetilde{\upsilon}$ coincide at $\mathcal{T}$ on the variables in $\bar{y}$, whence by the Coincidence Lemma $\langle \widetilde{\upsilon}, \mathcal{T} \rangle \vDash \psi(\bar{y})$.

Finally, it is straightforwardly checked that the rules of inference preserve the validity of the axioms of **FK**. ⊣

We now come to the promised examples. In the following pictures, a world is represented by a square box with individuals being represented by a bullet if they exist at that world and by a circle otherwise.

FIGURE 1 Refuting the Barcan Formula and its Converse

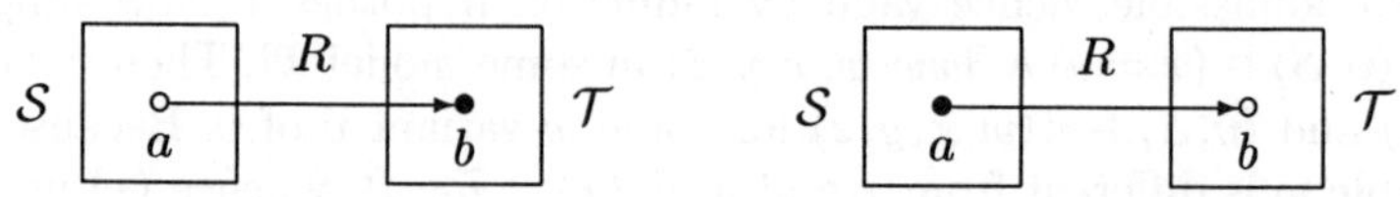

In Figure 1, the objects a and b are related via a CE-Relation R. The model on the left refutes the Barcan Formula $\Diamond \exists x(x \doteq x) \to \exists x \Diamond (x \doteq x)$ and also the Necessity of Fictionality $\neg E!(x) \to \Box \neg E!(x)$. Correspondingly, the model on the right refutes the Converse Barcan Formula and the Necessity of Existence. In Figure 2, the left model refutes the Ne-

FIGURE 2 Fusing and Splicing of Objects

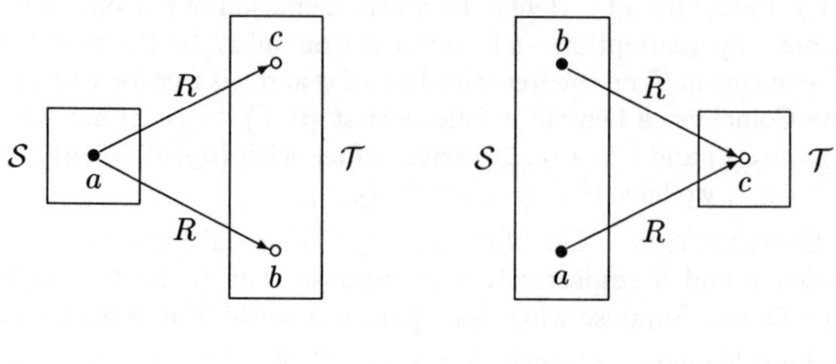

cessity of Identity $(x \doteq y) \to \Box(x \doteq y)$, whereas the model on the right refutes the Necessity of Distinctness $(x \neq y) \to \Box(x \neq y)$. The former also relates to "Pierre's Puzzle" (Kripke 1979). Think of world $\mathcal{S}$ as the actual world and of world $\mathcal{T}$ as Pierre's "World of Beliefs". Then Pierre's seemingly inconsistent beliefs about the actual city denoted in the real world by constants like `London` and `Londres` can be consistently modelled as a de re belief about two *distinct* (fictional) objects in his Belief-World.

Note that these examples are quite simple in that they don't make use of the feature of counterpart-frames to possibly posess many distinct counterpart-relations between two given worlds. Yet, despite first appearance, this feature is not eliminable and actually one of the basic ingredients of all generalizations of standard Kripke-semantics. This is underlined by the fact that one can easily construct second-order closed modal predicate logics that are frame-complete only with respect to frames having at least two counterpart-relations between any two worlds[3] and will also be apparent from the construction of canonical models, which we come to next.

4 Canonical Models

Definition 4.1 (Free Henkin-Types) A maximally consistent set of modal formulae Δ is called a *free Henkin-type* if it also satisfies the following condition:

(H) For every $\exists$-formula $\exists x\phi(x,\bar{z}) \in \Delta$ there is a variable y such that $\phi(y,\bar{z}) \in \Delta$, $E!(y) \in \Delta$ and y is free for x in $\phi(x,\bar{z})$.

The set of all free Henkin-types with respect to the logic L will be denoted by $FHen_L$.

In the sequel, we often use substitutions $f : Var \longrightarrow Var$ that are totally defined and injective. We will call them *faithful substitutions* and denote the set of all these functions by $\mathbb{F}$.

Lemma 4.2 (Existence of Free Henkin-Types) *For every consistent set Δ and any two* finite *sequences $y_1, \dots, y_m$ and $z_1, \dots, z_m$ of m pairwise distinct variables there is a faithful substitution $f \in \mathbb{F}$ such that $f(y_i) = z_i$ for $i = 1, \dots, m$, and a free Henkin-type Γ such that*

$$\Delta^f := \{\phi^f(\bar{v}) : \phi(\bar{u}) \in \Delta \textit{ and } f(\bar{u}) = \bar{v}\} \subseteq \Gamma,$$

where $f(\bar{u}) = \bar{v}$ abbreviates $f(u_i) = v_i, i = 1, \dots, n$.

Proof. Fix an enumeration $\pi : \omega \to \mathbb{ML}$. Let $\langle \Gamma_{-1}, \langle \Gamma_i : i \in \omega \rangle \rangle$ be a sequence defined in the following way. Because the given lists of variables

[3]For an example confer Kracht and Kutz 2001.

are finite, there are $k, l \in \omega$, such that $\{y_1, \ldots, y_m\} \subset \{x_1, \ldots, x_k\}$ and $\{z_1, \ldots, z_m\} \subset \{x_1, \ldots, x_l\}$. Now let f be defined as follows:

$$f(x) = \begin{cases} z_i & : \text{ if } x = y_i \text{ (for } i \in \{1, \ldots, m\}) \\ x_{l+2i+1} & : \text{ if } x = x_i \text{ and } x \neq y_i \text{ for } i = 1, \ldots, m \end{cases}$$

Since the y_i are all distinct, this (total) function is faithful. In particular, $range(f) \subset \{z_1, \ldots, z_n\} \cup \{x_{l+2i+1} : i \in \omega\}$, hence infinitely many variables (namely those of the form $x_{l+2\cdot(i+1)}$ for $i \in \omega$) are *not* within the range of f. Put $\Gamma_{-1} := \Delta^f$. Define Γ_n inductively as follows:

(i) If $\Gamma_{n-1} \cup \{\pi(n)\}$ is L-consistent and $\pi(n)$ is not an $\exists$-formula, put $\Gamma_n := \Gamma_{n-1} \cup \{\pi(n)\}$.

(ii) If $\Gamma_{n-1} \cup \{\pi(n)\}$ is L-consistent and $\pi(n) = \exists x \phi(x, x_{i_1}, \ldots, x_{i_j})$ is an $\exists$-formula, put $\Gamma_n := \{\phi(x_i, x_{i_1}, \ldots, x_{i_j}), E!(x_i), \pi(n)\} \cup \Gamma_{n-1}$, where $x_i \in$ *Var* is the first variable in the standard-enumeration, which is not under the (free or bound) variables occurring in $\Gamma_{n-1} \cup \{\pi(n)\}$. (Alternatively, choose x_i as $x_{l+2(n+1)}$.)

(iii) Finally, if $\Gamma_{n-1} \cup \{\pi(n)\}$ is L-inconsistent, we define $\Gamma_n := \Gamma_{n-1} \cup \{\neg\pi(n)\}$. (Notice that $\neg\pi(n)$ is then not an $\exists$-formula, even if it might be L-equivalent to one.)

Γ_n is easily seen to be consistent. Now set $\Gamma := \bigcup_{i \in \{-1\} \cup \omega} \Gamma_i$. This is a complete type. Moreover, by construction Γ is a free Henkin-type and in particular $\Delta^f \subset \Gamma$. $\dashv$

Definition 4.3 (Canonical Universe of Discourse) Let Δ be a free Henkin-type. Define an equivalence relation $\sim_\Delta$ on variables by putting $x \sim_\Delta y$ iff $(x \doteq y) \in \Delta$. Put $[x]_\Delta := \{y : y \sim_\Delta x\}$. The universe of the structure $\mathcal{S}_\Delta$ is then defined as: $U_\Delta := \{[x_i]_\Delta : x_i \in Var\}$ and the domain of quantification as $D_\Delta := \{[x_i]_\Delta^\sim : x_i \in Var \text{ and } E!(x_i) \in \Delta\}$.

Definition 4.4 (Canonical Interpretation) Let Δ be a free Henkin-type, R an n-place relation symbol and $y_1, \ldots, y_n$ n variables. We define an n-place relation R^Δ on the n-dimensional cartesian product of U_Δ by setting:

$$\langle [y_1]_\Delta, \ldots, [y_n]_\Delta \rangle \in R^\Delta :\Longleftrightarrow R(y_1, \ldots, y_n) \in \Delta.$$

The *canonical interpretation* is then $I_\Delta(R) := R^\Delta \subseteq U^n$.

It is easily shown that this is well-defined. Next we define the notion of a canonical structure.

Definition 4.5 (Canonical Structures) If Δ is a free Henkin-type, we define the *canonical structure* $\mathcal{S}_\Delta$ as the triple $\langle U_\Delta, D_\Delta, I_\Delta \rangle$. The class $\{\mathcal{S}_\Delta : \Delta \in FHen_L\}$ of all canonical structures is then denoted by $\mathcal{W}_L$.

Definition 4.6 (Canonical Valuation) Let Δ be a free Henkin-type and let $\mathcal{S}_\Delta$ be the associated canonical structure. Then the *canonical valuation* $v_\Delta : Var \longrightarrow U_\Delta$ is defined by $v_\Delta(x_i) := [x_i]_\Delta$.

Definition 4.7 (Canonical Counterpart-Relations) Let f be any faithful substitution and Δ and Γ be two free Henkin-types. We then define the expression $C_f := \{\langle [x]_\Delta, [y]_\Gamma\rangle : f(x) = y\}$ and finally put:

$$(\Delta^\Box)^f := \{\phi^f(\bar{z}) \ : \ \Box\phi(\bar{y}) \in \Delta \text{ and } f(y_i) = z_i, i = 1, \ldots, n\}.$$

Finally define: $\mathcal{S}_\Delta \xrightarrow{C_f} \mathcal{S}_\Gamma :\Longleftrightarrow (\Delta^\Box)^f \subset \Gamma$ (or briefly $\Delta \xrightarrow{C_f} \Gamma$). The class C_L of all *canonical counterpart-rclations* of the canonical modal structure is then given by:

$$C_L := \{C_f : f \in \mathbb{F},\ \Delta \xrightarrow{C_f} \Gamma \textit{ and } \Delta, \Gamma \in FHen_L\}.$$

By constructing appropriate free Henkin-types, it is easily seen that canonical counterpart-relations are in general neither functional nor injective or surjective etc. We now define the notion of a canonical modal structure and give the central result in Theorem 4.9.

Definition 4.8 (Canonical Modal Structure/Model) The *canonical modal structure* $\mathfrak{f}_L$ is given by the pair $\langle \mathcal{W}_L, C_L\rangle$. The *canonical modal model* is then $\mathfrak{M}_L = \langle \mathfrak{f}_L, v_L\rangle$, where v_L denotes the canonical valuation.

Theorem 4.9 (Fundamental Theorem) *Let L be any modal predicate logic, ϕ be an arbitrary modal formula, $\mathfrak{f}_L$ the canonical modal structure, $\mathcal{S}_\Delta$ any world and v_L the canonical valuation. Then*

$$\langle \mathfrak{f}_L, v_L, \mathcal{S}_\Delta\rangle \vDash \phi \Longleftrightarrow \phi \in \Delta.$$

Proof. The proof is by induction on the construction of ϕ. We perform only the problematic steps.

(i) Let $\phi = \exists x\psi(x, \bar{y})$. $\langle \mathfrak{f}_L, v_L, \mathcal{S}_\Delta\rangle \vDash \exists x\psi(x, \bar{y}) :\Longleftrightarrow$ *there exists an existential x-variant $\widetilde{v}$ such that $\langle \widetilde{v}, \mathcal{S}_\Delta\rangle \vDash \psi(x, \bar{y})$*. By definition this is the case if and only if there is a valuation $\widetilde{v}$ with $\widetilde{v}_\Delta(x) = [z]_\Delta \in D_\Delta$ such that $\langle v_x^{[z]_\Delta}, \mathcal{S}_\Delta\rangle \vDash \psi(x, \bar{y})$. Here the valuations v and $v_x^{[z]_\Delta}$ coincide at $\mathcal{S}_\Delta$ on all variables save possibly x and $v_\Delta(z) = [z]_\Delta = v_x^{[z]_\Delta}(x)$. Now let $\widetilde{\psi}(x, \bar{y})$ be a bound variant of $\psi(x, \bar{y})$, such that z is free for x in $\widetilde{\psi}(x, \bar{y})$. By the Coincidence Lemma it follows that $\langle v, \mathcal{S}_\Delta\rangle \vDash$

$\widetilde{\psi}(z, \bar{y})$ *and* $[z]_\Delta \in D_\Delta$. But by assumption and definition this means that there is a variable z such that $E!(z) \in \Delta$ and $\widetilde{\psi}(z, \bar{y}) \in \Delta$. Because $E!(z) \wedge \widetilde{\psi}(z, \bar{y}) \to \exists z \widetilde{\psi}(z, \bar{y}) \in L$ and Δ is a complete type it follows that $\exists z \psi(z, \bar{y}) \in \Delta$. But then, since complete types are closed under renaming of bound variables, $\exists x \psi(x, \bar{y})$ is in Δ as well. Conversely, assume that $\exists x \psi(x, \bar{y}) \in \Delta$. Then, since Δ is a free Henkin-type, there is a variable z such that $E!(z) \in \Delta$ and $\psi(z, \bar{y}) \in \Delta$. Now we can reverse the last argument. This proves part (i).

(ii) Assume that $\phi(x_1, \dots, x_n) = \Diamond \psi(x_1, \dots, x_n)$ and that the claim holds for ψ.

Assume first, that $\langle v, \mathcal{S}_\Delta \rangle \vDash \Diamond \psi(x_1, \dots, x_n)$. Then, by definition, there is a free Henkin-type Γ and a $C_f \in \mathfrak{Co}_{\Delta,\Gamma}$, such that $(\Delta^\Box)^f \subset \Gamma$. Furthermore there are variables $y_1, \dots, y_n$, such that $\langle [x_i]_\Delta, [y_i]_\Gamma \rangle \in C_f$ and $\langle \mathcal{S}_\Gamma, v^{[y_i]_\Gamma}_{x_i} \rangle \vDash \psi(x_1, \dots, x_n)$. But then there are variables $u_i \in [x_i]_\Delta$ and $v_i \in [y_i]_\Gamma$ such that $f(u_i) = v_i$ $(i = 1, \dots, n)$. Then, in particular, $[u_i]_\Delta = [x_i]_\Delta$ and $[v_i]_\Gamma = [y_i]_\Gamma$ and hence $\langle [u_i], [v_i] \rangle \in C_f$ for $i = 1, \dots, n$. It follows that $\langle v^{[v_i]_\Gamma}_{x_i}, \mathcal{S}_\Gamma \rangle \vDash \psi(x_1, \dots, x_n)$. The variables v_i might not be free for x_i in $\psi(x_1, \dots, x_n)$. Therefore let g be a faithful substitution such that $g(x_i) = v_i$ (and g arbitrary otherwise). Then $\langle v^{[v_i]_\Gamma}_{x_i} \circ g^{-1}, \mathcal{S}_\Gamma) \rangle \vDash \psi^g(v_1, \dots, v_n)$, i.e. $\langle v_L, \mathcal{S}_\Gamma \rangle \vDash \psi^g(v_1, \dots, v_n)$. By assumption it follows that

$$\psi^g(v_1, \dots, v_n) \in \Gamma \qquad (*).$$

Now assume that $\Diamond \psi(x_1, \dots, x_n) \notin \Delta$. Because $[x_i]_\Delta = [u_i]_\Delta$ for $i = 1, \dots, n$, it follows that $(x_i \doteq u_i) \in \Delta$ for $i = 1, \dots, n$. Moreover, let h be faithful with $h(x_i) = u_i$ for $i = 1, \dots, n$. It follows that $\Diamond \psi(x_1, \dots, x_n) \in \Delta \iff \Diamond \psi^h(u_1, \dots, u_n) \in \Delta$, so $\Diamond \psi^h(u_1, \dots, u_n) \notin \Delta$.

Then $\Box \neg \psi^h(u_1, \dots, u_n) \in \Delta$ and since $(\Delta^\Box)^f \subset \Gamma$ we also deduce that we have $(\neg \psi^h)^f(f(u_1), \dots, f(u_n)) \in \Gamma$, i.e. that

$$\neg \psi^{f \circ h}(v_1, \dots, v_n) \in \Gamma \qquad (**).$$

But since the formulae $(*)$ and $(**)$ differ only in the choice of bound variables, this immediately yields a contradiction.

Conversely, assume that $\Diamond \psi(x_1, \dots, x_n) \in \Delta$. Then there is a possible world Γ and a canonical counterpart-relation C_f in $\mathfrak{Co}_{\Delta,\Gamma}$, such that $\psi^f(y_1, \dots, y_n) \in \Gamma$. Then $(\Delta^\Box)^f \subset \Gamma$, where $f(x_i) = y_i$ for $i = 1, \dots, n$. Hence $\Delta \xrightarrow{C_f} \Gamma$ and $\langle [x_i]_\Delta, [y_i]_\Gamma \rangle \in C_f$ for $i = 1, \dots, n$. By assumption it follows that $\langle v, \mathcal{S}_\Gamma \rangle \vDash \psi^f(y_1, \dots, y_n)$ and hence $\langle v \circ f, \mathcal{S}_\Gamma \rangle \vDash \psi(x_1, \dots, x_n)$. (Here $(v \circ f)_\Theta = v$ for all $\Theta \neq \Gamma$, and $(v \circ f)_\Gamma := v_\Gamma \circ f$.) But then there is a $\langle x_1, .., x_n \rangle$-variant $\widetilde{v}$ of v, namely $\widetilde{v}_\Gamma(x_i) = [f(x_i)]_\Gamma =$

$[y_i]_\Gamma$, such that by the Coincidence Lemma $\langle \widetilde{v}, \mathcal{S}_\Gamma \rangle \vDash \psi(x_1, \ldots, x_n)$ and $\langle v_\Delta(x_i), \widetilde{v}_\Gamma(x_i) \rangle \in C_f$. Hence $\langle v, \mathcal{S}_\Delta \rangle \vDash \Diamond\psi(x_1, \ldots, x_n)$, which completes the proof of the theorem. ⊣

As usual, $\mathfrak{M} \vDash \phi$ (where $\mathfrak{M} = \langle \mathfrak{f}, v \rangle$) if for all worlds $\mathcal{S}$, $\langle \mathfrak{f}, v, \mathcal{S} \rangle \vDash \phi$. Put $Th(\mathfrak{M}) := \{\phi \in \mathbb{ML} : \mathfrak{M} \vDash \phi\}$. Furthermore, $\mathfrak{f} \vDash \phi$ if for all models $\mathfrak{M}$ based on $\mathfrak{f}$ (i.e. for all valuations in $\mathfrak{f}$), $\mathfrak{M} \vDash \phi$ holds. Finally, $\mathfrak{F} \vDash \phi$ if for all interpretations I in $\mathfrak{F}$, $\mathfrak{f} \vDash \phi$, where $\mathfrak{f} = \langle \mathfrak{F}, I \rangle$. On the basis of this it is not hard to show the following.

Theorem 4.10 (Canonical Model Theorem) *Let L be a modal predicate logic. Then $Th(\mathfrak{M}_L) = L$. In particular, the canonical valuation is dispensable, i.e. $L = Th(\mathfrak{f}_L)$.*

However, this only establishes completeness with respect to models and not frames.

5 Canonicity and Frame-Completeness

Definition 5.1 (Frame-Completeness) We say that a modal predicate logic L is *frame-complete*, if there is a class $\mathbb{K}$ of modal frames such that $L = \bigcap_{\mathfrak{F} \in \mathbb{K}} Th(\mathfrak{F})$.

Definition 5.2 (Canonicity) A logic L is said to be *canonical*, if it is valid on the canonical frame $\mathfrak{F}_L$, i.e. if $\mathfrak{F}_L \vDash L$.

The following are now immediate.

Theorem 5.3 *Every canonical logic is frame-complete.*

Proof. Immediately from Theorem 4.10. ⊣

Corollary 5.4 *The logic* **FK** *is frame-complete with respect to the class $\mathfrak{Fr}\mathfrak{K}$ of all frames.*

Proof. By 3.5 and 4.10. ⊣

It is possible to establish a correspondence between properties of Kripke-structures and certain axioms in the usual way but also between properties of the C-relations and axioms. Examples are the functionality or injectivity of all CE-relations of a model which correspond respectively to the validity of the axiom-schemes $(x \doteq y) \to \Box(x \doteq y)$ (Necessity of Identity) and $(x \not\doteq y) \to \Box(x \not\doteq y)$ (Necessity of Distinctness). If the former axiom-schema is denoted by (N) then specializing on **QS4N** (where 'Q' indicates that we also specialized to classical predicate logic), we obtain the C-sets of (Ghilardi 1991). We further illustrate this situation and the kind of correspondence that obtains by the following examples.

Definition 5.5 (Prominent Schemes) Let x and y be variables and let ϕ be a modal formula:

(NoI)	$(x \doteq y) \to \Box(x \doteq y)$	(Necessity of Identity)
(NoD)	$(x \neq y) \to \Box(x \neq y)$	(Necessity of Distinctness)
(NoE)	$E!(x) \to \Box E!(x)$	(Necessity of Existence)
(NoF)	$\neg E!(x) \to \Box\neg E!(x)$	(Necessity of Fictionality)
(BF)	$\forall x \Box\phi \to \Box\forall x\phi$	(Barcan-formulae)
(CBF)	$\Box\forall x\phi \to \forall x\Box\phi$	(Converse Barcan-formulae)
(T)	$\Box\phi \to \phi$	(T-Schema)
(4)	$\Box\phi \to \Box\Box\phi$	(4-Schema)

In the following we define various properties of frames and state some soundness and canonicity-results.

Definition 5.6 (Properties of Frames)

(a) A frame $\mathfrak{F}$ is called *functional*, if all $C \in C_{\mathfrak{F}}$ are functional.

(b) $\mathfrak{F}$ is called *injective*, if all $C \in C_{\mathfrak{F}}$ are injective.

(c) $\mathfrak{F}$ is called *existentially faithful*, if for all $\mathcal{S}$ and $\mathcal{T}$ and all $C \in \mathfrak{Co}_{\mathcal{S},\mathcal{T}}$: If $\mathbf{a} \in D_{\mathcal{S}}$ and $\langle \mathbf{a}, \mathbf{b} \rangle \in C$ then $\mathbf{b} \in D_{\mathcal{T}}$.

(d) We call $\mathfrak{F}$ *existentially friendly*, if for all worlds $\mathcal{S}$ and $\mathcal{T}$ such that $\mathfrak{Co}_{\mathcal{S},\mathcal{T}} \neq \emptyset$: If $\mathbf{b} \in D_{\mathcal{T}}$, then there is a $\mathbf{a} \in D_{\mathcal{S}}$ and a $C \in \mathfrak{Co}_{\mathcal{S},\mathcal{T}}$, such that $\langle \mathbf{a}, \mathbf{b} \rangle \in C$.

(e) If for all possible worlds $\mathcal{S}$ and $\mathcal{T}$ it holds that $C(U_{\mathcal{S}} - D_{\mathcal{S}}) \subseteq U_{\mathcal{T}} - D_{\mathcal{T}}$ (meaning that for all $C \in C_{\mathcal{S},\mathcal{T}} : \langle \mathbf{a}, \mathbf{b} \rangle \in C$ and $\mathbf{a} \in U_{\mathcal{S}} - D_{\mathcal{S}}$ implies that $\mathbf{b} \in U_{\mathcal{T}} - D_{\mathcal{T}}$), then $\mathfrak{F}$ is called *fictionally faithful.*

(f) A frame $\mathfrak{F}$ is said to be *locally reflexive*, if for every world $\mathcal{S}$, every natural number n and every n-tuple $\langle \mathbf{a}_1, \dots, \mathbf{a}_n \rangle$ of elements from $U_{\mathcal{S}}$ there is a relation $C \in \mathfrak{Co}_{\mathcal{S},\mathcal{S}}$ such that $C \supset \{\langle \mathbf{a}_i, \mathbf{a}_i \rangle : i = 1, \dots, n\}$.

(g) $\mathfrak{F}$ is then called *reflexive*, if it is locally-reflexive and for every world $\mathcal{S}$ and every n-tuple in $U_{\mathcal{S}}$ the same $C \in \mathfrak{Co}_{\mathcal{S},\mathcal{S}}$ can be chosen, i.e. $C \supset \{\langle \mathbf{a}, \mathbf{a} \rangle : \mathbf{a} \in U_{\mathcal{S}}\}$.

(h) $\mathfrak{F}$ is called *locally transitive*, if for every pair $\mathcal{S} \xrightarrow{C} \mathcal{T}$ and $\mathcal{T} \xrightarrow{\widehat{C}} \mathcal{R}$, every $n \in \mathbb{N}$ and every triple $\langle \mathbf{a}_1, \dots, \mathbf{a}_n \rangle$, $\langle \mathbf{b}_1, \dots, \mathbf{b}_n \rangle$ and $\langle \mathbf{c}_1, \dots, \mathbf{c}_n \rangle$ of n-tuples from $U_{\mathcal{S}}$, $U_{\mathcal{T}}$ and $U_{\mathcal{R}}$ respectively, such that $\langle \mathbf{a}_i, \mathbf{b}_i \rangle \in C$ and $\langle \mathbf{b}_i, \mathbf{c}_i \rangle \in \widehat{C}$ $(i = 1, \dots, n)$ there is a relation $\widetilde{C} \in \mathfrak{Co}_{\mathcal{S},\mathcal{R}}$, such that $\widetilde{C} \supset \{\langle \mathbf{a}_i, \mathbf{c}_i \rangle : i = 1, \dots, n\}$.

(i) $\mathfrak{F}$ is then called *transitive*, if for each pair $\mathcal{S} \xrightarrow{C} \mathcal{T}$ and $\mathcal{T} \xrightarrow{\widehat{C}} \mathcal{R}$ there is a relation $\mathcal{S} \xrightarrow{\widetilde{C}} \mathcal{R}$ in $\mathfrak{Co}_{\mathcal{S},\mathcal{R}}$, such that $\widetilde{C} \supset C \circ \widehat{C}$.

Notice in passing that the standard modal language that is assumed in this paper cannot distinguish between, for instance, transitivity and local transitivity because—when "moving" to another world—we can only "speak" about finitely many individuals at a time.

Now by showing that the corresponding property of frames holds on the canonical frame of a logic, we easily obtain e.g. the following completeness results.

Theorem 5.7

(a) The logic **FK** + (NoI) *is frame-complete with respect to the class of all functional frames.*

(b) **FK** + (NoD) *is frame-complete with respect to the class of all injective frames.*

(c) **FK** + (NoE) *is frame-complete with respect to the class of all existentially faithful frames.*

(d) **FK** + (CBF) *is frame-complete with respect to the class of all existentially friendly frames.*

(e) **FK** + (NoF) *is frame-complete with respect to the class of all fictionally faithful frames.*

(f) **FK** + (T) *is frame-complete with respect to the class of all locally reflexive frames.*

(g) **FK** + (4) *is frame-complete with respect to the class of all locally transitive frames.*

Definition 5.8 ((Semi-) categorial frames) A frame $\mathfrak{F}$ is said to be *semi-categorial*, if it is reflexive and transitive. Moreover, we say that $\mathfrak{F}$ is *categorial*, if $\mathfrak{F}$ is functional and semi-categorial.

As has already been noted, categorial frames that are also frames for classical predicate logic are semantically equivalent to C-sets as e.g. defined in (Ghilardi 1991). That is to say, a formula is valid on all C-sets if and only if it is valid on all categorial frames for classical predicate logic, i.e. where $U_{\mathcal{S}} = D_{\mathcal{S}}$ for all worlds $\mathcal{S}$. Now it was shown in (Ghilardi 1992) that all (standard) quantified extensions of canonical propositional modal logics above **S4** are frame-complete with respect to functor-semantics. Hence the semantics given in this paper allows for a wide class of frame-complete logics.

6 General Frames

In this last paragraph we want to sketch how to obtain general frame-completeness by enriching the notion of a modal frame by suitable algebras of 'admissible interpretations'. The situation is therefore quite analogous to the propositional case, where we have a natural possible worlds semantics, which is highly incomplete unless an algebraic component is added. The algebraic concept of a complex algebra as proposed below is in particular a straightforward extension of the concept of a boolean algebra with operators as known from propositional modal logic.

Definition 6.1 (Complex Algebras) Suppose $\mathfrak{F} = \langle U, C\rangle$ is a modal frame, where $U = \{\langle U_{\mathcal{S}}, D_{\mathcal{S}}\rangle : \mathcal{S} \in W\}$ is the associated family of universes. An **n**-*set* is a set of elements of the form $\langle \mathbf{a}, \mathcal{S}\rangle$, where $\mathcal{S} \in W$ and $\mathbf{a} \in (U_{\mathcal{S}})^n$. If A is an n-set, put

$$\blacklozenge A := \{\langle \mathbf{b}, \mathcal{S}\rangle \in (U_{\mathcal{S}})^n \times \{\mathcal{S}\} : \text{ there is a } \langle \mathbf{a}, \mathcal{T}\rangle \in A \text{ and a } C \in \mathfrak{Cos}_{\mathcal{S},\mathcal{T}} \text{ such that } \langle \mathbf{b}, \mathbf{a}\rangle \in C\} .$$

If $\sigma : \{1, \dots, m\} \to \{1, \dots, n\}$ is a map and $\mathbf{a}$ an n-tuple, let $\sigma(\mathbf{a})$ denote the m-tuple $\langle a_{\sigma(1)}, a_{\sigma(2)}, \dots, a_{\sigma(m)}\rangle$. In an m-set A and a map $\sigma : \{1, \dots, m\} \to \{1, \dots, n\}$ are given, put

$$\widehat{\sigma}(A) := \{\langle \mathbf{a}, \mathcal{S}\rangle \in (U_{\mathcal{S}})^n \times \{\mathcal{S}\} : \langle \sigma(\mathbf{a}), \mathcal{S}\rangle \in A\} .$$

Furthermore, if $j \in \omega$, A is an n-set and $j \leq n$, define the operation $\mathbb{E}_j$ as follows:

$$\mathbb{E}_j(A) := \{\langle a_1, \dots, a_{j-1}, c, a_{j+1}, \dots, a_n, \mathcal{S}\rangle \in (U_{\mathcal{S}})^n \times \{\mathcal{S}\} : c \in U_{\mathcal{S}} \text{ and there is a } b \in D_{\mathcal{S}}, \text{ such that } \langle a_1, \dots, a_{j-1}, b, a_{j+1}, \dots, a_n, \mathcal{S}\rangle \in A\}$$

Finally, if $i, j, n \in \omega$ with $i, j \leq n$ and A is an n-set, define the operation $id_{i,j}$ by

$$id_{i,j}(A) := \{\langle \mathbf{a}, \mathcal{S}\rangle \in (U_{\mathcal{S}})^n \times \{\mathcal{S}\} : \langle \mathbf{a}, \mathcal{S}\rangle \in A \text{ and } a_i = a_j\}.$$

A *complex algebra* $\mathbb{G}_n$ *of type n based on* $\mathfrak{F}$ is defined as a family of n-sets which is closed under all boolean operations and also under the operations $\blacklozenge$, $\mathbb{E}_j$ for every $j \in \omega$ and also under $id_{i,j}$ for every pair $i, j \in \omega$. A *complex algebra based on* $\mathfrak{F}$ is then defined as a sequence $\mathbb{G} = \langle \mathbb{G}_n : n \in \omega\rangle$ in which $\mathbb{G}_n$ are complex algebras of type n $(i \in \omega)$ and $\bigcup_{i\in\omega} \mathbb{G}_i$ is also closed under the operation $\widehat{\sigma}$ for every σ.

Definition 6.2 (General Frames) A *general frame* $\gamma\mathfrak{F}$ is a pair $\langle \mathfrak{F}, \mathbb{G}\rangle$, where $\mathfrak{F}$ is a modal frame and $\mathbb{G}$ is a complex algebra based on $\mathfrak{F}$. An *interpretation* I in $\mathfrak{F}$ is called *admissible*, if for every n-place relation symbol P, $I(P) \in \mathbb{G}_n$. A *model* based on a general frame $\langle \mathfrak{F}, \mathbb{G}\rangle$ is a triple $\langle \mathfrak{F}, \mathbb{G}, I \rangle$, where I is an admissible interpretation. A formula is then said to be *valid* on a general frame $\gamma\mathfrak{F}$, if it holds in every model based on $\gamma\mathfrak{F}$.

The reader will immediately notice the close connection between these models and the cartesian metaframes of Shehtman and Skvortsov (1993) or the general cartesian metaframes of Shirasu (1998). The connections between these different types of models will be examined in full detail in the sequel to this paper.

Now, if a model $\langle \mathfrak{F}, \mathbb{G}, I \rangle$ based on the general frame $\langle \mathfrak{F}, \mathbb{G} \rangle$ is given, we assign to each natural number n and each formula ϕ whose *free* variables are contained in $\bar{x} = \{x_1, \ldots, x_n\}$ and are exactly $\sigma(\bar{x}) = \langle x_{\sigma(1)}, \ldots, x_{\sigma(m)} \rangle$ (where $\sigma : \{1, \ldots, m\} \to \{1, \ldots, n\}$) an n-set $\widehat{I}(\phi, n)$ inductively as follows. Let $U^n := \bigcup_{\mathcal{S} \in W} U^n_{\mathcal{S}} \times \{\mathcal{S}\}$. Now set

$$\begin{array}{lcl} \widehat{I}(P^m(\sigma(\bar{x})), n) & := & \widehat{\sigma}(I(P^m)) \\ \widehat{I}(x_i \doteq x_j, n) & := & id_{i,j}(U^n) \\ \widehat{I}(\neg\phi, n) & := & U^n - \widehat{I}(\phi, n) \\ \widehat{I}(\phi_1 \wedge \phi_2, n) & := & \widehat{I}(\phi_1, n) \cap \widehat{I}(\phi_2, n) \end{array}$$

$$\begin{array}{lcl} \widehat{I}(\Diamond\phi, n) & := & \blacklozenge\widehat{I}(\phi, n) \\ \widehat{I}(\exists x_j.\phi, n) & := & \mathbb{E}_j(\widehat{I}(\phi, n)) \end{array}$$

It follows that for every formula ϕ, whose free variables are contained in $\{x_1, \ldots, x_n\}$, $\widehat{I}(\phi) \in \mathbb{G}_n$. The *canonical general* **L**-*frame* is then defined as follows. The set of worlds and relations is as defined above. Likewise, I_L is as previously defined on atomic predicates. We now define for every n a complex algebra of type n over U^n by means of the canonical interpretation as follows. Put

$$\mathbb{G}_n := \{\widehat{I_L}(\phi, n) : FV(\phi) \subseteq \{x_1, \ldots, x_n\}\} \, .$$

where $FV(\phi)$ denotes the set of variables occurring free in ϕ. It is easily shown that this is a general frame, which we call the *canonical general frame*. Still, as might be observed, the theory of a general frame is *not always* closed under second-order substitutions and hence not a modal predicate logic in the usual sense. However, the theory of the canonical general frame is closed under these substitutions and this suffices to establish the following completeness result.

In the following theorem, an L-frame is understood to be a modal frame $\mathfrak{F}$, such that $\mathfrak{F} \models L$.

Theorem 6.3 (Completeness) *If L is a modal predicate logic and $\mathbb{K}$ is the class of all general L-frames, then L is sound and complete with respect to the class $\mathbb{K}$, i.e. $L = \bigcap_{\gamma\mathfrak{F} \in \mathbb{K}} Th(\gamma\mathfrak{F})$.*

Proof. Let $\gamma\mathfrak{F}_L$ be the canonical general frame and suppose that there is a $\phi \in L$, such that $\gamma\mathfrak{F}_L \nvDash \phi$. Then there is an admissible interpretation I, a possible world $\mathcal{S}_\Delta$ and a valuation β such that $\langle \mathcal{S}_\Delta, I, \beta \rangle \nvDash \phi$. Because L is closed under first-order substitutions we can wlog assume that β is the canonical valuation v_Δ. Let $P_1^{n_1}, \ldots, P_m^{n_m}$ be all the relation symbols occurring in ϕ. Since I is admissible, $I(P_i^{n_i}, \Delta) = \widehat{I_L}(\psi_i) \in \mathbb{G}_{n_i}$ for $i = 1, \ldots, m$. In particular $FV(\psi_i) \subseteq \{x_1, \ldots, x_{n_i}\}$. Since $\phi \in L$, it follows by closure under second-order substitutions that $\widetilde{\phi} = (\psi_1/P_1^{n_1}, \ldots, \psi_m/P_m^{n_m})\phi \in L$ as well. But since $I(P_i^{n_i}) = \widehat{I_L}(\psi_i)$ it follows by induction that $\langle \mathcal{S}_\Delta, I_L \rangle \nvDash \widetilde{\phi}$, which contradicts $Th(\mathfrak{f}_L) = L$. This shows that the canonical general frame $\gamma\mathfrak{F}$ is an L-frame. Now, completeness follows immediately by appealing to the Canonical Model Theorem. ⊣

7 Conclusions and Further Work

We have introduced a semantics that combines the mathematical generality of functor semantics, metaframes or hyperdoctrines with aspects of Counterpart Theory. By constructing a canonical model for each modal predicate logic L, we have shown completeness with respect to modal structures. Furthermore, by defining appropriate conditions on counterpart relations, many interesting frame-completeness results are easily established. By enriching the notion of a modal frame by a suitable complex algebra of 'admissible interpretations', completeness with respect to general frames is obtained. Finally, the use of free logic does not only allow for an easy refutation of formulae like the Converse Barcan Formula, it is also the starting point for a proper treatment of non-denoting singular terms, talk of fictional objects or an investigation of different theories of definite descriptions in modal contexts. Thus, the move to a weaker base in free logic broadens the scope of modal predicate logics to be investigated considerably.

Since we have not considered these last topics in detail here, this is one of the subjects to be investigated in more detail in the future. Also, the exact connection between the different proposed semantics is to be spelled out. It should also be investigated how the semantics works with richer modal languages that use—besides an existence predicate—e.g. an actuality operator, universal modalities or lambda abstraction as discussed in (Fitting and Mendelsohn 1998). As a last point in case we would like to point to the problem of investigating the question which (philosophical) notion of object or individual is presupposed by the kind of semantics we have discussed in this paper. We believe that modal predicate logic in itself does not fix the notion of an object uniquely, since very different semantics involving quite distinct notions of an individual can give rise to general completeness results.

Acknowledgements

We would like to thank two anonymous referees for their helpful comments.

References

Bencivenga, Ermanno. 1986. Free Logics. In *Handbook of Philosophical Logic, Volume III: Alternatives to Classical Logic*, ed. D. Gabbay and F. Guenthner. 373–426. Kluwer Academic Publishers, Dordrecht.

Bencivenga, Ermanno, Karel Lambert, and Robert K. Meyer. 1982. The Ineliminability of E! in Free Quantification Theory Without Identity. In *Journal of Philosophical Logic Vol. 11.* 229–231.

Fitting, Melvin. 2001. Types, Tableaus and Gödels God. Unpublished Manuscript, `http://comet.lehman.edu/fitting/`.

Fitting, Melvin, and Richard L. Mendelsohn. 1998. First-Order Modal Logic. Kluwer Academic Publishers, Dordrecht.

Garson, James W. 1991. Applications of Free Logic to Quantified Intensional Logic. In *Philosophical Applications of Free Logic*, ed. Karel Lambert. 111–144. Oxford University Press, New York, Oxford.

Ghilardi, Silvio. 1991. Incompleteness Results in Kripke Semantics. In *Journal of Symbolic Logic, Nr. 2.* 516–538.

Ghilardi, Silvio. 1992. Quantified Extensions of Canonical Propositional Intermediate Logics. In *Studia Logica, Nr. 51.* 195–214.

Hintikka, Jaakko. 1969. Existential Presuppositions and Their Elimination. In *Models for Modalities. Selected Essays.* 23–44. D. Reidel Publishing Company, Dordrecht.

Hughes, G.E., and M.J. Cresswell. 1996. In *A New Introduction to Modal Logic.* Routledge, London.

Kracht, Marcus. 1999. Tools and Techniques in Modal Logic. Studies in Logic and the Foundations of Mathematics, Vol. 142, Elsevier Science Ltd., Amsterdam.

Kracht, Marcus, and Oliver Kutz. 2001. The Semantics of Modal Predicate Logic II. Modal Individuals Revisited. Unpublished Manuscript, `http://www.math.fu-berlin.de/~kracht/`.

Kripke, Saul A. 1979. A Puzzle About Belief. In *Meaning and Use*, ed. A. Margalit. 239–283. D. Reidel Publishing Co., Dordrecht, Boston.

Kutz, Oliver. 2000. Kripke–Typ Semantiken für die modale Prädikatenlogik. Master's thesis, Humboldt–Universität zu Berlin.

Kutz, Oliver. 2001. Some Remarks on Modal Predicate Logics. Unpublished Manuscript, `http://www.informatik.uni-leipzig.de/~kutz/`.

Lambert, Karel. 1997. Free Logics: Their Foundations, Character, and Some Applications Thereof. Academia Verlag, Sankt Augustin.

Lewis, David. 1968. Counterpart Theory and Quantified Modal Logic. In *Journal of Philosophy, Vol. 65.* 113–126. repr. in *The Possible and the Actual*, ed. Michael J. Loux, Ithaca 1979; also in: David Lewis, *Philosophical papers 1*, Oxford 1983.

Shirasu, Hiroyuki. 1998. Duality in Superintuitionistic and Modal Predicate Logics. In *Advances in Modal Logic, Volume I*, ed. Marcus Kracht et. al. 223–236. CSLI Publications, Stanford.

Skvortsov, D.P., and V.B. Shehtman. 1993. Maximal Kripke–Type Semantics for Modal and Superintuitionistic Predicate Logics. In *Annals of Pure and Applied Logic, Nr. 63.* 69–101.

van Benthem, Johan. 1993. Beyond Accessibility: Functional Models for Modal Logic. In *Diamonds and Defaults: Studies in Pure and Applied Intensional Logic*, ed. Maarten de Rijke. 1–18. Kluwer Academic Publishers, Dordrecht.

Marcus Kracht
Fachbereich Mathematik und Informatik
Institut für Mathematik II
Freie Universität Berlin
Arnimallee 3
D-14195 Berlin, Germany
E-mail: `kracht@math.fu-berlin.de`

Oliver Kutz
Institut für Informatik
Universitaet Leipzig
Augustus-Platz 10-11
04109 Leipzig , Germany
E-mail: `kutz@informatik.uni-leipzig.de`

17

S5 × S5 × S5 Lacks the Finite Model Property

AGNES KURUCZ

ABSTRACT. It follows from algebraic results of Maddux that every multi-modal logic L such that $[\mathbf{S5}, \mathbf{S5}, \ldots, \mathbf{S5}] \subseteq L \subseteq \mathbf{S5}^n$ is undecidable, whenever $n \geq 3$. This implies that the product logic $\mathbf{S5} \times \mathbf{S5} \times \mathbf{S5}$ does not have the product finite model property. Here we answer a question of Gabbay and Shehtman by showing that $\mathbf{S5} \times \mathbf{S5} \times \mathbf{S5}$ also lacks the 'real' finite model property (fmp). We prove that every logic L from the above interval lacks the fmp. (In algebraic setting: If V is a variety of n-dimensional diagonal-free cylindric algebras which contains all the representables then V does not have the finite algebra property.)

1 Introduction and the result

Multi-modal logics are of growing importance in many areas of computer science, artificial intelligence, knowledge representation and reasoning, and linguistics. In this paper we discuss the finite model property of *n-modal logics*: propositional multi-modal logics having finitely many unary modal operators $\Diamond_0, \ldots, \Diamond_{n-1}$ (and their duals $\Box_0, \ldots, \Box_{n-1}$). Formulas of this language, using propositional variables from some fixed countable set P, are called *n-modal formulas*. For each natural number $n > 1$, the well-known (cf. Gabbay and Shehtman 1998) n-modal logic

$$[\underbrace{\mathbf{S5}, \mathbf{S5}, \ldots, \mathbf{S5}}_{n}]$$

is the smallest set of n-modal formulas which

(1) is closed under the rules of Substitution, Modus Ponens, and Necessitation $A/\Box_i A$, for $i < n$;

and contains, for all $i, j < n$,

Advances in Modal Logic, Volume 3
F. Wolter, H. Wansing, M. de Rijke, and M. Zakharyaschev, eds.

(2) all propositional tautologies and formulas of the form

$$\Box_i(p \to q) \to (\Box_i p \to \Box_i q);$$

(3) the **S5**-axioms for $\Box_i$: $\Box_i p \to p$, $p \to \Box_i \Diamond_i p$, $\Box_i p \to \Box_i \Box_i p$;

(4) $\Box_i \Box_j p \leftrightarrow \Box_j \Box_i p$.

Special frames for n-modal logics are the *product frames*. Given Kripke frames $\mathfrak{F}_0 = \langle W_0, R_0 \rangle, \ldots, \mathfrak{F}_{n-1} = \langle W_{n-1}, R_{n-1} \rangle$, their product is defined to be the relational structure

$$\mathfrak{F}_0 \times \cdots \times \mathfrak{F}_{n-1} = \langle W_0 \times \cdots \times W_{n-1}, \bar{R}_0, \ldots, \bar{R}_{n-1} \rangle$$

where, for each $i < n$, $\bar{R}_i$ is the following binary relation on the Cartesian product $W_0 \times \cdots \times W_{n-1}$:

$$\langle u_0, \ldots, u_{n-1} \rangle \bar{R}_i \langle v_0, \ldots, v_{n-1} \rangle \quad \text{iff} \quad u_i R_i v_i \quad \text{and} \quad u_k = v_k, \quad \text{for } k \neq i.$$

Given Kripke complete unimodal logics L_i $(i < n)$, define the *product logic* $L_0 \times \cdots \times L_{n-1}$ as the set of all n-modal formulas which are valid in those product frames $\langle W_0, R_0 \rangle \times \cdots \times \langle W_{n-1}, R_{n-1} \rangle$ where, for each $i < n$, $\langle W_i, R_i \rangle$ is a frame for L_i. This way one can obtain new logics, e.g., for $n \geq 3$, the product logic $\mathbf{S5}^n$—being non-finitely axiomatizable by (Johnson 1969)—is different from $[\mathbf{S5}, \mathbf{S5}, \ldots, \mathbf{S5}]$. Though, it is routine to check that $[\mathbf{S5}, \mathbf{S5}, \ldots, \mathbf{S5}] \subseteq \mathbf{S5}^n$ holds. Also, it is not hard to see that $\mathbf{S5}^n$ is already determined by products of rooted (i.e., universal) **S5**-frames.

From now on, we always let $n \geq 3$. An n-modal logic L has the *finite model property* (*fmp*) if for any n-formula $\varphi \notin L$ there is a finite model $\mathfrak{M}$ such that $\mathfrak{M} \models L$ and $\mathfrak{M} \not\models \varphi$. Note that it is the same as requiring the existence of a finite frame $\mathfrak{F}$ with $\mathfrak{F} \models L$ and $\mathfrak{F} \not\models \varphi$ (see e.g. Chagrov and Zakharyaschev 1997, Thm.8.47). L has the *product fmp* if for any $\varphi \notin L$ there is a finite product frame such that $\mathfrak{F} \models L$ and $\mathfrak{F} \not\models \varphi$. Of course, the product fmp implies the fmp, but not the other way round, e.g., $\mathbf{K}^n$ has the fmp (see Gabbay and Shehtman 1998), but lacks the product fmp, for $n \geq 3$ (see Hirsch et al. to appear).

It follows from the algebraic results of (Maddux 1980) that every n-modal logic between $[\mathbf{S5}, \mathbf{S5}, \ldots, \mathbf{S5}]$ and $\mathbf{S5}^n$ is undecidable (see also section 2 below for the technique of Maddux 1980). This implies that the finitely axiomatizable logic $[\mathbf{S5}, \mathbf{S5}, \ldots, \mathbf{S5}]$ does not have the fmp. Since $\mathbf{S5}^n$ is recursively enumerable (see, e.g., Henkin et al. 1985) and, by the finite axiomatizability of **S5**, finite product frames for $\mathbf{S5}^n$ are recursively enumerable, it also follows that $\mathbf{S5}^n$ lacks the product fmp. However, even if a finite frame for $\mathbf{S5}^n$ is a p-morphic image of some product frame for $\mathbf{S5}^n$, this product frame cannot necessarily be chosen finite (see Hirsch et al. to appear for a counterexample). Thus, in case of $\mathbf{S5}^n$, the lack of fmp does not follow in an obvious way from the lack of product fmp.

Our main result is the following theorem:

Theorem 1.1 *Let L be a logic such that*

$$[\mathbf{S5}, \mathbf{S5}, \ldots, \mathbf{S5}] \subseteq L \subseteq \mathbf{S5}^n.$$

Then L lacks the finite model property, whenever $n \geq 3$.

As a corollary we obtain a negative answer to question Q21 of Gabbay and Shehtman (1998):

Corollary 1.2 $\mathbf{S5}^3$ *does not have the fmp.*

However, it remains open whether product logics like, e.g., $\mathbf{K4}^3$, $\mathbf{S4}^3$, $\mathbf{K4} \times \mathbf{S4} \times \mathbf{S5}$ have the fmp. Answering these questions may give some insight for intriguing open decision problems of two-dimensional products of transitive logics, such as $\mathbf{S4} \times \mathbf{S4}$, $\mathbf{K4} \times \mathbf{S4}$, $\mathbf{K4} \times \mathbf{K4}$. These logics are known to be finitely axiomatizable (see Gabbay and Shehtman 1998), thus if they had the fmp then they would be decidable. Note that these logics lack the product fmp.

2 Modal algebras

In this section we reformulate and prove Theorem 1.1 in an algebraic setting, see Theorem 2.1 below. Similarly to the unimodal case (see e.g. Chagrov and Zakharyaschev 1997), an *n-modal algebra* is a Boolean algebra with n unary normal operators, that is, a structure

$$\mathfrak{A} = \langle A, +^{\mathfrak{A}}, \cdot^{\mathfrak{A}}, -^{\mathfrak{A}}, 1^{\mathfrak{A}}, 0^{\mathfrak{A}}, \diamond_i^{\mathfrak{A}} \rangle_{i<n},$$

where $\langle A, +^{\mathfrak{A}}, \cdot^{\mathfrak{A}}, -^{\mathfrak{A}}, 1^{\mathfrak{A}}, 0^{\mathfrak{A}} \rangle$ is a Boolean algebra and, for all $a, b \in A$, $i < n$,

$$\diamond_i^{\mathfrak{A}}(a +^{\mathfrak{A}} b) = \diamond_i^{\mathfrak{A}} a +^{\mathfrak{A}} \diamond_i^{\mathfrak{A}} b \text{ and } \diamond_i^{\mathfrak{A}} 0^{\mathfrak{A}} = 0^{\mathfrak{A}}.$$

A *valuation to* $\mathfrak{A}$ is a function v mapping n-modal formulas to elements of $\mathfrak{A}$ which is defined the usual natural way: it turns the propositional connectives to the Booleans and the modal operators to themselves. An n-modal formula φ is said to be *true* in an *algebraic model* $\langle \mathfrak{A}, v \rangle$ if $v(\varphi) = 1^{\mathfrak{A}}$ holds. We say that φ is *valid* or *holds* in algebra $\mathfrak{A}$ (in symbols: $\mathfrak{A} \models \varphi$) if φ is true in all algebraic models of form $\langle \mathfrak{A}, v \rangle$. For any normal n-modal logic L, $\mathsf{Alg}L$ is the class of all n-modal algebras validating all formulas of L.

The class $\mathsf{Alg}[\mathbf{S5}, \mathbf{S5}, \ldots, \mathbf{S5}]$ is a class well-known in algebraic logic, it is the class Df_n of *diagonal-free cylindric algebras of dimension n*

(see Andréka et al. to appear, Henkin et al. 1985). In other words, an n-dimensional diagonal-free cylindric algebra is an n-modal algebra where (3)–(4) of the previous section hold. To obtain $\mathsf{AlgS5}^n$, observe that every Kripke frame $\mathfrak{F} = \langle W, R_0, \ldots, R_{n-1}\rangle$ for an n-modal logic gives rise to the n-modal algebra $\mathfrak{A}(\mathfrak{F})$ of all subsets of W where, for every $X \subseteq W$, $i < n$,

$$\Diamond_i^{\mathfrak{A}(\mathfrak{F})} X = \{w \in W \ : \ \exists u \in X \text{ with } wR_iu\}.$$

This way every product $\mathfrak{F}$ of rooted (i.e., universal) **S5**-frames leads to a diagonal-free cylindric algebra $\mathfrak{A}(\mathfrak{F})$ whose elements are all subsets of some Cartesian product $W_0 \times \cdots \times W_{n-1}$ and, for any such subset X, $i < n$,

$$\Diamond_i^{\mathfrak{A}(\mathfrak{F})} X = \{\langle w_0, \ldots, w_{n-1}\rangle : \exists u \in W_i \text{ with } \langle w_0, \ldots, w_{i-1}, u, w_{i+1}, \ldots, w_{n-1}\rangle \in X\}.$$

It is easy to see that, in general, an n-modal formula is valid in a frame $\mathfrak{F}$ iff it is valid in the algebra $\mathfrak{A}(\mathfrak{F})$. Therefore, $\mathsf{AlgS5}^n$ is the variety generated by the above kind of diagonal-free cylindric algebras. This variety is called in the algebraic logic literature the variety RDf_n of *representable diagonal-free cylindric algebras of dimension* n (see Andréka et al. to appear, Henkin et al. 1985).

A variety V of n-modal algebras has the *finite algebra property* if V is generated by its finite members. In other words, for all n-modal formulas φ, if $\mathfrak{A} \not\models \varphi$ for some $\mathfrak{A} \in V$ then there is some finite $\mathfrak{B} \in V$ with $\mathfrak{B} \not\models \varphi$. It is well-known (see e.g. Chagrov and Zakharyaschev 1997) that a logic L has the fmp iff $\mathsf{Alg}L$ has the finite algebra property. Now we are ready to reformulate Theorem 1.1 for diagonal-free cylindric algebras. Note that related results about cylindric algebras are in (Németi 1984).

Theorem 2.1 *Let $n \geq 3$, and V be a variety with $\mathsf{RDf}_n \subseteq V \subseteq \mathsf{Df}_n$. Then V does not have the finite algebra property. Namely, there is an n-modal formula φ such that*

- *$\mathfrak{A} \models \varphi$, for every finite $\mathfrak{A} \in \mathsf{Df}_n$, and*
- *there is some $\mathfrak{B} \in \mathsf{RDf}_n$ with $\mathfrak{B} \not\models \varphi$.*

Proof. Consider the following quasi-equation (that is, a conjunction of finitely many equations implying one equation) Q^* of the language of semigroups (we use $\circ$ for multiplication):

$$[(x = e\circ x = x\circ e) \wedge (y = e\circ y = y\circ e) \wedge (e = e\circ e) \wedge (x\circ y = e)] \rightarrow (y\circ x = e).$$

Claim 2.2 *Q^* fails in some infinite semigroup.*

Proof. For instance, consider the monoid of $\omega \to \omega$ functions with composition. It is routine to check that the function $m \mapsto m+1$ has an inverse from one side but not from the other. $\dashv$

Claim 2.3 *Q^* holds in every finite semigroup.*

Proof. Assume that the antecedent of Q^* holds in some finite semigroup $\mathfrak{S}$. Since $\mathfrak{S}$ is finite, the subsemigroup of $\mathfrak{S}$ generated by x, y and e is a finite monoid with identity element e. Then in this finite monoid $x^k = x^\ell$ must hold, for some natural numbers $k < \ell$. Therefore, by x having a right-inverse, $x^{\ell-k} = e$ holds (with $\ell - k \geq 1$). Thus

$$y \circ x = x^{\ell-k} \circ y \circ x = x^{\ell-k-1} \circ x \circ y \circ x = x^{\ell-k-1} \circ x = x^{\ell-k} = e$$

holds. $\dashv$

Next, using the technique of (Maddux 1980), we interpret semigroups in diagonal-free cylindric algebras. Take some $\mathfrak{A} \in \mathsf{Df}_n$, and $d_0, d_1 \in A$. For any $a, b \in A$, define

$$a \circ^{\mathfrak{A}}_{d_0 d_1} b \;=\; \Diamond^{\mathfrak{A}}_2 \left[\Diamond^{\mathfrak{A}}_1 (d_0 \cdot^{\mathfrak{A}} \Diamond^{\mathfrak{A}}_2 a) \cdot^{\mathfrak{A}} \Diamond^{\mathfrak{A}}_0 (d_1 \cdot^{\mathfrak{A}} \Diamond^{\mathfrak{A}}_2 b)\right].$$

Given propositional variables $r_0, r_1 \in P$ and n-modal formulas ψ, χ, the n-modal formula $\psi \circ_{r_0 r_1} \chi$ is defined analogously, by taking

$$\psi \circ_{r_0 r_1} \chi \;=\; \Diamond_2 \left[\Diamond_1 (r_0 \wedge \Diamond_2 \psi) \wedge \Diamond_0 (r_1 \wedge \Diamond_2 \chi)\right].$$

Now fix some propositional variables $r_0, r_1 \in P$. For any term τ of the language of semigroups, define inductively an n-modal formula τ^+ as follows:

- for any variable x, let $x^+ = p_x$, where $p_x \in P$ and $p_x \neq r_0, r_1$, and
- $(\sigma \circ \varrho)^+ = \sigma^+ \circ_{r_0 r_1} \varrho^+$.

For any quasi-equation Q of the language of semigroups of form

$$[(\tau_1 = \sigma_1) \wedge \cdots \wedge (\tau_m = \sigma_m)] \;\to\; (\tau_0 = \sigma_0),$$

let φ_Q be the following n-modal formula:

$$\Box \left[(\tau_1^+ \leftrightarrow \sigma_1^+) \wedge \cdots \wedge (\tau_m^+ \leftrightarrow \sigma_m^+) \wedge (r_0 \leftrightarrow \Diamond_0 r_0) \wedge (r_1 \leftrightarrow \Diamond_1 r_1)\right] \to$$
$$\to \left[(\tau_0^+ \circ_{r_0 r_1} \Diamond_1 q) \leftrightarrow (\sigma_0^+ \circ_{r_0 r_1} \Diamond_1 q)\right],$$

where $\Box \chi$ abbreviates $\Box_0 \Box_1 \ldots \Box_{n-1} \chi$, and q is a fresh propositional variable.

Lemma 2.4 (Maddux 1980) *Let $\mathfrak{A} \in \mathsf{Df}_n$, and $d_0, d_1 \in A$ such that $\Diamond^{\mathfrak{A}}_0 d_0 = d_0$ and $\Diamond^{\mathfrak{A}}_1 d_1 = d_1$. For any $a, b \in A$, define*

$$a \sim^{\mathfrak{A}}_{d_0 d_1} b \iff (\forall w \in A)\; a \circ^{\mathfrak{A}}_{d_0 d_1} \Diamond^{\mathfrak{A}}_1 w = b \circ^{\mathfrak{A}}_{d_0 d_1} \Diamond^{\mathfrak{A}}_1 w.$$

Then the following hold:

(i) $\sim^{\mathfrak{A}}_{d_0 d_1}$ *is a congruence of* $\langle A, \circ^{\mathfrak{A}}_{d_0 d_1}\rangle$, *that is, for any* $a, b, c \in A$, *if* $a \sim^{\mathfrak{A}}_{d_0 d_1} b$ *then*

$$(a \circ^{\mathfrak{A}}_{d_0 d_1} c) \sim^{\mathfrak{A}}_{d_0 d_1} (b \circ^{\mathfrak{A}}_{d_0 d_1} c) \quad \textit{and} \quad (c \circ^{\mathfrak{A}}_{d_0 d_1} a) \sim^{\mathfrak{A}}_{d_0 d_1} (c \circ^{\mathfrak{A}}_{d_0 d_1} b).$$

(ii) *The quotient algebra* $\langle A, \circ^{\mathfrak{A}}_{d_0 d_1}\rangle / \sim^{\mathfrak{A}}_{d_0 d_1}$ *is a semigroup.*

(iii) *If* $\mathfrak{A}$ *is a simple algebra then, for any quasi-equation* Q *of the language of semigroups,*

$$\langle A, \circ^{\mathfrak{A}}_{d_0 d_1}\rangle / \sim^{\mathfrak{A}}_{d_0 d_1} \models Q \ \textit{ implies } \ \mathfrak{A} \models \varphi_Q.$$

Now recall the quasi-equation Q^* above.

Corollary 2.5 φ_{Q^*} *holds in every finite* $\mathfrak{A} \in \mathsf{Df}_n$.

Proof. Assume that there is some finite $\mathfrak{A} \in \mathsf{Df}_n$ and $d_0, d_1 \in A$ such that

$$\mathfrak{A} \not\models \varphi_{Q^*}[r_0/d_0, r_1/d_1] \,.$$

It is well-known (see Henkin et al. 1985) that every subdirectly irreducible algebra in Df_n is simple, so we may assume that $\mathfrak{A}$ is simple. Then $\Diamond^{\mathfrak{A}}_0 d_0 = d_0$ and $\Diamond^{\mathfrak{A}}_1 d_1 = d_1$ must hold, otherwise the antecedent of φ_{Q^*} would be false. Thus, by Lemma 2.4 above, Q^* fails in the finite semigroup $\langle A, \circ^{\mathfrak{A}}_{d_0 d_1}\rangle / \sim^{\mathfrak{A}}_{d_0 d_1}$, contradicting Claim 2.3. ⊣

Lemma 2.6 φ_{Q^*} *fails in some (infinite)* $\mathfrak{B} \in \mathsf{RDf}_n$.

Proof. This is the similar to the argument in (Maddux 1980), but we sketch it for completeness. By Claim 2.2, there is some (infinite) semigroup $\mathfrak{S} = \langle S, \circ\rangle$ where φ fails. We may assume that $\mathfrak{S}$ is a monoid. (If not then choose some $e \notin S$ and take the monoid $\mathfrak{S}^+ = \langle S \cup \{e\}, \circ'\rangle$, where $x \circ' y = x \circ y$ whenever $x, y \in S$ and $x \circ' e = e \circ' x = x$, for every $x \in S \cup \{e\}$.) Now consider the diagonal-free cylindric algebra $\mathfrak{A}_S \in \mathsf{RDf}_n$ of all subsets of S^n, and take the *diagonal elements* D_{02} and D_{12} of $\mathfrak{A}_S$ defined by

$$D_{i2} = \{\langle s_0, \ldots, s_{n-1}\rangle \in S^n \ : \ s_i = s_2\} \quad (i = 0, 1).$$

Then $\Diamond^{\mathfrak{A}_S}_0 D_{12} = D_{12}$ and $\Diamond^{\mathfrak{A}_S}_1 D_{02} = D_{02}$ hold. Define a map h by taking, for any $s \in S$,

$$h(s) = \{\langle s_0, \ldots, s_{n-1}\rangle \in S^n \ : \ s_1 = s_0 \circ s\} \,.$$

It is easy to check that h is one-one and, for all $s, t \in S$,

$$h(s \circ t) = h(s) \circ^{\mathfrak{A}_S}_{D_{12} D_{02}} h(t)$$

hold. (This h is in fact the Cayley representation of the monoid $\mathfrak{S}$, taken in the first two coordinates of $\mathfrak{A}_S$.) Now, in order to show that φ_{Q^*} fails in $\mathfrak{A}_S$, it is enough to prove that $h(s) \not\sim^{\mathfrak{A}_S}_{D_{12}D_{02}} h(t)$ whenever $s \neq t$. To this end, take some $s \neq t \in S$ and let

$$w = \{\langle s_0, \ldots, s_{n-1}\rangle \in S^n \ : \ s_0 = s\} .$$

Let e denote the identity element of $\mathfrak{G}$. Then clearly $\Diamond_1^{\mathfrak{A}_S} w = w$, and any sequence of form $\langle e, \ldots\rangle$ belongs to $h(s) \circ^{\mathfrak{A}_S}_{D_{12}D_{02}} w$ but not to $h(t) \circ^{\mathfrak{A}_S}_{D_{12}D_{02}} w$, showing that they are different. ⊣

Finally, Corollary 2.5 and Lemma 2.6 prove Theorem 2.1. ⊣

Acknowledgments

Research supported by UK EPSRC grant GR/L85978 and by Hungarian National Foundation for Scientific Research grant T30314. Thanks to Robin Hirsch and Gábor Sági for helpful discussions.

References

Andréka, H., I. Németi, and I. Sain. to appear. Algebraic logic. In *Handbook of Philosophical Logic*, ed. D. M. Gabbay and F. Guenthner. Kluwer Academic Publishers, 2nd edition.

Chagrov, A., and M. Zakharyaschev. 1997. *Modal logic.* Oxford Logic Guides, No. 35. Oxford University Press.

Gabbay, D.M., and V.B. Shehtman. 1998. Products of modal logics, part I. *Logic Journal of the IGPL* 6(1):73–146.

Henkin, L., J.D. Monk, and A. Tarski. 1985. *Cylindric Algebras, Part II.* North Holland.

Hirsch, R., I. Hodkinson, and A. Kurucz. to appear. On modal logics between **K** × **K** × **K** and **S5** × **S5** × **S5**. *Journal of Symbolic Logic.* Available at `http://www.doc.ic.ac.uk/~kuag/k3undec.ps`.

Johnson, J.S. 1969. Nonfinitizability of classes of representable polyadic algebras. *Journal of Symbolic Logic* 34:344–352.

Maddux, R. 1980. The equational theory of CA_3 is undecidable. *Journal of Symbolic Logic* 45:311–316.

Németi, I. 1984. Neither the variety of cylindric algebras nor the variety of representable cylindric algebras is generated by its finite members for dimensions greater than 2. Manuscript.

Agnes Kurucz
Department of Computer Science
King's College
Strand, London WC2R 2LS, UK
E-mail: `kuag@doc.ic.ac.uk, kurucz@dcs.kcl.ac.uk`
URL: `http://www.doc.ic.ac.uk/~kuag/`

18

The Complexity of Reasoning with Boolean Modal Logics

CARSTEN LUTZ AND ULRIKE SATTLER

ABSTRACT. In this paper, we investigate the complexity of reasoning with various Boolean Modal Logics. The main results are that (i) adding negation of modal parameters to (multi-modal) **K** makes reasoning ExpTime-complete and (ii) adding atomic negation and conjunction to **K** even yields a NExpTime-complete logic. The last result is relativized by the fact that it depends on an infinite number of modal parameters to be available. If the number of modal parameters is bounded, full Boolean Modal Logic becomes ExpTime-complete.

1 Motivation

Since Modal Logics are an extension of Propositional Logic, they provide Boolean operators for constructing complex formulae. However, most Modal Logics do not admit Boolean operators for constructing *complex modal parameters* to be used in the box and diamond operators. This asymmetry is not present in Boolean Modal Logics, in which box and diamond quantify over arbitrary Boolean combinations of atomic modal parameters; see Gargov and Passy 1987. Boolean Modal Logics have been considered in various forms and contexts:
1. "Pure" Boolean Modal Logic has been studied in Gargov and Passy 1987. Negation and intersection of modal parameters occur in some variants of Propositional Dynamic Logic, see, e.g., Danecki 1984, Harel 1984, Passy and Tinchev 1991.
2. The modal box operator can be thought of as expressing necessity. More precisely, when employing the usual Kripke Semantics, $\Box\varphi$ holds at a world w iff w' being accessible from w implies that φ holds

Advances in Modal Logic, Volume 3
F. Wolter, H. Wansing, M. de Rijke, and M. Zakharyaschev, eds.

at w'. Given this, it is obviously quite natural to define a symmetric operator $\boxdot$ (sometimes called "window operator") such that $\boxdot\varphi$ holds at a world w iff φ holding at a world w' implies that w' is accessible from w. Obviously, the window operator can be thought of as expressing sufficiency. Logics with this operator were investigated from different viewpoints by, e.g., Humberstone, Gargov et al., and Goranko Humberstone 1983, Gargov et al. 1987, Goranko 1987, Goranko 1990. If negation of modal parameters is available, the window operator comes for free since we can write $\boxdot_R\varphi$ as $[\neg R]\neg\varphi$.

3. There are several Description Logics that provide "negation of roles" which corresponds to the negation of modal parameters, see, e.g. Hustadt and Schmidt 2000. Union and intersection of modal parameters are also considered in Description Logics and other KR formalisms, as is the window operator; see Givan et al. 1991, Lutz and Sattler 2000a.

Although—as we just argued—logics involving Boolean operators on modal parameters or the window operator are widely used, to the best of our knowledge, complexity results for this class of logics have never been obtained. In this paper, we close the gap and determine the complexity of the satisfiability and validity problems for many Boolean Modal Logics. In the first part of this paper (Sections 2 and 3), we investigate the logic $\mathbf{K}_\omega$ ($\mathbf{K}$ with a countably infinite number of accessibility relations) enriched with negation of modal parameters and show that the afore mentioned inference problems are ExpTime-complete using an automata-theoretic approach. We then demonstrate the generality of our approach by extending this result to the logic $(\mathbf{K}_\omega \otimes \mathbf{K4}_\omega)^\neg$, i.e., to the fusion of $\mathbf{K}_\omega$ with $\mathbf{K4}_\omega$ enriched with negation on relations. In the second part of this paper (Sections 4 and 5), we add other Boolean operators on roles. In doing so, one has the choice to either restrict negation to atomic relations or to allow for full negation of relations.

We give a complete list of complexity results for the logics obtained in this way, the central result being that the combination of (atomic) negation with intersection yields a logic whose inference problems are NExpTime-complete. The lower bound is obtained by a reduction of a NExpTime-complete variant of the domino problem. The mentioned result obviously implies that full Boolean Modal Logic $\mathbf{K}_\omega^{\neg,\cap,\cup}$ is also NExpTime-complete. However, the lower bound crucially depends on the number of relations to be unbounded. Inspired by this observation, in Section 5, we supplement our result by showing that, for any fixed finite number of relations, full Boolean Modal Logic is ExpTime-complete. The upper bound is proved by a reduction to multi-modal $\mathbf{K}$ (with finitely many relations) enriched with the universal modality.

To complete our investigation, in Section 6 we show that $\mathbf{K}_\omega$ with union and intersection of roles and without negation is of the same complexity as pure $\mathbf{K}_\omega$, i.e., PSpace-complete. Summing up, we thus have tight complexity bounds for $\mathbf{K}_\omega$ extended with any combination of Boolean operators on roles. This paper is accompanied by a technical report which contains all proofs and technical details (Lutz and Sattler 2000b).

2 Preliminaries

We define syntax and semantics of $\mathbf{K}_\omega^\neg$, introduce looping automata, and discuss some model- and complexity-theoretic properties of $\mathbf{K}_\omega^\neg$.

Definition 2.1 Given a countably infinite set of *propositional variables* Φ and a countably infinite set of *atomic modal parameters* $R_1, R_2, \ldots$, the set of $\mathbf{K}_\omega^\neg$*-formulae* is the smallest set that (i) contains the propositional variables in Φ, (ii) is closed under Boolean connectives $\wedge$, $\vee$, and $\neg$, and (iii) if it contains φ, then it also contains $\langle R_i \rangle \varphi$, $[R_i]\varphi$, $\langle \neg R_i \rangle \varphi$, and $[\neg R_i]\varphi$ for $i \geq 1$. The set of $\mathbf{K}_\omega^\neg$-modal parameters is the smallest set containing all atomic modal parameters and their negations (i.e., expressions of the form $\neg R_i$).

$\mathbf{K}_\omega^\neg$ semantics is given by *Kripke structures* $\mathcal{M} = \langle W, \pi, \mathcal{R}_1, \ldots \rangle$, where W is a set of *worlds*, π is a mapping from the set of propositional variables into sets of worlds (i.e., for each $p \in \Phi$, $\pi(p)$ is the set of worlds in which p holds), and $\mathcal{R}_i$ is a binary relation on the worlds W, the so-called *accessibility relation* for the atomic modal parameter R_i.

The semantics is then given as follows, where, for a $\mathbf{K}_\omega^\neg$-formula φ, a Kripke structure $\mathcal{M}$, and a world $w \in W$; the expression $\mathcal{M}, w \models \varphi$ is read as "φ holds in $\mathcal{M}$ in world w".

$\mathcal{M}, w \models p$	iff	$w \in \pi(p)$ for $p \in \Phi$
$\mathcal{M}, w \models \varphi_1 \wedge \varphi_2$	iff	$\mathcal{M}, w \models \varphi_1$ and $\mathcal{M}, w \models \varphi_2$
$\mathcal{M}, w \models \varphi_1 \vee \varphi_2$	iff	$\mathcal{M}, w \models \varphi_1$ or $\mathcal{M}, w \models \varphi_2$
$\mathcal{M}, w \models \neg\varphi$	iff	$\mathcal{M}, w \not\models \varphi$
$\mathcal{M}, w \models \langle R_i \rangle \varphi$	iff	$\exists w' \in W$ with $(w, w') \in \mathcal{R}_i$ and $\mathcal{M}, w' \models \varphi$
$\mathcal{M}, w \models [R_i]\varphi$	iff	$\forall w' \in W$, if $(w, w') \in \mathcal{R}_i$, then $\mathcal{M}, w' \models \varphi$
$\mathcal{M}, w \models \langle \neg R_i \rangle \varphi$	iff	$\exists w' \in W$ with $(w, w') \notin \mathcal{R}_i$ and $\mathcal{M}, w' \models \varphi$
$\mathcal{M}, w \models [\neg R_i]\varphi$	iff	$\forall w' \in W$, if $(w, w') \notin \mathcal{R}_i$, then $\mathcal{M}, w' \models \varphi$

A $\mathbf{K}_\omega^\neg$-formula φ is *satisfiable* iff there is a Kripke structure $\mathcal{M}$ with a set of worlds W and a world $w \in W$ such that $\mathcal{M}, w \models \varphi$. Such a structure is called a *model* of φ. Two $\mathbf{K}_\omega^\neg$-formulae φ and ψ are *equivalent* (written $\varphi \equiv \psi$) iff $\mathcal{M}, w \models \varphi \iff \mathcal{M}, w \models \psi$ for all Kripke structures $\mathcal{M}$ with set of worlds W and all worlds $w \in W$. Let R be a modal parameter. We write $\mathcal{M}, (w, w') \models R$ to express that (i) $(w, w') \in \mathcal{R}_i$ if R is an atomic modal parameter R_i and (ii) $(w, w') \notin \mathcal{R}_i$ if $R = \neg R_i$ for an atomic modal parameter R_i.

Throughout this paper, we denote modal parameters by R and S. For the sake of brevity, we will often omit the word "modal" when talking about modal parameters. As usual, we write $\varphi \to \psi$ for $\neg\varphi \vee \psi$ and $\varphi \leftrightarrow \psi$ for $(\varphi \to \psi) \wedge (\psi \to \varphi)$. The semantics of the window operator discussed in the motivation can formally be defined as follows:

$$\mathcal{M}, w \models \boxminus_{R_i} \varphi \quad \text{iff} \quad \text{for all } w' \in W, \text{ if } \mathcal{M}, w' \models \varphi, \text{ then } (w, w') \in \mathcal{R}_i$$

It is easy to see that $\boxminus_{R_i} \varphi \equiv [\neg R_i]\neg\varphi$, and, hence, the window operator is expressible in $\mathbf{K}_\omega^\neg$.

It is not hard to see that satisfiability of $\mathbf{K}_\omega^\neg$-formulae is ExpTime-hard and in NExpTime: (i) the logic $\mathbf{K}^u$, i.e., uni-modal $\mathbf{K}$ enriched with the universal modality, is a fragment of $\mathbf{K}_\omega^\neg$: Just replace

- every occurrence of $[u]\,\varphi$ by $[R]\,\varphi \wedge [\neg R]\,\varphi$ and
- every occurrence of $\langle u\rangle\,\varphi$ by $\langle R\rangle\,\varphi \vee \langle\neg R\rangle\,\varphi$

where $[u]$ and $\langle u\rangle$ denote the universal modality, and R is an arbitrary atomic modal parameter. This translation may clearly lead to an exponential blowup in the formula. However, in the class of formulae used to prove the ExpTime-hardness of $\mathbf{K}^u$ in Spaan 1993, $[u]$ occurs only once, and $\langle u\rangle$ does not occur. In this case, the translation is linear, and, thus, satisfiability of $\mathbf{K}_\omega^\neg$-concepts is ExpTime-hard; (ii) when using the standard translation of modal formulae into first order formulae (see, e.g, Blackburn et al. 2001, van Benthem 1983), $\mathbf{K}_\omega^\neg$-formulae are translated to first-order formulae with at most 2 variables. Since L^2, the two-variable fragment of first-order logic, is decidable in NExpTime (Grädel et al. 1997), this implies that satisfiability of $\mathbf{K}_\omega^\neg$-formulae is also in NExpTime. However, these two complexity bounds are obviously not tight. One main contribution of this paper is to give an ExpTime-algorithm for the satisfiability of $\mathbf{K}_\omega^\neg$-formulae, thus tightening the complexity bounds.

For devising a satisfiability algorithm, it is interesting to know what kind of models need to be considered. In Gargov et al. 1987, it is proved that $\mathbf{K}_\omega^\neg$ has the finite model property. $\mathbf{K}_\omega^\neg$ does not have the tree model property since, e.g., the formula $p \wedge [\neg R]\neg p$ has no tree model. However, we will show that there exists a one-to-one correspondence between models and so-called Hintikka-trees which we then use to decide satisfiability (and thus validity) of $\mathbf{K}_\omega^\neg$-formulae. We do this by building, for each $\mathbf{K}_\omega^\neg$-formula φ, a looping automaton $\mathcal{A}_\varphi$ which accepts the empty (tree-) language iff φ is unsatisfiable. Hence we introduce trees, looping automata, and the language they accept here.

Definition 2.2 Let M be a set and $k \geq 1$. A *k-ary M-tree* is a mapping $T : \{1, \ldots, k\}^* \mapsto M$ that labels each node $\alpha \in \{1, \ldots, k\}^*$ with $T(\alpha) \in M$. Intuitively, the node αi is the i-th child of α. We use ϵ to denote the empty word (corresponding to the root of the tree).

A *looping automaton* $\mathcal{A} = (Q, M, I, \Delta)$ for k-ary M-trees is defined by a set Q of states, an alphabet M, a subset $I \subseteq Q$ of initial states, and a transition relation $\Delta \subseteq Q \times M \times Q^k$. A *run* of $\mathcal{A}$ on an M-tree T is a mapping $r : \{1, \ldots, k\}^* \mapsto Q$ with $(r(\alpha), T(\alpha), r(\alpha 1), \ldots, r(\alpha k)) \in \Delta$ for each $\alpha \in \{1, \ldots, k\}^*$.

A looping automaton accepts all those M-trees for which a run exists, i.e., the language $\mathcal{L}(\mathcal{A})$ of M-trees accepted by $\mathcal{A}$ is

$$\mathcal{L}(\mathcal{A}) = \{T \mid \text{There is a run from } \mathcal{A} \text{ on } T\}.$$

Since looping automata are special Büchi automata, emptiness of their language can effectively be tested using the well-known (quadratic) emptiness test for Büchi-automata Vardi and Wolper 1986. However, for looping automata, this algorithm can be specialized into a simpler (linear) one.

3 Negation of Modal Parameters

We show that satisfiability of $\mathbf{K}_\omega^\neg$-formulae is decidable in exponential time. For this purpose, we first abstract from models of $\mathbf{K}_\omega^\neg$-formulae to Hintikka-trees, and then show how to construct a looping automaton that accepts exactly Hintikka-trees.

Notation: We assume all formulae to be in negation normal form (NNF), i.e., negation occurs only in front of atomic parameters and propositional variables. Each formula can easily be transformed into an equivalent one in NNF by pushing negation inwards, employing de Morgan's law and the duality between $[R]$ and $\langle R \rangle$ and between $[\neg R]$ and $\langle \neg R \rangle$. We use $\bar{\varphi}$ to denote the NNF of $\neg \varphi$.

Since we treat modalities with negated and unnegated modal parameters symmetrically, we introduce the notion

$$\langle \bar{R} \rangle \varphi = \begin{cases} \langle \neg R \rangle \varphi & \text{if } R \text{ is atomic,} \\ \langle S \rangle \varphi & \text{if } R = \neg S \text{ for some atomic parameter } S \end{cases}$$

and analogously $[\bar{R}]\, \varphi$. Let $\mathsf{cl}(\varphi)$ denote the set of φ's subformulae and the NNFs of their negations, i.e.,

$$\mathsf{cl}(\varphi) := \{\psi \mid \psi \text{ is a subformula of } \varphi \text{ or } \psi = \bar{\rho} \text{ for a subformula } \rho \text{ of } \varphi\}.$$

Obviously, the cardinality of $\mathsf{cl}(\varphi)$ is linear in the length of φ. We assume that diamond-formulae $\langle R \rangle \psi$ in $\mathsf{cl}(\varphi)$ are linearly ordered, and that $\mathcal{D}(i)$ yields the i-th diamond-formula in $\mathsf{cl}(\varphi)$.

Definition 3.1 Hintikka-set and Hintikka-tree

Let φ be a $\mathbf{K}_\omega^\neg$-formula and k the number of diamond-formulae in $\mathsf{cl}(\varphi)$. A set $\Psi \subseteq \mathsf{cl}(\varphi)$ is a *Hintikka-set* iff it satisfies the following conditions:

(H1) if $\varphi_1 \wedge \varphi_2 \in \Psi$, then $\{\varphi_1, \varphi_2\} \subseteq \Psi$,
(H2) if $\varphi_1 \vee \varphi_2 \in \Psi$, then $\{\varphi_1, \varphi_2\} \cap \Psi \neq \emptyset$, and
(H3) $\{\psi, \bar{\psi}\} \not\subseteq \Psi$ for all $\mathbf{K}_\omega^\neg$-formulae ψ.

A k-ary $2^{\mathsf{cl}(\varphi)}$-tree T is a *Hintikka-tree for* φ iff $T(\alpha)$ is a Hintikka-set for each node α in T, and T satisfies, for all nodes $\alpha, \beta \in \{1, \ldots, k\}^*$, the following conditions:

(H4) $\varphi \in T(\epsilon)$,
(H5) if $\{\langle R \rangle \psi, [R]\rho_1, \ldots, [R]\rho_m\} \subseteq T(\alpha)$ and $\mathcal{D}(i) = \langle R \rangle \psi$, then $\{\psi, \rho_1, \ldots, \rho_m\} \subseteq T(\alpha i)$
(H6) if $\mathcal{D}(i) \notin T(\alpha)$, then $T(\alpha i) = \emptyset$,
(H7) if $[R]\rho \in T(\alpha)$, then $\rho \in T(\beta)$, $\bar{\rho} \in T(\beta)$, or $T(\beta) = \emptyset$, and
(H8) if $\{[R]\rho, [\bar{R}]\psi\} \subseteq T(\alpha)$ and $\bar{\rho} \in T(\beta)$, then $\psi \in T(\beta)$.

For **(H5)**, **(H7)**, and **(H8)**, recall that R denotes an atomic parameter or the negation of an atomic parameter. The following lemma shows the connection between models and Hintikka trees.

Lemma 3.2 *A $\mathbf{K}_\omega^\neg$-formula φ is satisfiable iff φ has a Hintikka-tree.*

Thus, we have that Hintikka-trees are appropriate abstractions of models of $\mathbf{K}_\omega^\neg$-formulae. Hintikka-trees enjoy the nice property that they are trees, and we can thus define, for a $\mathbf{K}_\omega^\neg$-formula φ, a tree-automaton $\mathcal{A}_\varphi$ that accepts exactly the Hintikka-trees for φ.

Definition 3.3 For a $\mathbf{K}_\omega^\neg$-formula φ with k diamond-formulae in $\mathsf{cl}(\varphi)$, the looping automaton $\mathcal{A}_\varphi = (Q, 2^{\mathsf{cl}(\varphi)}, \Delta, I)$ is defined as follows:

- Let
$$P = \{\{[R]\psi, [\bar{R}]\rho\} \mid [R]\psi, [\bar{R}]\rho \in \mathsf{cl}(\varphi)\},$$
$$S = \{[R]\psi \mid [R]\psi \in \mathsf{cl}(\varphi)\},$$
Q is the set of all those elements (Ψ, p, s) of
$$\{\Psi \in 2^{\mathsf{cl}(\varphi)} \mid \Psi \text{ is a Hintikka-set}\} \times 2^P \times 2^S$$
satisfying the following conditions:
  1. if $\{[R]\rho, [\bar{R}]\psi\} \in p$ and $\bar{\rho} \in \Psi$, then $\psi \in \Psi$,

2. if $[R]\,\rho \in s$, then $\Psi = \emptyset$ or $\{\rho, \bar{\rho}\} \cap \Psi \neq \emptyset$,
3. if $[R]\,\rho \in \Psi$, then $[R]\,\rho \in s$, and
4. if $\{[R]\,\rho, [\bar{R}]\,\psi\} \subseteq \Psi$, then $\{[R]\,\rho, [\bar{R}]\,\psi\} \in p$.

- $I = \{(\Psi, p, s) \mid \varphi \in \Psi\}$.
- $((\Psi, p, s), \Psi', (\Psi_1, p_1, s_1), \ldots, (\Psi_k, p_k, s_k)) \in \Delta$ iff

  $\Psi = \Psi', \quad p_i = p, \quad s_i = s$ for all $1 \leq i \leq k$, and
  $\mathcal{D}(i) = \langle R\rangle\,\psi \in \Psi$ implies $\psi \in \Psi_i$ and $\rho \in \Psi_i$ for each $[R]\,\rho \in \Psi$
  $\mathcal{D}(i) = \langle R\rangle\,\psi \notin \Psi$ implies $\Psi_i = \emptyset$.

Note that, since $\mathcal{A}_\varphi$ is a looping automata, every run is accepting. As a consequence of the following lemma and Lemma 3.2, we can reduce satisfiability of $\mathbf{K}_\omega^\neg$-formulae to the emptyness of the language accepted by looping automata.

Lemma 3.4 *T is a Hintikka-tree for a $\mathbf{K}_\omega^\neg$-formula φ iff $T \in \mathcal{L}(\mathcal{A}_\varphi)$.*

Obviously, the cardinality of $\mathsf{cl}(\varphi)$ is linear in the length of φ. Hence, by definition of $\mathcal{A}_\varphi$, the cardinality of each component of $\mathcal{A}_\varphi$ is exponential in the length of φ, and thus the size of $\mathcal{A}_\varphi$ is also exponential in the length of φ. This fact together with Lemma 3.2, Lemma 3.4, and the fact that emptiness of the language accepted by a looping automaton $\mathcal{A}_\varphi$ can be tested in time linear in the size of $\mathcal{A}_\varphi$ (Vardi and Wolper 1986) implies that satisfiability of $\mathbf{K}_\omega^\neg$-concepts is in ExpTime. We already noted in Section 2 that satisfiability of $\mathbf{K}_\omega^\neg$-concepts is also ExpTime-hard and, hence, we obtain the following theorem:

Theorem 3.5 *Satisfiability of $\mathbf{K}_\omega^\neg$-formulae is ExpTime-complete.*

Is our approach still of use if we replace $\mathbf{K}_\omega$ by some logic with a restricted class of frames? In the following, we perform a case study by extending the presented algorithm to deal with $(\mathbf{K}_\omega \otimes \mathbf{K4}_\omega)^\neg$-formulae, where $(\mathbf{K}_\omega \otimes \mathbf{K4}_\omega)^\neg$ denotes the fusion of $\mathbf{K}_\omega$ and $\mathbf{K4}_\omega$ enriched with the negation of modal parameters. More precisely, $(\mathbf{K}_\omega \otimes \mathbf{K4}_\omega)^\neg$ provides two disjoint sets of atomic modal parameters $R_1, R_2, \ldots$ and $S_1, S_2, \ldots$, where the latter are called *transitive modal parameters*. Moreover, accessibility relations corresponding to transitive modal parameters are required to be transitive. Apart from demonstrating the generality of our approach, the logic $(\mathbf{K}_\omega \otimes \mathbf{K4}_\omega)^\neg$ is very natural if viewed as a Description Logic (Lutz and Sattler 2000b, Sattler 1996).

To define Hintikka trees for $(\mathbf{K}_\omega \otimes \mathbf{K4}_\omega)^\neg$, we introduce counterparts for **(H5)** and **(H8)** which deal with transitive modal parameters.

Definition 3.6 A $(\mathbf{K}_\omega \otimes \mathbf{K4}_\omega)^\neg$-Hintikka-tree is a Hintikka-tree as in Definition 3.1 extended with the following two conditions:

(H5b) if, for a trans. parameter S_j, we have $\{\langle S_j\rangle \psi, [S_j]\rho_1, \ldots, [S_j]\rho_m\} \subseteq T(\alpha)$ and $\mathcal{D}(i) = \langle S_j\rangle \psi$, then $\{\psi, \rho_1, \ldots, \rho_m, [S_j]\rho_1, \ldots, [S_j]\rho_m\} \subseteq T(\alpha i)$.

(H8b) if, for a trans. parameter S_j, we have $\{[S_j]\psi, [\neg S_j]\rho\} \subseteq T(\alpha)$ and $\bar{\rho} \in T(\beta)$, then $\{[S_j]\psi, \psi\} \subseteq T(\beta)$.

We can now "lift" Lemma 3.2 to the $(\mathbf{K}_\omega \otimes \mathbf{K4}_\omega)^\neg$ case.

Lemma 3.7 *A* $(\mathbf{K}_\omega \otimes \mathbf{K4}_\omega)^\neg$*-formula* φ *is satisfiable iff* φ *has a* $(\mathbf{K}_\omega \otimes \mathbf{K4}_\omega)^\neg$*-Hintikka-tree.*

It remains to construct a looping automaton that accepts exactly the Hintikka-trees for a given $(\mathbf{K}_\omega \otimes \mathbf{K4}_\omega)^\neg$-formula φ. This construction is the same as the one in Defintion 3.3, with an additional fifth condition in the definition of Q as a translation of **(H8b)**, and an additional implication in the definition of Δ as a translation of **(H5b)**.

Definition 3.8 Let $\mathcal{A}_\varphi = (Q, 2^{\mathsf{cl}(\varphi)}, \Delta, I)$ be the looping automaton corresponding to a $(\mathbf{K}_\omega \otimes \mathbf{K4}_\omega)^\neg$-formula φ as defined in Lemma 3.3. Define a new looping automaton $\mathcal{A}'_\varphi := (Q', 2^{\mathsf{cl}(\varphi)}, \Delta', I)$ by setting

- Q' to the maximal subset of Q such that, for all $(\Psi, p, s) \in Q'$, if $\{[S_j]\psi, [\neg S_j]\rho\} \in p$ and $\bar{\rho} \in \Psi$ for a transitive parameter S_j, then $\{\psi, [S_j]\psi\} \in \Psi$.
- Δ' to the maximal subset of Q such that, for all
$$((\Psi, p, s), \Psi', (\Psi_1, p_1, s_1), \ldots, (\Psi_k, p_k, s_k)) \in \Delta',$$
if $\mathcal{D}(i) = \langle S_j\rangle \psi \in \Psi$ for a transitive parameter S, then $[S_j]\rho \in \Psi_i$ for each $[S_j]\rho \in \Psi$.

Proving an analogon to Lemma 3.4, we obtain the following theorem:

Theorem 3.9 *Satisfiability of* $(\mathbf{K}_\omega \otimes \mathbf{K4}_\omega)^\neg$*-formulae is ExpTime-complete.*

4 Adding Intersection and Union of Modal Parameters

In this section, we investigate the complexity of adding intersection and union of modal parameters to the logic $\mathbf{K}_\omega^\neg$. In doing this, one has the choice to either restrict the applicability of negation to atomic modal parameters or allowing for full negation w.r.t. modal parameters. In the latter case, adding union is obviously equivalent to adding intersection or both. We start with the smallest extension, i.e., we add either intersection or union on modal parameters while restricting negation to atomic parameters.

Definition 4.1 A $\mathbf{K}_\omega^{(\neg),\cup}$-formula ($\mathbf{K}_\omega^{(\neg),\cap}$-formula) is a $\mathbf{K}_\omega^{\neg}$-formula which, additionally, allows for modal parameters of the form $S_1 \cup \cdots \cup S_k$ ($S_1 \cap \cdots \cap S_k$), where each S_i is an atomic or a negated atomic parameter. The semantics of the new modal operators is defined as follows:

$$\begin{array}{lll}
\mathcal{M}, w \models \langle S_1 \cup \cdots \cup S_k \rangle \varphi & \text{iff} & \exists w' \in W \text{ with } \mathcal{M}, (w, w') \models S_i \text{ for} \\
& & \text{some } i \in \{1, \ldots, k\} \text{ and } \mathcal{M}, w' \models \varphi \\
\mathcal{M}, w \models [S_1 \cup \cdots \cup S_k] \varphi & \text{iff} & \forall w' \in W, \text{ if } \mathcal{M}, (w, w') \models S_i \text{ for} \\
& & \text{some } i \in \{1, \ldots, k\}, \text{ then } \mathcal{M}, w' \models \varphi \\
\mathcal{M}, w \models \langle S_1 \cap \cdots \cap S_k \rangle \varphi & \text{iff} & \exists w' \in W \text{ with } \mathcal{M}, (w, w') \models S_i \\
& & \text{for all } 1 \le i \le k \text{ and } \mathcal{M}, w' \models \varphi \\
\mathcal{M}, w \models [S_1 \cap \cdots \cap S_k] \varphi & \text{iff} & \forall w' \in W, \text{ if } \mathcal{M}, (w, w') \models S_i \\
& & \text{for all } 1 \le i \le k, \text{ then } \mathcal{M}, w' \models \varphi
\end{array}$$

Let us first investigate the logic $\mathbf{K}_\omega^{(\neg),\cup}$. It is not hard to see that

$$\begin{array}{rcl}
[S_1 \cup \cdots \cup S_k] \varphi & \equiv & [S_1] \varphi \wedge \cdots \wedge [S_k] \varphi \text{ and} \\
\langle S_1 \cup \cdots \cup S_k \rangle \varphi & \equiv & \langle S_1 \rangle \varphi \vee \cdots \vee [S_k] \varphi,
\end{array}$$

i.e., satisfiability of $\mathbf{K}_\omega^{(\neg),\cup}$-formulae can be reduced to satisfiability of $\mathbf{K}_\omega^{\neg}$-formulae. However, this naive reduction might lead to an exponential blow-up of the formula. In order to avoid this blow-up, we can proceed as follows to transform a $\mathbf{K}_\omega^{(\neg),\cup}$-formula ψ into an equivalent $\mathbf{K}_\omega^{\neg}$-formula $\widehat{\psi}$ whose length is linear in the length of ψ: As the first step, recursively apply the following substitutions to ψ from the inside to the outside (i.e., no union on modal parameters occurs in φ)

$$\begin{array}{rcl}
[S_1 \cup \cdots \cup S_k] \varphi & \rightsquigarrow & [S_1] p_\varphi \wedge \cdots \wedge [S_k] p_\varphi \text{ and} \\
\langle S_1 \cup \cdots \cup S_k \rangle \varphi & \rightsquigarrow & \langle S_1 \rangle p_\varphi \vee \cdots \vee [S_k] p_\varphi
\end{array}$$

where p_φ is a new propositional variable. Call the result of these substitutions ψ'. Secondly, use a new modal parameter R and define

$$\widehat{\psi} \quad := \quad \psi' \wedge \bigwedge_{p_\varphi \text{ occurs in } \psi'} [R](p_\varphi \leftrightarrow \varphi) \wedge [\neg R](p_\varphi \leftrightarrow \varphi)$$

It can easily be seen that this gives the following result.

Theorem 4.2 *Satisfiability of* $\mathbf{K}_\omega^{(\neg),\cup}$*-formulae is ExpTime-complete.*

Next, we show that satisfiability of $\mathbf{K}_\omega^{(\neg),\cap}$-formulae is NExpTime-hard. The proof is given by a reduction of a NExpTime-complete variant of the well-known, undecidable domino problem.

A domino problem (Berger 1966, Knuth 1968) is given by a finite set of *domino types*. All domino types are of the same size, each type has a quadratic shape and colored edges. Of each type, an unlimited number of dominoe is available. The problem in the original domino problem is to arrange these dominoe to cover the plane without holes or overlapping such that adjacent dominoe have identical colors on their touching edges (rotation of the dominoe is not allowed). In the NExpTime-complete variant of the domino problem that we use, the task is not to tile the whole plane, but to tile a $2^{n+1} \times 2^{n+1}$-torus, i.e., a $2^{n+1} \times 2^{n+1}$-rectangle whose edges are "glued" together. See, e.g., Berger 1966, Knuth 1968 for undecidable versions of the domino problem and Börger et al. 1997 for bounded variants.

Definition 4.3 Let $\mathcal{D} = (D, H, V)$ be a *domino system*, where D is a finite set of *domino types* and $H, V \subseteq D \times D$ represent the horizontal and vertical matching conditions. For $s, t \in \mathbb{N}$, let $U(s,t)$ be the torus $\mathbb{Z}_s \times \mathbb{Z}_t$, where $\mathbb{Z}_n$ denotes the set $\{0, \ldots, n-1\}$. Let $a = a_0, \ldots, a_{n-1}$ be an n-tuple of dominoe (with $n \leq s$). We say that $\mathcal{D}$ *tiles* $U(s,t)$ *with initial condition* a iff there exists a mapping $\tau : U(s,t) \rightarrow D$ such that, for all $(x,y) \in U(s,t)$:

- if $\tau(x,y) = d$ and $\tau(x \oplus_s 1, y) = d'$, then $(d, d') \in H$
- if $\tau(x,y) = d$ and $\tau(x, y \oplus_t 1) = d'$, then $(d, d') \in V$
- $\tau(i, 0) = a_i$ for $0 \leq i < n$.

where $\oplus_n$ denotes addition modulo n. Such a mapping τ is called a *solution* for $\mathcal{D}$ w.r.t. a.

The following is a consequence of Theorem 6.1.2 in Börger et al. 1997 (see also Lutz and Sattler 2000b).

Theorem 4.4 *There exists a domino system $\mathcal{D}$ such that the following is a NExpTime-hard problem: Given an initial condition $a = a_0 \cdots a_{n-1}$ of length n, does $\mathcal{D}$ tile the torus $U(2^{n+1}, 2^{n+1})$ with initial condition a?*

We reduce the NExpTime-complete variant of the domino problem from Theorem 4.4 to the satisfiability of $\mathbf{K}_\omega^{(\neg),\cap}$-formulae. Given a domino system $\mathcal{D} = (D, H, V)$ and an initial condition $a = a_0, \ldots, a_{n-1}$, we define a reduction formula $\varphi_{(\mathcal{D},a)}$ such that $\varphi_{(\mathcal{D},a)}$ is satisfiable iff $\mathcal{D}$ tiles the torus $U(2^{n+1}, 2^{n+1})$ with initial condition a. The subformulae of the reduction formula

$$\varphi(\mathcal{D}, a) = \mathit{Count}_x \wedge \mathit{Count}_y \wedge \mathit{Stable} \wedge \mathit{Unique} \wedge \mathit{Tiling} \wedge \mathit{Init}$$

$$\begin{aligned}
Count_x &= [u]\Big[\bigwedge_{k=0}^{n}\Big(\big(\bigwedge_{j=0}^{k-1} x_j\big) \to (x_k \leftrightarrow [R_x]\neg x_k)\Big) \wedge \\
&\qquad \bigwedge_{k=0}^{n}\Big(\big(\bigvee_{j=0}^{k-1} \neg x_j\big) \to (x_k \leftrightarrow [R_x]\, x_k)\Big) \wedge \langle R_x\rangle \mathsf{true}\Big] \\
Stable &= [u]\Big[\bigwedge_{k=0}^{n}(x_k \to [R_y]\, x_k) \wedge \bigwedge_{k=0}^{n}(\neg x_k \to [R_y]\neg x_k) \wedge \\
&\qquad \bigwedge_{k=0}^{n}(y_k \to [R_x]\, y_k) \wedge \bigwedge_{k=0}^{n}(\neg y_k \to [R_x]\neg y_k)\Big] \\
Unique &= [u]\Big[\bigwedge_{k=0}^{n}\Big((x_k \to [\neg R_k]\neg x_k) \wedge (\neg x_k \to [\neg R_k]\, x_k)\Big) \wedge \\
&\qquad \bigwedge_{k=0}^{n}\Big((y_k \to [\neg S_k]\neg y_k) \wedge (\neg y_k \to [\neg S_k]\, y_k)\Big) \wedge \\
&\qquad \bigwedge_{d\in D} p_d \to [R_0 \cap \cdots \cap R_n \cap S_0 \cap \cdots \cap S_n]\, p_d\Big] \\
Tiling &= [u]\Big[\big(\bigvee_{d\in D} p_d\big) \wedge \bigwedge_{d\in D}\ \bigwedge_{d'\in D\setminus\{d\}} \neg(p_d \wedge p_{d'}) \wedge \\
&\qquad \bigwedge_{d\in D} p_d \to \big([R_x] \bigvee_{(d,d')\in H} p_{d'}\big) \wedge \\
&\qquad \bigwedge_{d\in D} p_d \to \big([R_y] \bigvee_{(d,d')\in G} p_{d'}\big)\Big] \\
Init &= \bigwedge_{k=0}^{n}(\neg x_i \wedge \neg y_i) \wedge p_{w_0} \wedge [R_x]\, p_{w_1} \wedge \cdots \wedge [R_x]^{n-1} p_{w_{n-1}}
\end{aligned}$$

FIGURE 1 Sub-formulae of $\varphi_{(\mathcal{D},a)}$ for $\mathcal{D} = (D, H, V)$ and $a = a_0, \ldots, a_{n-1}$.

can be found in Figure 1, where $Count_y$ is $Count_x$ with R_x replaced by R_y, x_j by y_j, and x_k by y_k. In this figure, $[u]\varphi$ is an abbreviation for $[R]\,\varphi \wedge [\neg R]\,\varphi$, where R is an arbitrary atomic modal parameter. Obviously, in each model of $[u]\varphi$, each world satisfies φ. In *Init*, we write $[R]^n\varphi$ to denote the n-fold nesting of $[R]$. The strategy of the reduction is to define the reduction formula $\varphi_{(\mathcal{D},a)}$ such that, for every model $\mathcal{M}$ of $\varphi_{(\mathcal{D},a)}$ with set of worlds W,

1. there exists a propositional variable p_d for every domino type $d \in D$ such that each $w \in W$ is in the extension of p_d for exactly one $d \in D$ (first line of *Tiling*),
2. for each point (i, j) in the torus $U(2^{n+1}, 2^{n+1})$, there exists a corresponding set of worlds $\{w_1, \ldots, w_k\} \subseteq W$ with $k \geq 1$ and a $d \in D$ such that all $w_1, \ldots, w_k$ are in the extension of p_d (*Count*$_x$, *Count*$_y$, *Stable*, and *Unique* formulae),
3. the horizontal and vertical conditions V and H are satisfied w.r.t. sets of worlds representing points in the plane (second and third line of *Tiling*), and
4. the initial condition is satisfied (*Init*).

Properties 1, 3, and 4 are enforced in a standard way using $\mathbf{K}$ formulae. Property 2, however, needs some explanation. Usually, domino-reductions axiomatize a "grid" in order to capture the structure of the torus. As Property 2 indicates, we employ a different strategy: Each world in each model of $\varphi_{(\mathcal{D},a)}$ corresponds to a point (i, j) in the torus. The number i is binarily encoded by the propositional variables $x_0, \ldots, x_n$ while the number j is encoded by the propositional variables $y_0, \ldots, y_n$. We use standard binary incrementation modulo 2^{n+1} to ensure that, for every world w corresponding to a position (i, j), there exists a world w_1 such that w_1 corresponds to $(i \oplus_{2^{n+1}} 1, j)$ and $\mathcal{M}, (w, w_1) \models R_x$, and a world w_2 such that w_2 corresponds to $(i, j \oplus_{2^{n+1}} 1)$ and $\mathcal{M}, (w, w_2) \models R_y$. The *Count*$_x$ and *Count*$_y$ formulae encode the incrementation of the one dimension while the *Stable* formula ensures that the other dimension does not change. It remains to guarantee that every two worlds corresponding to the same position are labeled with the same domino. This task is accomplished by the *Unique* formula which is the only one to use conjunction of modal parameters and the only one to use negation for a purpose different from expressing the universal modality. In order to understand the *Unique* formula, it may be helpful to read subformulae of the form $[\neg R]\neg\varphi$ as $\boxminus_R \varphi$.

Proposition 4.5 *A domino system $\mathcal{D}$ tiles the torus $U(2^{n+1}, 2^{n+1})$ with initial condition $a = a_0, \ldots, a_{n-1}$ iff $\varphi_{(\mathcal{D},a)}$ is satisfiable.*

Together with Theorem 4.4, we obtain a NExpTime lower bound for $\mathbf{K}_\omega^{(\neg),\cap}$-formulae. The corresponding upper bound follows from the fact that the translation of $\mathbf{K}_\omega^{\neg}$-formulae to L^2-formulae mentioned in Section 2 can also be applied to $\mathbf{K}_\omega^{(\neg),\cap}$-formulae.

Theorem 4.6 *Satisfiability of $\mathbf{K}_\omega^{(\neg),\cap}$-formulae is NExpTime-complete.*

5 Full Boolean Modal Logic

In this section, we investigate the complexity of full Boolean Modal Logic. Let us start with introducing this logic formally.

Definition 5.1 A *complex modal parameter* is a Boolean formula of atomic modal parameters. We use $\mathbf{K}_\omega^{\neg,\cap,\cup}$ to denote the extension of $\mathbf{K}_\omega$ with complex modal parameters. Let $\mathcal{M} = \langle W, \pi, \mathcal{R}_1, \dots \rangle$ be a Kripke structure, and S a (possibly complex) modal parameter. Then the extension $\mathcal{E}(S)$ is inductively defined as follows:

if $S = R_i$ (i.e., S is atomic)	then $\mathcal{E}(S) = \mathcal{R}_i$
if $S = \neg S'$	then $\mathcal{E}(S) = (W \times W) \setminus \mathcal{E}(S')$
if $S = S_1 \cap S_2$	then $\mathcal{E}(S) = \mathcal{E}(S_1) \cap \mathcal{E}(S_2)$
if $S = S_1 \cup S_2$	then $\mathcal{E}(S) = \mathcal{E}(S_1) \cup \mathcal{E}(S_2)$

The semantics of formulae is extended as follows:

$$\mathcal{M}, w \models \langle S \rangle \varphi \quad \text{iff} \quad \exists w' \in W \text{ with } (w, w') \in \mathcal{E}(S) \text{ and } \mathcal{M}, w' \models \varphi$$
$$\mathcal{M}, w \models [S] \varphi \quad \text{iff} \quad \forall w' \in W, \text{ if } (w, w') \in \mathcal{E}(S), \text{ then } \mathcal{M}, w' \models \varphi$$

We write $\mathcal{M}, (w, w') \models S$ iff $(w, w') \in \mathcal{E}(S)$.

From Theorem 4.6 and the standard tranlation of $\mathbf{K}_\omega^{\neg,\cap,\cup}$ into L^2, we easily obtain the following result:

Theorem 5.2 *Satisfiability of* $\mathbf{K}_\omega^{\neg,\cap,\cup}$*-formulae is NExpTime-complete.*

However, it is interesting to note that the NExpTime reduction used to prove Theorem 4.6 crucially depends on the fact that an infinite number of modal parameters is available: Since the size of the torus to be tiled is not bounded, there exists no upper bound for the number of the R_i and S_i parameters used for the reduction either. Although Boolean Modal Logics usually provide an infinite number of modal parameters (see, e.g., Gargov and Passy 1987), the question whether NExpTime-hardness can still be obtained if only a bounded number of modal parameters is available is natural. In the remainder of this section, we answer this question by showing that satisfiability and validity of $\mathbf{K}_m^{\neg,\cap,\cup}$, i.e., full Boolean Modal Logic with a fixed number m of modal parameters, is ExpTime-complete. The upper bound is proved by a reduction to multi-modal **K** enriched with the universal modality.

We show that satisfiability of $\mathbf{K}_m^{\neg,\cap,\cup}$-formulae can be reduced to satisfiability of $\mathbf{K}_n^u$-formulae (i.e., formulae of multi-modal **K** enriched with the universal modality) by giving a series of polynomial reduction steps. We do not introduce $\mathbf{K}_n^u$ formally but refer the reader to, e.g., Spaan 1993. The following notions are central to several of the reduction steps.

Definition 5.3 A Kripke structure $\mathcal{M} = \langle W, \pi, \mathcal{R}_1, \ldots \mathcal{R}_m \rangle$ is called *simple* iff we have $\mathcal{R}_i \cap \mathcal{R}_j = \emptyset$ for all $1 \leq i < j \leq m$. $\mathcal{M}$ is called *complete* iff, for all $w, w' \in W$, there exists a unique i with $1 \leq i \leq m$ such that $(w, w') \in \mathcal{R}_i$. A formula (of any logic defined in this paper) is called *s-satisfiable* iff it has a model which is a simple Kripke structure. Similarly, a formula is called *c-satisfiable* iff it has a model which is a complete Kripke structure.

Note that every complete Kripke structure is also simple. We now describe the reduction steps in detail. Let φ be a $\mathbf{K}_m^{\neg,\cap,\cup}$-formula whose satisfiability is to be decided and let $R_1, \ldots, R_m$ be the modal parameters of $\mathbf{K}_m^{\neg,\cap,\cup}$.

Step 1. Convert all modal parameters in φ to disjunctive normal form using a truth table and one disjunct for each line in the truth table that yields *true*. If the "empty disjunction" is obtained when converting a modal parameter S, then replace every occurrence of $\langle S \rangle \psi$ with *false* and every occurrence of $[S]\psi$ with *true*. Call the result of the conversion φ_1. The length of φ_1 is linear in the length of φ since the number m of atomic modal parameters is fixed (and the conversion can be done in linear time). It is easy to see that φ_1 is satisfiable iff φ is satisfiable.

Since the conversion to DNF was done using a truth table, each disjunct occurring in a modal parameter in φ_1 is a *relational type*, i.e., of the form

$$S_1 \cap \cdots \cap S_m \text{ with } S_i = R_i \text{ or } S_i = \neg R_i \text{ for } 1 \leq i \leq m.$$

Let Γ be the set of all relational types. As is easily seen, if $\mathcal{M}, (w, w') \models S$ for some Kripke structure $\mathcal{M}$ with set of worlds W, $w, w' \in W$, and $S \in \Gamma$, then, for every atomic modal parameters R_i, this determines whether $\mathcal{M}, (w, w') \models R_i$ holds. Hence, for every $w, w' \in W$, we have $\mathcal{M}, (w, w') \models S$ for exactly one $S \in \Gamma$.

Step 2. We reduce satisfiability of $\mathbf{K}_m^{\neg,\cap,\cup}$-formulae of the form of φ_1 (i.e, the modal parameters are in DNF and hence $\cup$ does not appear nested inside other operators) to the satisfiability of $\mathbf{K}_m^{(\neg),\cap}$-formulae in which all modal parameters are relational types. It is not hard to see that this can be done as in Section 4, where $\mathbf{K}_\omega^{(\neg),\cup}$ is reduced to $\mathbf{K}_\omega^{\neg}$: In the reduction, just replace the formula $[R](p_\varphi \leftrightarrow \varphi) \wedge [\neg R](p_\varphi \leftrightarrow \varphi)$ with $\bigwedge_{S \in \Gamma} [S](p_\varphi \leftrightarrow \varphi)$.[1] The reduction can again be done in linear time since m is fixed. The $\mathbf{K}_m^{(\neg),\cap}$-formula obtained by converting φ_1 is called φ_2.

[1]This reduction ensures that all modal parameters in the resulting formula are relational types.

Step 3. We reduce satisfiability of $\mathbf{K}_m^{(\neg),\cap}$-formulae of the form of φ_2 to c-satisfiability of $\mathbf{K}_{2^m}$-formulae. Set $n := 2^m$ and let $K_1, \ldots, K_n$ be the atomic modal parameters of the logic $\mathbf{K}_n$. Let r be some bijection between Γ and the set $\{K_1, \ldots, K_n\}$. The formula φ_3 is obtained from φ_2 by replacing each element S of Γ in φ_2 with $r(S)$. Considering the special syntactic form of φ_2 and the definitions of Γ and of c-satisfiability, it is easy to see that φ_2 is satisfiable iff φ_3 is c-satisfiable. Furthermore, the reduction is obviously linear. Note that using 2^m instead of m modal parameters does not spoil the reduction since, ultimately, our reduction goes to satisfiability of multi-modal **K** enriched with the universal modality, and this logic is known to be in ExpTime for *any* fixed number of modalities (Spaan 1993).

Step 4. We reduce c-satisfiability of $\mathbf{K}_n$-formulae to s-satisfiability of $\mathbf{K}_n^u$-formulae. W.l.o.g., we assume φ_3 to be in negation normal form. Define φ_4 as the conjunction of φ_3 with the formula

$$\chi := [u]\Big(\bigwedge_{\psi_1,\ldots,\psi_n \text{ subformulae of } \varphi_3} [K_1]\psi_1 \wedge \cdots \wedge [K_n]\psi_n \rightarrow [u](\psi_1 \vee \cdots \vee \psi_n)\Big)$$

Note that the length of φ_4 is polynomial in the length $|\varphi_3|$ of φ_3: The number of subformulae of φ_3 is bounded by $|\varphi_3|$; hence, χ consists of at most $|\varphi_3|^\ell$ conjuncts, where ℓ is a constant since the number of modal parameters is fixed. Let us prove that φ_3 is c-satisfiable iff φ_4 is s-satisfiable. The "only if" direction is straightforward: Let $\mathcal{M}$ be a complete model for φ_3. Obviously, $\mathcal{M}$ is also simple. Moreover, using the fact that $\mathcal{M}$ is complete, it is straightforward to check that $\mathcal{M}$ is a model for φ_4. It remains to prove the "if" direction. Let $\mathcal{M} = \langle W, \pi, \mathcal{K}_1, \ldots, \mathcal{K}_n\rangle$ be a simple model for φ_4. We first show that

Claim. For each $w, w' \in W$, there exists an ℓ with $1 \leq \ell \leq n$ such that, for all subformulae ψ of φ_3, $\mathcal{M}, w \models [K_\ell]\psi$ implies $\mathcal{M}, w' \models \psi$.

Assume to the contrary that, for some $w, w' \in W$, there exist no ℓ as in the claim. Hence, for each i with $1 \leq i \leq n$, there exists a subformula ρ_i of φ_3 such that $\mathcal{M}, w \models [K_i]\rho_i$ and $\mathcal{M}, w' \not\models \rho_i$. Since $\mathcal{M}$ is a model for χ, we clearly have

$$\mathcal{M}, w \models [K_1]\rho_1 \wedge \cdots \wedge [K_n]\rho_n \rightarrow [u](\rho_1 \vee \cdots \vee \rho_n).$$

This is obviously a contradiction to the fact that $\mathcal{M}, w \not\models \rho_1 \vee \cdots \vee \rho_n$ which proves the claim.

Extend the Kripke structure $\mathcal{M}$ to $\mathcal{M}' = \langle W, \pi, \mathcal{K}_1', \ldots, \mathcal{K}_n'\rangle$ as follows: For any $w, w' \in W$ with $(w, w') \notin \mathcal{K}_i$ for all i with $1 \leq i \leq n$, augment $\mathcal{K}_\ell$ with the tuple (w, w'), where ℓ is as in the claim. Obviously, $\mathcal{M}'$ is

complete. It is now a matter of routine to prove that $\mathcal{M}, w \models \psi$ implies $\mathcal{M}', w \models \psi$ for all subformulae ψ of φ_3. The proof is by induction over the structure of ψ. The only interesting case is:
$\psi = [K_i]\psi'$. Let $(w, w') \in \mathcal{K}'_i$. We need to show that $\mathcal{M}', w' \models \psi'$. First assume that $(w, w') \in \mathcal{K}_i$. Since $\mathcal{M}, w \models \psi$, this implies $\mathcal{M}, w' \models \psi'$. By induction, we have $\mathcal{M}', w' \models \psi'$ and are done. Now assume $(w, w') \in \mathcal{K}'_i \setminus \mathcal{K}_i$. By definition of $\mathcal{K}'_i$, we have that $\mathcal{M}, w \models [K_i]\rho$ implies $\mathcal{M}, w' \models \rho$ for all subformulae ρ of φ_3. Since ψ is a subformula of φ_3, we have $\mathcal{M}, w' \models \psi'$. It remains to apply the induction hypothesis.

Since $\mathcal{M}$ is a model for φ_4, we have that $\mathcal{M}'$ is a model for φ_3. ❑

Step 5. It remains to prove that s-satisfiability of $\mathbf{K}^u_n$-formulae is decidable in ExpTime. This is, however, an easy consequence of the facts that satisfiability of $\mathbf{K}^u_n$-formulae is in ExpTime and that $\mathbf{K}^u_n$ has the tree model property: since every tree model is obviously simple, satisfiability coincides with s-satisfiability.

The sequence of reductions given above yields an ExpTime upper bound for the satisfiability of $\mathbf{K}^{\neg,\cap,\cup}_m$-formulae. Since the lower bound for $\mathbf{K}^{\neg}_\omega$ already holds if we have only a single modal parameter available (again, see Spaan 1993), we obtain the following theorem.

Theorem 5.4 *Satisfiability of $\mathbf{K}^{\neg,\cap,\cup}_m$-formulae (i.e., $\mathbf{K}^{\neg,\cap,\cup}_\omega$ with a bounded bumber of modal parameters) is ExpTime-complete.*

6 Boolean Modal Logics without Negation

So far, we have only considered logics with negation of modal parameters. We will complete our investigation by showing that adding intersection and union of modal parameters does not increase the complexity of $\mathbf{K}_\omega$ (and thus neither the complexity of $\mathbf{K}_m$ is increased by this extension). The fact that the extension of $\mathbf{K}_\omega$ with intersection of modal parameters (i.e., $\mathbf{K}^{\cap}_\omega$) is still in PSpace is an immediate consequence of PSpace-completeness of the Description Logic $\mathcal{ALCR}$ (Donini et al. 1991) and the fact that $\mathcal{ALCR}$ is a notational variant of $\mathbf{K}^{\cap}_\omega$ (Schild 1991). Moreover, it is folklore that $\mathbf{K}_\omega$ extended with union of modal parameters (i.e., $\mathbf{K}^{\cup}_\omega$) is also in PSpace (however, the reduction from Section 4 cannot be applied since the universal modality is not available). For both union and intersection, we go into more detail.

With $\mathbf{K}^{\cap,\cup}_\omega$, we denote the variant of $\mathbf{K}^{\neg,\cap,\cup}_\omega$ obtained by disallowing the use of negation of modal parameters. In the following, we will present a slight extension of the standard PSpace tableau algorithm for $\mathbf{K}$, $\mathbf{K}$-*World* (Ladner 1977), to decide satisfiability of $\mathbf{K}^{\cap,\cup}_\omega$-formulae. Please note that we cannot adapt the reduction from the previous section since the disjunctive normal form of a complex modal parameter can yield an exponential blow-up if the number of boolean parameters is not bounded. When started with an input formula φ, **K**-World decides φ's satisfiability by recursively searching a finite tree-model of φ in a depth-first manner. For each world w in this tree model, it checks whether the set Δ of formulae that w must satisfy is not contradictory. Then, for each $\Diamond\psi$ in Δ, **K**-World is called recursively with ψ and all ρ with $\Box\rho$ in Δ.

To extend **K**-World to $\mathbf{K}_\omega^{\cap,\cup}$, it is comfortable to view the semantics of roles in a different way. For S a complex modal parameter and s a set of atomic modal parameters, we say $s \models S$ iff s, when viewed as the valuation that maps each $R_i \in s$ to true and each $R_j \notin s$ to false, evaluates the Boolean expression S to true. Then $(w, w') \in \mathcal{E}(S)$ iff there is a set s of atomic modal parameters with $s \models S$ and $(w, w') \in \mathcal{R}_i$ for each $R_i \in s$. The only modifications to **K**-World concern the recursive calls for diamond formulae which are more elaborate in the presence of complex modal parameters. For each $\langle S \rangle \psi$ in the set Δ of formulae currently considered, we guess an s with $s \models S$, and then consider ψ together with all ρ where $[S']\rho$ is in Δ and $s \models S'$.

For the sake of a succinct presentation, we assume the input formula φ to contain no disjunction and no diamond-formulae. For Δ and S sets of $\mathbf{K}_\omega^{\cap,\cup}$-formulae where S is closed under subformulae and single negations, $\mathbf{K}_\omega^{\cap,\cup}$-World$(\Delta, S)$ returns true iff

- Δ is a maximally propositionally consistent subset of S, i.e.,
  - $\Delta \subseteq S$,
  - for each $\neg\psi \in S$, $\psi \in \Delta$ iff $\neg\psi \notin \Delta$, and
  - for each $\psi_1 \wedge \psi_2 \in S$, $\psi_1 \wedge \psi_2 \in \Delta$ iff $\psi_1 \in \Delta$ and $\psi_2 \in \Delta$.
- For each subformula $\neg [S] \psi \in \Delta$, there exists a set s of modal parameters with $s \models S$ and a set $\Delta_{\psi,s}$ such that
  - $\neg\psi \in \Delta_{\psi,s}$,
  - for each S' and ρ, if $[S']\rho \in \Delta$ and $s \models S'$, then $\rho \in \Delta_{\psi,s}$,
  - $\mathbf{K}_\omega^{\cap,\cup}$–$\mathsf{World}(\Delta_{\psi,s}, S')$ returns true, where S' is the closure under subformulae and single negation of $\{\rho \mid [S']\rho \in \Delta$ and $s \models S'\} \cup \{\neg\psi\}$.

Let $\mathsf{cl}(\varphi)$ be the smallest set of formulae containing φ that is closed under subformulae and single negation. The proof that a $\mathbf{K}_\omega^{\cap,\cup}$-formula φ is satisfiable iff there exists a $\Delta \subseteq \mathsf{cl}(\varphi)$ with $\varphi \in \Delta$ such that

$$\mathbf{K}_\omega^{\cap,\cup}\text{-World}(\Delta, \mathsf{cl}(\varphi)\})$$

returns true is analogous to the one for **K**-World. Just like **K**-World, $\mathbf{K}_\omega^{\cap,\cup}$-World runs in PSpace (since PSpace = NPSpace (Savitch 1970), the additional non-deterministic guessing of the set of modal parameters s does not matter). Moreover, $\mathbf{K}$ is known to be PSpace-hard (Ladner 1977), and we thus have the following result.

Theorem 6.1 *Satisfiability of* $\mathbf{K}_\omega^{\cup,\cap}$*-formulae is PSpace-complete.*

7 Conclusion

We have given a complete picture of the complexity of Boolean Modal Logics, both with and without a bound on the number of modal parameters. The results for (fragments of) Boolean Modal Logic with an unbounded number of modal parameters are summarised in Figure 2, showing known results in grey.

NExpTime-hardness of $\mathbf{K}_\omega^{(\neg),\cap}$ was rather surprising since so far, intersection of atomic modal parameters (not of chainings/composition of modal parameters) is mostly considered to be "harmless" w.r.t. complexity. Interestingly, we were able to show that, if a bound m is imposed on the number of atomic modal parameters, then full Boolean Modal Logic $\mathbf{K}_m^{\neg,\cap,\cup}$ becomes ExpTime-complete. For this proof, we did not use the automata-based approach because we considered that extending it to take care of complex modal parameters was more involved than the reduction to $\mathbf{K}_n^u$ that we used.

As future work, it may be interesting to extend our techniques to more expressive logics. For example, one may consider arbitrary combinations of the Boolean operators on modal parameters with composition and converse. Several results for such logics are known from the area of Propositional Dynamic Logics (PDL). For example, Harel proves that PDL extended with negation of modal parameters is undecidable using a reduction to the equivalence problem for relation algebra (Harel 1984). It is not hard to see that a similar reduction (of the equivalence problem of boolean algebras of relations with composition only, see, e.g., Andreka et al. 2001) can be used to show that Boolean Modal Logic extended with composition of modal parameters is undecidable. On the contrary, it follows from Danecki's results on PDL with

	no negation	atomic negation	full negation
–	PSpace-compl.	ExpTime-compl.	
∪	PSpace-compl.	ExpTime-compl.	NExpTime-compl.
∩	PSpace-compl.	NExpTime-compl.	NExpTime-compl.
∩ and ∪	PSpace-compl.	NExpTime-compl.	NExpTime-compl.

FIGURE 2 Complexity of $\mathbf{K}_\omega$ extended with various role constructors.

intersection that $\mathbf{K}_{\omega}^{\cap,\cup}$ extended with composition is decidable in double ExpTime (Danecki 1984). As we demonstrated by extending our results to $(\mathbf{K}_{\omega} \otimes \mathbf{K4}_{\omega})^{\neg}$, our automata-based approach to proving ExpTime-bounds can be considered flexible. As a first step towards more expressive logics, we hope that our approach can be "married" with the standard automata-based decidability procedure for PDL thus yielding a decidability result for PDL extended with atomic negation of modal parameters.

Acknowledgements

The authors would like to thank Franz Baader, Stephane Demri, Marcus Kracht, Agnes Kurucz, and Maarten Marx for fruitful discussions. The first author was supported by the DFG Project BA1122/3-1 "Combinations of Modal and Description Logics".

References

Andreka, H., I. Nemeti, and I. Sain. 2001. Algebraic Logic. In *Handbook of Philosophical Logic*. D. Reidel Publishing Co., 2 edition. Forthcoming.

Berger, R. 1966. The undecidability of the dominoe problem. *Mem. Amer. Math. Soc.* 66.

Blackburn, P., M. de Rijke, and Y. Venema. 2001. *Modal Logic*. Cambridge: Cambridge University Press. Forthcoming. Preprint available at http://www.mlbook.org/.

Börger, E., E. Grädel, and Y. Gurevich. 1997. *The Classical Decision Problem*. Perspectives in Mathematical Logic. Springer-Verlag.

Danecki, S. 1984. Nondeterministic Propositional Dynamic Logic with intersection is decidable. In *Proc. of the 5th Symp. on Computation Theory*, LNCS, Vol. 208, 34–53. Springer-Verlag.

Donini, F., M. Lenzerini, D. Nardi, and W. Nutt. 1991. The Complexity of Concept Languages. In *Proc. of KR-91*. Boston, MA, USA.

Gargov, G., and S. Passy. 1987. A note on Boolean modal logic. In *Mathematical Logic and Applications*. Plenum Press.

Gargov, G., S. Passy, and T. Tinchev. 1987. Modal Environment for Boolean Speculations. In *Mathematical Logic and Applications*. Plenum Press.

Givan, R., D. McAllester, and S. Shalaby. 1991. Natural Language Based Inference Procedures Applied to Schubert's Steamroller. In *Proc. of the 9th Nat. Conf. on Artificial Intelligence*. MIT Press.

Goranko, V. 1987. Completeness and incompleteness in the bimodal base L($R, -R$). In *Proc. of the Conf. on Mathematical Logic "Heyting '88", Chaika, Bulgaria*. Plenum Press.

Goranko, V. 1990. Modal Definability in Enriched Languages. *Notre Dame Journal of Formal Logic* 31(1):81–105.

Grädel, E., P. Kolaitis, and M. Vardi. 1997. On the Decision Problem for Two-Variable First-Order Logic. *Bulletin of Symbolic Logic* 3:53–69.

Harel, D. 1984. Dynamic Logic. In *Handbook of Philosophical Logic, Volume II.* D. Reidel Publishers.

Humberstone, I. L. 1983. Inaccessible Worlds. *Notre Dame Journal of Formal Logic* 24(3):346–352.

Hustadt, U., and R. A. Schmidt. 2000. Issues of Decidability for Description Logics in the Framework of Resolution. In *Automated Deduction in classical and non-classical logic*, LNAI, Vol. 1761, 191–205. Springer-Verlag.

Knuth, D.E. 1968. *The Art of computer programming.* Addison Wesley Publ. Co., Reading, Massachussetts.

Ladner, R. E. 1977. The computational complexity of provability in systems of modal propositional logic. *SIAM J. of Computing* 6(3):467–480.

Lutz, C., and U. Sattler. 2000a. Mary likes all Cats. In *Proc. of DL2000.* CEUR-WS, No. 33. Proceedings online available from http://SunSITE.Informatik.RWTH-Aachen.DE/Publications/CEUR-WS/Vol-33/.

Lutz, C., and U. Sattler. 2000b. The Complexity of Reasoning with Boolean Modal Logics. LTCS-Report 00-02. Germany: LuFG Theoretical Computer Science, RWTH Aachen. See http://www-lti.informatik.rwth-aachen.de/Forschung/Reports.html.

Passy, S., and T. Tinchev. 1991. An Essay in Combinatory Dynamic Logic. *Information and Computation* 93(2).

Sattler, U. 1996. A Concept Language Extended with Different Kinds of Transitive Roles. In *20. Deutsche Jahrestagung für Künstliche Intelligenz*, ed. G. Görz and S. Hölldobler, LNAI, Vol. 1137. Springer Verlag.

Savitch, W. J. 1970. Relationsship between nondeterministic and deterministic tape complexities. *J. of Computer and System Science* 4:177–192.

Schild, K. 1991. A Correspondence Theory for Terminological Logics: Preliminary Report. In *Proc. of IJCAI-91*, 466–471. Sydney.

Spaan, E. 1993. *Complexity of Modal Logics.* Doctoral dissertation, University of Amsterdam.

van Benthem, J. F. A. K. 1983. *Modal Logic and Classical Logic.* Naples: Bibliopolis.

Vardi, M. Y., and P. Wolper. 1986. Automata-theoretic techniques for modal logic of programs. *Journal of Computer and System Sciences* 32:183–221.

Carsten Lutz and Ulrike Sattler
LuFG Theoretical Computer Science, RWTH Aachen
Ahornstr. 55, 52074 Aachen, Germany
E-mail: {lutz,sattler}@cs.rwth-aachen.de
URL: http://www-lti.informatik.rwth-aachen.de/~clu/
URL: http://www-lti.informatik.rwth-aachen.de/ti/uli-en.html

19

Projective Beth's Properties in Infinite Slice Extensions of the Modal Logic K4

LARISA MAKSIMOVA

ABSTRACT. We consider interrelation of various versions of the Beth property with interpolation and amalgamation. In particular, projective Beth's property is weaker than interpolation on the family of normal extensions of K4. We prove that for each infinite-slice extension of K4 with projective Beth's property its reflexive fragment is contained in Grz.2. This is a generalisation of a similar result proved by the author in 1991b for interpolation property.

The famous Beth theorem on implicit definability (Beth 1953) states that in the classical first order logic, if a relation is implicitly definable in some theory then it is explicitly definable. The notions of implicit and explicit definability can be defined for many logics and in various ways.

We consider several versions of the Beth property in normal modal logics, in particular, projective Beth's properties defined in Barwise and Feferman 1985. Some strong necessary conditions for projective Beth properties in extensions of K4 will be found. We focus on infinite-slice logics whose frames admit infinite chains. We prove that for each infinite-slice extension of K4 with projective Beth's property its reflexive fragment is contained in the logic Grz.2. It follows that, for instance, logics of linear frames, of binary trees, of finite width and so on do not possess projective Beth's property.

These statements are generalisations of our results on interpolation in Maksimova 1991b. One can prove that in classical model theory the

Advances in Modal Logic, Volume 3
F. Wolter, H. Wansing, M. de Rijke, and M. Zakharyaschev, eds.

projective Beth property is equivalent to interpolation. On the contrast, we show that for modal logics extending K4, and even S4, the global version PB2 of projective Beth property is weaker than interpolation properties. A logic of infinite slice is found which has PB2 and does not possess interpolation.

Note that Grz.2 is the greatest among the infinite-slice logics in NE(S4) that have the interpolation property (Maksimova 1982). Also in the family of infinite-slice extensions of the provability logic GL, which have interpolation property, there exists the greatest logic $G\gamma$ (Maksimova 1988). In the present paper we extend these results to the family of infinite-slice logics that possess projective Beth properties.

In Maksimova 1988, Maksimova 1991b we used an equivalence of interpolation in logics to amalgamation properties of varieties of modal algebras. Here we show that we can take some restricted form of the amalgamation property for our proofs, and prove that this restricted amalgamation property is implied by projective Beth properties.

1 Definitions

If $\mathbf{p}$ is a list of variables, let $A(\mathbf{p})$ denote a formula whose all variables are in $\mathbf{p}$, and $\mathcal{F}(\mathbf{p})$ the set of all such formulas.

Let L be a logic, $\vdash_L$ consequence relation in L. Suppose that $\mathbf{p}$, $\mathbf{q}$, $\mathbf{q}'$ are disjoint lists of variables that do not contain x and y, $\mathbf{q}$ and $\mathbf{q}'$ are of the same length, and $A(\mathbf{p},\mathbf{q},x)$ is a formula. We say that the logic L possesses *the projective Beth property PB1*, whenever $\vdash_L A(\mathbf{p},\mathbf{q},x)\&A(\mathbf{p},\mathbf{q}',y) \rightarrow (x \leftrightarrow y)$ implies $\vdash_L A(\mathbf{p},\mathbf{q},x) \rightarrow (x \leftrightarrow B(\mathbf{p}))$ for some formula $B(\mathbf{p})$. Further, L has *the projective Beth property PB2*, if $A(\mathbf{p},\mathbf{q},x), A(\mathbf{p},\mathbf{q}',y) \vdash_L x \leftrightarrow y$ implies $A(\mathbf{p},\mathbf{q},x) \vdash_L x \leftrightarrow B(\mathbf{p})$ for some $B(\mathbf{p})$; L possesses *the Beth property B1*, whenever $\vdash_L A(\mathbf{p},x)\&A(\mathbf{p},y) \rightarrow (x \leftrightarrow y)$ implies $\vdash_L A(\mathbf{p},x) \rightarrow (x \leftrightarrow B(\mathbf{p}))$ for a suitable formula $B(\mathbf{p})$; L has *the Beth property B2* if $A(\mathbf{p},x), A(\mathbf{p},y) \vdash_L x \leftrightarrow y$ implies $A(\mathbf{p},x) \vdash_L x \leftrightarrow B(\mathbf{p})$ for a formula $B(\mathbf{p})$.

It is evident that B1 is a special case of PB1, and B2 of PB2.

The Beth properties are closely connected with *the interpolation properties CIP* and *IPD* defined as follows (where the lists $\mathbf{p},\mathbf{q},\mathbf{r}$ are disjoint):

CIP. If $\vdash_L A(\mathbf{p},\mathbf{q}) \rightarrow B(\mathbf{p},\mathbf{r})$, then there exists a formula $C(\mathbf{p})$ such that $\vdash_L A(\mathbf{p},\mathbf{q}) \rightarrow C(\mathbf{p})$ and $\vdash_L C(\mathbf{p}) \rightarrow B(\mathbf{p},\mathbf{r})$.

IPD. If $A(\mathbf{p},\mathbf{q}) \vdash_L B(\mathbf{p},\mathbf{r})$, then there exists a formula $C(\mathbf{p})$ such that $A(\mathbf{p},\mathbf{q}) \vdash_L C(\mathbf{p})$ and $C(\mathbf{p}) \vdash_L B(\mathbf{p},\mathbf{r})$.

2 Modal logics

The most known modal logics such as Lewis' systems S4 and S5, Grzegorczyk's logic Grz, the logic GL of provability, the logic K4 and the least normal modal logic K have the properties CIP and PB1. We consider propositional modal logics, and we mean that the language contains at least one propositional constant $\top$ ("true") or $\bot$ ("false").

A *normal modal logic* is any set of modal formulas containing all the tautologies of the two-valued propositional logic and the axioms $\Box(A \to B) \to (\Box A \to \Box B)$, and closed under the inference rules R1: $A, A \to B/B$ and R2: $A/\Box A$, and the substitution rule. The set of normal modal logics containing a logic L is denoted by $NE(L)$.

Let L be a normal modal logic, $\vdash_L$ deducibility in L by the rules R1 and R2. Let us define $[*]A = A\&\Box A$.

We bring denotations for some known logics in $NE(K)$:

$$
\begin{aligned}
&\text{K4} = \text{K}+(\Box A \to \Box\Box A),\\
&\text{S4} = \text{K4} + (\Box A \to A),\\
&\text{S4.1.2} = \text{S4} + (\Diamond\Box p \leftrightarrow \Box\Diamond p),\\
&\text{S5} = \text{S4} + (p \to \Box\Diamond p),\\
&\text{Grz} = \text{S4} + (\Box(\Box(p \to \Box p) \to p) \to p),\\
&\text{Grz.2} = \text{Grz} + (\Diamond\Box p \to \Box\Diamond p),\\
&\text{Grz.3} = \text{Grz} + (\Box(\Box p \to \Box q) \vee \Box(\Box q \to \Box p)),\\
&\text{GL} = \text{K4} + (\Box(\Box p \to p) \to \Box p),\\
&\text{G.3} = \text{GL} + (\Box([*]p \to q) \vee \Box([*]q \to p)),\\
&\text{K4.3} = \text{K4} + (\Box([*]p \to q) \vee \Box([*]q \to p)).
\end{aligned}
$$

Interrelations of interpolation with Beth properties are given by the following

Proposition 2.1 *In the family of normal modal logics*

a) PB1 is equivalent to each of the properties CIP and B1;

b) PB1 implies conjunction B2 and IPD but the converse does not hold;

c) conjunction of B2 and IPD implies PB2;

d) PB2 implies B2 but the converse is not true.

Proof. Proposition was stated in Maksimova 1998. The proof is presented in Maksimova 1999a, Theorem 2.4. In that paper, an algebraic proof for c) was found. Here we prove it by syntactic method.

c) Let L have B2 and IPD. Suppose

$$A(\mathbf{p}, \mathbf{q}, x), A(\mathbf{p}, \mathbf{q}', y) \vdash_L x \leftrightarrow y.$$

Then by Deduction Lemma we have

$$A(\mathbf{p},\mathbf{q},x) \vdash_L [n]A(\mathbf{p},\mathbf{q}',y) \to (x \leftrightarrow y)$$

for some n, where $[0]A \rightleftharpoons A$, $[n+1]A \rightleftharpoons A \& \Box[n]A$. By IPD, there exists a formula $B(\mathbf{p},x)$ such that

$$A(\mathbf{p},\mathbf{q},x) \vdash_L B(\mathbf{p},x)$$

and

$$B(\mathbf{p},x) \vdash_L [n]A(\mathbf{p},\mathbf{q}',y) \to (x \leftrightarrow y).$$

It follows that

$$B(\mathbf{p},x), A(\mathbf{p},\mathbf{q}',y) \vdash_L (x \leftrightarrow y),$$

and

$$A(\mathbf{p},\mathbf{q}',y) \vdash [m]B(\mathbf{p},x) \to (x \leftrightarrow y)$$

for some m.

Again by IPD, there exists $B'(\mathbf{p},y)$ such that

$$A(\mathbf{p},\mathbf{q}',y) \vdash B'(\mathbf{p},y) \text{ and } B'(\mathbf{p},y) \vdash [m]B(\mathbf{p},x) \to (x \leftrightarrow y).$$

So we get

$$A(\mathbf{p},\mathbf{q},x) \vdash_L B(\mathbf{p},x) \& B'(\mathbf{p},x)$$

and

$$B'(\mathbf{p},y), B(\mathbf{p},x) \vdash_L (x \leftrightarrow y),$$

hence

$$B(\mathbf{p},x) \& B'(\mathbf{p},x), B(\mathbf{p},y) \& B'(\mathbf{p},y) \vdash_L (x \leftrightarrow y).$$

By B2, there exists a formula $C(\mathbf{p})$ such that

$$B(\mathbf{p},x) \& B'(\mathbf{p},x) \vdash_L (x \leftrightarrow C(\mathbf{p})).$$

As a consequence, we have

$$A(\mathbf{p},\mathbf{q},x) \vdash_L (x \leftrightarrow C(\mathbf{p})).$$

Thus, L has PB2. $\dashv$

It follows immediately from this proposition that PB2 is weaker than PB1 on the class of modal logics. In general, IPD does not imply PB2. In Maksimova 1992 we proved that all normal extensions of the modal logic K4 possess the property B2. Therefore, the proposition 2.1 c implies

Proposition 2.2 *Let L be a normal modal logic containing K4. If L has IPD then it has PB2.*

We state that the converse is not true.

Theorem 2.3 *There exists a logic in NE(K4), and even in NE(S4), which has PB2 but does not possess IPD.*

Proof. As an example, one can take a logic $L = S4.1.2 \cap S5$. It follows from the results of Maksimova 1979 that L does not possess IPD. In order to prove that L has PB2, we use the well-known correspondence between logics in $NE(S4)$ and varieties of topoboolean algebras. Theorem 4.2 below gives an algebraic equivalent of PB2, and our statement follows from Theorem 4.4. ⊣

3 Logics of infinite slice

Let L be in $NE(K4)$. One can associate with L its *reflexive fragment* $r(L)$ defined as follows:

$$r(L) \rightleftharpoons \{A | f(A) \in L\},$$

where $f(A)$ is defined inductively:
$f(\bot) = \bot$, $f(p) = p$ for any variable p,
$f(A \to B) = f(A) \to f(B)$, $f(\Box A) = [*]f(A)$.

If L is in $NE(K4)$, then $r(L)$ is in $NE(S4)$; in particular, $r(K4) = S4$ (Meskhi 1978), $r(GL) = Grz$ (Goldblatt 1978), $r(G.3) = Grz.3$. Obviously, $r(L) = L$ for $L \in NE(S4)$.

A classification of logics in $NE(K4)$ by slices is defined as follows.

A logic L in $NE(K4)$ is *of finite slice,* if it contains a formula φ_n for some n, where
$\varphi_0 = \bot$, $\varphi_{n+1} = p_{n+1} \to \Box(\Box\neg p_{n+1} \to \varphi_n)$;
L is of *infinite slice,* if φ_n is not in L for any n.

It is well known that the logic K4 can be characterized by transitive Kripke frames. For a transitive frame $\mathbf{W} = (W, R)$ and $x \in W$, we denote $W^x = \{y | y = x \text{ or } xRy\}$. The frame $\mathbf{W}$ is *initial* if $W = W^x$ for some x in W. A sequence $x_1, x_2, \ldots$ of elements of a frame (W, R) is *an R-chain,* if $x_n R x_{n+1}$ and not $x_{n+1} R x_n$ for any n. For any frame $W = (W, R)$, its *depth* $d(W)$ is defined as supremum of lengths of finite R-chains in W. For each x in W, let $d(x) = d(W^x)$.

Remark that L is of infinite slice if and only if there are R-chains of any finite length in suitable L-frames.

Now we can formulate the main results of the paper. For the proofs we essentially use an algebraic approach given in Section 4.

Theorem 3.1 *Let $L \in NE(K4)$ be a logic of infinite slice which has the projective Beth property PB2. Then its reflexive fragment $r(L)$ is contained in Grz.2 = Grz+ ($\Diamond\Box p \to \Box\Diamond p$).*

Proof. It is well known that there is a duality between $NE(K4)$ and the family of varieties of transitive algebras (see Section 4). Then we apply Corollary 5.9(i). ⊣

As PB1 is stronger than PB2, the theorem 3.1 holds also for PB1. In Maksimova 1991b we proved the analogous result for IPD. Since PB2 is weaker than IPD in $NE(K4)$, Theorem 3.1 is a generalisation of Theorem 4 in Maksimova 1991b.

Evidently, $r(L) = L$ for all L in $NE(S4)$. Further, Grz.2 itself has CIP, so it possesses PB2, and we get

Corollary 3.2 *The family of infinite-slice logics in $NE(S4)$ possessing PB2 has the greatest element, namely the logic Grz.2.*

We obtain an analogous result for extensions of the provability logic GL. It was proved in 1988 that there is a logic which has the Craig interpolation property CIP, and is the greatest among the logics of infinite slice in $NE(GL)$ possessing IPD (Maksimova 1988, theorem 3.5).

This logic $G\gamma$ is defined as follows:

$$G\gamma \rightleftharpoons GL + \{\gamma_n(x) | n \geq 0\},$$

where $\gamma_n(x) \rightleftharpoons \beta_0(x) \& \ldots \& \beta_n(x)$,

$\beta_n(x) \rightleftharpoons [*](\alpha_n \to x) \vee [*](\alpha_n \to \neg x)$,

$\alpha_n \rightleftharpoons \Box^{n+1}\bot \& \neg\Box^n\bot$.

We have shown that $r(G\gamma)$=Grz.2. Furthermore, $G\gamma$ does not possess the finite model property and is not finitely axiomatizable. On the other hand, it is decidable and Kripke-complete (Maksimova 1988). This logic can be characterized by the class of so-called *ω-linear* GL-frames whose elements of finite depth form a chain. Remind that the class of all GL-frames consists of transitive and irreflexive frames whithout infinitely increasing chains (Segerberg 1971).

Now we can replace IPD by PB2.

Theorem 3.3 *The logic $G\gamma$ is the greatest among the logics of infinite slice in NE(GL) possessing PB2.*

Proof. Since $G\gamma$ has CIP, it has PB2. On the other hand, we show that each infinite-slice logic with PB2 in $NE(GL)$ is contained in $G\gamma$. For the proof, we use duality between the set of normal extensions of GL and the family of varieties of diagonalizable algebras (see Section 4). The statement follows immediately from Corollary 5.9(ii). ⊣

In addition, we note that there exist proper extensions of $G\gamma$ whose reflexive fragment is Grz.2 (see Maksimova 1988).

Note, that L in $NE(GL)$ is a logic of infinite slice iff L is contained in $G.3$. A logic L in $NE(S4)$ is of infinite slice iff L is contained in Grz.3. So we have the following

Corollary 3.4 *Let $L \in NE(K4)$ have PB2.*

(i) If $S4 \subseteq L \subseteq$ Grz.3 then L is contained in Grz.2.

(ii) If $GL \subseteq L \subseteq G.3$ then L is contained in $G\gamma$.

Recall that a logic L in NE(K4) is *of finite width* if it contains a formula

$$\Diamond p_0 \& \ldots \& \Diamond p_n \to \bigvee_{0 \leq i \neq j \leq n} \Diamond(p_i \& (p_j \vee \Diamond p_j))$$

for some n. If L is of finite width then the cardinality of anti-chains in initial L-frames may be bounded by some natural number. In particular, all logics in $NE(K4.3)$ are of the width 1.

From Theorem 3.1 we obtain one more

Corollary 3.5 *(i) If L in NE(K4) is a logic of infinite slice and of finite width then L does not possess PB2.*

(ii) For any $n < \omega$, the logic of n-ary transitive and irreflexive finite trees does not have PB2.

Proof. One can show that for each of these logics, its reflexive fragment is not contained in Grz.2. ⊣

4 Algebraisation

In our proofs we use an algebraic approach. First we find algebraic equivalents for the projective Beth properties. For properties B1 and B2, and also for CIP and IPD it was done in Maksimova 1991a.

It is well known that there exists a one-to-one correspondence between normal modal logics and varieties of modal algebras. *A modal algebra* is an algebra $\mathbf{A} = (A, \to, \neg, \Box)$ that is a boolean algebra with respect to $\to$ and $\neg$ and, moreover, satisfies the conditions $\Box\top = \top$ and $\Box(x \to y) \leq \Box x \to \Box y$. An algebra is *non-degenerate* if $\bot \neq \top$. A modal algebra $\mathbf{A}$ is called *transitive* if it satisfies an inequality $\Box x \leq \Box\Box x$. A transitive algebra is called *a topoboolean algebra*, or *interior algebra* if it satisfies $\Box x \leq x$; a *diagonalizable* algebra if it satisfies $\Box(\Box x \to x) = \Box x$. It is well known that the modal logic K4 can be characterized by the variety of all transitive algebras, S4 by the variety of topoboolean algebras, GL by diagonalizable algebras.

If A is a formula, $\mathbf{A}$ a modal algebra, then $\mathbf{A} \models A$ denotes that the identity $A = \top$ is satisfied in $\mathbf{A}$. We write $\mathbf{A} \models L$ instead of $(\forall A \in L)(\mathbf{A} \models A)$. We denote $V(L) = \{\mathbf{A} | \mathbf{A} \models L\}$. Each normal modal logic L is characterized by the variety $V(L)$. If $V = V(L)$ is a variety of transitive algebras, we define $r(V) = V(r(L))$.

A variety V is called ***locally finite,*** if any finitely generated algebra in V is finite. We proved in Maksimova 1991b

Theorem 4.1 *For any logic L in NE(K4) the following are equivalent:*
(i) L is a logic of finite slice,
(ii) r(L) is a logic of finite slice,
(iii) V(L) is locally finite,
(iv) the one-generated free algebra in V(L) is finite.

In Maksimova 1991a we found an algebraic equivalent of Craig's interpolation property. Namely, a normal modal logic L satisfies CIP if and only if its associated variety $V(L)$ has the super-amalgamation property SAP; therefore B1 and PB1 are also equivalent to the super-amalgamation property of the corresponding variety. Further, IPD is equivalent to the amalgamation property AP.

We say that a class K has the Amalgamation Property if it satisfies the following condition for any algebras $\mathbf{A}$,$\mathbf{B}$,$\mathbf{C}$ in K:

(AP) for any monomorphisms $\beta : \mathbf{A} \to \mathbf{B}$ and $\gamma : \mathbf{A} \to \mathbf{C}$ there exists an algebra $\mathbf{D}$ in K and monomorphisms $\delta : \mathbf{B} \to \mathbf{D}$, $\varepsilon : \mathbf{C} \to \mathbf{D}$, such that $\delta\beta = \varepsilon\gamma$.

If, in addition, $\delta(\mathbf{B}) \cap \varepsilon(\mathbf{C}) = \delta\beta(\mathbf{A})$ holds, then the class K is said to have Strong Amalgamation Property ($StrAP$). At last, the class K is said to have Super-Amalgamation Property ($SupAP$) if for any algebras $\mathbf{A}$,$\mathbf{B}$,$\mathbf{C}$ in K the condition AP is satisfied and, moreover, for any $x \in \mathbf{B}$, $y \in \mathbf{C}$ the following equivalencies are fulfilled:

$$\delta(x) \leq \varepsilon(y) \text{ iff } (\exists z \in \mathbf{A})(x \leq \beta(z) \& \gamma(z) \leq y),$$

$$\delta(x) \geq \varepsilon(y) \text{ iff } (\exists z \in \mathbf{A})(x \geq \beta(z) \& \gamma(z) \geq y).$$

Now we consider so-called *Strong Epimorphisms Surjectivity (SES)*:

For each $\mathbf{A}, \mathbf{B}$ in K, for every monomorphism $\alpha : \mathbf{A} \to \mathbf{B}$ and for every $x \in \mathbf{B} - \alpha(\mathbf{A})$ there exist $\mathbf{C} \in K$ and homomorphisms $\beta : \mathbf{B} \to \mathbf{C}$, $\gamma : \mathbf{B} \to \mathbf{C}$ such that $\beta\alpha = \gamma\alpha$ and $\beta(x) \neq \gamma(x)$.

In Maksimova 1998 we obtained

Theorem 4.2 *Let L be a normal modal logic. Then L possesses the projective Beth property PB2 if and only if $V(L)$ has the property SES.*

The detailed proof of this theorem is presented in Maksimova 1999a. In addition, we bring out a useful criterion found in that paper. Recall that a modal algebra is *finitely indecomposable* if it can not be represented as a finite subdirect product of its proper quotient algebras. A transitive algebra is finitely indecomposable if anf only if it satisfies the condition:
$[*]x \vee [*]y = \top \Rightarrow ([*]x = \top \text{ or } [*]y = \top)$.

Theorem 4.3 *(Maksimova 1999a, Theorem 3.8) For any normal modal logic L the following are equivalent:*

1) L has the property PB2;

2) for any finitely generated and finitely indecomposable $\mathbf{A}, \mathbf{B}, \mathbf{C}$ in $V(L)$, for every monomorphisms $\beta : \mathbf{A} \to \mathbf{B}$, $\gamma : \mathbf{A} \to \mathbf{C}$ and for any $b \in \mathbf{B}, c \in \mathbf{C}$, if $\neg(\exists z \in \mathbf{A})(b = \beta(z) \& \gamma(z) = c)$ then there exist a subdirectly irreducible algebra $\mathbf{D}$ in $V(L)$ and homomorphisms $\delta : \mathbf{B} \times \mathbf{C} \to \mathbf{D}$, $\varepsilon : \mathbf{B} \times \mathbf{C} \to \mathbf{D}$ such that $\delta(b, c) \neq \varepsilon(b, c)$ and $\delta(\beta(z), \gamma(z)) = \varepsilon(\beta(z), \gamma(z))$ for all z in $\mathbf{A}$.

Using this criterion, we prove

Theorem 4.4 *There exists a variety of topoboolean algebras with SES and without AP.*

Proof. Take $L = L_1 \cap L_2$, where L_1=S4.1.2 and L_2=S5, and consider the variety $V(L)$. It follows from Maksimova 1979 that $V(L)$ does not possess the amalgamation property. Remind that the logics S4.1.2 and S5 have CIP and hence PB2. By Theorem 4.2, both $V(L_1)$ and $V(L_2)$ have SES. We apply Theorem 4.3 to prove that $V(L)$ has SES.

Let us take any finitely indecomposable algebras $\mathbf{A}, \mathbf{B}, \mathbf{C}$ in $V(L)$ and monomorphisms $\beta : \mathbf{A} \to \mathbf{B}$, $\gamma : \mathbf{A} \to \mathbf{C}$. Assume $b \in \mathbf{B}, c \in \mathbf{C}$ and $\neg(\exists z \in \mathbf{A})(b = \beta(z) \& \gamma(z) = c)$.

The variety $V(L)$ is determined by the identities for topoboolean algebras plus
$\Box(\Diamond\Box p \leftrightarrow \Box\Diamond p) \vee \Box(q \to \Box\Diamond q) = \top$.

One can easily see that each finitely indecomposable algebra in $V(L)$ is in $V(L_1)$ or in $V(L_2)$. Thus the algebras $\mathbf{B}$ and $\mathbf{C}$ belong to $V(L_1) \cup V(L_2)$ as they are finitely indecomposable. If both $\mathbf{B}$ and $\mathbf{C}$ are in $V(L_1)$ then by Theorem 4.3 one can find a subdirectly irreducible algebra $\mathbf{D}$ in $V(L_1)$ (and so in $V(L)$) and homomorphisms $\delta : \mathbf{B} \times \mathbf{C} \to \mathbf{D}$, $\varepsilon : \mathbf{B} \times \mathbf{C} \to \mathbf{D}$ such that $\delta(b,c) \neq \varepsilon(b,c)$ and $\delta(\beta(z), \gamma(z)) = \varepsilon(\beta(z), \gamma(z))$ for all z in $\mathbf{A}$ because $V(L_1)$ has *SES*. The similar is true for $\mathbf{B}, \mathbf{C} \in V(L_2)$.

Now let $\mathbf{B} \in V(L_1)$ and $\mathbf{C} \in V(L_2)$. Since $\mathbf{B}$ and $\mathbf{C}$ are non-degenerate and $\mathbf{A}$ is embeddable into both $\mathbf{B}$ and $\mathbf{C}$, we conclude that $\mathbf{A} \models p \leftrightarrow \Box p$. Indeed, we have $S5 \vdash p \to \Box\Diamond p$, $S4.1.2 \vdash \Box\Diamond p \leftrightarrow \Diamond\Box p$, $S5 \vdash \Diamond\Box p \to \Box p$, and $S4 \vdash \Box p \to p$, so $(S4.1.2 + S5) \vdash p \leftrightarrow \Box p$. As $\mathbf{A}$ is finitely indecomposable, it is the two-element topoboolean algebra $\{\bot, \top\}$.

Remind that $V(L_1)$ has SES. If $b \in (\mathbf{B} - \beta(\mathbf{A}))$ then there exist $\mathbf{D} \in V(L_1)$ and homomorphisms $h : \mathbf{B} \to \mathbf{D}$, $g : \mathbf{B} \to \mathbf{D}$ such that $h\beta = g\beta$ and $h(b) \neq g(b)$. One can take $\mathbf{D}$ subdirectly irreducible. Then we set $\delta(x,y) = h(x)$ and $\varepsilon(x,y) = g(x)$ for all $x \in \mathbf{B}, \mathbf{y} \in \mathbf{C}$. If $c \in (\mathbf{C} - \gamma(\mathbf{A}))$, we act by analogy.

Now we consider the case $b, c \in \mathbf{A}$. Remember that $\neg(\exists z \in \mathbf{A})(b = \beta(z) \& \gamma(z) = c)$. Then we have $b = \bot, c = \top$ or $b = \top, c = \bot$.

Since $\mathbf{B}$ is a subdirectly irreducible algebra in $V(S4.1.2)$, for each x in $\mathbf{B}$ we have $\Diamond\Box x = \Box\Diamond x = \top$ or $\Box\Diamond\neg x = \Diamond\Box\neg x = \top$. We define $f(x) = \top$ if $\Diamond\Box x = \top$, and $f(x) = \bot$ otherwise. One can show that f is a homomorphism from $\mathbf{B}$ into the two-element topoboolean algebra that is, in turn, a subalgebra of $\mathbf{C}$. Let us set

$$\delta(x,y) = f(x) \text{ and } \varepsilon(x,y) = y \text{ for all } x \in \mathbf{B}, \mathbf{y} \in \mathbf{C}.$$

Then δ and ε are homomorphisms from $\mathbf{B} \times \mathbf{C}$ into $\mathbf{C}$, and $\delta(\bot, \top) = \bot \neq \varepsilon(\bot, \top)$, $\delta(\top, \bot) = \top \neq \varepsilon(\top, \bot)$, so $\delta(b,c) \neq \varepsilon(b,c)$. Obviously, $\delta(z,z) = \varepsilon(z,z)$ for $z \in \mathbf{A}$. ⊣

From the statements of Section 2 and also from the results obtained in Maksimova 1991a we can conclude the following.

1. In the family of all normal modal logics, PB1 is equivalent to B1 and to the super-amalgamation property of the corresponding variety; *SupAP* is stronger than *StrAP*, which is equivalent to the conjunction of *AP* and B2. Further, *StrAP* implies *SES* that is equivalent to PB2; *AP* does not imply *SES*.

2. Each normal extension of the logic K4 possesses the property B2, therefore, *AP* is equivalent to *StrAP* and implies *SES* in varieties of transitive algebras. *SES* does not imply *AP*.

3. By the proposition 2.1 a, the number of normal extensions of the logic S4 possessing the property PB1 is finite (Maksimova 1979). There are only finitely many varieties of topoboolean algebras with the amalgamation property. We do not know, what is the number of logics in $NE(S4)$ having the property PB2 and how many varieties of topoboolean algebras have the property SES.

5 Restricted Amalgamation Property

Now we find a necessary condition for SES in varieties of transitive algebras.

We say that an element a of a transitive algebra $\mathbf{A}$ is an *opremum* of $\mathbf{A}$ if $a \neq \top$, $[*]a = a$ and $(\forall x \in \mathbf{A})(x = \top$ or $[*]x \leq a)$. Remind that a transitive algebra is subdirectly irreducible iff it has an opremum; it is easy to see that an opremum is unique.

Say that a monomorphism $\alpha : \mathbf{A} \to \mathbf{B}$ of two subdirectly irreducible transitive algebras is *o-preserving* if it maps the opremum of $\mathbf{A}$ to the opremum of $\mathbf{B}$. Let us define the *Restricted Amalgamation Property (RAP)* of a class K as follows:

(RAP) for any o-preserving monomorphisms $\beta : \mathbf{A} \to \mathbf{B}$ and $\gamma : \mathbf{A} \to \mathbf{C}$ of subdirectly irreducible algebras in K there exist an algebra $\mathbf{D}$ in K and monomorphisms $\delta : \mathbf{B} \to \mathbf{D}$, $\varepsilon : \mathbf{C} \to \mathbf{D}$, such that $\delta\beta = \varepsilon\gamma$.

It is evident that AP implies implies RAP, and for any variety K of transitive algebras, RAP is equivalent to the property:

for each subdirectly irreducible $\mathbf{A}, \mathbf{B}, \mathbf{C} \in K$, such that $\mathbf{A}$ is a common subalgebra of $\mathbf{B}$ and $\mathbf{C}$ and all three algebras have the same opremum, there exist an algebra $\mathbf{D}$ in K and monomorphisms $\delta : \mathbf{B} \to \mathbf{D}$, $\varepsilon : \mathbf{C} \to \mathbf{D}$, such that $\delta(x) = \varepsilon(x)$ for all $x \in \mathbf{A}$.

One can easily prove

Lemma 5.1 *For any variety K of transitive algebras, RAP is equivalent to the property:*

for any o-preserving monomorphisms $\beta : \mathbf{A} \to \mathbf{B}$ and $\gamma : \mathbf{A} \to \mathbf{C}$ of subdirectly irreducible algebras in K there exist a subdirectly irreducible algebra $\mathbf{D}$ in K and o-preserving monomorphisms $\delta : \mathbf{B} \to \mathbf{D}$, $\epsilon : \mathbf{C} \to \mathbf{D}$, such that $\delta\beta = \epsilon\gamma$.

Theorem 5.2 *If a variety of transitive algebras has SES then it has RAP.*

A similar statement was proved for varieties of pseudoboolean algebras as a part of Theorem 3.3 in Maksimova 1999b. We use the same ideas. First we need some lemmas.

Lemma 5.3 *Let* $\mathbf{A}$ *bc a transitive algebra,* $a, b \in \mathbf{A}$ *and* $a = [*]a$, $b = [*]b$. *If* $a \not\leq b$ *then there exist a subdirectly irreducible* $\mathbf{C}$ *with an opremum* ω *and a homomorphism* $f : \mathbf{A} \to \mathbf{C}$ *such that* $f(a) = \top$ *and* $f(b) = \omega$.

Proof. By analogy with that of Lemma 2.3 in Maksimova 1999b. ⊣

Lemma 5.4 *Let* $\mathbf{A}$ *be a subdirectly irreducible transitive algebra with an opremum* ω, h *a homomorphism from* $\mathbf{A}$ *into* $\mathbf{B}$ *such that* $h(\omega) < \top$. *Then* h *is a monomorphism.*

Proof. If $x, y \in \mathbf{A}$, $x \neq y$ then $[*](x \leftrightarrow y) \leq \omega$, $[*](h(x) \leftrightarrow h(y)) = h([*](x \leftrightarrow y)) \leq h(\omega) < \top$, $h(x) \neq h(y)$. ⊣

Lemma 5.5 *Let* $\mathbf{B}, \mathbf{C}$ *be transitive algebras,* $h : \mathbf{B} \times \mathbf{C} \to \mathbf{D}$ *a homomorphism into a subdirectly irreducible transitive algebra* $\mathbf{D}$. *Then just one of the following conditions is fulfilled:*

1) $h(\top, \bot) = \top$, $h(x, y) = h(x, \bot)$ *for all* $y \in \mathbf{C}$, *and* $\alpha(x) = h(x, \bot)$ *is a homomorphism from* $\mathbf{B}$ *into* $\mathbf{D}$,

2) $h(\bot, \top) = \top$, $h(x, y) = h(\bot, y)$ *for all* $x \in \mathbf{B}$, *and* $\beta(y) = h(\bot, y)$ *is a homomorphism from* $\mathbf{C}$ *into* $\mathbf{D}$.

Proof. By analogy with Lemma 2.7 in Maksimova 1999b. ⊣

Now we are in position to prove Theorem 5.2.

Proof. of Theorem 5.2. Assume a variety V of transitive algebras has SES. We take subdirectly irreducible $\mathbf{A}, \mathbf{B}, \mathbf{C}$ in V such that $\mathbf{A}$ is a common subalgebra of $\mathbf{B}$ and $\mathbf{C}$, and all three algebras have the same opremum ω. We consider

$$\mathbf{A}_0 = \{(x, x) | x \in \mathbf{A}\}$$

that is a subalgebra of the product $\mathbf{B}_0 = \mathbf{B} \times \mathbf{C}$. Then an element $(\top, \omega)$ is in $\mathbf{B}_0 - \mathbf{A}_0$.

By SES there exist an algebra $\mathbf{D}_1 \in V$ and homomorphisms $g_1 : \mathbf{B}_0 \to \mathbf{D}_1$, $h_1 : \mathbf{B}_0 \to \mathbf{D}_1$ such that $g_1(z, z) = h_1(z, z)$ for all $z \in \mathbf{A}$, and $g_1(\top, \omega) \neq h_1(\top, \omega)$. Suppose for determiness that $g_1(\top, \omega) \not\leq h_1(\top, \omega)$. Due to the lemma 5.3 there exist a subdirectly irreducible transitive algebra $\mathbf{D}$ and a homomorphism $f : \mathbf{D}_1 \to \mathbf{D}$ such that $f(g_1(x, y)) = \top$, $f(h_1(x, y))$ is the opremum of $\mathbf{D}$. Then $g = fg_1$, $h = fh_1$ are homomorphisms into $\mathbf{D}$, $g(z, z) = h(z, z)$ for all $z \in \mathbf{A}$, $g(\top, \omega), = \top$ and $h(\top, \omega)$ is the opremum of $\mathbf{D}$.

We have $h(\top,\bot) \leq h(\top,\omega) < \top$; by Lemma 5.5 we get $h(\bot,\top) = \top$, $h(x,y) = h(\bot,y)$ for all $x \in \mathbf{B}$, and $\beta(y) = h(\bot,y)$ is a homomorphism from $\mathbf{C}$ into $\mathbf{D}$. Moreover,
$\beta(\omega) = h(\bot,\omega) = h(\omega,\omega) = h(\top,\omega) < \top$, i.e. $\beta(\omega) < \top$.
So β is a monomorphism from $\mathbf{C}$ into $\mathbf{D}$ by Lemma 5.4.

Now we apply the lemma 5.5 to g. We have $g(\top,\omega) = \top$, $g(\omega,\omega) = h(\omega,\omega) = h(\top,\omega) < \top$, therefore, $g(\top,\omega) \neq g(\omega,\omega)$. Hence the case 2 of Lemma 5.5 can not be satisfied for g, so the case 1 holds, i.e. $g(\top,\bot) = \top$, $g(x,y) = g(x,\bot)$ for all $y \in \mathbf{C}$, and $\alpha(x) = g(x,\bot)$ is a homomorphism from $\mathbf{B}$ into $\mathbf{D}$. We get $\alpha(\omega) = g(\omega,\bot) = g(\omega,\omega) = h(\omega,\omega) < \top$, so α is a monomorphism from $\mathbf{B}$ into $\mathbf{D}$. At last, for $z \in \mathbf{A}$ we have $\alpha(z) = g(z,\bot) = g(z,z) = h(z,z) = h(\bot,z) = \beta(z)$. It completes the proof. ⊣

So we see that in any variety V of transitive algebras, SES implies some restricted form of the amalgamation property for the class of subdirectly irreducible algebras in V. It follows from the theorem 2.3 that neither RAP nor SES do imply AP in varieties of transitive algebras.

Now we formulate our main results in terms of varieties of transitive algebras. By close inspection of difficult algebraic constructions found in Maksimova 1991b we prove the following

Theorem 5.6 *Let V be any variety of modal transitive algebras which possesses the restricted amalgamation property and is not locally finite. Then*

(i) $r(V)$ contains $V(Grz.2)$;

(ii) if V consists of topoboolean algebras then V contains $V(Grz.2)$;

(iii) if V is a variety of diagonalizable algebras then V contains $V(G\gamma)$ that has $SupAP$.

Proof. (i) In Maksimova 1991b (Theorem 4) we proved that $r(V) \supseteq V(Grz.2)$, whenever V is non-locally-finite variety of transitive algebras with the amalgamation property. A technical algebraic proof is given there in detail. One should carefully follow that proof and apply Lemma 4.3 in order to show that one can take RAP instead of AP in all special technical constructions used in the proof. In particular, one can show that all algebras can be chosen subdirectly irreducible and all monomorphisms o-preserving.

(ii) Immediately from (i) since $r(V) = V$ for varieties of topoboolean algebras.

(iii) By revising the proof of Theorem 3.5 in Maksimova 1988.

For every Kripke frame $\mathbf{W} = (W,R)$, we denote $\mathbf{W}^+ = (W^+, \to, \bot, \Box)$, where $(W^+, \to, \bot)$ is the boolean algebra of all subsets of W and

$\Box X = \{x | \forall y(xRy \Rightarrow y \in X\}$. A transitive algebra $\mathbf{W}^+$ is subdirectly irreducible iff $\mathbf{W}$ is *initial*, i.e. it has an initial element a such that aRx for all $x \in W - \{a\}$.

For an initial infinite ω-linear frame $\mathbf{W} = (W, R)$, define $Fin(W)$ to be the set of elements of finite depth, and $Inf(W) = W - Fin(W)$. The set $Fin(W)$ is isomorphic to the set N of natural numbers with the relation $xRy \Leftrightarrow y < x$.

It follows from Theorem 2.5 in Maksimova 1988 that the logic $G\gamma$ is complete with respect to the class of initial ω-linear frames whose elements of infinite depth form a finite (irreflexive and transitive) tree. We state a strengthening of Proposition 3.3 in Maksimova 1988:

Lemma 5.7 *Let V be an infinite-slice variety of diagonalizable algebras with RAP. Then the algebra $\mathbf{W}^+$ is in V for each initial ω-linear frame $\mathbf{W}$ such that $Inf(W)$ is a finite tree.*

In order to prove this lemma, we use the same constructions as in Maksimova 1988. We note that all algebras needed in the proof are subdirectly irreducible and all embeddings are o-preserving.

The lemma 5.7 immediately implies our statement (iii) by Theorem 4.1. ⊣

From Theorems 5.6 and 5.2 we immediately obtain

Corollary 5.8 *Let V be any variety of transitive algebras which has SES and is not locally finite. Then $r(V)$ contains $V(Grz.2)$.*

In addition, we have

Corollary 5.9 *Let $L \in NE(K4)$ be a logic af infinite slice which has the projective Beth property PB2. Then*

(i) $V(r(L)) \supseteq V(Grz.2)$;

(ii) if $L \supseteq GL$ then $V(L) \supseteq V(G\gamma)$.

Proof. By Theorem 4.2, $V(L)$ has SES, so it possesses RAP by Theorem 5.2. Due to Theorem 4.1, $V(L)$ is not locally finite. Then $V(r(L)) = r(V(L)) \supseteq V(Grz.2)$ by Theorem 5.6(i), and so we obtain (i). Also (ii) follows from Theorem 5.6(iii). ⊣

Acknowledgements

The research was supported by Russian Foundation for Basic Research, grant no. 99-01-00600, and by Russian Ministry of Education, Program "Universities of Russia", 2000.

References

Barwise, J., and S. Feferman (ed.). 1985. *Model-Theoretic Logics.* Springer-Verlag, N.Y.

Beth, E. 1953. On Padoa's method in the theory of definitions. *Indagationes Math.* 15(4):330–339.

Goldblatt, R. 1978. Arithmetical necessity, provability and intuitionistic logic. *Theoria* 44:38–46.

Maksimova, L. 1998. Explicit and Implicit Definability in Modal and Related Logics. *Bulletin of Section of Logic* 27:36–39.

Maksimova, L.L. 1979. Interpolation theorems in modal logics and amalgamated varieties of topoboolean algebras. *Algebra and Logic* 18(5):556–586.

Maksimova, L.L. 1982. Failure of the Interpolation Property in Modal Companions of Dummett's Logic. *Algebra and Logic* 21(6):690–694.

Maksimova, L.L. 1988. Interpolation in infinite slice extensions of the provability logic. *Algebra and Logic* 27(5):581–604.

Maksimova, L.L. 1991a. The Beth properties, interpolation, and amalgamability in varieties of modal algebras. *Doklady Akademii Nauk SSSR* 319(6):1309–1312.

Maksimova, L.L. 1991b. Interpolation in modal logics of infinite slice containing K4. *Siberian Advances in Mathematics* 1(3):52–74.

Maksimova, L.L. 1992. An analog of Beth's theorem in normal extensions of the modal logic K4. *Siberian Mathematical Journal* 33:118–130.

Maksimova, L.L. 1999a. Projective Beth's properties in modal and superintuitionistic logics. *Algebra and Logic* 38(3):316–333.

Maksimova, L.L. 1999b. Superintuitionistic logics and the projective Beth property. *Algebra and Logic* 38(6):680–696.

Meskhi, V. 1978. Algebraic analysis of modal fragments of tense logics. In *Logic, Semantics, Methodology.* 113–124. Metsniereba, Tbilissi.

Segerberg, K. 1971. *An Essay in Classical Modal Logic.* Uppsala University.

Larisa Maksimova
Institute of Mathematics,
Siberian Branch of Russian Acad. Sci.
630090, Novosibirsk, Russia
E-mail: lmaksi@math.nsc.ru

20

Outline of a Logic of Action

KRISTER SEGERBERG

ABSTRACT. A semantics for action logic is presented, which combines elements from three sources: von Wright's philosophy of action, dynamic logic, and action logic in the style of Scott, Chellas, and Belnap.

1 Autobiographical remarks

At the AiML meeting in Leipzig, the author's address bore the title "Remarks on the prospects of dynamic doxastic logic". An overview of a project in three steps was given: to present (i) a logic of action, (ii) a limited deontic logic, and (iii) a dynamic deontic logic. This project was meant to integrate several different traditions: under (i), to combine action logic in the Scott/Chellas/Belnap tradition with Pratt's dynamic logic; under (ii) to adapt ordinary deontic logic to the new environment described under (i); and (iii) to implement in this deontic context the AGM paradigm, especially the ideas emanating from Alchourrón. This was a vast undertaking, impossible (as it turned out) to cover satisfactorily in one lecture. It will not surprise readers that the scope of this paper is narrower: what follows is only a slightly elaborated version of the first part of the address. The plan of the paper is as follows. In Section 2 we go through, in an informal manner, notions that are important for what follows, and they are then given a formal definition. The next several sections describe further formal notions, syntactic and semantic (the most fundamental ones in Section 3, temporal in Section 4, dynamic in Section 5, and a novel kind in Section 6). The theory presented is related to some previous theories in Section 7. In Section 8 we sketch some easy ways to extend our theory and in Section 9 some that are more problematic.

Advances in Modal Logic, Volume 3
F. Wolter, H. Wansing, M. de Rijke, and M. Zakharyaschev, eds.

2 Informal motivation and formal groundwork

Agents function in an environment which, for want of a better word, we will call *the world.* The world is always in exactly one of a definite number of possible ways, which we will call ***possible real states.*** Actions in the world – actions modifying the environment, ***real actions*** – may be seen as changing one state to another (one extreme or degenerate case being the null change, when there is an action but no change). This analysis is inspired by and largely in agreement with Georg Henrik von Wright's views as expressed in (von Wright 1963) and elsewhere.[1]

1. *States, propositions, events.* Let U be a nonempty set (the *universe* of (*total real*) *states*). A path in U is a nonempty sequence of elements that is either finite (of type m, for some positive natural number m) or infinite (of type ω or $^*\omega$ or $^*\omega + \omega$).[2] If p is a path we write $p(\star)$ for its first element, if there is one, and $p(\sharp)$ for its last element, if there is one. If p and q are paths such that $p(\sharp)$ and $q(\star)$ exist and $p(\sharp) = q(\star)$, then we consider that the concatenation pq of p and q, in that order, exists (and is a path). For example, if $p = (u_0, \ldots, u_{m-1})$ and $q = (v_0, \ldots, v_{n-1})$ and $u_{m-1} = v_0 = w$, then $pq = (u_0, \ldots, u_{m-2}, w, v_1, \ldots, v_{n-1})$. (A formal definition of concatenation of paths is omitted.) Furthermore, we consider that $(pq)r = p(qr)$, if p, q, and r are paths such that this equation is meaningful, and we will usually refer to that entity as just pqr.

A (*real*) *proposition* in U is a set of states (even though not every subset of U need be a proposition). Normally, the set of propositions is closed under certain operations.

An *event* (or *event type*) in U is a set of finite paths (even though not every set of finite paths need be an event). If a is an event, then every element $p \in a$ is a (*possible*) *realization* of a. If a and b are events, then the *concatenation* ab of a and b, in that order, is defined as the set $\{ pq : p \in a \;\&\; q \in b \;\&\; p(\sharp) = q(\star) \}$. Note that concatenation is associative. An individual event is a pair (a, p) such that a is an event type and p is a possible realization.

[1]To be really interesting, a theory of action should portray agents as able to collect and evaluate information, to hold preferences, to reason, to form intentions, to act on intentions, and so on. The agents appearing in this paper lack all those abilities; at least, the present version of our theory has nothing to say about them. (The topic of this paper is an old interest of the author. For previous efforts, which may throw some light on what is done here, see, for example, (Segerberg 1985), (Segerberg 1988) and (Segerberg 1992).)

[2]We denote by ω (the order type of) the set of nonnegative integers, by $^*\omega$ (the order type of) the set of negative integers.

2. *Agents, actions, acting. Agents* are identified by nonnegative numbers. (In certain applications one of these numbers may be assigned to Nature, which may sometimes be considered as a degenerate kind of agent.) The set of agents may have a certain structure, as indeed the agents themselves; see Section 9 below. Following von Wright, we take the view that an action is an event brought about by the agent: every (possible) *individual* (*real*) *action* is a triple (i, a, p) such that i is an agent, a is an event, and p is an element of a; thus (a, p) is an individual event. But not every triple (i, a, p) counts as an individual action. For every event it makes sense to ask if the agent is able to bring it off. The answer is affirmative if and only if the agent has a *way* or *means* of realizing it – some kind of method, procedure or (in the case of humans) mode of behaviour. We shall think of i's way or ways of bringing off an event a as a certain subset of a. Thus (i, a, p) is a (possible) individual action if and only if p is a path in a and i has a way of bringing a about that could result in p. In that case we say that i is the *author* of (i, a, p) and that p is the *trace*. Note that traces of actions are always finite.

3. *Histories.* A (*possible*) *history* is a certain kind of sequence of individual actions that is either finite (of type m, for some positive natural number m) or infinite (of type ω or $^*\omega$ or $^*\omega+\omega$). A necessary condition is that (i, a, p) and (j, b, q) occur in a history, the former immediately preceding the latter, only if $p(\sharp) = q(\star)$. A schematic portrait of a history is given in Fig. 1:

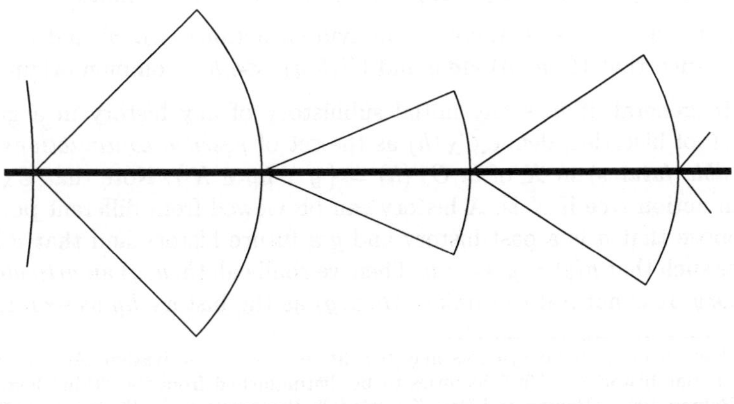

By the *trace* of a history h, in symbols $tr(h)$, we mean the concatenation, in the order defined by h, of the sequences of traces of the elements of h. Notice that, strictly speaking, $((i, ab, pq))$ and $((i, a, p), (i, b, q))$ are different histories – one of type 1, the other of type 2. They have the same trace, though, and are obviously closely related. If h is a history we shall write $h(\star)$ for its first element, if there is one, and $h(\sharp)$ for its last element, if there is one; a history of the former kind is a (possible) *past* (history), a history of the latter kind a (possible) *future* (history). If h and g are histories such that $h(\sharp) = (i, a, p)$ and $g(\star) = (j, b, q)$ exist and, furthermore, $p(\sharp) = q(\star)$, then we consider that the concatenation hg of h and g, in that order, exists (and is a history). (A formal definition of concatenation of histories is omitted.) As with paths, we consider that the operation of concatenation is associative: that $h(gf) = (hg)f$ whenever f, g, and h are histories such that this condition is meaningful. By courtesy we admit the empty history, in symbols $\emptyset$, which has no agent, takes no time and leaves no trace: for all histories h, $h\emptyset = \emptyset h = h$. We say that g is an *initial subhistory* of h, in symbols $g \lhd\!\blacktriangleleft h$, if there is some history f, not necessarily nonempty, such that $h = gf$. Similarly, we say that g is a *final subhistory* of h, in symbols $g \blacktriangleleft\!\lhd h$, if there is some history f, not necessarily nonempty, such that $h = fg$.[3] Note that if hg and $h'g'$ are histories such that $hg = h'g'$, then h is an initial subhistory of h' if and only if g' is a final subhistory of g: $h \lhd\!\blacktriangleleft h'$ if and only if $g' \blacktriangleleft\!\lhd g$.

Let us say that a set T of histories is an *action tree* if

1. for all f, g, h, if $g \in T$ and $fgh \in T$ then $f = h = \emptyset$ ("maximality"),
2. if $hg \in T$ and $((i, a, p)) \lhd\!\blacktriangleleft g$ then, for all actions (i, a, q), there is some g' such that $((i, a, q)) \lhd\!\blacktriangleleft g'$ and $hg' \in T$ ("fullness"),
3. for all g, $h \subset T$ there are individual actions (i, a, p) and (i, b, q) such that $((i, a, p)) \lhd\!\blacktriangleleft g$ and $((i, b, q)) \lhd\!\blacktriangleleft h$ ("common origin").

In general, if h is any initial subhistory of any history in a given set X of histories, define $C_X(h)$ as the set of *possible continuations* (or *possible futures*) in X of h: $C_X(h) = \{ g : hg \in X \}$. Note that $C_X(h)$ is an action tree if X is. A history can be viewed from different points. Suppose that h is a past history and g a future history and that u is a state such that $h(\sharp) = g(\star) = u$. Then we shall call (h, u, g) an *articulated history*. It is natural to think of (h, u, g) as the history hg as seen from

---

[3]The unconventional symbols may benefit from some motivation. According to Fig. 1, our histories – "fat" histories to be distinguished from the "thin" histories of (Belnap 1991),(Belnap and Perloff 1988),(Chellas 1969) and (Thomason 1970) – can be thought of as being of the shape $\ldots \lhd\lhd\lhd \ldots$. This suggests writing $\lhd$ for actions and $\lhd\!\lhd$ for histories. Furthermore, in the symbol $\lhd\!\blacktriangleleft$ an initial subhistory is highlighted with the remainder (possibly empty) left in darkness; in the case of $\blacktriangleleft\!\lhd$ it is the other way around.

u: h as the past, g as the future, and u as the present.[4] Special cases arise when either h or g is empty: $(\emptyset, u, g)$ is a future, $(h, u, \emptyset)$ a past, both seen from u, which is thus a starting point in the former case and an end point in the latter.

4. *Action structures.* The previous considerations lead up to and motivate the following definition: $\mathfrak{A} = (U, \mathfrak{prop}, \mathfrak{ev}, \mathfrak{ag}, \mathfrak{wam})$ is an action structure if

1. U (the *universe*) is a set of points,
2. $\mathfrak{prop}$ (the set of (*real*) *propositions*) is a set of subsets of U, closed under intersection, union, and complement,
3. $\mathfrak{ev}$ (the set of (*real*) *events*) is a set of sets of finite sequences of elements of U, closed under concatenation,
4. $\mathfrak{ag}$ (the set of *agents*) is a set of nonnegative numbers,
5. $\mathfrak{wam}$ (the *ways-and-means function*) is a function assigning to each $i \in \mathfrak{ag}$ and event $a \in \mathfrak{ev}$ a subset $\mathfrak{wam}(i, a) \subseteq a$,
6. $\mathfrak{ev}$ includes the ranges of the D and E, which are defined on $\mathfrak{ag} \times \mathfrak{prop}$ and $\mathfrak{prop}$, respectively: for all $i \in \mathfrak{ag}$ and $P \in \mathfrak{prop}$,

$$\mathsf{D}(i, P) = \{\, p : \exists a(p \in \mathfrak{wam}(i, a) \ \&\ \forall q(q \in a \Rightarrow q(\sharp) \in P))\,\},$$

$$\mathsf{E}P = \{\, p : \exists a(a \in \mathfrak{ev} \ \&\ p \in a \ \&\ p(\sharp) \in P)\,\}.$$

It is clear that the concepts discussed above are well-defined in relation to $\mathfrak{A}$. In particular, an individual action is a triple (i, a, p) where $i \in \mathfrak{ag}$, $a \in \mathfrak{ev}$ and $p \in \mathfrak{wam}(i, a)$. We write $\mathfrak{hist}$ for the set of all histories in $\mathfrak{A}$.

5. *Typographical conventions.* In the following section we will discuss some object languages that suit this kind of structure. To make the presentation less cumbersome we adopt some typographical conventions: if nothing else is said, m, n will stand for nonnegative integers, f, g, h for histories, i, j, k for agents, a, b, c for events, p, q, r for paths, u, v, w for states, φ, ψ, ϑ for formulae, α, β, γ for terms. Furthermore, when we discuss triples of type (i, a, p) we assume that they represent individual actions.

3 Boolean connectives and term-forming operators

Our notion of an object language depends on a fixed set I of nonnegative numbers. We assume that our object language includes the following: (i) denumerably many propositional letters, (ii) denumerably many event letters, (iii) Boolean connectives (the set of which is truth-functionally

[4] However, it must be borne in mind that the state u might occur more than once in the trace of hg.

complete), (iv) the term-forming binary term-operator ; (composition), and (v) the term-forming propositional operators ∂_i (for all $i \in I$) and ε. (We assume that all these elements are distinct.) Formulae made up of only propositional letters and Boolean connectives are called *purely Boolean.* By convention, ∂_i and ε operate only on purely Boolean formulae. The following (heuristical, informal) readings are suggested:

$\alpha\,;\beta$	"α (immediately) followed by β",
$\partial_i\varphi$	"i's bringing about that φ",
$\varepsilon\varphi$	"the coming about that φ".

Let $\mathfrak{A} = (U, \mathfrak{prop}, \mathfrak{ev}, \mathfrak{ag}, \mathfrak{wam})$ be an action structure. A *valuation* in $\mathfrak{A}$ is a function V that assigns a proposition $||\mathsf{P}||_V$ in $\mathfrak{prop}$ to each propositional letter P and an event $||\mathsf{e}||_V$ in $\mathfrak{ev}$ to each event letter e. (For simplicity we will drop the subscript when it is clear what the valuation is.) An *action model* is a pair $(\mathfrak{A}, V)$ where $\mathfrak{A}$ is an action structure and V is a valuation in $\mathfrak{A}$. A valuation can be extended, in the first place to all purely Boolean formulae:

$$\begin{aligned} ||\varphi \wedge \psi|| &= ||\varphi|| \cap ||\psi||, \\ ||\varphi \vee \psi|| &= ||\varphi|| \cup ||\psi||, \\ ||\neg\varphi|| &= U - ||\varphi||, \end{aligned}$$

and similarly for all other Boolean connectives; in the second place to all terms:

$$\begin{aligned} ||\alpha\,;\beta|| &= ||\alpha||||\beta||, \\ ||\partial_i\varphi|| &= \mathsf{D}_i||\varphi||, \\ ||\varepsilon\varphi|| &= \mathsf{E}||\varphi||. \end{aligned}$$

In this way, the valuation assigns a proposition to each purely Boolean formula and an event to each term.

The central theme of the paper is truth- and meaning-conditions for formulae and terms. These conditions will be relative in two ways: with respect to articulated histories and with respect to a set of histories. Thus if (h, u, g) is an articulated history and H is a set of histories such that $hg \in H$, then we wish to define the *truth* of every formula φ with respect to (h, u, g) and H, in symbols

$$(h, u, g) \models_H \varphi,$$

and the *meaning* of every term α with respect to (h, u, g) and H, in symbols

$$||\alpha||_{(h,u,g),H}.$$

In this paper the truth-conditions for purely Boolean formulae will always be the same, as will the meaning-conditions for terms:

$$(h, u, g) \models_H \varphi \text{ iff } u \in \|\varphi\|, \text{ if } \varphi \text{ is purely Boolean,}$$

$$\|\alpha\|_{(h,u,g),H} = \|\alpha\|, \text{ if } \alpha \text{ is a term.}$$

The truth-conditions for the Boolean connectives extend the corresponding conditions given above: for all formulae φ and ψ, not necessarily purely Boolean,

$$(h, u, g) \models_H (\varphi \wedge \psi) \text{ iff } (h, u, g) \models_H \varphi \text{ and } (h, u, g) \models_H \psi,$$

$$(h, u, g) \models_H (\varphi \vee \psi) \text{ iff } (h, u, g) \models_H \varphi \text{ or } (h, u, g) \models_H \psi,$$

$$(h, u, g) \models_H \neg\varphi \text{ iff not } (h, u, g) \models_H \varphi,$$

and so on. A formula is *valid* in $\mathfrak{A}$ if true for all valuations with respect to all sets H of histories and all articulated histories (h, u, g) such that $hg \in H$.

4 Temporal operators

Traditional operators suitable for discrete time are equally useful in the present context. We shall be content to introduce just a few: [F], [P], [NEXT:φ], and [LAST:φ] (where, in the last two, φ is any formula). The following readings are convenient:

[F]ϑ	"always in the future, ϑ",
[P]ϑ	"always in the past, ϑ",
[NEXT:φ]ϑ	"next time that φ, ϑ",
[LAST:φ]ϑ	"last time that φ, ϑ".

Formal truth-conditions:

$$(h, u, g) \models_H [\mathrm{F}]\vartheta \text{ iff } \forall h', g', v((hg = h'g' \;\&\; h \lhd\!\!\blacktriangleleft h' \;\&\; v = tr(h')(\sharp) = tr(g')(\star)) \Rightarrow (h', v, g') \models_H \vartheta),$$

$$(h, u, g) \models_H [\mathrm{P}]\vartheta \text{ iff } \forall h', g', v((hg = h'g' \;\&\; h \blacktriangleleft\!\!\lhd h' \;\&\; v = tr(h')(\sharp) = tr(g')(\star)) \Rightarrow (h', v, g') \models_H \vartheta),$$

$$(h, u, g) \models_H [\mathrm{NEXT}{:}\varphi]\vartheta \text{ iff } \forall h', g', v((hg = h'g' \;\&\; h \lhd\!\!\blacktriangleleft h' \;\&\; v = tr(h')(\sharp) = tr(g')(\star)) \Rightarrow (h', v, g') \not\models_H \varphi), \text{ or else } \exists h', g', v((hg = h'g' \;\&\; h \lhd\!\!\blacktriangleleft h' \;\&\; v = tr(h')(\sharp) = tr(g')(\star) \;\&\; (h', v, g') \models_H \varphi \;\&$$

$$(h', v, g') \models_H \vartheta) \;\&\; \forall h'', g'', w((hg = h''g'' \;\&\; h \lhd\!\!\blacktriangleleft h'' \;\&\; w = tr(h'')(\sharp) = tr(g'')(\star) \;\&\; (h'', w, g'') \models_H \varphi) \Rightarrow h' \lhd\!\!\blacktriangleleft h'')),$$

$$(h, u, g) \models_H [\text{LAST}{:}\varphi]\vartheta \text{ iff } \forall h', g', v((hg = h'g' \;\&\; g \blacktriangleleft\!\!\lhd g' \;\&\; v = tr(h')(\sharp) = tr(g')(\star)) \Rightarrow (h', v, g') \not\models_H \varphi), \text{ or else } \exists h', g', v((hg = h'g' \;\&\; g \blacktriangleleft\!\!\lhd g' \;\&\; v = tr(h')(\sharp) = tr(g')(\star) \;\&\; (h', v, g') \models_H \varphi \;\&\; (h', v, g') \models_H \vartheta) \;\&\; \forall h'', g'', w((hg = h''g'' \;\&\; g \blacktriangleleft\!\!\lhd g'' \;\&\; w = tr(h'')(\sharp) = tr(g'')(\star) \;\&\; (h'', w, g'') \models_H \varphi) \Rightarrow g' \blacktriangleleft\!\!\lhd g'')).$$

Let us call the preceding operators *purely temporal*: their truth-conditions involve only one history at a time, albeit seen from different viewpoints. Formulae involving only operators that are either Boolean or purely temporal are also called purely temporal.

Our final operator, [H], is not purely temporal. However, as it involves all future histories (simultaneously, seen from the present), we nevertheless include it here as a distant relative:

$[\text{H}]\vartheta$ "no matter what the future may hold, ϑ".

Truth-condition:

$$(h, u, g) \models_H [\text{H}]\vartheta \text{ iff } \forall f(f \in C_H(h) \Rightarrow (h, u, f) \models_H \vartheta).$$

5 Dynamic operators

We introduce a number of new formula forming term operators which come in two families, one whose members are forward-looking: does_i, reals_i, occs, (for all $i \in I$) and one whose members are backward-looking: done_i, $\mathsf{realled}_i$, occed (for all $i \in I$). The following informal readings are helpful:

$\mathsf{does}_i\alpha$ "i will just do α",

$\mathsf{done}_i\alpha$ "i has just done α",

$\mathsf{reals}_i\alpha$ "i will just realize α",

$\mathsf{realled}_i\alpha$ "i has just realized α",

$\mathsf{occs}\alpha$ "α will just occur",

$\mathsf{occed}\alpha$ "α has just occurred".

The following truth-conditions are made with respect to a certain action model. We will assume that the corresponding set $\mathfrak{ag}$ of agents and the object language suit one other in the sense that $\mathfrak{ag} = I$. There are then the following new truth-conditions. Forward-looking operators:

$$(h,u,g) \models_H \mathsf{does}_i\alpha \text{ iff } \exists g', m, a_0, \ldots, a_{m-1}, p_0, \ldots, p_{m-1}(g' \vartriangleleft\!\blacktriangleleft g \ \& \ m > 0 \ \& \ g' = ((i,a_0,p_0), \ldots, (i,a_{m-1},p_{m-1})) \ \& \ a_0 \ldots a_{m-1} = \|\alpha\|),$$

$$(h,u,g) \models_H \mathsf{reals}_i\alpha \text{ iff } \exists g', m, a_0, \ldots, a_{m-1}, p_0, \ldots, p_{m-1}(g' \vartriangleleft\!\blacktriangleleft g \ \& \ m > 0 \ \& \ g' = ((i,a_0,p_0), \ldots, (i,a_{m-1},p_{m-1})) \ \& \ p_0 \ldots p_{m-1} \in \|\alpha\|),$$

$$(h,u,g) \models_H \mathsf{occs}\alpha \text{ iff } \exists g', m, i_0, \ldots, i_{m-1}, a_0, \ldots, a_{m-1}, p_0, \ldots, p_{m-1} (g' \vartriangleleft\!\blacktriangleleft g \ \& \ m > 0 \ \& \ g' = ((i_0,a_0,p_0), \ldots \ldots, (i_{m-1},a_{m-1},p_{m-1})) \ \& \ p_0 \ldots p_{m-1} \in \|\alpha\|).$$

Backward-looking operators:

$$(h,u,g) \models_H \mathsf{done}_i\alpha \text{ iff } \exists h', m, a_0, \ldots, a_{m-1}, p_0, \ldots, p_{m-1}(h' \blacktriangleleft\!\vartriangleleft h \ \& \ m > 0 \ \& \ h' = ((i,a_0,p_0), \ldots, (i,a_{m-1},p_{m-1})) \ \& \ a_0 \ldots a_{m-1} = \|\alpha\|),$$

$$(h,u,g) \models_H \mathsf{realled}_i\alpha \text{ iff } \exists h', m, a_0, \ldots, a_{m-1}, p_0, \ldots, p_{m-1}(h' \blacktriangleleft\!\vartriangleleft h \ \& \ m > 0 \ \& \ h' = ((i,a_0,p_0), \ldots, (i,a_{m-1},p_{m-1})) \ \& \ p_0 \ldots p_{m-1} \in \|\alpha\|),$$

$$(h,u,g) \models_H \mathsf{occed}\alpha \text{ iff } \exists h', m, i_0, \ldots, i_{m-1}, a_0, \ldots, a_{m-1}, p_0, \ldots, p_{m-1} (h' \blacktriangleleft\!\vartriangleleft h \ \& \ m > 0 \ \& \ h' = ((i_0,a_0,p_0), \ldots \ldots, (i_{m-1},a_{m-1},p_{m-1})) \ \& \ p_0 \ldots p_{m-1} \in \|\alpha\|).$$

Examples of valid formulae:

$$\mathsf{does}_i\alpha \supset \mathsf{reals}_i\alpha,$$

$$\mathsf{reals}_i\alpha \supset \mathsf{occs}\alpha,$$

$$[\text{NEXT}{:}\mathsf{done}_i\partial_i\varphi]\varphi,$$

$$[\text{NEXT}{:}\mathsf{realled}_i\partial_i\varphi]\varphi,$$

$$[\text{NEXT}{:}\mathsf{occed}\varepsilon\varphi]\varphi.$$

Furthermore, the following are valid in action structures that satisfy the condition that, for all numbers $m > 1$ and nonempty events a, b, $c_0, \ldots, c_{m-1}$,

$$ab = c_0 \ldots c_{m-1} \Rightarrow \exists i(0 < i < m \;\&\; c_0 \ldots c_{i-1} \in a \;\&\; c_i \ldots c_{m-1} \in b):$$

$$\mathsf{does}_i(\alpha;\beta) \equiv (\mathsf{does}_i\alpha \land [\mathrm{NEXT}{:}\mathsf{done}_i\alpha]\mathsf{does}_i\beta),$$

$$\mathsf{reals}_i(\alpha;\beta) \equiv (\mathsf{reals}_i\alpha \land [\mathrm{NEXT}{:}\mathsf{realled}_i\alpha]\mathsf{reals}_i\beta),$$

$$\mathsf{occs}(\alpha;\beta) \equiv (\mathsf{occs}\alpha \land [\mathrm{NEXT}{:}\mathsf{occed}\alpha]\mathsf{occs}\beta).$$

6 The focus operator

Sometimes we want to focus on a certain subset of possible futures. Then the operator $[\mathrm{H}{:}\varphi]$ is useful, where φ is any formula:

$[\mathrm{H}{:}\varphi]\vartheta$ "given that φ, ϑ".

Truth-condition:

$$(h,u,g) \models_H [\mathrm{H}{:}\varphi]\vartheta \text{ iff } \forall f(f \in C_{H'}(h) \Rightarrow (h,u,g) \models_{H'} \vartheta)$$
$$\text{where } H' = \{\, hg' \in H : (h,u,g') \models_H \varphi \,\}.$$

The focus operator is evidently a kind of conditional. Note that the schema

$$[\mathrm{H}]\vartheta \equiv [\mathrm{H}{:}\top]\vartheta$$

is valid. Thus the operator [H] has the same effect as $[\mathrm{H}{:}\top]$. In other words, the operator [H] is definable in terms of the focus operator. However, the converse is not true. What is true is that the schema

$$[\mathrm{H}{:}\varphi]\vartheta \equiv [\mathrm{H}](\varphi \supset \vartheta)$$

is valid for large classes of formulae φ and ϑ, for example, if φ and ϑ are purely temporal. But there are instances of the last schema that are not valid, for example, if $\varphi = \langle \mathrm{F} \rangle \mathsf{P}$ and $\vartheta = [\mathrm{H}]\langle \mathrm{F} \rangle \mathsf{P}$, where P is a propositional letter.[5]

7 Comparison with other theories

The modelling in this paper may be regarded as an attempt, inspired by Georg Henrik von Wright's philosophy of action, to combine Vaughan Pratt's dynamic logic with the tradition in the logic of action of which the current work of Nuel Belnap and his collaborators is the most mature expression. In this section we review some operators in the literature

[5] We use the convention that the dual of a box operator $[\ldots]$ is the corresponding diamond operator $\langle \ldots \rangle$. In other words, $\langle \ldots \rangle\vartheta$ is always logically equivalent to $\neg[\ldots]\neg\vartheta$, no matter what formula ϑ is.

which are more or less closely related to some we have defined here and others that it would be easy to define.

Our notion of an action structure is very general. Too general! In a more extensive study it would be necessary to discuss what restrictions to place on it. Here, in order to facilitate comparison of our theory with other theories in the literature we shall assume that all sets of histories dealt with below are action trees.

First, consider Pratt's operator $[\alpha]$ with the reading "after every terminating computation according to the program α" (Pratt 1977). The definition is obvious:

$$[\alpha]\vartheta =_{\mathrm{df}} [\mathrm{H{:}occs}\alpha][\mathrm{NEXT{:}occed}\alpha]\vartheta.$$

Second, the operators [F] and [P] are of course the operators Arthur Prior wrote G and H (his F and P correspond to our ⟨F⟩ and ⟨P⟩, respectively). Third, our operator [H] is the same as the operator written $\boxdot$ in (Chellas 1969) and L in (Thomason 1970). Chellas referred to $\boxdot$ as a form of historical or physical necessity, saying that "$\boxdot\varphi$ expresses the truth of φ at all courses of events historically identical with the given one". [6] Thomason described L as an operator of unavoidability or inevitability: "a thing is inevitable if it is the case with respect to all alternative futures". Where $[\mathrm{H}]\varphi$ is true with respect to a set H and a history (h, u, g) it would be convenient to borrow (and slightly modify) a phrase from Belnap (Belnap 1991) and say that "φ is settled true for (h, u, g)".[7]

Finally two operators from the Chellas/Belnap tradition. Chellas's operator Δ_i in (Chellas 1969) with the reading "i sees to it that" is approximated by the definition

$$\Delta_i\varphi =_{\mathrm{df}} \mathsf{realled}_i\partial_i\varphi$$

Belnap's and Perloff's related but more sophisticated operator $stit_i$, introduced in (Belnap and Perloff 1988), is similarly approximated by the following two (nonequivalent) definitions:[8]

$$stit_i\varphi =_{\mathrm{df}} \mathsf{realled}_i\partial_i\varphi \land [\mathrm{LAST}{:}\mathsf{reals}_i\partial_i\varphi]\langle\mathrm{H}\rangle\mathsf{reals}_i\partial_i\neg\varphi,$$

$$stit_i\varphi =_{\mathrm{df}} \mathsf{realled}_i\partial_i\varphi \land [\mathrm{LAST}{:}\mathsf{reals}_i\partial_i\varphi]\langle\mathrm{H}\rangle\mathsf{occs}\varepsilon\neg\varphi.$$

Because of the difference regarding iteration – our operators take only purely Boolean formulae as arguments – all three approximations are by necessity imperfect.

---

[6] Notation slightly changed.

[7] Notice that it would not do to borrow Belnap's phrase – "φ is settled true at u" – without the modification.

[8] Belnap's and Perloff's own notation differs from the one used here.

8 Extensions of the theory

The theory outlined in this paper can be extended in a number of ways. In this section we discuss some ways in which the expressibility of the object language can be increased.

1. *The regular operations.* In this paper, the three regular operations, introduced by S. C. Kleene in (Kleene 1956), are called *sum*, *composition* and *Kleene's star operation* and are denoted by + and ; and $*$. Composition is already part of our theory. In order to provide for the other two, we must first assume that the set of events of every action structure is closed under set-theoretic union and *finite repetition*, the latter defined by the condition

$$a* = \{ p_0 \dots p_{m-1} \ : \ m > 0 \ \& \ \forall i < m (p_i \in a) \} \cup \{ (u) \ : \ u \in U \}.$$

The symbols + and $*$ are added to the object language along with the following informal readings:

$\alpha + \beta$ "α or β",

$\alpha *$ "α some number of times (including zero)".

The definition of valuation in Section 3 would then have to be augmented by two clauses:

$$\|\alpha + \beta\| = \|\alpha\| \cup \|\beta\|,$$

$$\|\alpha *\| = (\|\alpha\|)^*.$$

There are two closely related actions which must not be confused with those just defined. Let us introduce yet a new symbol $\oplus$ and write $\alpha \oplus \beta$ for the action that is done by either first choosing α over β and then doing α or else first choosing β over α and then doing β. That this operation is different from the sum operation is seen by the fact that, informally, we expect

$$\mathfrak{wam}(i, \alpha + \beta) \supseteq \mathfrak{wam}(i, \alpha) \cup \mathfrak{wam}(i, \beta),$$

$$\mathfrak{wam}(i, \alpha \oplus \beta) = \mathfrak{wam}(i, \alpha) \cup \mathfrak{wam}(i, \beta),$$

to hold and that the inclusion in the former condition would be proper in certain action structures. Thus, in the case of $\oplus$, real action is preceded by a "mental act", a choice between α and β. As a first approximation we could work with the definition just given for $\oplus$. But in order to do full justice to intuition, our semantics would have to be extended in a significant way in order to provide for "mental acts". (One way would be to introduce "states of mind" to supplement the states of the world.)

Similar remarks hold in relation to Kleene's star operation. Let $\circledast$ be a new symbol and write $\alpha\circledast$ for the action that is done by first choosing a nonnegative integer and then doing α that number of times. Informally,

$$\mathfrak{wam}(i, \alpha*) \supseteq \bigcup\{\, \mathfrak{wam}(i, \alpha^n) \; : \; n \geqslant 0 \,\},$$

$$\mathfrak{wam}(i, \alpha\circledast) = \bigcup\{\, \mathfrak{wam}(i, \alpha^n) \; : \; n \geqslant 0 \,\}$$

where again the inclusion in the former condition may be proper.

2. *Product.* In propositional dynamic logic as originally formulated there is no operation corresponding to set-theoretical intersection. The absence of such an operator reflects the fact that running two programs at the same time on the same machine is either not possible or else raises the difficult notion of concurrency. However, admitting intersection into the present framework presents no problem. All it takes is to adopt the condition that if a and b are events, then so is $a \cap b$; introduce into the object language a new binary term-forming term operator, $\bullet$ say; and add the condition

$$\|\alpha \bullet \beta\| = \|\alpha\| \cap \|\beta\|.$$

to the semantic conditions in Section 3. It seems logical that if the operator interpreted by union is called sum, we should call the new operator, which is interpreted by intersection, *product.* Readers are warned not confuse what we call product with composition, which is referred to as "product" by some authors.

3. *Negative action.* By contrast with intersection, the question concerning complement is intricate and involves much extratheoretical consideration: Do we humans really think in terms of complements? Does the analysis of human languages suggest that we do? To persist in the jargon of this paper, is it not the case that the choice between two actions a and $U - a$ is often a choice between a and some action b that is a proper subset of $U - a$? Before these questions have been answered, this author feels a certain unease about the unrestricted acceptance of closure under complement.

But even in the absence of such closure can we handle some forms of negative action. One is by pushing the negative element up a level. Consider an action such that it makes sense to think that an agent should perform it. Then it also makes sense to think that he should not perform it. Let us contrast doing with *avoiding* and realizing with *omitting*: an agent may do an action or avoid it or neither; and he may realize an action or omit it or neither. Suppose we enrich our object language with four new families of proposition-forming term-operators: avoids_i, $\mathsf{avoided}_i$, omits_i and $\mathsf{omitted}_i$ (for all $i \in I$) with informal readings

$$\begin{array}{ll}
\mathsf{avoids}_i\alpha & \text{“}i\text{ will just avoid }\alpha\text{”},\\
\mathsf{avoided}_i\alpha & \text{“}i\text{ has just avoided }\alpha\text{”},\\
\mathsf{omits}_i\alpha & \text{“}i\text{ will just omit }\alpha\text{”},\\
\mathsf{omitted}_i\alpha & \text{“}i\text{ has just omitted }\alpha\text{”}.
\end{array}$$

The corresponding truth-conditions are

$$\begin{aligned}
(h,u,g) \models_H \mathsf{avoids}_i\alpha \text{ iff } & \exists a,p(((i,a,p)) \lhd\!\!\blacktriangleleft g) \;\&\; \forall g',m,a_0,\dots,a_{m-1},\\
& p_0,\dots,p_{m-1}(g' \lhd\!\!\blacktriangleleft g \;\&\; m>0 \;\&\; g' = ((i,a_0,p_0),\dots\\
& \dots,(i,a_{m-1},p_{m-1})) \Rightarrow a_0\dots a_{m-1} \neq \|\alpha\|),\\
(h,u,g) \models_H \mathsf{avoided}_i\alpha \text{ iff } & \exists a,p(((i,a,p)) \blacktriangleleft\!\!\lhd h) \;\&\; \forall h',m,a_0,\dots,a_{m-1},\\
& p_0,\dots,p_{m-1}(h' \blacktriangleleft\!\!\lhd h \;\&\; m>0 \;\&\; h' = ((i,a_0,p_0),\dots\\
& \dots,(i,a_{m-1},p_{m-1})) \Rightarrow a_0\dots a_{m-1} \neq \|\alpha\|),\\
(h,u,g) \models_H \mathsf{omits}_i\alpha \text{ iff } & \exists a,p(((i,a,p)) \lhd\!\!\blacktriangleleft g) \;\&\; \forall g',m,a_0,\dots,a_{m-1},\\
& p_0,\dots,p_{m-1}(g' \lhd\!\!\blacktriangleleft g \;\&\; m>0 \;\&\; g' = ((i,a_0,p_0),\dots\\
& \dots,(i,a_{m-1},p_{m-1})) \Rightarrow p_0\dots p_{m-1} \notin \|\alpha\|),\\
(h,u,g) \models_H \mathsf{omitted}_i\alpha \text{ iff } & \exists a,p(((i,a,p)) \blacktriangleleft\!\!\lhd h) \;\&\; \forall h',m,a_0,\dots,a_{m-1},\\
& p_0,\dots,p_{m-1}(h' \blacktriangleleft\!\!\lhd h \;\&\; m>0 \;\&\; h' = ((i,a_0,p_0),\dots\\
& \dots,(i,a_{m-1},p_{m-1})) \Rightarrow p_0\dots p_{m-1} \notin \|\alpha\|).
\end{aligned}$$

The words “avoid” and “omit” have been chosen over such words as “refrain” and “forebear”as there seems to be a certain consensus among action logicians that the latter two should be reserved for intentional action; in this paper the question of intention has not been addressed. Notice that all instances of $\neg(\mathsf{does}_i\alpha \wedge \mathsf{avoids}_i\alpha)$ and of $\neg(\mathsf{reals}_i\alpha \wedge \mathsf{omits}_i\alpha)$ are valid but not all instances of $\mathsf{does}_i\alpha \vee \mathsf{avoids}_i\alpha$ or of $\mathsf{reals}_i\alpha \vee \mathsf{omits}_i\alpha$.

We close this section by mentioning two further actions that may also be regarded as negative, although in a different sense.
One is what is termed “skipping” in the literature: when the action that the agent performs consists in staying with *status quo* (see, for example, (Goldblatt 1987), page 87). For the other we suggest the term “slipping” (if there had been a one-word verb in English like the German *schweigen* or the Swedish *tiga* we could have used that instead). We define two events, the “diagonal event” and the “singleton event”:

$$\begin{aligned}
diag &= \{\, p : \exists u(p = (u,u)) \,\},\\
sing &= \{\, p : \exists u(p = (u)) \,\}.
\end{aligned}$$

Introduce the new constants skips_i and $\mathsf{skipped}_i$, slips_i and $\mathsf{slipped}_i$ (for all $i \in I$). There are the following truth-conditions:

$$(h,u,g) \models_H \mathsf{skips}_i \text{ iff } \exists g'(g' \lhd\!\!\blacktriangleleft g \;\&\; g' = ((i, diag, (u,u))),$$

$$(h,u,g) \models_H \mathsf{skipped}_i \text{ iff } \exists h'(h' \blacktriangleleft\!\!\lhd h \;\&\; h' = ((i, diag, (u,u))),$$

$$(h,u,g) \models_H \mathsf{slips}_i \text{ iff } \exists g'(g' \lhd\!\!\blacktriangleleft g \;\&\; g' = ((i, sing, (u))),$$

$$(h,u,g) \models_H \mathsf{slipped}_i \text{ iff } \exists h'(h' \blacktriangleleft\!\!\lhd h \;\&\; h' = ((i, sing, (u))).$$

It is clear that skipping and slipping are two different actions since the former changes the trace of the past history but the latter does not. Both are negative in the sense that the current state is the same before and after the action.

9 Formal limitations of the present theory

Among many formal limitations, we will single out three, which fall under three different headings: the structure of the set of agents, the conception of order, and the possibility of interruption. (Arguably the most important limitation of the theory presented here — that nothing is said about application — is not formal. This is a topic much neglected by philosophical logicians, and, alas, it will be neglected here as well. Readers are referred to (Segerberg 1992) as a surrogate, especially the section on state spaces.)

1. *Agent structure.* We have said nothing about interaction. Obviously, if two agents act at the same time it is possible that the outcome will be influenced by both; the result of my action may depend on your simultaneous action. In fact, this is the point of social action; it is also why game theory is nontrivial and why concurrent action is a challenge to theories of computation. To extend our theory in such a way as to represent this feature requires care; here we confine ourselves to a few remarks. One possible way of extending the theory would be to replace the ways-and-means function $\mathfrak{wam}$ defined in Section 2 by a function $\mathfrak{wam}^*$: while the former is defined on $\mathfrak{ag} \times \mathfrak{ev}$, let the new function be defined on $\wp(\mathfrak{ag}) \times \mathfrak{ev}$, subject to the conditions that, for all C, $D \subseteq \mathfrak{ag}$ and all $a \in \mathfrak{ev}$,

$$\mathfrak{wam}^*(C,a) \subseteq a,$$

$$\mathfrak{wam}^*(\emptyset,a) = \emptyset,$$

$$\text{if } C \subseteq D \text{ then } \mathfrak{wam}^*(C,a) \subseteq \mathfrak{wam}^*(D,a).$$

The new function is a generalization of the old one if, for all agents $i \in \mathfrak{ag}$ and $a \in \mathfrak{ev}$, $\mathfrak{wam}(i,a) = \mathfrak{wam}^*(\{i\},a)$. The difference between noncooperative action and cooperative action, for given C and a, is the difference between $\bigcup\{\mathfrak{wam}^*(\{i\},a) : i \in C\}$ and $\mathfrak{wam}^*(C,a)$. Intuitively, one would not be surprised to learn, in a particular case, that the former quantity were a proper subset of the latter.

2. *Order.* Our modelling is action driven, not time driven. In other words, time (or, perhaps better: order) is generated by action. Since paths are seen as sequences of states, there is an order implicit in our theory that is discrete. But discreteness, while reasonable in many contexts, is not a necessary feature.

In order to sketch a more general approach, let T be either the set of integers or the set of rational numbers or the set of real numbers. A *prepath* is a function into U whose domain is a closed interval of T. A *path* is the range of a prepath. If p is a path and f a prepath such that $p = \mathit{range}(f)$, then we say that f is a *carrier* of p. In order to clarify the nature of paths we introduce an equivalence relation $\sim$ between prepaths:

$$f \sim g \text{ iff } \forall r, r', s, s'((r \leqslant s \;\&\; r' \leqslant s' \;\&\; \mathit{dom}(f) = [r,s] \;\&\; \mathit{dom}(g) = [r',s'] \;\&\; s - r = s' - r') \Rightarrow \forall t((t \in \mathit{dom}(f) \Rightarrow f(t) = g(t - r + r')) \;\&\; (t \in \mathit{dom}(g) \Rightarrow g(t) = f(t - r' + r)))).$$

It follows that two prepaths f and g are carriers of the same path if and only if they are equivalent; that is,

$$\mathit{range}(f) = \mathit{range}(g) \text{ iff } f \sim g.$$

The first and last elements of a path p are defined as follows. Take any carrier f of p. Assume that $\mathit{dom}(f) = [r,s]$. Then

$$p(\star) = f(r),$$

$$p(\sharp) = f(s).$$

It is clear that this definition is correct (in the sense that it does not depend on the particular carrier chosen). The *concatenation* of two paths p and q such that $p(\sharp) = q(\star)$ is defined in the obvious manner: Suppose that f and g are carriers of p and q, respectively. Assume that $\mathit{dom}(f) = [r,s]$ and $\mathit{dom}(g) = [r',s']$. Define $pq = \mathit{range}(h)$, where h is the prepath defined on the closed interval $[r+r', s+s']$ and, for all $t \in [r+r', s+s']$,

$$h(t) = \begin{cases} f(t - r'), & \text{if } r + r' \leqslant t \leqslant s + r', \\ g(t - s), & \text{if } s + r' \leqslant t \leqslant r' + s'. \end{cases}$$

Note that, for $t = s + r'$, $f(t - r') = f(s) = p(\sharp)$ and $g(t - s) = g(r') = q(\star)$. Again, it is clear that the definition is correct.

The theory presented here can now be developed as before. In particular, events are sets of paths and histories are sequences of individual actions. (It does not seem important to generalize the notion of a history to allow that between two points of time there can be infinitely many actions. But those who think otherwise will be able to carry out such a generalization.)

3. *Interrupted action.* In game theory, the nodes of game trees (other than the end nodes) have an agent assigned to them. In our action trees, which are special sets of histories, the situation is similar, yet slightly different. The similarity is that if (h, u, g) is an articulated history and $g \neq \emptyset$, then there is an agent i and an initial subhistory (i, a, p) of g; in that sense, the agent i is assigned to the present u (which can be thought of as a node). The difference is that nothing prevents another articulated history (h, u, g') from belonging to the same action tree, where g' contains an initial subhistory (j, b, q) and j is an agent other than i. Therefore, instead of comparing our action trees to game trees it may be a better idea to compare them to game trees with information sets.

The question of interrupted action is somewhat related to this feature but goes beyond the modelling as presented here. In real life it is not uncommon that an extended action begun by an agent is thwarted before it is completed. In order to capture this feature, we might redefine the notion of an individual action: say that a (*possibly interrupted*) *individual action* is a triple (i, a, p) such that i is an agent, a is an event, p is a path, and there is a path p' such that $pp' \in \mathfrak{wam}(i, a)$. A special case is when $p'(\star) = p'(\sharp)$ and therefore p' is of type 1; in this case $p \in \mathfrak{wam}(i, a)$, and we say that (i, a, p) is *complete*. An individual action that is not complete is said to be *incomplete* or *interrupted*. The new concept is a generalization of the old one inasmuch as individual actions of the old kind may be identified with complete individual actions of the new kind.

Similarly we redefine histories as sequences of possibly interrupted indivudual actions. If formerly our histories could be thought of as in Fig.1 above, now they should be thought of as in Fig. 2.

10 Postscript: Further questions and answers

In the hope of clarifying what the present theory offers and does not offer, and where there is room for improvement, this section has been added. *

1.*I would agree with the author that "action in the world ... may be seen as changing one state to another", provided that by "state" is*

*A thought-provoking referee's report arrived late in the preparation of this paper — too late to be taken into account in the main text. I should like to thank the referee for his perceptive and useful remarks, which are quoted here in a slightly edited form.

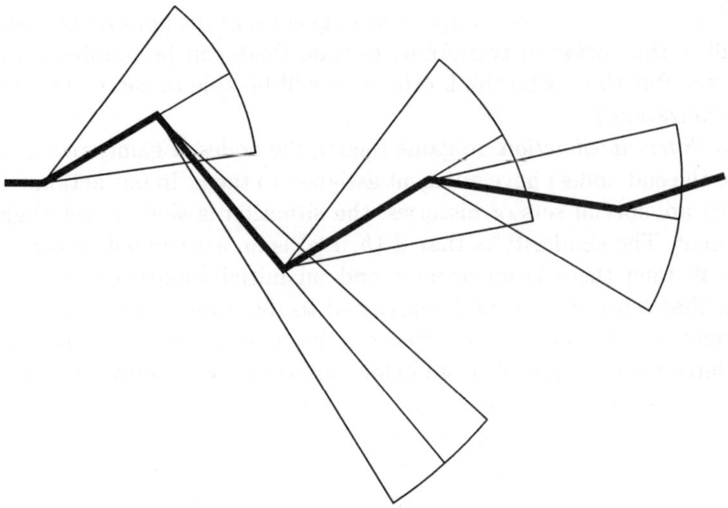

meant "state of affairs". But if by "state" is meant "total state", which is how the author uses the term, then the author's thesis is too strong.

It is the former reading that von Wright has in mind: "The event, for example, which we call the opening of a window, consists in a change or transition from a state of affairs when this window is closed, to a state of affairs when it is open." (von Wright 1963, p. 28) And, yes, there is an important sense in which, normally, an agent does not have it in his power to make the world change from one total state to another. Yet the world is always in some total state or other, so there is also another sense in which any change in the world consists in the transition from one total state to another.

There is a familiar distinction between event type and event token (and between action type and action token), a distinction that is supposed to be clear. However, it is not entirely clear how to implement it with respect to the modelling presented in this paper. Perhaps one can think of event tokens as finite paths and event types as sets of finite paths. Nevertheless, in our modelling an individual event involves both an event type and an event token. To defend this choice, suppose that we have an action model in which a is the action of opening a certain window. Saying with respect to a certain history that a took place on a certain occasion when the world was in total state u can be expressed by saying that, at a certain point in the history, the event a (or the event $\{ p \in a : p(\star) = u \}$) was realized. But suppose now that, in this particular case, it is important to find out what "really happened" — say the police are interested what the agent "really did". Detectives may be able to work out — with respect to the chosen action model, which should be of a kind suited to represent the kind of information the police are interested in — that a certain path p was the path that realized a. The pair (a, p) then represents the individual event that took place. If it is determined that i was the agent of (a, p), then (i, a, p) represents the individual action that took place.

2.*In the paper, propositions are considered as sets of total sets. This is correct, for example, with respect to the proposition expressed by the sentence "The wine in the cup is red". But the proposition expressed by the sentence "Socrates is dying" cannot be represented by a set of total states.*

In the paper, the (undeclared) intention is that all formulae express propositions and, in particular, that purely Boolean ones express real propositions (perhaps "stative propositions" would have been a better term). It is only the latter that are identified with sets of total states.

The referee's remarks highlight the importance of always referring to a particular, fixed state space (universe of total states). On the one hand, "The wine in the cup is red" can be evaluated only in universes where this sentence is meaningful. On the other hand, as von Wright himself remarks, what is represented as a process in one modelling (for example, the running or not-running of a machine) can be represented as a state of affairs in another. Thus it is possible, if perhaps not very profitable, to consider state spaces in which "Socrates is dying" is interpreted as a set of total states.

This said, it must be admitted that the referee has pointed to an interesting problem: how, given a state space, to represent processes. This is in fact an old problem that still awaits a satisfactory solution. (For some previous efforts, see (Segerberg 1992) and (Segerberg 1994).)

3.*The ways-and-means function is static. This assumption calls for comment.*

There is certainly a sense in which the ways-and-means function is static: what the agent can do does not change over time. In another sense, it is not static: what the agent can do in one state he may not be able to do in another. Let $\mathfrak{wam}^\circ(i, u, a)$ be a new function defined by the condition

$$\mathfrak{wam}^\circ(i, u, a) = \{\, p \in \mathfrak{wam}(i, a) \; : \; p(\star) = u \,\}.$$

Then $\mathfrak{wam}^\circ(i, u, a)$ represents the ways in which i is or would be able to achieve a in u. It is clear that in general we may have states u and v such that $\mathfrak{wam}^\circ(i, u, a) \neq \mathfrak{wam}^\circ(i, v, a)$. Note that

$$\mathfrak{wam}(i, a) = \bigcup \{\, \mathfrak{wam}^\circ(i, u, a) \; : \; u \in U \,\}.$$

Say that m is a *method* if m is a (computable?) function from U to the set of sets of finite paths in U such that, for all u and p, $p \in m(u)$ only if $p(\star) \in u$. Let us write $\mathfrak{meth}$ for the set of methods. We might postulate a function $\mathfrak{rep}$ defined on the set $\mathfrak{ag}$ of agents such that, for each $i \in \mathfrak{ag}$, $\mathfrak{rep}(i)$ is a set of methods (thus a subset of $\mathfrak{meth}$) called the *repertoire* of i. We could then define

$$\mathfrak{wam}^\circ(i,u,a) = \bigcup\{\, m(u) \,:\, m \in \mathfrak{rep}(i)\,\}.$$

This conceptual framework is more expressive than that presented in the paper. For example, we might say that i has the *ability* to do a if $\mathfrak{wam}(i,a) \neq \emptyset$, the *opportunity* to do a at u if, for some method m, $m(u) \subseteq a$, and the *power* to do a at u if, for some method $m \in \mathfrak{rep}(i)$, $m(u) \subseteq a$. In this terminology, power implies ability and opportunity. But the converse does not hold: if the agent has a method that works in some states but not in the actual one, then (according to this terminology) he would have the ability and the opportunity but not the power to do the action in question.

Coming back to the referee's question, is it a limitation that $\mathfrak{rep}(i)$ is fixed? Yes, if we want to model phenomena such as learning, forgetting, injury or, in a deontic context, the acquisition or loss of legal competence. But extending our theory in this respect has to await another occasion.

4. *What constitutes an action as an action of the agent? In particular, what is the relationship between i and a in an individual action (i,a,p)?*

5. *Is it possible, within the author's theory, to express the fundamental idea that an agent is free to 'realize' his action?*

It is a tacit assumption in this paper that in order to understand change one has to assume the existence of some sort of mechanism or regularity governing the change. And it is an assumption underlying the logic of action that what agents "do" can make a difference: each agent can have what, for want of a better name, we might call an *input*, at least at times (at all times, if the possibility of a "null input" is admitted). Let us assume that there is a function from the set $\mathfrak{ag}$ of agents to a given set of inputs; and let us write $\mathfrak{menu}(i)$ for the value of this function for the argument i — agent i's *menu* of possible inputs. (Imagine that each agent has a keyboard of buttons of which he is to press one, with the system going into computation after all inputs have been made. Or imagine that for each agent there is a list of code words of which he is to select one to give to an umpire, who then proceeds to carry out a routine identifiable by the totality of codewords provided by the agents.) With n agents there are n inputs, so it is natural to represent the total contribution of the agents — their total input — as a vector $\bar{x} = \langle x_0, \ldots, x_{n-1}\rangle$. We may think of the system as an automaton Γ

that, given a particular input $\bar{x}$ and a particular total state u, delivers a unique path

$$p = \Gamma(\bar{x}, u).$$

Let z be a possible input of agent i. Define

$$\Delta_i(z, u) = \{ p : \exists \bar{x}(x_i = z \Rightarrow p = \Gamma(\bar{x}, u) \}.$$

Then $\Delta_i(z, u)$ may be regarded as the event corresponding to i's making input z in total state u. This can be related to our previous theory by the condition

$$\mathfrak{rep}(i) = \{ m \in \mathfrak{meth} : \exists u \exists z(z \in \mathfrak{menu}(i) \ \& \ m(u) = \Delta_i(z, u)) \},$$

whence

$$\mathfrak{wam}^\circ(i, u, a) = \bigcup \{ \Delta_i(z, u) : \exists z(z \in \mathfrak{menu}(i) \ \& \ \Delta_i(z, u) \subseteq a) \}.$$

Notice that, in our theory, freedom belongs at the level of inputs: on each occasion each agent is supposed to be free to choose ("at will") from his menu of possible inputs (although he is not free to choose the menu itself). If our mechanism is deterministic, then what indeterminism there is in our theory will be of an epistemic rather than an ontological nature.

6. *The author defines histories as sequences of actions. The basic and certainly correct idea behind this is that what is done by all agents (including Nature) constitutes and determines the real outcome, the real world, the real history. But, according to the author's definition, no two agents can begin an action at the same instant. This is not very plausible!*

The referee is right. A wider definition of 'history' can be given along the following lines. Suppose – another idealization – that what happens proceeds in a discrete fashion: as described above, the agents furnish a total input, whereupon the system, represented by the function Γ, computes a path; the procedure is perhaps repeated. A history may then be represented as a sequence of pairs $(\bar{x}, u)$ where $\bar{x}$ is the total input of the agents and u is the total state of the world. An obvious condition is that if $(\bar{y}, v)$ is the immediate successor of $(\bar{x}, u)$, then v is the last element of the path $\Gamma(\bar{x}, u)$.

Technically, histories are needed in order to define truth-conditions. The following conditions (in which previous terminology has been extrapolated in an obvious fashion) are in the spirit of Section 5:

$$\begin{aligned}(h, u, g) \models \mathsf{does}_i\alpha \text{ iff } & \exists g', m, \bar{x}_0, \ldots, \bar{x}_{m-1}, u_0, \ldots, u_{m-1}(m > 0 \ \& \\ & u_0 = u \ \& \ g = (\bar{x}_0, u_0) \ldots (\bar{x}_{m-1}, u_{m-1})g' \ \& \\ & \Delta(x_{0,i}, u_0) \ldots \Delta(x_{m-1,i}, u_{m-1}) = \|\alpha\|),\end{aligned}$$

$$(h,u,g) \models \text{reals}_i\alpha \text{ iff } \exists g', m, \bar{x}_0, \ldots, \bar{x}_{m-1}, u_0, \ldots, u_{m-1}(m > 0 \;\&\; u_0 = u \;\&\; g = (\bar{x}_0, u_0)\ldots(\bar{x}_{m-1}, u_{m-1})g' \;\&\; \Gamma(\bar{x}_0, u_0)\ldots\Gamma(\bar{x}_{m-1}, u_{m-1}) \in \|\alpha\|).$$

*

The paper was an attempt to construct a conceptual framework that is reasonably simple and yet reasonably adequate, in the sense of being rich enough for an interesting logic of action to be developed. Pushed by the referee, we have now extended our theory step by step, at each step compromising on simplicity for the sake of increased adequacy. No doubt the referee would be able to pose further questions, which would force further extensions, which would give rise to still further questions, and so on.

The problem of striking a balance between adequacy and simplicity remains open, and there is room for many solutions.

References

Belnap, N. 1991. Backwards and forwards in the logic of agency. *Philosophy and Phenomenological Research*, vol. 51. 777–807.

Belnap, N., and M. Perloff. 1988. Seeing to it that: a canonical form for agentives. *Theoria*, vol. 54. 175–199.

Chellas, B.F. 1969. *The logical form of imperatives.* Perry Lane Press, Stanford.

Goldblatt, R. 1987. *Logics of time and computation.* CSLI, Stanford.

Kleene, S.C. 1956. Representation of events in nerve nets and finite automata. In *Annals of Mathematics Studies*, no. 34, ed. C.E. Shannon and J. McCarthy. 3–41.

Pratt, V. 1977. Semantical considerations on Floyd-Hoare logic. In *Proceedings of the 17th IEEE Symposium on Foundations of Computer Science.* 109–121.

Segerberg, K. 1985. Routines. *Synthese*, vol. 65. 185–210.

Segerberg, K. 1988. Talking about action. *Studia Logica*, vol. 48. 347–352.

Segerberg, K. 1992. Representing facts. In *Knowledge, belief and strategic interaction*, ed. Cristina Bicchieri and Maria Luisa Dalla Chiara. 239–256. Cambridge University Press.

Segerberg, K. 1994. A festival of facts. *Logic and Logical Philosophy*, vol. 2. 239–256.

Thomason, R.H. 1970. Indeterminist time and truth-value gaps. *Theoria*, vol. 36. 264–281.

von Wright, G.H. 1963. *Norm and action: a logical enquiry.* Routledge & Kegan Paul, London.

Krister Segerberg
Department of Philosophy, Uppsala University
Villavagen 5
S-752 36 Uppsala, Sweden
E-mail: `krister.segerberg@filosofi.uu`

21

Belief, Names, and Modes of Presentation

RUILI YE AND MELVIN FITTING

ABSTRACT. **We develop an axiom system for a first-order logic of belief with names, in which we emphasize the sense and reference distinction, and make use of the notion of modes of presentation in interpreting *de dicto* beliefs. We show the system to be sound and complete, and discuss its philosophical implications.**

Introduction

In investigating the nature of equality in (Frege 1993), Gottlob Frege introduced the notion of *modes of presentation*, suggesting that, beside reference, a name also has a *sense* containing a "mode of presentation" of the reference. For example, 'Phosphorus' and 'Hesperus' have the same reference, the planet Venus, but they have different senses, each containing a mode of presentation of the planet: being seen in the morning, and being seen in the evening. In philosophical discussions, modes of presentation have often been used in interpreting belief reports (*e.g.*, (Engel 1991), (Forbes 1990), (Salmon and Soames 1988), (Schiffer 1990), (Schiffer 1992), (Schiffer 1994), (Schiffer 1995)). In this paper we develop a logic of belief with names, **FMP**(Frege's Modes of Presentation), in which modes of presentation play a central role in the possible-world semantics. We modify Frege's original notion in some aspects: (*i*) Modes of presentation are associated with *names* rather than objects; (*ii*), they are believer-dependent; (*iii*) as "senses" of names, they are relevant only in belief contexts. An axiom system is developed, and its completeness is proved.

Advances in Modal Logic, Volume 3
F. Wolter, H. Wansing, M. de Rijke, and M. Zakharyaschev, eds.

1 Overview

1.1 Belief Representation

Suppose Jones says, "Phosphorus is a planet, Hesperus is not," and we conclude: "Jones does not believe that Hesperus is a planet." This sentence is true under a *de dicto* reading, that is, when it is regarded as being about the *proposition* (*dictum*) "Hesperus is a planet", since what Jones said indicates that he does not believe that "Hesperus is a planet" is true, even though he believes that "Phosphorus is a planet" is true, which, however, is a different proposition. Another possible reading, the *de re* reading, of the sentence is to regard it as being about the *thing* (*res*) called 'Hesperus', so what the sentence says is that *of* the thing named 'Hesperus', Jones does not believe that it is a planet. But this contradicts the fact that Jones said "Phosphorus is a planet", since Phosphorus *is* Hesperus, and anything that is true *of* Phosphorus ought to be true *of* Hesperus, making it reasonable to conclude that Jones believes of Hesperus that it is a planet. *De re* beliefs concern the reference of a name, while *de dicto* beliefs concern the sense of a name.

In **FMP** we make a distinction not only between *de re* and *de dicto* beliefs, but also, among *de dicto* beliefs, the senses associated with the same name by different believers. The *de re* and *de dicto* readings of the sentence "i believes that a is φ" (where i is a believer, a is a name, and φ is some property) are represented, respectively, by $B_i\varphi(a)$ ("Of the thing named 'a', i believes that it is φ") and $B_i\langle x.\varphi\rangle(a)$ ("i believes that a (as he perceives it) is φ")[1]. The explicit occurrence of 'a' in $B_i\langle x.\varphi\rangle(a)$ indicates that 'a' has the sense or mode of presentation that i associates with it. For a different believer j, the 'a' in $B_j\langle x.\varphi\rangle(a)$ is associated with the sense or mode of presentation of 'a' for j. In the above example about Jones, $B_{Jones}\langle x.Planet(x)\rangle(Phos)$ and $B_{Jones}\langle x.Planet(x)\rangle(Hes)$ are two different sentences which may express quite different propositions.

Iterated beliefs such as "Smith believes that Jones believes that Phosphorus is a planet" are ambiguous in ordinary languages: it is not clear whether the name 'Phosphorus' has the sense that Smith associates with it or the one that Jones associates with it. In **FMP**, the two cases are distinctively represented by $B_{Smith}\langle x.B_{Jones}Planet(x)\rangle(Phos)$ and $B_{Smith}B_{Jones}\langle x.Planet(x)\rangle(Phos)$.
A third possible reading of the sentence, when 'Phosphorus' is used as *de re*, is simply expressed as $B_{Smith}B_{Jones}Planet(Phos)$.
The notation extends to cases when multiple names are involved; *e.g.*, $B_i\langle x.\langle y.\langle z.\varphi\rangle(c)\rangle(b)\rangle(a)$ (abbreviated $B_i\langle xyz.\varphi\rangle(a,b,c)$), or

[1]The notation is similar to the λ abstraction notation used in, *e.g.*, (Fitting 1993) and (Fitting and Mendelsohn 1998)

$B_i\langle x.B_j\langle y.\varphi\rangle(b)\rangle(a)$. Sense is irrelevant for an occurrence of a name that is not in the scope of the abstract (*e.g.*, the 'a' in $B_i\varphi(a)$): the name occurs *de re*, and the *reference* of the name is what is involved. In $B_i\langle x.B_j\langle y.\varphi(x,y,c)\rangle(b)\rangle(a)$, for example, a has the sense that i associates with the name a, while b has the sense j associates with the name b, and c occurs *de re*. In $B_i\langle x.\varphi(a,x)\rangle(a)$, the first occurence of a is *de re*, the second *de dicto*.

1.2 The Role of Modes of Presentation in FMP

It is natural to think of modes of presentation as descriptions, represented by predicates. For technical simplification in this work in progress, we abstract descriptions to *sets of objects*. Intuitively, the members of a set representing a description are the objects satisfying that description. The mode of presentation of a name a for a believer i at a possible world Γ (written $m(a,i,\Gamma)$) can be thought of as a set consisting of just the objects satisfying the description that i associates with a. This set may be of any size, depending on how many objects satisfy the description.

To say that a believer i has a *de re* belief about a name a is to say that the belief i has about a is about the *object designated by a*. A *de dicto* belief of i about a, on the other hand, is about what i sees through the lens of the mode of presentation of 'a'; therefore, *de dicto* beliefs are associated with (and only with) modes of presentation: i believes that a (as he perceives the name) is φ if and only if everything that fits the mode of presentation of a for i is believed by him to be φ.

2 The Syntax of FMP

The language $\mathcal{L}_{\mathbf{FMP}}$ consists of the following:

1. a countable set of *constant symbols* (names): $C = \{a_1, a_2, \dots\}$;
2. a countable set of *variables*: $V = \{x_1, x_2, \dots\}$;
3. a countable set $\mathcal{P}$ of predicates: let $\mathcal{P}^n$ denote the set of n-place predicates, where $n = 1, 2, \dots$, with $E \in \mathcal{P}^1$ as the existential predicate. Then $\mathcal{P} = \bigcup_n \mathcal{P}^n$;
4. the equality sign '$=$';
5. propositional connectives: $\neg$ (negation), $\wedge$ (conjunction);
6. quantifier: $\exists$;
7. modal operators: B_i ("i believes"), $B_j, \dots$;
8. punctuation: '(', ')', '⟨', '⟩', ',', and '.'.

Informally, constants and variables are often written as $a, b, \dots$, and $x, y, \dots$, resp., and predicates may be written in a way that suits the context. Propositional connectives $\vee$, $\supset$, and $\equiv$, and the universal quantifier $\forall$ are defined as customary.

A *term* is a constant or a variable. (There are no function symbols.)

Definition 2.1 *Atomic formulas* are all expressions of the form $P(t_1, \dots, t_n)$ or $t_1 = t_2$, where $P \in \mathcal{P}^n$ and $t_1, \dots, t_n$ are terms. The set, Φ, of *formulas* and the set, Ψ, of *pseudo-formulas* are defined by a mutual recursion:

1. The set of formulas, Φ, is the smallest set such that
   (1) all atomic formulas are in Φ; and
   (2) if $\varphi_1, \varphi_2 \in \Phi$, and $\psi \in \Psi$, then $\neg\varphi_1, \varphi_1 \wedge \varphi_2, (\exists x)\varphi_1$ and $B_i\psi$ are in Φ, where $i \in I = \{1, 2, \dots\}$, x is a variable.
2. The set of pseudo-formulas, Ψ, is the smallest set such that
   (1) $\Phi \subseteq \Psi$; and
   (2) if $\psi \in \Psi$, $x \in V$, $a \in C$, then $\langle x.\psi\rangle(a) \in \Psi$, provided x has not "represented" any constant in ψ, and a has not "acquired a sense" in ψ, where by "representing a constant" and "acquiring a sense" we mean the following:
   i) for any formula φ, no variable represents any constant in φ, and no constant acquires a sense in φ;
   ii) for $\psi \in \Psi$, the variables which represent constants in $\langle x.\psi\rangle(a)$ are those which have represented constants in ψ, together with x, and the constants which acquire senses in $\langle x.\psi\rangle(a)$ are those which have acquired senses in ψ, together with a.

Roughly speaking, anything that can immediately follow the belief operator B_i is a pseudo-formula: both P and $\langle x.P\rangle(a)$ are pseudo-formulas, but while the former is a formula, the latter is not.

Definition 2.2 $\langle x_1 \dots x_n.\varphi\rangle(a_1, \dots, a_n)$ abbreviates $\langle x_1.\langle x_2.\dots\langle x_n.\varphi\rangle(a_n)\rangle\dots(a_2)\rangle(a_1)$. The occurrences of the $a_1, \dots, a_n$ in such positions are called *Fregean occurrences*.

Definition 2.3 (Nesting Complexity) The *nesting complexity* of a pseudo-formula ψ is defined recursively as follows:
(1) The nesting complexity of a formula is 0;
(2) The nesting complexity of $\langle x.\varphi\rangle(a)$ is one more than the nesting complexity of φ.

Definition 2.4 (θ–Formula) A *θ–formula* is any formula of the form $\neg B_i\langle x.\psi\rangle(a)$ for some constant a and some pseudo-formula ψ.

By a *sublanguage* of $\mathcal{L}_{\mathbf{FMP}}$ we mean a language all of whose constant symbols and predicate symbols are in $\mathcal{L}_{\mathbf{FMP}}$. The next definition introduces a shorthand for an expression used in one of the axioms; its intuitive meaning will be explained in Remark 5.1 below, following the axioms.

Definition 2.5 Let $\theta = \neg B_i\langle x.\psi\rangle(a)$, $\mathcal{L}$ be a sublanguage of $\mathcal{L}_{\mathbf{FMP}}$ in which θ is written, and z be a variable, then

$$\theta^{\mathcal{L}}(z) \stackrel{df}{=} \{(\forall\vec{y})(B_i\langle x.\varphi'\rangle(a) \supset B_i\varphi'(z)) \wedge E(z) \wedge \neg B_i\psi(z) \mid \varphi' \text{ is a formula of } \mathcal{L},\ \vec{y} \text{ includes all the variables of } \mathcal{L} \text{ occurring in } \varphi'\}.$$

3 The Semantics of FMP

Definition 3.1 (FMP Model) An **FMP** *model* is a tuple

$$M = \langle G, I, R_1, \ldots, R_{|I|}, D, \sigma, m, \pi\rangle,$$

where

1. G is a non-empty set of possible worlds.
2. I is the set of believers.
3. $R_i \subseteq G^2$, for each $i \in I$, is an accessibility relation, which is serial, transitive and Euclidean (this condition can be relaxed, and the system can be modified to deal with other propositional attitudes such as knowledge).
4. For each $\Gamma \in G$, $D(\Gamma)$ is the domain at Γ (by stipulating world-relative domains, actualist quantification is being used). D is monotonic: if $\Gamma R_i \Delta$, then $D(\Gamma) \subseteq D(\Delta)$. For convenience, we shall refer to the union of the domains of all worlds of M, as the *universe* of M, denoted by $\mathcal{D}_M$.
5. $\sigma : C \times G \mapsto \mathcal{D}_M$, the *designation function*, assigns to each name at each possible world an object in the universe: $\sigma(a, \Gamma) \in \mathcal{D}_M$ (a name, therefore, need not designate an object in the domain of quantification, hence we have adopted a free-logic approach). σ is subject to the *rigidity* condition: for any constant a, any Γ, and any Δ such that $\Gamma R_i \Delta$ for some i, $\sigma(a, \Gamma) = \sigma(a, \Delta)$.
6. $m : C \times I \times G \mapsto 2^{\mathcal{D}_M}$, the *sense function*, is a function such that for each $a \in C, i \in I, \Gamma \in G$, $m(a, i, \Gamma) \subseteq D(\Gamma)$; it represents the *mode of presentation* of the name a for the believer i at the world Γ.
7. For each $\Gamma \in G$ and each $n = 1, 2, \ldots$, $\pi : (\mathcal{P}^n - \{E\}) \times \Gamma \mapsto 2^{\mathcal{D}_M{}^n}$ is the interpretation of atomic formulas at each world.

Remark 3.2 The rigidity condition on σ reflects the view that names are rigid designators (cf. (Kripke 1972)); that is, though a believer's perception of a name may vary from world to world, the actual reference of a name remains the same under any circumstance, or $\sigma(a,\Gamma) = \sigma(a,\Delta)$ for any Γ, Δ. For our purpose, the weaker rigidity condition ("local rigidity" in (Fitting and Mendelsohn 1998)) in the definition above suffices.

Definition 3.3 Let $M = \langle G, I, R_1, \ldots, R_{|I|}, D, \sigma, m, \pi\rangle$ be an **FMP** model. An *assignment* μ is a mapping: $V \mapsto \mathcal{D}_M$.
Using the same symbol μ, we define, for each $x \in V$, $a \in C$, $\Gamma \in G$:

$$\mu(x,\Gamma) \stackrel{df}{=} \mu(x); \;\; \mu(a,\Gamma) \stackrel{df}{=} \sigma(a,\Gamma).$$

Given an **FMP** model M, a possible world Γ in M, an assignment μ, and a formula φ, the expression

$$M, \Gamma, \mu \models_{\mathbf{FMP}} \varphi.$$

expresses that φ is true at Γ under μ. M is generally omitted unless confusion may arise.

Definition 3.4 (Satisfaction) The *satisfaction* relation $\models_{\mathbf{FMP}}$ is defined as follows.

Atom For atomic formulas: if $t_1, \ldots, t_n$ are terms, then
$\Gamma, \mu \models_{\mathbf{FMP}} P(t_1, \ldots, t_n)$ iff $\langle \mu(t_1,\Gamma), \ldots, \mu(t_n,\Gamma)\rangle \in \pi(P,\Gamma)$, where $P \in \mathcal{P}^n - \{E\}$;
$\Gamma, \mu \models_{\mathbf{FMP}} E(t)$ iff $\mu(t,\Gamma) \in D(\Gamma)$;
$\Gamma, \mu \models_{\mathbf{FMP}} t_1 = t_2$ iff $\mu(t_1,\Gamma) = \mu(t_2,\Gamma)$.

Neg $\Gamma, \mu \models_{\mathbf{FMP}} \neg\varphi$ iff $\Gamma, \mu \not\models_{\mathbf{FMP}} \varphi$.

Conj $\Gamma, \mu \models_{\mathbf{FMP}} \varphi \wedge \psi$ iff $\Gamma, \mu \models_{\mathbf{FMP}} \varphi$ and $\Gamma, \mu \models_{\mathbf{FMP}} \psi$.

Quant $\Gamma, \mu \models_{\mathbf{FMP}} (\exists x)\varphi$ iff $\Gamma, \mu \models_{\mathbf{FMP}} \varphi[x:o]$[2] for some $o \in D(\Gamma)$.

B-r [3] $\Gamma, \mu \models_{\mathbf{FMP}} B_i\varphi$ iff for every Δ such that $\Gamma R_i \Delta$, $\Delta, \mu \models_{\mathbf{FMP}} \varphi$, where $\varphi \in \Phi$.

B-d $\Gamma, \mu \models_{\mathbf{FMP}} B_i\langle x.\psi\rangle(a)$ iff for every $o \in m(a,i,\Gamma)$,
$\Gamma, \mu \models_{\mathbf{FMP}} B_i\psi[x:o]$, where $\psi \in \Psi$.

Definition 3.5 A formula φ is *valid* in an **FMP** model M, written $M \models_{\mathbf{FMP}} \varphi$, iff for every assignment μ, $M, \mu \models_{\mathbf{FMP}} \varphi$. φ is **FMP** *valid* (written $\models_{\mathbf{FMP}} \varphi$) iff it is valid in every **FMP** model.

[2]The notation "$\Gamma, \mu \models \varphi[x:o]$" is a shorthand for: "$\Gamma, \mu' \models \varphi$, where μ' is an assignment like μ except that it assigns the object o to the free variable x".

[3]Because of rigidity, **B-r** gives $B_i\varphi(a)$ (where $\varphi(a) \in \Phi$) a *de re* interpretation.

4 Philosophical Discussions

4.1 A Closer Look at Modes of Presentation in FMP

A mode of presentation is what a name means to a believer. Since a person may well commit an error as to the actual reference of a name, and the error may have nothing to do with the actual reference, there is no necessary connection between the mode of presentation and the reference of a name (in belief context). Moreover, while a name can have at most one reference in **FMP**, it is possible for the mode of presentation of a name to be satisfied by more than one object in reality, therefore, a mode of presentation is a *set* of object(s). For example, a person who does not know that the British Queen has three sons may think of Charles as "the son of the British Queen"; the mode of presentation of "Charles" for him would be *Is_Son_of_British_Queen*, in its predicate form. Since the Queen in fact has three sons, the set representing this mode of presentation is a set with three members, *viz.,* the three sons. If, on the other hand, one (mistakenly) thinks of "Charles" as "the oldest son of George V", then the mode of presentation of "Charles" would be the singleton consisting of Edward VIII. Of course, for a person who thinks, rightly, of "Charles" as "the oldest son of the British Queen", the mode of presentation of "Charles" is the singleton containing Prince Charles, the actual reference of the name.

The difference between *de re* and *de dicto* beliefs is that the former is directly about a thing, regardless what it is called (it may not even have a name), whereas the latter is by way of mode of presentation, through which one thinks of a *name* (the name may not even denote anything). In **FMP**, *de dicto* belief about *a* is in effect a *de re* belief about members of the set representing the mode of presentation of '*a*'.

Example 4.1 Imagine in the late 1950's, Fred heard much talk about *Principia Mathematica* (*PM*), which seemed to him to be always associated with the name "Russell". At the same time, he knew there was someone named "Lord Russell of Liverpool", so, without a second thought, he took Lord Russell of Liverpool to be the (sole) author of *PM*[4]. The mode of presentation of "Lord Russell of Liverpool" for Fred, then, is "Is_Author_of_*PM*" (in its predicate form), and in **FMP**, it is represented by a set consisting of the two objects having the property, *viz.,* Bertrand Russell and A. N. Whitehead. Any *de dicto* belief that Fred had about Lord Russell of Liverpool was in fact a *de re* belief about B. Russell and A. N. Whitehead. Convinced that he knew enough about

[4]The confusion soon lead to a letter to the London *Times* in which the two Russells stated "neither of us is the other."

PM, albeit all from hearsay, Fred wrote a review, thinking he was reviewing a book by Lord Russell of Liverpool (book-reviewing being his sole means of support, he could not afford the volumes and so never saw the actual names of the authors). The authors of *PM* then wrote a reply: B. Russell wrote the first part, and Whitehead, the second. For some reason, their names didn't appear with the reply, though it's clear that it was written by whoever wrote the book. Fred, of course, took it to be a reply from Lord Russell of Liverpool. Both parts of the reply impressed Fred with great courtesy, so he concluded that Lord Russell of Liverpool was a courteous man. But while the first part impressed him as being exceptionally witty, he found the second part not so at all. As a result, on the one hand, Fred had the *de dicto* belief that Lord Russell of Liverpool was a courteous man—because he had the *de re* belief about both members of its mode of presentation that it was courteous; on the other hand, he did not have the *de dicto* belief that Lord Russell of Liverpool was a witty man, for, *of* at least one member of its mode of presentation, Fred did not have the *de re* belief that it was witty.

It's worth mentioning that from the inconsistency between the two parts of the reply in regard to wittiness, Fred would not infer that the reply was written by two different people. In his mind, he had only "Lord Russell of Liverpool", and the inconsistency only results in Fred's not having the *de dicto* belief about Lord Russell of Liverpool's being witty. It was in terms of names, not actual objects, that Fred carried out his thinking. His opinions were formed through the lens of modes of presentation: actual objects were in a sense beyond his reach. One talks, at the *de dicto* level, only about the name "Lord Russell of Liverpool", not the real objects B. Russell and Whitehead; using the name "Lord Russell of Liverpool" one cannot express the distinction between these two objects at the *de re* level. There is, for example, no question that Fred has the *de dicto* belief that Lord Russell of Liverpool is Lord Russell of Liverpool, but that is not to be confused with the fact that Fred has the *de re* belief that B. Russell and A. N. Whitehead are two distinct objects. The *de dicto* belief about the self-identity of Lord Russell of Liverpool was due to the *de re* belief about the self-identity of each member of its mode of presentation (how could Fred deny the fact that each object is nothing other than itself? He need not know the name of an object in order to form such a belief; in fact, an object need not even have a name).

4.2 Contradictions

When one has a contradictory mode of presentation for a name '*a*' *e.g.*, 'round-square', it seems reasonable to suppose that he would believe anything about a, *e.g.*, that a is round, that a is not round, a is triangle, a exists and a does not exist, ... The **FMP** semantics reflects this: if $m(a, i, \Gamma)$ is the empty set, by **B-d**, regardless what φ is, $B_i\langle x.\varphi\rangle(a)$ would always be true: one believes, *de re*, everything about all the members of the empty set. It follows that one believes, *de dicto*, everything about a, the round-square. However, it does not follow from this that one believes everything about everything: $B_i\langle x.Round\rangle(a) \wedge B_i\langle x.\neg Round\rangle(a)$ is of an entirely different nature than $B_i\varphi \wedge B_i\neg\varphi$, so an explosion of beliefs is not a consequence.

4.3 Each Name is *Believed* to Refer to a Unique Object

Though a mode of presentation $m(a, i, \Gamma)$ may have more than one member, it does not follow that i *believes* there is more than one thing called 'a'. In fact, the following is a validity in **FMP**: for any i and any a,

$$\models_{\mathbf{FMP}} B_i\langle x.(\exists y)(\forall z)(z = x \equiv z = y)\rangle(a)$$

(i believes that 'a' names a unique thing).

4.4 Identities

De re and de dicto beliefs about identities

1. $B_i\langle xy.x = y\rangle(a, b)$ expresses "i believes that a is b."
   By **B-d**, this is true at a world Γ iff either both $m(a, i, \Gamma)$ and $m(b, i, \Gamma)$ are the empty set, (*e.g.*, when both modes of presentation of a and b for i at Γ are of contradictory sort), or they are the same singleton (when they are met uniquely by the same object). This is a *de dicto* belief about both 'a' and 'b'.

2. $B_i\langle x.x = b\rangle(a)$ expresses "of the thing named b, i believes that a is that thing."
   By **B-d**, this is true at a world Γ iff $m(a, i, \Gamma) = \{\sigma(b, \Gamma)\}(= \{\sigma(b, \Delta)\}$ for any Δ such that $\Gamma R_i \Delta)$; *i.e.*, iff the reference of b uniquely satisfies the mode of presentation of 'a' for i at Γ. In this case, it is a *de dicto* belief about 'a', and a *de re* belief about 'b'.

The following are validities in FMP:

(1) $\models_{\mathbf{FMP}} B_i\langle x.x = x\rangle(a)$.

(2) $\models_{\mathbf{FMP}} B_i\langle x.\varphi\rangle(a) \wedge B_i\langle x.x = b\rangle(a) \supset B_i\varphi(b)$.

(3) $\models_{\mathbf{FMP}} B_i\langle x.\varphi\rangle(a) \wedge B_i\langle xy.x = y\rangle(a, b) \supset B_i\langle x.\varphi\rangle(b)$.

(4) $\models_{\mathbf{FMP}} B_i\langle x.\varphi\rangle(a) \wedge B_i\langle x.\neg\varphi\rangle(b) \supset B_i\langle xy.x \neq y\rangle(a, b)$, where x has at least one free occurrence in φ.

4.5 Iterated Belief Sentences

Example 4.2 Imagine a situation in which Smith believes that Cicero is Tully, and he heard Jones saying: "Cicero was a Roman orator, but Tully was not." To Smith, then, Jones believes that *Tully* (as *Smith* understands the name) was a Roman orator as well. The situation can be expressed by

$$B_S\langle xy.x = y\rangle(Cic, Tul) \wedge B_S\neg B_J\langle x.\varphi\rangle(Tul) \wedge B_S\langle x.B_J\varphi(x)\rangle(Tul),$$

where φ stands for the predicate "Is Roman orator". The first conjunct says "Smith believes that Cicero was Tully", the second conjunct says "Smith believes that Jones does not believe that Tully (as *Jones* understands the name Tully) was a Roman orator, and the third conjunct says "Smith believes that Jones believes that Tully (as *Smith* understands the name Tully) was a Roman orator".

5 Axioms and Rules of Inference

In this and the next section, φ, ψ are formulas, a, b are constants, and t, t_1, t_2 are terms, unless indicated otherwise.

5.1 Axioms

FMP 0 All tautologies

FMP 1 Substitution Principle: $t_1 = t_2 \supset (\varphi \supset \psi)$, where ψ is a formula resulting from free substitution (Kleene 1952) of one or more occurrences of t_1 in the formula φ by t_2, with the restriction that, if t_1 is a constant, then all the occurrences of t_1 in φ which are being substituted by t_2 are non-Fregean occurrences (cf. Definition 2.2).

FMP 2 $(\forall x)E(x)$

FMP 3 $(\forall x)\varphi(x) \wedge E(t) \supset \varphi(t)$

FMP 4 $x = x$

FMP b1 $(\forall x)B_iE(x)$

FMP b2 $B_i(\varphi \supset \psi) \supset (B_i\varphi \supset B_i\psi)$

FMP b3 $B_i\varphi \supset \neg B_i\neg\varphi$

FMP b4 $B_i\varphi \supset B_iB_i\varphi$

FMP b5 $\neg B_i\varphi \supset B_i\neg B_i\varphi$

FMP b6a $x = y \supset B_i(x = y)$

FMP b6b $x \neq y \supset B_i(x \neq y)$

FMP bs1 $B_i\langle xy.\varphi\rangle(a, b) \equiv$
$(\forall x)\{(\forall y)[B_i\langle x.\varphi\rangle(a) \supset B_i\varphi] \supset B_i\langle y.\varphi\rangle(b)\}$, where $\varphi \in \Psi$

FMP bs2 $\theta \supset (\exists w) \bigwedge \theta_0^{\mathcal{L}}(w)$, for *any* finite subset $\theta_0^{\mathcal{L}}(w)$ of $\theta^{\mathcal{L}}(w)$, where $\mathcal{L}$ is any sublanguage of $\mathcal{L}_{\mathbf{FMP}}$ in which θ is written, w is a variable not in $\mathcal{L}$, and $\bigwedge \theta_0^{\mathcal{L}}(w)$ is the conjunction of all members of $\theta_0^{\mathcal{L}}(w)$.

Remark 5.1 1. With **FMP b3, b4** and **b5,** this is a **KD45** system.

2. Intuitively, **FMP bs1** says, *e.g.*, Smith believes that LBJ killed JFK (as he understands these two names), just in case that, for anyone x such that whenever Smith believes that LBJ killed any person, then he believes that x killed the same person—for any such x, Smith believes that x killed JFK.
3. **FMP bs2** says, if i does not believe that a (as perceived by i) is ψ, then there is something, w, such that whenever i has any (*de dicto*) belief about a, i has the same belief about w, yet i does not believe that w is ψ. Here w represents something that satisfies the mode of presentation of a for i, and the axiom **FMP bs2** essentially says that if i doesn't believe that a is ψ, then there is something satisfying the mode of presentation of a for i, *of* which i does not believe that it is ψ.

5.2 Rules of Inference

R1 Modus Ponens: $\dfrac{\varphi,\ \varphi \supset \psi}{\psi}$.

R2 Universal Generalization: $\dfrac{\varphi \supset \psi(z)}{\varphi \supset (\forall w)\psi(w)}$, where $\psi(z)$ is the result of replacing all free occurrences of w in $\psi(w)$ by occurrences of z, and z does not occur free in φ or $\psi(w)$.

R3 *De Re* Belief Generalization: $\dfrac{\varphi}{B_i\varphi}$.

R4 $\dfrac{\varphi(\vec{x})}{\varphi(\vec{a})}$, where $\vec{x}$ occur free in φ, and $\varphi(\vec{a})$ is the result of replacing $\vec{x}$ by $\vec{a}$.

6 Some Theorems and Derived Rules of Inference of FMP

Theorem 6.1 *(1)* $\vdash_{\mathbf{FMP}} B_i(\varphi \wedge \psi) \equiv B_i\varphi \wedge B_i\psi$.

(2) $\vdash_{\mathbf{FMP}} a = b \supset B_i(a = b)$; $\vdash_{\mathbf{FMP}} a \neq b \supset B_i(a \neq b)$.

(3) $\vdash_{\mathbf{FMP}} (\forall x)B_i\varphi \supset B_i\langle x.\varphi\rangle(a)$; $\vdash_{\mathbf{FMP}} (\forall \vec{x})B_i\varphi \supset B_i\langle \vec{x}.\varphi\rangle(\vec{a})$.

(4) $\vdash_{\mathbf{FMP}} B_i\langle x.\varphi \supset \psi\rangle(a) \supset (B_i\langle x.\varphi\rangle(a) \supset B_i\langle x.\psi\rangle(a))$.

(5) $\vdash_{\mathbf{FMP}} B_i\langle x.\varphi \wedge \psi\rangle(a) \equiv B_i\langle x.\varphi\rangle(a) \wedge B_i\langle x.\psi\rangle(a)$.

(6) $\vdash_{\mathbf{FMP}} B_i\langle x.\varphi\rangle(a) \supset B_i\langle x.B_i\varphi\rangle(a)$, *where* $\varphi \in \Psi$

(7) **(Converse Barcan Formula)** $\vdash_{\mathbf{FMP}} B_i(\forall x)\varphi \supset (\forall x)B_i\varphi$.

Derived Rules:

DR1 $\dfrac{B_i\varphi}{B_i\langle x.\varphi\rangle(a)}$

DR2 *De Dicto* Belief Generalization: $\dfrac{\varphi}{B_i\langle x.\varphi\rangle(a)}$

DR3 $\dfrac{\bigwedge\theta_0^{\mathcal{L}}(z) \supset \chi}{\theta \supset \chi}$, where z is a variable not in $\mathcal{L}$, $\theta_0^{\mathcal{L}}(z)$ is any finite subset of $\theta^{\mathcal{L}}(z)$, and $\bigwedge\theta_0^{\mathcal{L}}(z)$ is the conjunction of all members of $\theta_0^{\mathcal{L}}(z)$.

7 Soundness

Theorem 7.1 (Soundness) *The axiom system of* **FMP** *is sound with respect to the* **FMP** *semantics.*

The proof is as usual, though lengthy, and is omitted.

8 Completeness

The completeness is shown in two steps: first, we construct a (*non-normal*) canonical model for $\mathcal{L}_{\mathbf{FMP}}$; then we show that this model can be "converted" into an *equivalent* normal one. From these the completeness result follows.

Definition 8.1 An **FMP** model M is *normal* if for any world Γ in that model, $\pi(=,\Gamma)$, the interpretation of '=' at Γ, is the equality relation on the universe of that model: $\pi(=,\Gamma) = \{\langle o, o\rangle \mid o \in \mathcal{D}_M\}$ (where G is the set of all possible worlds of the model).

Definition 8.2 Two **FMP** models M and M' are *equivalent* if, for every closed formula φ, $M \models_{\mathbf{FMP}} \varphi$ iff $M' \models_{\mathbf{FMP}} \varphi$.

8.1 The Construction of a (Non-Normal) Canonical Model

$B-$Completeness and $B-$Saturatedness

Lemma 8.3 *Let* $\theta = \neg B_i\langle x.\psi\rangle(a)$ *(where* ψ *is a formula). If* S *is a consistent set of a sublanguage* $\mathcal{L}$ *of* $\mathcal{L}_{\mathbf{FMP}}$ *containing* θ*, and* z *is a variable not in* $\mathcal{L}$*, then* $S \cup \theta^{\mathcal{L}}(z)$ *is consistent.*

Proof. Suppose $S \cup \theta^{\mathcal{L}}(z)$ is inconsistent, then there is a finite subset S_0 of S, and a finite subset $\theta_0^{\mathcal{L}}(z)$ of $\theta^{\mathcal{L}}(z)$, such that $S_0 \cup \theta_0^{\mathcal{L}}(z)$ is inconsistent; *i.e.*,

$$\vdash_{\mathbf{FMP}} \bigwedge\theta_0^{\mathcal{L}}(z) \supset \neg\bigwedge S_0,$$

where $\bigwedge S_0$ is the conjunction of all formulas in S_0, and $\bigwedge \theta_0^{\mathcal{L}}(z)$ is the conjunction of all members of $\theta_0^{\mathcal{L}}(z)$. Since z is not in $\mathcal{L}$, by **DR3**,

$$\vdash_{\mathbf{FMP}} \theta \supset \neg \bigwedge S_0,$$

hence $\theta \cup S_0$ is inconsistent. Therefore S, with θ in it, is inconsistent. ⊣

Definition 8.4 (Goodness) Let S be a set of formulas and a be a constant in some sublanguage $\mathcal{L}$ of $\mathcal{L}_{\mathbf{FMP}}$. A term t is (a,i,S)-*good* w.r.t. $\mathcal{L}$ if
(1) $E(t) \in S$, and
(2) for every formula φ of $\mathcal{L}$, $B_i\langle x.\varphi\rangle(a) \supset B_i\varphi(t) \in S$.

Definition 8.5 (Witness) Let $\theta = \neg B_i\langle x.\psi\rangle(a)$ and S be a set of formulas of some sublanguage $\mathcal{L}$ of $\mathcal{L}_{\mathbf{FMP}}$ containing θ. If there is some term t such that $\neg B_i\psi(t) \in S$, we say t is a *witness* to θ w.r.t. S. If, in addition, t is (a,i,S)-good w.r.t. $\mathcal{L}$, we say t is a *good witness* to θ w.r.t. S and $\mathcal{L}$.

Definition 8.6 (B–Completeness) A set S of formulas of a sublanguage $\mathcal{L}$ of $\mathcal{L}_{\mathbf{FMP}}$ is *B-complete* w.r.t. $\mathcal{L}$ if every θ-formula in S has a good witness w.r.t. S and $\mathcal{L}$.

Definition 8.7 (E–Completeness) [5] A set S of formulas of a sublanguage $\mathcal{L}$ of $\mathcal{L}_{\mathbf{FMP}}$ is *E-complete* in $\mathcal{L}$ if for every formula of the form $(\exists x)\varphi(x)$ in S, there is a term t of $\mathcal{L}$ such that $\varphi(t) \in S$.

Definition 8.8 (B–Saturatedness w.r.t. a Language) A set S is *B-saturated* w.r.t. a sublanguage $\mathcal{L}$ of $\mathcal{L}_{\mathbf{FMP}}$ if it meets all the following three conditions:
(1) S is maximal consistent w.r.t. $\mathcal{L}$,
(2) S is E-complete in $\mathcal{L}$,
(3) S is B-complete w.r.t. $\mathcal{L}$.

Lemma 8.9 *Every consistent set S in a sublanguage $\mathcal{L}$ of $\mathcal{L}_{\mathbf{FMP}}$ can be extended to a* B*-saturated set w.r.t. an extension $\mathcal{L}^+$ of $\mathcal{L}$.*

Proof. Let there be an infinite supply of variables not in $\mathcal{L}$. Let $\mathcal{L}^+$ be the language enlarged from $\mathcal{L}$ with the new variables. Arrange all formulas of $\mathcal{L}^+$ in some order $\varphi_1, \varphi_2, \varphi_3, \ldots$. Let $S_0 = S$. For any positive integer n, if $S_{n-1} \cup \{\varphi_n\}$ is inconsistent, then let $S_n = S_{n-1}$; otherwise:
(a) If φ_n is not of the form $(\exists x)\psi$ or a θ-formula, let $S_n = S_{n-1} \cup \{\varphi_n\}$;

[5] cf. (Smullyan 1968). Also called *ω–completeness* in, *e.g.*, (Garson 1984).

(b) If φ_n is $(\exists x)\psi(x)$, let $S_n = S_{n-1} \cup \{(\exists x)\psi(x)\} \cup \{E(z)\} \cup \{\psi(z)\}$, where z is a variable not occurring in S_{n-1};
(c) If φ_n is θ, let $S_n = S_{n-1} \cup \{\theta\} \cup \theta^{\mathcal{L}}(z)$, where z is a variable not occurring in S_{n-1}.
Since there is an infinite supply of new variables, (b) and (c) can always be done, and by the usual argument about E-completeness and Lemma 8.3, each time (a), (b) or (c) is carried out, consistency is retained. Clearly, $S = S_0 \subseteq S_1 \subseteq S_2 \subseteq \ldots$. Finally, let

$$S^* = S_0 \cup S_1 \cup S_2 \cup S_3 \cup \ldots.$$

Then S^* is maximal consistent and E-complete in $\mathcal{L}^+$, by the usual argument about E-completeness. We claim that S^* is also B-complete w.r.t. $\mathcal{L}^+$. To show this, suppose $\theta \in S^*$, where $\theta = \neg B_i\langle x.\psi\rangle(a)$ (we need to show that θ has a good witness w.r.t. S^* and $\mathcal{L}^+$). Then $\theta = \varphi_k$ for some k, so by the construction of the S_n's, $\theta \in S_k$, and

(1) $$\theta^{\mathcal{L}}(z) \subseteq S_k \subseteq S^*$$

(where z is a variable not occurring in S_{k-1}). By Definition 2.5, (1) implies

$$\bigwedge_{\varphi'}(\forall\vec{y})(B_i\langle x.\varphi'\rangle(a) \supset B_i\varphi'(z)) \wedge E(z) \wedge \neg B_i\psi(z) \in S^*,$$

for the conjunction $\bigwedge_{\varphi'}$ of members φ' of any *finite* subset of the set of formulas of $\mathcal{L}$; since S^* is maximal consistent, this implies

(2) $$\neg B_i\psi(z) \in S^*,$$

(3) $$E(z) \in S^*,$$

and for every formula φ' of $\mathcal{L}$,

(4) $$(\forall\vec{y})(B_i\langle x.\varphi'\rangle(a) \supset B_i\varphi'(z)) \in S^*;$$

from (4), for every formula φ' of $\mathcal{L}$ and every term $t_1, t_2, \ldots,$

(5) $$B_i\langle x.\varphi'\rangle(a) \supset B_i\varphi'(z) \in S^*,$$

where every occurrence of the element y_j of $\vec{y}$ in φ' is substituted by t_j. Because all the formulas of $\mathcal{L}^+ - \mathcal{L}$ are a result of the introduction of the new variables, for every formula $\varphi(z_{i_1}, \cdots, z_{i_m})$ of $\mathcal{L}^+ - \mathcal{L}$, there is a corresponding formula $\varphi(y_{i_1}, \cdots, y_{i_m})$ of $\mathcal{L}$ which is one of the φ''s. By (5), when $y_{i_1}, \ldots, y_{i_m}$ in $\varphi(y_{i_1}, \cdots, y_{i_m})$ are substituted by $z_{i_1}, \ldots, z_{i_m}$, we must have $B_i\langle x.\varphi(z_{i_1}, \cdots, z_{i_m})\rangle(a) \supset B_i\varphi(z) \in S^*$. This means that not only when φ' is a formula of $\mathcal{L}$, but also when φ' is a formula of $\mathcal{L}^+ - \mathcal{L}$, the following is true: $B_i\langle x.\varphi'\rangle(a) \supset B_i\varphi'(z) \in S^*$. Consequently, for every formula φ of $\mathcal{L}^+$,

(6) $$B_i\langle x.\varphi\rangle(a) \supset B_i\varphi(z) \in S^*.$$

By (2), z is a witness to θ w.r.t. S^*. By (3) and (6), z is (a, i, S^*)-good w.r.t. $\mathcal{L}^+$. Therefore z is a good witness to θ w.r.t. S^* and $\mathcal{L}^+$.

We conclude that every θ-formula in S^* has a good witness w.r.t. S^* and $\mathcal{L}^+$. So S^* is B-complete w.r.t. $\mathcal{L}^+$. Therefore S^* is $B-$saturated w.r.t. $\mathcal{L}^+$. $\dashv$

Definition 8.10 (**$B-$Saturatedness**) A set is *B-saturated* if it is B-saturated with respect to some language $\mathcal{L}$.

Definition 8.11 For any set S of formulas, $S^{-B_i} \overset{df}{=} \{\varphi \mid B_i\varphi \in S \text{ and } \varphi \in \Phi\}$.

Lemma 8.12 *If S is a B-saturated set, and $\neg B_i\psi \in S$, where $\psi \in \Phi$, then $S^{-B_i} \cup \{\neg\psi\}$ is consistent.*

Proof. Similar reasoning as in, *e.g.*, (Hughes and Cresswell 1996). $\dashv$

The Canonical Model M^*

We now construct the model

$$M^* = \langle G^*, I, R_1^*, \dots, R_{|I|}^*, D^*, \sigma^*, m^*, \pi^* \rangle :$$

1. G^* is the collection of all B-saturated sets of formulas (for each $\Gamma \in G^*$, $\mathcal{L}_\Gamma$ denotes the smallest language in which the formulas in Γ are written);
2. For any $i \in I$ and any $\Gamma, \Delta \in G^*$, $\Gamma R_i^* \Delta$ iff $\Gamma^{-B_i} \subseteq \Delta$;
3. For any $\Gamma \in G^*$, $D^*(\Gamma) = \{t \mid t \text{ is a term of } \mathcal{L}_\Gamma \text{ and } E(t) \in \Gamma\}$;
4. For each $a \in C$, $\Gamma \in G^*$, $\sigma^*(a, \Gamma) = a$;
5. For any $a \in C$, $i \in I, \Gamma \in G^*$, $m^*(a, i, \Gamma) =$ $\{t \mid t$ is a term of $\mathcal{L}_\Gamma$ and t is (a, i, Γ)-good w. r. t. $\mathcal{L}_\Gamma\}$;
6. $\langle t_1, \dots, t_n\rangle \in \pi^*(P, \Gamma)$ iff $P(t_1, \dots, t_n) \in \Gamma$ (where $P \in \mathcal{P}^n$);
7. $\langle t_1, t_2\rangle \in \pi^*(=, \Gamma)$ iff $t_1 = t_2 \in \Gamma$

Let ρ be an assignment such that for each variable x, $\rho(x) = x$.

Proposition 8.13 *(1) For every term t and every $\Gamma \in G^*$, $\rho(t, \Gamma) = t$. (2) For any $\Gamma \in G^*$ and any instance of $\neg B_i\langle x.\psi\rangle(a) \in \Gamma$, if t is a good witness to the latter w.r.t. Γ and $\mathcal{L}_\Gamma$, then $t \in m^*(a, i, \Gamma)$.*

Lemma 8.14 *M^* is an* **FMP** *model (except that it is non-normal).*

Lemma 8.15 *For any $i \in I$, $a \in C$, $\Gamma \in G^*$ and any $\varphi \in \Psi$,*
$B_i\langle x.\varphi\rangle(a) \in \Gamma$ iff for every term $t \in m^*(a, i, \Gamma)$, $B_i\varphi(t) \in \Gamma$.

Proof. By induction on the nesting complexity of φ.

1. Base: φ is a formula (*i.e.*, of nesting complexity 0):
   ($\Longrightarrow$) Suppose $B_i\langle x.\varphi\rangle(a) \in \Gamma$. Let $t \in m^*(a,i,\Gamma)$. By the definition of m^*, t is (a,i,Γ)-good w.r.t. $\mathcal{L}_\Gamma$, so by Definition 8.4, $B_i\langle x.\varphi\rangle(a) \supset B_i\varphi(t) \in \Gamma$. Since $B_i\langle x.\varphi\rangle(a) \in \Gamma$, and Γ is maximal consistent, $B_i\varphi(t) \in \Gamma$. So for every $t \in m^*(a,i,\Gamma)$, $B_i\varphi(t) \in \Gamma$.
   ($\Longleftarrow$) Suppose now $B_i\langle x.\varphi\rangle(a) \notin \Gamma$, then $\neg B_i\langle x.\varphi\rangle(a) \in \Gamma$. Since Γ is B-saturated, $\neg B_i\langle x.\varphi\rangle(a)$ has a good witness t_0 w.r.t. Γ and $\mathcal{L}_\Gamma$ such that $\neg B_i\varphi(t_0) \in \Gamma$, *i.e.*, $B_i\varphi(t_0) \notin \Gamma$. By Part (2) of Proposition 8.13, $t_0 \in m^*(a,i,\Gamma)$. Thus for some t (namely t_0) such that $t \in m^*(a,i,\Gamma)$, $B_i\varphi(t) \notin \Gamma$.
2. Induction: φ is $\langle y.\psi\rangle(b)$, where ψ is a pseudo-formula of nesting complexity k:
   By Definition 2.2 and Axiom **FMP bs1**, $B_i\langle x.\langle y.\psi\rangle(b)\rangle(a) \in \Gamma$ iff $(\forall x)\{(\forall y)[B_i\langle x.\psi\rangle(a) \supset B_i\psi] \supset B_i\langle y.\psi\rangle(b)\} \in \Gamma$
   iff for every term $t \in D^*(\Gamma)$, either

   $$\neg(\forall y)[B_i\langle x.\psi\rangle(a) \supset B_i\psi(t,y)] \in \Gamma \tag{7}$$

   or

   $$B_i\langle y.\psi(t,y)\rangle(b) \in \Gamma. \tag{8}$$

   By $E-$completeness, (7) means that there is some term $\hat{t}$ such that $E(\hat{t}) \in \Gamma$ and

   $$B_i\langle x.\psi(x,\hat{t})\rangle(a) \in \Gamma \tag{9}$$

   and

   $$\neg B_i\psi(t,\hat{t}) \in \Gamma. \tag{10}$$

   By the inductive hypothesis, (9) is true iff for every $t_1 \in m^*(a,i,\Gamma)$,

   $$B_i\psi(t_1,\hat{t}) \in \Gamma. \tag{11}$$

   Now, in (10), since t is any member of $D^*(\Gamma)$, and $m^*(a,i,\Gamma) \subseteq D^*(\Gamma)$, we have: for every $t_1 \in m^*(a,i,\Gamma)$,

   $$\neg B_i\psi(t_1,\hat{t}) \in \Gamma, \tag{12}$$

   contradicting (11). So if t is any $t_1 \in m^*(a,i,\Gamma)$, then instead of (7), (8) is true: for every $t_1 \in m^*(a,i,\Gamma)$, $B_i\langle y.\psi(t_1,y)\rangle(b) \in \Gamma$, *i.e.*, $B_i\varphi(t_1) \in \Gamma$.

$\dashv$

Corollary 8.16 *For any $i \in I$, $a_1,\ldots,a_n \in C$, $\Gamma \in G^*$, and any formula φ, $B_i\langle x_1\ldots x_n.\varphi\rangle(a_1,\ldots,a_n) \in \Gamma$ iff for every t_j such that $t_j \in m^*(a_j,i,\Gamma)$ $(j = 1,\ldots,n)$, $B_i\varphi(t_1,\ldots,t_n) \in \Gamma$.*

Lemma 8.17 (Main Lemma) *For any* $\Gamma \in G^*$ *and any formula* φ *in* $\mathcal{L}_{\mathbf{FMP}}$, $\mathcal{M}^*, \Gamma, \rho \models_{\mathbf{FMP}} \varphi \Longleftrightarrow \varphi \in \Gamma$.

Proof. By induction on the complexity of φ. With simple cases omitted, let's consider only the case when $\varphi = B_i\langle x.\psi\rangle(a)$, where ψ is a pseudo-formula: $M^*, \Gamma, \rho \models_{\mathbf{FMP}} B_i\langle x.\psi\rangle(a)$
iff for every $o \in m^*(a, i, \Gamma)$, $M^*, \Gamma, \rho \models_{\mathbf{FMP}} B_i\psi[x : o]$ (by **B-d**)
iff for every $t \in m^*(a, i, \Gamma)$, $M^*, \Gamma, \rho \models_{\mathbf{FMP}} B_i\psi[x : t]$
iff for every $t \in m^*(a, i, \Gamma)$, $M^*, \Gamma, \rho \models_{\mathbf{FMP}} B_i\psi(t)$ (provable by induction on the complexity of ψ)
iff for every t such that $t \in m^*(a, i, \Gamma)$, $B_i\psi(t) \in \Gamma$ (by the inductive hypothesis)
iff $B_i\langle x.\psi\rangle(a) \in \Gamma$ (by Lemma 8.15). $\dashv$

8.2 Converting M^* to a Normal Model

Converting M^* to a Special Model

Definition 8.18 Given an **FMP** model M (not necessarily normal), two possible worlds Γ and Δ of M are said to be *directly connected* if $\Gamma R_i \Delta$ or $\Delta R_i \Gamma$ for some i. Γ and Δ are *connected* if there are $\Sigma_1, \ldots \Sigma_k$ ($k \geq 1$) such that Σ_1 is Γ, Σ_k is Δ, and for every j such that $1 \leq j < k$, Σ_j and Σ_{j+1} are directly connected.

Proposition 8.19 *For any* **FMP** *model* M, *connectedness is an equivalence relation on its set of possible worlds.*

Definition 8.20 (Special Model) A (not necessarily normal) **FMP** model $M = \langle G, I, R_1, \ldots, R_{|I|}, D, \sigma, m, \pi\rangle$ is a *special model* if for every $\Gamma, \Delta \in G$ and every $o, o' \in \mathcal{D}_M$, $\langle o, o'\rangle \in \pi(=, \Gamma)$ iff $\langle o, o'\rangle \in \pi(=, \Delta)$.

Lemma 8.21 *Every model* M *in which the Axioms* **FMP b6a** *and* **FMP b6b** *are valid has an equivalent special model* M'*; in particular, there is a special* **FMP** *model equivalent to* M^*.

Converting a Special Model to a Normal Model

Definition 8.22 Given a special model M, we write $o \approx_M o'$ if, for some possible world Γ of M, $\langle o, o'\rangle \in \pi(=, \Gamma)$.

Proposition 8.23 *For any special model* $M = \langle G, I, R_1, \ldots, R_{|I|}, D, \sigma, m, \pi\rangle$,
(1) $\approx_M$ *is an equivalence relation.*
(2) If $o_1, \ldots, o_n, o'_1, \ldots, o'_n \in D(\Gamma)$, *and* $o_1 \approx_M o'_1, \ldots, o_n \approx_M o'_n$, *then for any* $P \in \mathcal{P}_n$, $\langle o_1, \ldots, o_n\rangle \in \pi(P, \Gamma)$ *iff* $\langle o'_1, \ldots, o'_n\rangle \in \pi(P, \Gamma)$.

Lemma 8.24 *(1) For every non-normal special model M, there is an equivalent normal model $\overline{M}$.*
(2) For every non-normal special **FMP** *model M, there is an equivalent normal* **FMP** *model $\overline{M}$.*

8.3 The Completeness Result

Theorem 8.25 (Completeness) *The axiom system of* **FMP** *is complete with respect to the* **FMP** *semantics.*

Proof. Follows from Lemma 8.17, Lemma 8.21 and Lemma 8.24. ⊣

9 Concluding Remarks

A notion frequently used in philosophical discussions, modes of presentation did not (as far as we know) appear in previous logical treatments of names, belief and the sense and reference distinction (*e.g.*, (Asher 1986), (Church 1951), (Taschek 1996)). The present approach shows that the notion can be built into a formal system of belief and used to interpret belief statements, reflecting the fact that a person's (*de dicto*) belief involving a name depends on his perception of the name involved. The combination of modal operators with the abstraction notation gives **FMP** a certain expressive power. The approach does not, however, solve the problem of logical omniscience.

The present formalization is a basis for a more natural semantics, which we hope to develop, in which modes of presentation are formalized by *formulas*, representing descriptions. For some name 'a' and believer i, then, the mode of presentation of 'a' for i will be a formula $m(a,i)(.)$, with a free variable, and for a pseudo-formula ψ, $\Gamma, \mu \models B_i\langle x.\psi\rangle(a)$ iff

$$\Gamma, \mu \models B_i(\forall x)[m(a,i)(x) \supset \psi(x)]$$

(i believes that a is ψ just in case that he believes that anything satisfying the description he associates with 'a' ($m(a,i)(x)$) is ψ). With two (or more) names, $\Gamma, \mu \models B_i\langle xy.\psi\rangle(a,b)$ iff

$$\Gamma, \mu \models B_i(\forall x)(\forall y)[m(a,i)(x) \wedge m(b,i)(y) \supset \psi(x,y)].$$

To make it more natural, the new semantics will assume a constant domain instead of the monotonic varying domain.

Acknowledgements

We thank Dr. Larry Moss for providing very helpful comments on some preliminary versions of this work. Thanks also to the anonymous referees for their helpful comments and suggestions.

References

Asher, N. 1986. Belief in Discourse Representation Theory. *Journal of Philosophical Logic* 15:127–189.

Church, A. 1951. A Formulation of the Logic of Sense and Denotation. In *Structure, Method and Meaning Essays in Honor of Henry M. Sheffer*, ed. P. Henle, H. M. Kallen, and S. K. Langer. 3–24. Liberal Arts Press.

Engel, P. 1991. *The Norm of Truth—An Introduction to the Philosophy of Logic.* University of Toronto Press.

Fitting, M. C. 1993. Basic Modal Logic. In *Handbook of Logic in Artificial Intelligence and Logic Programming*, ed. D. M. Gabbay, C. J. Hogger, and J. A. Robinson. 368–448. Oxford University Press.

Fitting, M. C., and R. Mendelsohn. 1998. *First-Order Modal Logic.* Kluwer.

Fitting, Melvin C. 1996. *First-Order Logic and Automated Theorem Proving.* Springer-Verlag. First edition, 1990.

Forbes, G. 1990. The Indispensability of *Sinn*. *The Philosophical Review* 99(4):535–563.

Frege, G. 1993. On Sense and Reference. In *Meaning and Reference*, ed. A. W. Moore. 23–42. Oxford University Press.

Garson, J. 1984. Quantification in Modal Logic. In *Handbook of Philosophical Logic, Vol. II*, ed. D. Gabbay and F. Guenthner. 249–307. Kluwer.

Hughes, G., and M. Cresswell. 1996. *A New Introduction to Modal Logic.* Routledge.

Kleene, S. C. 1952. *Introduction to Metamathematics.* North-Holland.

Kripke, S. 1972. *Naming and Necessity.* Harvard University Press.

Salmon, N., and S. Soames. 1988. *Propositions and Attitudes.* Oxford University Press.

Schiffer, S. 1990. The Mode-of-Presentation Problem. In *Propositional Attitudes—the role of content in logic, language and mind*, ed. C. Anderson and J. Owen. 249–268. Stanford: CSLI.

Schiffer, S. 1992. Belief Ascription. *Journal of Philosophy* 89(10):499–521.

Schiffer, S. 1994. A Paradox of Meaning. *Noûs* 279–324.

Schiffer, S. 1995. Descriptions, Indexicals, and Belief Reports: Some Dilemmas. *Mind* 104:107–131.

Smullyan, R. 1968. *First-Order Logic.* Springer-Verlag.

Taschek, W. 1996. Opacity in Context: On Ascribing Beliefs. Paper presented at the Eastern Division APA meetings, Atlanta, Georgia.

Ruili Ye
Mathematics Department
John Jay College, CUNY
445 West 59 Street
New York, NY 10019
E-mail: `bonbistro@yahoo.com`

Melvin Fitting
Dept. of Mathematics and Computer Science
Lehman College, CUNY
250 Bedford Park Boulevard West
Bronx, NY 10468-1589
E-mail: fitting@lehman.cuny.edu
URL: http://comet.lehman.cuny.edu/fitting

Subject Index

Author Index